MODULES OVER ENDOMORPHISM RINGS

This is an extensive synthesis of recent work in the study of endomorphism rings and their modules, bringing together direct sum decompositions of modules, the class number of an algebraic number field, point set topological spaces, and classical noncommutative localization.

The main idea behind the book is to study modules G over a ring R via their endomorphism ring $\text{End}_R(G)$. The author discusses a wealth of results that classify G and $\text{End}_R(G)$ via numerous properties, and in particular results from point set topology are used to provide a complete characterization of the direct sum decomposition properties of G.

For graduate students this is a useful introduction, while the more experienced mathematician will discover that the book contains results that are not otherwise available. Each chapter contains a list of exercises and problems for future research, which provide a springboard for students entering modern professional mathematics.

THEODORE G. FATICONI is Professor in the Mathematics Department at Fordham University, New York.

ENCYCLOPEDIA OF MATHEMATICS AND ITS APPLICATIONS

All the titles listed below can be obtained from good booksellers or from Cambridge University Press. For a complete series listing visit
http://www.cambridge.org/uk/series/sSeries.asp?code=EOM

Modules over Endomorphism Rings

THEODORE G. FATICONI

Fordham University

CAMBRIDGE UNIVERSITY PRESS
Cambridge, New York, Melbourne, Madrid, Cape Town, Singapore,
São Paulo, Delhi, Dubai, Tokyo

Cambridge University Press
The Edinburgh Building, Cambridge CB2 8RU, UK

Published in the United States of America by Cambridge University Press, New York

www.cambridge.org
Information on this title: www.cambridge.org/9780521199605

First published 2010

Printed in the United Kingdom at the University Press, Cambridge

A catalogue record for this publication is available from the British Library

ISBN 978-0-521-19960-5 Hardback

To my wife Barbara Jean
who helped me read our book of life.

Contents

Preface

The chapters in this book are from papers published or submitted to peer-reviewed journals. These papers were written by the author during the calendar years 2006–2008.

There is a simple example that motivates the point of view of this text. Let $\mathbf{k}$ be a field, let V be an n-dimensional $\mathbf{k}$-vector space for some integer $n > 0$, and let $E = \mathrm{Mat}_n(\mathbf{k})$ denote the ring of $n \times n$-matrices over $\mathbf{k}$. Fix an ordered basis β for V and let $[\mathbf{v}]_\beta$ denote the vector representation for $\mathbf{v}$ relative to β. Given $r \in E$ and $\mathbf{v} \in V$ then we define

$$r\mathbf{v} = r \cdot [\mathbf{v}]_\beta,$$

where $\cdot$ is the usual multiplication between the $n \times n$ matrix r and the column vector $[\mathbf{v}]_\beta$. This multiplication makes V a left E-module. Given a right ideal $I \subset E$ we define

$$IV = \left\{ \sum_i r_i \mathbf{v}_i \,\middle|\, \text{finitely many elements } r_i \in I \text{ and } \mathbf{v}_i \in V \right\}.$$

Then the assignment

$$I \longmapsto IV$$

defines a bijection between the set of right ideals of E and the set of $\mathbf{k}$-subspaces of V. Thus we can study some properties of V by studying the right ideals in the ring E. Notice that we have passed from a strictly additive setting into a setting that is additive and multiplicative. This gain in stucture improves our chances of solving certain problems concerning V.

There is little hope of generalizing the bijection $I \longmapsto IV$ to more general modules over associative rings without sacrificing something, so we hope for the best possible generalization. To find this generalization we will use elements from ring theory, module theory, and some elementary homology and homotopy theory of complexes over associative rings, and in at least a couple of instances we use some point set topology. More details follow.

Most modern research into direct sum decompositions of reduced torsion-free finite rank abelian groups (now called *rtffr groups*) begins with a study of projective modules over End(G). This stems from the Arnold–Lady theorem, which shows that direct summands of G^n correspond to projective direct summands of End(G)^n.

This method begins a study of modules G over a ring R and the endomorphism ring End$_R(G)$. We begin by constructing a category G-plex of what are called G-plexes. This category is category equivalent to **Mod**-End$_R(G)$, which makes G-plex the category to be studied if we wish to characterize G or End$_R(G)$. A duality is used to characterize End$_R(G)$-**Mod** in terms of a category G-coplex. Thus we can characterize the rings End$_R(G)$ whose properties are on the following list of properties of rings:

1. right or left hereditary
2. right or left Noetherian
3. right or left coherent
4. right or left FP-injective
5. right or left self-injective
6. right or left cogenerator
7. right PF rings
8. QF rings

We consider End$_R(G)$ and the left End$_R(G)$-module G, and we characterize several integers associated with rings and modules. Specifically we characterize the integers on the following list:

1. projective dimension of G
2. injective dimension of G
3. flat dimension of G
4. right or left global dimension of End$_R(G)$

Several properties are left as exercises.

One of the purposes of this book is to show that we can study groups locally isomorphic to G by studying invertible fractional right ideals of $E(G)$ where

$$E(G) = \text{End}(G)/\mathcal{N}(\text{End}(G)).$$

For example, let $n > 0$ be an integer, and given a commutative prime ring R, let Pic(R) denote the abelian group of isomorphism classes of invertible fractional right ideals of R. If G is a strongly indecomposable *rtffr* group and if $E(G)$ is commutative then the set of isomorphism classes (H) of groups H that are *locally isomorphic* to G^n is bijective with the finite abelian group Pic($E(G)$). In this setting we show that if H, K, and L are direct summands of G^n, then cancellation in the isomorphism $H \oplus K \cong H \oplus L$ can be viewed as cancellation of elements in the abelian group Pic($E(G)$).

This point of view gives a new insight into the problem of finding the class number $h(\mathbf{k})$ of the algebraic number field $\mathbf{k}$. For example, those $\mathbf{k}$ with $h(\mathbf{k}) = 1$ are classified in the following result. Let $\overline{E}$ denote the algebraic integers in **k**. Let $\Omega(\overline{E}) = \{$rtffr groups $G \mid \mathrm{End}(G) \cong \overline{E}\}$.

Theorem. *The following are equivalent for the algebraic number field* **k**.

1. $h(\mathbf{k}) = 1$.
2. *Each* $G \in \Omega(\mathbf{k})$ *has the power cancellation property. (For each integer* $m > 0$ *and group* H, $G^m \cong H^m$ *implies that* $G \cong H$.)
3. *Each group* H *that is locally isomorphic to* G *is isomorphic to* G.

Some research over the thirty-year period from 1970 to 2000 dealt with the rtffr groups G that were finitely generated left $\mathrm{End}(G)$-modules, or projective left $\mathrm{End}(G)$-modules, or that had right hereditary endomorphism ring. Our thirty-odd pages on this type of result give us a unified approach to these problems and extends existing results. Subsequently, we use the machinery developed in Chapter 9 to characterize the left $\mathrm{End}(G)$-module G that possesses some properties from the following list:

1. finitely generated
2. finitely presented
3. coherent
4. projective
5. quasi-projective
6. possesses a projective cover
7. cogenerator
8. generator
9. progenerator
10. quasi-progenerator
11. Noetherian

Let R be an associative ring with identity, let G be a right R-module, and let

$$\mathrm{End}_R(G)$$

denote the ring of R-endomorphisms of G. The module G is *self-small* if for each index set $\mathcal{I}$ and each R-module map $\phi : G \longrightarrow G^{(\mathcal{I})}$ there is a finite set $\mathcal{J} \subset \mathcal{I}$ such that $\phi(G) \subset G^{(\mathcal{J})}$. In other words there is a natural isomorphism

$$\mathrm{Hom}_R(G, G)^{(\mathcal{I})} \longrightarrow \mathrm{Hom}_R(G, G^{(\mathcal{I})}).$$

Let $\mathbf{P}(G) = \{$right R-modules $Q \mid Q \oplus Q' \cong G^{(\mathcal{I})}$ for some index set $\mathcal{I}$ and some right R-module $Q'\}$. A *G-plex* is a complex

$$Q = \cdots \xrightarrow{\delta_3} Q_2 \xrightarrow{\delta_2} Q_1 \xrightarrow{\delta_1} Q_0$$

Preface

with the properties that

1. $Q_k \in \mathbf{P}(G)$ for each $k \geq 0$ and
2. G has the following lifting property for each $k \geq 1$. Given a map $\phi : G \longrightarrow Q_k$ such that $\delta_k \phi = 0$ there is a map $\psi : G \longrightarrow Q_{k+1}$ such that $\phi = \delta_{k+1}\psi$ as in the commutative triangle

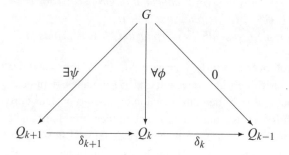

of right R-modules.

*The category of G-plexes G-***Plex** is the additive category whose objects are the G-plexes Q and whose morphisms are homotopy equivalence classes $[f]$ of chain maps

$$f : Q \longrightarrow Q'$$

between G-plexes Q and Q'.

If we let **Mod-**End$_R(G)$ denote the category of right End$_R(G)$-modules then the functor

$$h_G(\cdot) : G\text{-}\mathbf{Plex} \longrightarrow \mathbf{Mod}\text{-}\mathrm{End}_R(G)$$

sends $Q \in G$-**Plex** to the zeroth homology group of the complex

$$\mathrm{Hom}(G, Q) = \cdots \xrightarrow{\delta_2^*} \mathrm{Hom}_R(G, Q_1) \xrightarrow{\delta_1^*} \mathrm{Hom}_R(G, Q_0),$$

or in other words

$$h_G(Q) = \mathrm{coker}\ \delta_1^*.$$

Theorem. *Let G be a self-small right R-module. Then the additive functor*

$$h_G(\cdot) : G\text{-}\mathbf{Plex} \longrightarrow \mathbf{Mod}\text{-}\mathrm{End}_R(G)$$

is a category equivalence.

Thus the category of right End$_R(G)$-modules, **Mod-**End$_R(G)$, is characterized in terms of a category G-**Plex** in which G is a small projective generator.

One of the more attractive elements of this point of view is that it dualizes without too much effort. We will assume the set theoretic condition

(μ) measurable cardinals do not exist.

The assumption (μ) is true under Gödel's constructibility hypothesis. Under (μ) we can make a complete dualization of the above theorem. The right R-module G is *self-slender* if for each index set $\mathcal{I}$ and R-module map $\phi : G^{\mathcal{I}} \longrightarrow G$ there is a finite set $\mathcal{J} \subset \mathcal{I}$ such that

$$G^{\mathcal{I} \setminus \mathcal{J}} \subset \ker \phi.$$

Equivalently G is self-slender if for each index set $\mathcal{I}$ the canonical map

$$\text{Hom}_R(G, G)^{(\mathcal{I})} \longrightarrow \text{Hom}_R(G^{\mathcal{I}}, G)$$

is an isomorphism. Let

$$\mathcal{W} = W_0 \xrightarrow{\sigma_1} W_1 \xrightarrow{\sigma_2} W_2 \xrightarrow{\sigma_3} \cdots$$

be a complex of right R-modules. Then $\mathcal{W}$ is a *G-coplex* if W_k is a direct summand of a direct product of copies of G for each integer $k \geq 0$, and if it satisfies the lifting property that is dual to the lifting property satisfied by a G-plex. Define *the category of G-coplexes*, G-**Coplx**, to be that category whose objects are G-coplexes and whose maps are homotopy equivalence classes $[f]$ of chain maps f between G-coplexes. The functor

$$h^G(\cdot) : G\text{-}\mathbf{Coplx} \longrightarrow \text{End}_R(G)\text{-}\mathbf{Mod}$$

is defined by

$$h^G(\mathcal{W}) = \text{coker } \text{Hom}_R(\partial_1, G)$$

which is just the zeroth homology group of the complex of left $\text{End}_R(G)$-modules $\text{Hom}_R(\mathcal{W}, G)$.

Theorem. *Assume (μ) and let G be a self-slender right R-module. Then the additive functor*

$$h^G(\cdot) : G\text{-}\mathbf{Coplx} \longrightarrow \text{End}_R(G)\text{-}\mathbf{Mod}$$

is a category equivalence.

Consequently, we have characterized the category $\text{End}_R(G)$-**Mod** of *left* $\text{End}_R(G)$-modules in terms of the category G-**Coplx** in which G is a slender injective cogenerator. It is worth noting that if G is a reduced torsion-free finite rank abelian group then G is both self-small and self-slender. Thus for these groups we have a complete characterization of the right $\mathrm{End}_{\mathbf{Z}}(G)$-modules and the left $\mathrm{End}_{\mathbf{Z}}(G)$-modules by categories completely determined by G.

From these theorems we sample the existing module theoretic properties for G that can be characterized in terms of $\mathrm{End}_R(G)$, and we look at those properties of $\mathrm{End}_R(G)$ that can be characterized in terms of G. Specifically we characterize the homological dimensions of G as a left $\mathrm{End}_R(G)$-module, we characterize the global dimensions of $\mathrm{End}_R(G)$ in terms of G, and we consider ring theoretic properties for $\mathrm{End}_R(G)$. E.g. we determine when $\mathrm{End}_R(G)$ is left or right Noetherian, left or right coherent, right or left self-injective, a left or right cogenerator ring, a left or right PF ring, a *QF* ring, or a left or right *FP*-injective ring. We also characterize those C such that $\mathrm{Hom}_R(G, C)$ is a projective or an injective right $\mathrm{End}_R(G)$-module.

There are a few diagrams that illustrate a connection between G, $\mathrm{End}_R(G)$, homology, and point set topological spaces. This equivalence of ideas from different areas of mathematics is rare. Given a (not necessarily self-small) right R-module G there is a commutative diagram of categories and functors in which **M-spaces** denotes a category whose objects are point set topological spaces with specified homology groups. It is usual to call a space concentrated at some integer $k \geq 0$ a *Moore k-space*. Notice that the diagram 16.1 contains the left modules and the right modules over $\mathrm{End}_R(G)$, as well as the functors Tor* and Ext*.

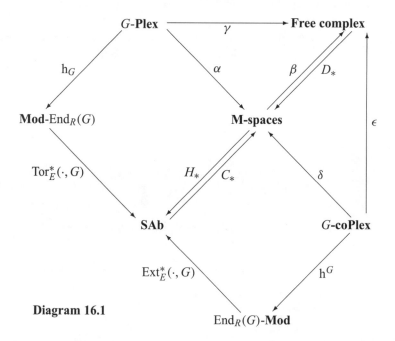

Diagram 16.1

Fix G. By letting X denote the topological space whose fundamental group is G, called an *Eilenberg–MacLane space*, we develop a commutative triangle (see Diagram 17.1) of categories and functors. When G is self-small this triangle consists of category equivalences between homology theory, modules over a ring, and a category of point set topology.

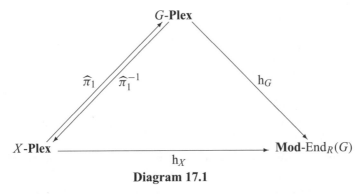

Diagram 17.1

The text ends with a chapter that combines noncommutative localization of rings with an additive functor $\mathbf{QH}_G(\cdot)$, and a new measurement of modules called *margimorphism* to give several right R-modules G possessing unique decompositions. For example, let Q_G denote the semi-primary classical right ring of quotients of the ring $\mathrm{End}_R(G)$. Then G is margimorphic to G' iff $\mathbf{QH}_G(G) \cong \mathbf{QH}_G(G')$ as right Q_G-modules. Furthermore, we prove that Q_G can be used to find a unique direct sum decomposition for G.

Theorem. *Suppose that* $\mathrm{End}_R(G)$ *possesses a semi-primary classical right ring of quotients* Q_G *such that* $Q_G/\mathcal{J}(Q_G)$ *is a product of division rings. Then G has a unique direct sum decomposition in the sense of the Azumaya–Krull–Schmidt theorem.*

Organization: Aside from a preliminary chapter, the book is in three parts:

1. A portion of the book is devoted to the study of a number-theoretic connection between G and $E(G)$. This includes an investigation into the algebraic number theory of algebraic number fields.
2. A portion of the book is devoted to the study of the module and ideal theoretic connections between G and $\mathrm{End}(G)$.
3. A portion of the book shows a categorical connection between G, $\mathrm{End}(G)$, and certain point set topological spaces.

Chapters 2–7 develop a method for utilizing the commutative property in $E(G)$ in discussing unique direct sum decompositions of G. These techniques characterize the class number of an algebraic number field. Chapter 8 uses analytic number theory to study groups G such that each group locally isomorphic to G is isomorphic to G. Chapter 9 gives the homological framework needed to study $\mathrm{End}(G)$ systematically, including the theorem giving the category equivalence $h_G : G\text{-}\mathbf{Plex} \longrightarrow \mathbf{Mod}\text{-}\mathrm{End}_R(G)$. Chapter 10 gives several hypotheses under which the tensor functor $\mathbf{T}_G$ with G is a category equivalence. In Chapter 11 we describe the ring $\mathrm{End}_R(G)$ with small right or left global dimension. Chapters 12–15 give characterizations of modules of the form $\mathrm{Hom}_R(G, C)$. Chapters 16 and 17 give the diagrams relating abelian groups, modules, $\mathrm{End}_R(G)$, and point set topology. These techniques characterize the class number of an algebraic number field. Chapter 18 is devoted to margimorphisms.

Each chapter ends with a number of exercises and the chapters themselves contain many statements of the type *the reader will prove that*... These are details or generalizations that I felt detracted from the discussion. The young ring or module theorist should attempt these exercises. Since examples guide our intuition and guide us to theorems, the reader should not be surprised at the number of examples used to motivate our discussions.

I would like to thank Fordham University for giving me the time and the resources needed to write this book during the years 2002–2007. I would also like to thank my colleagues for carefully reading this book in manuscript form and for their subsequent comments. Their e-mails helped me to polish soft points in the manuscript. I am especially grateful to the faculty and students at New Mexico State University at Las Cruces; to Professors D. Arnold, R. Mines, R. Hunter, E. A. Walker, C. Walker, F. Richman, C. Vinsonhaler, and W. Wickless who brought me to modern ideas for research in endomorphism rings, modules, and abelian groups; and to Professor C. Faith, who introduced me to much of the ring and module theory that the reader will find within.

Since my research style produces many TEX files and almost no paper files, I am both author and technical typist on this project. Any errors within are my responsibility.

1

Preliminary results

This text assumes that the reader is familiar with abelian groups and unital modules over associative rings with unity as contained in the texts by L. Fuchs [59] and F. W. Anderson and K. R. Fuller [7]. We will reference but not prove those results that we feel fall outside of the line of thought of this book. I suggest that you use this chapter as a reference and nothing more. Skim through this chapter. Do not attempt to plow through these results as though they were exercises.

1.1 Rings, modules, and functors

We will deal with several rings at once in our discussions so we will use more than one symbol to denote rings. Thus R and E denote rings. Given right R-modules G and H and an index set $\mathcal{I}$, let $c = \text{card}(\mathcal{I})$. For each $i \in \mathcal{I}$ let $G_i \cong G$. Then

$$G^{(c)} = G^{(\mathcal{I})} = \oplus_{i \in \mathcal{I}} G_i$$
$$G^c = G^{\mathcal{I}} = \prod_{i \in \mathcal{I}} G_i$$

are the usual *direct sum* and *direct product* of c copies of G. We say that G is *indecomposable* if $G \cong H \oplus K$ implies that $H = 0$ or $K = 0$. The abelian group G is said to be *strongly indecomposable* if each subgroup of finite index in G is an indecomposable group.

Let G and H be right R-modules. As usual

$$\text{End}_R(G)$$

is *the ring of R-endomorphisms $f : G \longrightarrow G$* and

$$\text{Hom}_R(G, H)$$

is the group of R-module homomorphisms $f : G \longrightarrow H$. We consider G as a left $\text{End}_R(G)$-module by setting $f \cdot x = f(x)$ for $f \in \text{End}_R(G)$ and $x \in G$. Also $\text{Hom}_R(G, H)$ is a right $\text{End}_R(G)$-module if we define $x \cdot f = x \circ f$ for each

$x \in \mathrm{Hom}_R(G, H)$ and $f \in \mathrm{End}_R(G)$. A right module over $\mathrm{End}_R(G)$ is called an *endomorphism module*. Specifically, $\mathrm{Hom}_R(G, H)$ is an endomorphism module.

The right R-module G is *self-small* if for each cardinal c the canonical injection

$$\mathrm{Hom}_R(G, G)^{(c)} \longrightarrow \mathrm{Hom}_R(G, G^{(c)})$$

is an isomorphism. Equivalently, given an index set $\mathcal{I}$ and an R-module map ϕ : $G \longrightarrow G^{(\mathcal{I})}$ there is a finite subset $\mathcal{J} \subset \mathcal{I}$ such that $\phi(G) \subset G^{(\mathcal{J})}$. Finitely generated modules are self-small, as are the abelian subgroups of finite-dimensional $\mathbf{Q}$-vector spaces. (Such groups are called *torsion-free finite rank groups* in the literature). The quasi-cyclic group $\mathbf{Z}(p^\infty)$ is not self-small for primes $p \in \mathbf{Z}$. (See [59].)

Let E be ring. *The Jacobson radical of E* is the ideal $\mathcal{J}(E)$ defined as follows.

$$
\begin{aligned}
\mathcal{J}(E) &= \cap\{M \mid M \subset E \text{ is a maximal right ideal }\} \\
&= \cap\{M \mid M \subset E \text{ is a maximal left ideal }\} \\
&= \{r \in E \mid 1 + rx \text{ is a unit in } E \text{ for each } x \in E\}
\end{aligned}
$$

In particular, if $J \subset E$ is a right ideal such that $J + \mathcal{J}(E) = E$ then $J = E$.

We say that E is a *local ring* if any of the following equivalent properties hold.

1. E possesses a unique maximal right ideal M.
2. $\mathcal{J}(E)$ is the unique maximal right ideal of E.
3. $u \in E$ is a unit of E iff $u \notin \mathcal{J}(E)$.

Nakayama's theorem 1.1. *Let M be a finitely generated right R-module and let $N \subset M$ be an R-submodule of M. If $N + M\mathcal{J}(R) = M$ then $M = N$.*

The right ideal $I \subset E$ is a *nil right ideal* if each $x \in I$ is *nilpotent*. That is, for each $x \in I$ there exists an integer n such that $x^n = 0$. The *nilradical of E* is the ideal $\mathcal{N}(E)$ that is defined as follows.

$$
\begin{aligned}
\mathcal{N}(E) &= \{x \in E \mid xE \text{ is a nilpotent right ideal in } E\} \\
&= \{x \in E \mid Ex \text{ is a nilpotent left ideal in } E\}
\end{aligned}
$$

Let $x \in E$. Since $x^n = 0$ implies that $1 - x$ is a unit in E (show that one, reader), $1 - xy$ is a unit of E for each $x \in \mathcal{N}(E)$ and $y \in E$. Thus

$$\mathcal{N}(E) \subset \mathcal{J}(E).$$

The right R-module P is *projective* if for each surjection $K \xrightarrow{f} L \to 0$ of right R-modules, each mapping $g : M \to L$ lifts to a mapping $h : M \to K$ such that $fh = g$. The free R-module F for some cardinal c is projective, as is any direct summand of F. In fact, every projective right R-module is a direct summand of a free right R-module.

An *idempotent* is an element $e \in E$ such that $e^2 = e$. Often we will avoid the term idempotent and just write $e^2 = e$. If $e^2 = e \in E$ then eE is a cyclic projective right E-module, and every cyclic projective has this form. Given a ring E and an $e^2 = e \in E$ then

$$eEe = \{exe \,|\, x \in E\}$$

is a ring with identity

$$1_{eEe} = e.$$

Suppose $S \subset R$ are rings. Then S is a *unital subring* of R if $1_S = 1_R$. Although $eEe \subset E$, eEe is not a unital subring of E unless $e = 1$. There are some relationships between eEe and E.

Lemma 1.2. [7, Proposition 5.9] *Let E be a ring and let $e^2 = e$. Then* $\mathrm{End}_E(eE) = eEe$.

The ring E is *semi-perfect* if

1. $E/\mathcal{J}(E)$ is semi-simple Artinian and
2. Given an $\bar{e}^2 = \bar{e} \in E/\mathcal{J}(E)$ there is an $e^2 = e \in E$ such that $\bar{e} = e + \mathcal{J}(E)$. That is, *idempotents lift modulo $\mathcal{J}(E)$.*

See [7, Chapter 7, §27] for a complete discussion of semi-perfect rings and their modules. Fields, local rings, and Artinian rings are semi-perfect rings. $\mathbf{Z}$ is not semi-perfect but the localization of $\mathbf{Z}$ at a prime p, $\mathbf{Z}_p$, is semi-perfect.

The next result follows from [7, Theorem 27.11].

Lemma 1.3. *Let E be a semi-perfect ring and let P be a projective right E-module.*

1. *P is indecomposable iff $\mathrm{End}_E(P)$ is a local ring.*
2. *P is a direct sum of cyclic right E-modules with local endomorphism rings.*

1.2 Azumaya–Krull–Schmidt theorem

Let G be a right R-module. The purpose of this section is to show that under some conditions direct sum decompositions of G are well behaved, in a sense that we will make precise below.

An *indecomposable decomposition* of G is a direct sum

$$G = G_1 \oplus \cdots \oplus G_t$$

for some integer $t > 0$ and indecomposable right R-modules $G_1, \ldots, G_t$.

We say that G has a *unique decomposition* if

1. G has an indecomposable decomposition $G \cong G_1 \oplus \cdots \oplus G_t$, and
2. given another indecomposable decomposition $G \cong H_1 \oplus \cdots \oplus H_s$ then $s = t$ and there is a permutation π of the subscripts $\{1, \ldots, t\}$ such that $G_i \cong H_{\pi(i)}$ for each $i = 1, \ldots, t$.

In this case we call $G_1 \oplus \cdots \oplus G_t$ *the unique decomposition of G*. Notice that the unique decomposition of G is necessarily indecomposable. Professional mathematicians believe that modules possessing a unique decomposition are rare.

The Azumaya–Krull–Schmidt theorem is the most referenced result on the subject of the existence of a unique decomposition for a module. A proof can be found in [7].

The Azumaya–Krull–Schmidt theorem 1.4. [7, Theorem 12.6] *Suppose that* $G = G_1 \oplus \cdots \oplus G_t$ *is an indecomposable decomposition of right R-modules such that* $\mathrm{End}_R(G_i)$ *is a local ring for each* $i = 1, \ldots, t$.

1. *The direct sum* $G = G_1 \oplus \cdots \oplus G_t$ *is a unique decomposition of G,*
2. *If* $G = H \oplus K$ *then there is an* $\mathcal{I} \subset \{1, \ldots, t\}$ *such that* $H \cong \oplus_{i \in \mathcal{I}} G_i$.
3. $G \cong H \oplus K \cong H \oplus N \Longrightarrow K \cong N$ *for any right R-modules H, K, and N.*

Thus direct sum decompositions of projective modules over semi-perfect rings are unique.

1.3 The structure of rings

We present in this section a few results on the structure of a ring and its modules. The first shows that if S is a commutative ring then an S-module M is 0 iff it is locally zero. Thus a function is an epimorphism (or a monomorphism) iff it is locally an epimorphism (or a monomorphism).

Theorem 1.5. [97, Theorem 3.80] *Let S be a commutative ring and let M be a finitely generated S-module. Then* $M = 0$ *iff* $M_I = 0$ *for each maximal ideal* $I \subset S$.

Corollary 1.6. [97, exercise 9.22] *Let S be a commutative ring and let M be a finitely presented S-module. Then M is projective (or a generator) iff* M_I *is projective (or a generator) for each maximal ideal* $I \subset S$.

The next two results give a direct sum decomposition of a finite-dimensional S-algebra, where S is a commutative ring.

Wedderburn's theorem 1.7. [9, Theorem 14.1] *Let A be an Artinian* **k**-*algebra over some field* **k**. *Then there is a semi-simple* **k**-*subalgebra B of A such that* $A = B \oplus \mathcal{N}(A)$ *as* **k**-*vector spaces.*

Theorem 1.8. [9, Beaumont–Pierce theorem, Theorem 14.2] *Let E be a rtffr ring. Then there is a semi-prime subring T of E such that* $T \oplus \mathcal{N}(E)$ *has finite index in E as groups.*

We will need to count the number of isomorphism classes of right ideals in a ring R. To do this we use a result due to Jordan and Zassenhaus.

Lemma 1.9. [94, Jordan–Zassenhaus lemma, Lemma 26.3] *Let E be a semi-prime rtffr ring. Then there are at most finitely many isomorphism classes of right ideals of E.*

It will be necessary to use the fact that each rtffr ring E is $\text{End}(G)$ for some rtffr group G. The results of Butler and Corner are most often referenced in this regard.

Butler's theorem 1.10. [46, Theorem I.2.6] *If E is an rtffr ring whose additive structure is a locally free abelian group then $E \cong \text{End}(G)$ for some group $E \subset G \subset QE$.*

Theorem 1.11. [46, Theorem F.1.1] *If E is an rtffr ring then there is an rtffr group G of rank $2 \cdot \text{rank}(E)$ such that $E \cong \text{End}(G)$.*

Theorem 1.12. [56] *If M is a countable reduced torsion-free left E-module and if $E = \{q \in QE \mid qM \subset M\}$ then there is a short exact sequence*

$$0 \longrightarrow M \longrightarrow G \longrightarrow QE \oplus QE \longrightarrow 0$$

of left E-modules such that $E \cong \text{End}(G)$.

1.4 The Arnold–Lady theorem

Let G be a right R-module, let **Mod**-R denote the category of right R-modules, let

$$\boxed{\textbf{Mod-}\text{End}_R(G) = \text{the category of right } \text{End}_R(G)\text{-modules}}$$

and let

$$\boxed{\text{End}_R(G)\textbf{-Mod} = \text{the category of left } \text{End}_R(G)\text{-modules.}}$$

While characterizing module theoretic properties of G in terms of $\text{End}_R(G)$ will not be easy, the theorem of Arnold–Lady shows us that we can characterize direct summands of G as projective $\text{End}_R(G)$-modules. With this tool we can characterize properties surrounding the direct sum decompositions of G in terms of direct sum decompositions of projective right $\text{End}_R(G)$-modules. Finitely generated projective modules, at least on the surface, seem to be easier to work with than more general modules.

Let

$$\boxed{\begin{array}{c} \mathbf{P}(G) = \{H \mid G^{(c)} \cong H \oplus K \text{ for some cardinal } c > 0 \\ \text{and some right } R\text{-module } K\}. \end{array}}$$

We consider $\mathbf{P}(G)$ to be a full subcategory of the category **Mod**-R of right R-modules. Similarly, given a ring E,

$$\mathbf{P}(E) = \text{category of projective right } E\text{-modules.}$$

Define additive functors

$$\mathbf{T}_G(\cdot) = \cdot \otimes_{\text{End}_R(G)} G \qquad \mathbf{H}_G(\cdot) = \text{Hom}_R(G, \cdot)$$

$$\mathbf{H}_G(\cdot) : \mathbf{Mod}\text{-}R \longrightarrow \mathbf{Mod}\text{-}\text{End}_R(G)$$

$$\mathbf{T}_G(\cdot) : \mathbf{Mod}\text{-}\text{End}_R(G) \longrightarrow \mathbf{Mod}\text{-}R.$$

That is, $\mathbf{H}_G(\cdot)$ takes right R-modules to right $\text{End}_R(G)$-modules, and $\mathbf{T}_G(\cdot)$ takes right $\text{End}_R(G)$-modules to right R-modules. Associated with $\mathbf{H}_G(\cdot)$ and $\mathbf{T}_G(\cdot)$ are the natural transformations

$$\Theta : \mathbf{T}_G \mathbf{H}_G(\cdot) \longrightarrow 1$$

$$\Psi : 1 \longrightarrow \mathbf{H}_G \mathbf{T}_G(\cdot)$$

defined by

$$\Theta_H(f \otimes x) = f(x)$$

$$\Psi_M(x)(\cdot) = \cdot \otimes x$$

for each $f \in \text{Hom}_R(G, M)$ and $x \in G$.

$$\mathbf{T}_G \circ \mathbf{H}_G(\cdot) = \text{Hom}_R(G, \cdot) \otimes_{\text{End}_R(G)} G.$$

$$\mathbf{H}_G \circ \mathbf{T}_G(\cdot) = \text{Hom}_R(G, \cdot_{\text{End}_R(G)} G).$$

A good exercise is to demonstrate that

$$\Theta_{\mathbf{T}_G(M)} \circ \mathbf{T}_G(\Psi_M) = 1_{\mathbf{T}_G(M)}$$

for each right $\text{End}_R(G)$-module M.

One of the themes in this text is to identify categories of modules $\mathcal{C}$ and $\mathcal{D}$ such that Θ_H and Ψ_M are isomorphisms for each $H \in \mathcal{C}$ and $M \in \mathcal{D}$. The first such result is the theorem of Arnold–Lady.

The Arnold–Lady theorem 1.13. [9, Theorem 7.21] *If G is a self-small right R-module then the functors*

$$\mathbf{H}_G(\cdot) : \mathbf{P}(G) \longrightarrow \mathbf{P}(\text{End}_R(G))$$

$$\mathbf{T}_G(\cdot) : \mathbf{P}(\text{End}_R(G)) \longrightarrow \mathbf{P}(G)$$

are inverse category equivalences.

Proof: For the sake of the argument let $E = \text{End}_R(G)$. Since Θ and Ψ are natural transformations $\Theta_{H \oplus K} = \Theta_H \oplus \Theta_K$ for right E-modules H and K, and $\Psi_{M \oplus N} = \Psi_M \oplus \Psi_N$ for right $\text{End}_R(G)$-modules M and N. Moreover, since G is self-small $\Theta_{G^{(I)}} = \oplus_I \Theta_G$. Thus given $H \oplus K \cong G^{(I)}$ we can prove that Θ_H is an isomorphism if we can prove that Θ_G is an isomorphism. Similarly, to show that Ψ_P is an isomorphism for each projective right E-module P, it suffices to show that Ψ_E is an isomorphism.

Consider the map

$$\Psi_E : E \longrightarrow \mathbf{H}_G \mathbf{T}_G(E).$$

Notice that $\mathbf{H}_G \mathbf{T}_G(E) \cong E$ with generator the map $f : G \longrightarrow \mathbf{T}_G(E)$ such that $f(x) = 1 \otimes x$ for each $x \in G$. Then $\Psi_E(1) = f$, which implies that Ψ_E is an isomorphism.

Recall that

$$\Theta_{\mathbf{T}_G(E)} \circ \mathbf{T}_G(\Psi_E) = 1_{\mathbf{T}_G(E)}.$$

Since Ψ_E, and so $\mathbf{T}_G(\Psi_E)$ are isomorphisms, it follows that $\Theta_{\mathbf{T}_G(E)} = \Theta_G$ is an isomorphism. Given our reductions the proof is complete.

Let

> $\mathbf{P}_o(G) = \{H \mid G^{(n)} \cong H \oplus H' \text{ for some integer } n > 0$
> and some right E-module $H'\}.$

Similarly, given a ring E,

> $\mathbf{P}_o(E) = $ the category of finitely generated projective
> right E-modules.

Notice the missing self-small hypothesis in the next result.

Theorem 1.14. [46, Arnold–Lady Theorem] *Let G be a right E-module. The functors*

$$\mathbf{H}_G(\cdot) : \mathbf{P}_o(G) \longrightarrow \mathbf{P}_o(\text{End}_R(G))$$

$$\mathbf{T}_G(\cdot) : \mathbf{P}_o(\text{End}_R(G)) \longrightarrow \mathbf{P}_o(G)$$

are inverse category equivalences.

Example 1.15. Let $G = \oplus_p \mathbf{Z}_p$ where p ranges over the primes in $\mathbf{Z}$. Then $\text{End}(G) = \prod_p \mathbf{Z}_p$ is a semi-hereditary ring and G is a projective ($= \text{flat}$) left $\text{End}(G)$-module. Let $I = \oplus_p \mathbf{Z}_p$ be the ideal in $\text{End}(G)$. Then I is a projective ideal in $\text{End}(G)$ that is not finitely generated and such that $\mathbf{T}_G(I) = IG = G$. Inasmuchas $I \neq \text{End}(G)$ we have shown that $\mathbf{T}_G(\cdot) : \mathbf{P}(\text{End}(G)) \longrightarrow \mathbf{P}(G)$ is not a category equivalence if the self-small hypothesis is deleted from Theorem 1.13.

Example 1.16. Let $G = \mathbf{Q}^{(\aleph_o)}$ and let $E = \text{End}(G)$. Then G is a cyclic projective left E-module so $\mathbf{T}_G(I) \cong IG$ for each right ideal $I \subset \text{End}(G)$. If we let $I = \{f \in A \mid f(G)$ has finite dimension$\}$ then $I \ncong \text{End}(G)$ while

$$\mathbf{T}_G(I) = IG = G = \mathbf{T}_G(\text{End}(G))$$

but $I \neq \text{End}(G)$. This is another example of the necessity of the self-small hypothesis in Theorem 1.13 even though G has a rather restricted left $\text{End}(G)$-module structure.

2

Class number of an abelian group

The study of direct sum decompositions of abelian groups is as old as the study of abelian groups. In this chapter we study the direct sum decompositions of reduced torsion-free finite rank abelian groups, and we show that the associated direct sum problems are equivalent to a pair of deep problems in algebraic number theory.

2.1 Preliminaries

Let G be a reduced torsion-free finite-rank group. Following [46] we write *rtffr* to abbreviate the string of hypotheses *reduced torsion-free finite rank*.

At all times in this text

$$E(G) = \mathrm{End}(G)/\mathcal{N}(\mathrm{End}(G)).$$

We say that G is *cocommutative* if $E(G)$ is a commutative ring. If G is a cocommutative strongly indecomposable rtffr group then $E(G)$ is an rtffr Noetherian commutative integral domain. For instance, it follows from a theorem of J. D. Reid's that strongly indecomposable rank two abelian groups are cocommutative. Any group whose quasi-endomorphism ring is the Hamiltonian quaternions is not a cocommutative group. One of the consequences of the work in this chapter is that cocommutative groups occur naturally and often.

As in [46] we write *locally isomorphic* instead of *nearly isomorphic*, [9], or *in the same genus class*, [94]. Thus, groups G and H are *locally isomorphic* if for each integer $n > 0$ there are maps $f : G \longrightarrow H$ and $g : H \longrightarrow G$ and an integer m such that m is relatively prime to n and $fg = m1_H$ and $gf = m1_G$. Lattices M and N over a semi-prime rtffr ring E are *locally isomorphic* if for each integer $n > 0$ there are E-module maps $f : M \longrightarrow N$ and $g : N \longrightarrow M$ and an integer m such that m is relatively prime to n and $fg = m1_N$ and $gf = m1_M$. The *class number* of X, $h(X)$, where X is either an rtffr group or a lattice, is the number of isomorphism classes of those Y that are locally isomorphic to X.

An important result due to R. B. Warfield, Jr. is the following.

Warfield's theorem 2.1. [9, Theorem 13.9] *Let M and N be rtffr modules over an rtffr ring E. Then M is locally isomorphic to N iff $M^n \cong N^n$ for some integer $n > 0$.*

Let $\mathbf{P}_o(G) = \{$groups $H \mid H \oplus H' = G^m$ for some group H' and some integer $m > 0\}$. We say that G satisfies the *power cancellation property* if $G^n \cong H^n$ for some group H and integer $n > 0$ implies that $G \cong H$. We say that G has *a Σ-unique decomposition* if G^n has a unique direct sum decomposition for each integer $n > 0$. Furthermore, G has *internal cancellation* if given $H, K, L \in \mathbf{P}_o(G)$ such that $H \oplus K \cong H \oplus L$ then $K \cong L$.

Let E be an *rtffr ring*, (i.e., a ring whose additive structure $(E, +)$ is an rtffr group). The semi-prime rtffr ring $\overline{E}$ is *integrally closed* if given a ring $\overline{E} \subset E' \subset \mathbf{Q}\overline{E}$ such that $E'/\overline{E}$ is finite then $\overline{E} = E'$.

Let (X) be the isomorphism class of X, and let

$$\Gamma(X) = \{(Y) \mid Y \text{ is locally isomorphic to } X\}.$$

Let $\mathbf{P}_o(E)$ be the set of finitely generated projective right E-modules. The local isomorphism class of X is denoted by $[X]$.

Lemma 2.2. *Let $P, Q \in \mathbf{P}_o(R)$, and suppose that $J \subset \mathcal{J}(R)$. Then $P/PJ \cong Q/QJ$ iff $P \cong Q$.*

Proof: Since P and Q are projective right R-modules, the isomorphism $P/PJ \cong Q/QJ$ lifts to a map $\phi : P \longrightarrow Q$ such that

$$\ker \phi \subset PJ \quad \text{and} \quad Q = \phi(P) + QJ.$$

By Nakayama's theorem 1.1, $Q = \phi(P)$. Since Q is a projective R-module, $P = U \oplus \ker \phi$ where $U \cong Q$. Furthermore, since $\ker \phi \subset PJ$, Nakayama's theorem shows us that $\ker \phi = 0$, whence $P \cong Q$. The converse is clear so the proof is complete.

Theorem 2.3. [19, Proposition 2.12] *Let R be an rtffr ring. The functor*

$$A_R(\cdot) : \mathbf{P}_o(R) \longrightarrow \mathbf{P}_o(R/\mathcal{N}(R))$$

defined by

$$A_R(\cdot) = \cdot \otimes_R R/\mathcal{N}(R)$$

is full. Furthermore, $A_R(\cdot)$ induces bijections of sets

1. $\{(P) \mid P \in \mathbf{P}_o(R)\} \longrightarrow \{(W) \mid W \in \mathbf{P}_o(R/\mathcal{N}(R))\}$, *and*
2. $\alpha_R : \{[P] \mid P \in \mathbf{P}_o(R)\} \longrightarrow \{[W] \mid W \in \mathbf{P}_o(R/\mathcal{N}(R))\}$.

Proof: Part 1 is true by [19, Proposition 2.12].

2. It is easily verified that α_R is well defined.

Say $W \in \mathbf{P}_o(R/\mathcal{N}(R))$. By part 1 there is a $P \in \mathbf{P}_o(R)$ such that $A_R(P) \cong W$. Thus α_G is a surjection. Say $\alpha_R[P] = \alpha_R[Q]$. Then $[P/P\mathcal{N}(R)] = [Q/Q\mathcal{N}(R)]$. By Warfield's theorem 2.1, $(P/P\mathcal{N}(R))^n \cong (Q/Q\mathcal{N}(R))^n$ for some integer $n > 0$.

Then $P^n/P^n\mathcal{N}(R) \cong Q^n/Q^n\mathcal{N}(R)$, so by Lemma 2.2, $P^n \cong Q^n$. Hence, by Warfield's theorem 2.1, $[P] = [Q]$, and so α_R is an injection. This proves part 2.

The next result, due to D.M. Arnold, follows immediately from [9, Theorem 9.6(a), Corollary 12.7(b)].

Theorem 2.4. *Let G be an rtffr group. There is a full additive functor*

$$A_G(\cdot) : \mathbf{P}_o(G) \longrightarrow \mathbf{P}_o(E(G))$$

defined by

$$A_G(\cdot) = \mathrm{Hom}(G, \cdot) \otimes_{\mathrm{End}(G)} E(G).$$

1. *Given $P \in \mathbf{P}_o(E(G))$ there is an $H \in \mathbf{P}_o(G)$, that is unique up to isomorphism, such that $A_G(H) \cong P$.*
2. *For $H \in \mathbf{P}_o(G)$, $A_G(\cdot)$ induces a bijection $\Gamma(H) \longrightarrow \Gamma(A_G(H))$.*

The next result follows immediately from [9, Theorem 9.8, and Corollaries 9.7 and 10.14].

Lemma 2.5. *If G is a cocommutative strongly indecomposable rtffr group then $E(G)$ is a Noetherian commmutative integral domain.*

The next result shows us that cocommutative rtffr groups are common enough.

Lemma 2.6. *Let G be a strongly indecomposable rtffr group. If $\mathrm{rank}(G)$ is square-free then $E(G)$ is a commutative Noetherian rtffr integral domain. That is, G is cocommutative .*

Proof: Suppose that G is strongly indecomposable of square-free rank. By the Wedderburn theorem 1.7 there is a semi-simple Artinian subalgebra $B \subset \mathbf{Q}\mathrm{End}(G)$ such that $\mathbf{Q}\mathrm{End}(G) = B \oplus \mathcal{N}(\mathbf{Q}\mathrm{End}(G))$ as B-modules. Since G is strongly indecomposable, B is a division algebra. Thus $\mathbf{Q}G = B^n$ for some integer $n \geq 1$. Let $\mathbf{k}$ be the center of B, and say that $\mathbf{k}$ has degree $\ell \geq 1$ over $\mathbf{Q}$. Since $\mathbf{k}$-$\dim(B) = c^2 \geq 1$ is a perfect square, [94], we have $\mathbf{Q}G = B^n = \mathbf{Q}^{c^2 n\ell}$ so that c^2 divides the square-free integer $\mathbf{Q}$-$\dim(\mathbf{Q}G) = \mathrm{rank}(G)$. Then $c = 1$, thus $B = \mathbf{k}$ is commutative, and therefore $B \cong \mathbf{Q}E(G)$ is commutative. Hence $E(G)$ is a Noetherian commutative rtffr integral domain. This completes the proof.

Let E be an rtffr commutative integral domain. We say that U is a *fractional ideal of E* if U is a finitely generated E-submodule of $\mathbf{Q}E$ In this case $\mathbf{Q}U = \mathbf{Q}E$. We say that U is *invertible* if $U^*U = E$ where $U^* = \{q \in \mathbf{Q}E \mid qU \subset E\}$. Given an rtffr commutative integral domain E, let

$$\boxed{\mathrm{Pic}(E)}$$

be the set of isomorphism classes (U) of invertible fractional ideals U of E. Then $\text{Pic}(E)$ is a finite abelian group if we define $(U)(W) = (UW)$ and if $(U)^{-1} = (U^*)$. $\text{Pic}(E)$ is finite by the Jordan–Zassenhaus lemma 1.9. Let

$$h(E) = \text{card}(\text{Pic}(E)).$$

I. Kaplansky [94, Theorem 38.13] shows that if E is a Noetherian commutative integral domain, (if, e.g., E is a commutative rtffr integral domain), then E satisfies the following property.

Property 2.7. Let $U_1, \ldots, U_t, W_1, \ldots, W_t$ be fractional ideals of the Noetherian commutative integral domain E. Then

$$U_1 \oplus \cdots \oplus U_s \cong W_1 \oplus \cdots \oplus W_t \quad \text{iff} \quad s = t \quad \text{and} \quad U_1 \cdots U_s \cong W_1 \cdots W_t.$$

For example, let E be a ring that satisfies Property 2.7, and let $U_1, \ldots, U_t$ be invertible fractional ideals of E. Then

$$U_1 \oplus \cdots \oplus U_t \cong E^{t-1} \oplus (U_1 \cdots U_t).$$

For example, by Lemma 2.6, Property 2.7 is satisfied by $E(G)$ for a strongly indecomposable rtffr group G of square-free rank.

Next we present several properties about the groups and rings that are quasi-equal to G and $E(G)$. These groups and associated integers are essential to the subsequent results.

Lemma 2.8. [9, Corollary 10.14] *Let E be an rtffr semi-prime ring. There is an integrally closed ring $\overline{E}$ such that $E \subset \overline{E}$ and such that $\overline{E}/E$ is a finite group. Thus there is an integer $n(E) = n > 0$ such that $n\overline{E} \subset E \subset \overline{E}$.*

Let E be a commutative rtffr integral domain. By the previous Lemma there is an integrally closed ring (a Dedekind domain) $\overline{E}$ such that $E \subset \overline{E}$ and $\overline{E}/E$ is a finite group. Thus there is an integer $n(E) = n > 0$ such that $n\overline{E} \subset E \subset \overline{E}$. The next lemma shows us that a similar arrangement occurs for rtffr groups.

Lemma 2.9. *Let G be an rtffr group. There is an rtffr group $\overline{G}$ and an integer $n(G) = n > 0$ such that $n\overline{G} \subset G \subset \overline{G}$, such that $nE(\overline{G}) \subset E(G) \subset E(\overline{G})$, and such that $E(\overline{G})$ is integrally closed.*

Proof: By the Beaumont–Pierce theorem 1.8 there is a semi-prime ring $T \subset \text{End}(G)$ such that $\text{End}(G) \doteq T \oplus \mathcal{N}(\text{End}(G))$. Then $\mathbf{Q}\text{End}(G) = B \oplus \mathcal{J}$ where $B = \mathbf{Q}T$ is a semi-simple subalgebra of $\mathbf{Q}\text{End}(G)$ and where $\mathcal{J} = \mathbf{Q}\mathcal{N}(\text{End}(G))$. We identify $E(G)$ with a full subring of B in the natural way.

$$E(G) \cong \{(x, 0) \in B \oplus \mathcal{J} \mid (x, y) \in \text{End}(G) \text{ for some } y \in \mathcal{N}(\text{End}(G))\}.$$

Then $T \subset E(G)$, and

$$E(G) \cong \mathrm{End}(G)/\mathcal{N}(\mathrm{End}(G)) \doteq T.$$

By Lemma 2.8 there is an integer $n = n(G) \neq 0$ and an integrally closed rtffr subring $\overline{T}$ of B such that

$$n\overline{T} \subset T \subset E(G) \subset \overline{T}.$$

Let $\overline{G} = \overline{T}G \subset \mathbf{Q}G$. Since $1 \in T$,

$$n\overline{G} = (n\overline{T})G \subset TG = G \subset \overline{G},$$

so that $\overline{G}/G$ is finite. Evidently $\overline{T} \subset \mathrm{End}(\overline{G})$ so that $\overline{T} \subset E(\overline{G})$. Because $G \doteq \overline{G}$, $E(\overline{G}) \doteq E(G) \doteq T \doteq \overline{T}$, so that $E(\overline{G})/\overline{T}$ is finite. Because $\overline{T}$ is integrally closed, $\overline{T} = E(\overline{G})$. Because $E(G) \subset \overline{T}$,

$$nE(\overline{G}) = n\overline{T} \subset E(G) \subset \overline{T} = E(\overline{G}),$$

which completes the proof.

The group $\overline{G}$ constructed above is called an *integral closure of G*. If $G = \overline{G}$ then we say that G is an *integrally closed group*.

Lemma 2.10. [9, Corollary 12.14(a)] *Let U be a fractional ideal in the commutative rtffr integral domain E, and let $n(E) = n > 0$ be an integer such that $n\overline{E} \subset E \subset \overline{E}$ for some integrally closed ring $\overline{E} \subset \mathbf{Q}E$. If $U + nE = E$ then U is an invertible ideal in E.*

The next result shows us that a projective module over a commutative rtffr integral domain decomposes completely.

Lemma 2.11. *Let E be a commutative rtffr integral domain. If P is a finitely generated projective right E-module then $P = U_1 \oplus \cdots \oplus U_t$ for some invertible fractional ideals $U_1, \ldots, U_t$ of E.*

Proof: Let E be an commutative rtffr integral domain, and let P be a finitely generated projective E-module. By Lemma 2.8 there is an integrally closed ring $\overline{E}$ and an integer $n = n(E) > 0$ such that $n\overline{E} \subset E \subset \overline{E}$. Consider E/nE. Since E/nE is finite, $E/nE = A_1 \times \cdots \times A_r$ for some integer $r > 0$ and some local commutative Artinian rings $A_1, \ldots, A_r$.

Furthermore, for each maximal ideal $M \subset E$, P_M is a (free) generator of the local ring E_M, so by the Local-Global Theorem 1.5, P is a generator of E. Thus P/nP is a finitely generated projective generator of E/nE. It follows that $P/nP = B_1 \oplus \cdots \oplus B_r$ where each B_i is a projective generator of A_i. Since each A_i is local there exists for each $i = 1, \ldots, r$ a surjection $f_i : B_i \longrightarrow A_i$. Consequently there is a surjection $f = \oplus_i f_i : P/nP \longrightarrow E/nE$. Since P is projective, f lifts to a map $\phi : P \longrightarrow E$ such that $\phi(P) + nE = E$. Let $U_1 = \phi(P)$ and let $P_2 = \ker \phi$. By Lemma 2.10, U_1 is an invertible fractional ideal of E, and therefore $P = U_1 \oplus P_2$. By an induction on the $E(G)$-rank of P, $P_2 = U_2 \oplus \cdots \oplus U_t$ for some invertible fractional ideals $U_2, \ldots, U_t$. This completes the proof.

Corollary 2.12. *Let G be a cocommutative strongly indecomposable rtffr group.*

1. *If $P \in \mathbf{P}_o(E(G))$ then $P = U_1 \oplus \cdots \oplus U_t$ for some integer $t > 0$ and some invertible fractional ideals $U_1, \ldots, U_t$ of $E(G)$.*
2. *If $H \in \mathbf{P}_o(G)$ then $H \cong H_1 \oplus \cdots \oplus H_t$ for some groups $H_1, \ldots, H_t$ such that each H_i is locally isomorphic to G.*

Proof: 1. By Lemma 2.5, $E(G)$ is a commutative Noetherian rtffr integral domain. Let $P \in \mathbf{P}_o(E(G))$. Then by Lemma 2.11, $P = U_1 \oplus \cdots \oplus U_t$ for some integer $t > 0$ and some invertible fractional ideals $U_1, \ldots, U_t$ of $E(G)$. This proves part 1.

2. Apply Theorem 2.4 to part 1. This completes the proof.

2.2 A functorial bijection

Let E be a commutative rtffr integral domain, and let $\mathrm{Pic}(E)$ be the set of isomorphism classes of invertible fractional ideals of E. Read [94, Corollary 38.12] to see that our definition of $\mathrm{Pic}(E)$ agrees with the traditional one.

Lemma 2.13. [46, Lemma 2.5.4] *Let E be an rtffr ring, and let M and N be finitely generated projective right E-modules. The following are equivalent.*

1. *M is locally isomorphic to N.*
2. *$M_p \cong N_p$ as right E_p-modules for each prime $p \in \mathbf{Z}$.*
3. *$M_n \cong N_n$ as right E_n-modules for each $n \in \mathbf{Z}$.*

Lemma 2.14. *Let E be a commutative Noetherian rtffr integral domain, and let I be a fractional ideal of E. Then I is invertible iff I is locally isomorphic to E.*

Proof: By Lemma 2.13, we need only prove that I is projective if I is either invertible or locally isomorphic to E.

If I is an invertible fractional ideal of the commutative ring E then $II^* = E$, so that I is a generator of E. Let

$$\mathrm{End}_E(I) = \mathcal{O}(I) = \{q \in \mathbf{Q}E \mid qI \subset I\}.$$

Then $E \subset \mathcal{O}(I)$ and I is projective over $\mathcal{O}(I)$. But also

$$\mathcal{O}(I) = \mathcal{O}(I)E = \mathcal{O}(I)II^* = II^* = E.$$

Thus I is projective over E.

If I is locally isomorphic to E then by Warfield's theorem 2.1, $I^n \cong E^n$ for some integer $n > 0$, and so I is a projective E-module. This completes the proof.

The following functorial bijection changes direct sums of rtffr groups into a product of elements in a finite multiplicative abelian group. This is different from the usual categorical methods used to study direct sum decompositions of torsion-free finite rank groups. See the Arnold–Lady theorem 1.13.

Lemma 2.15. *Let G be a cocommutative strongly indecomposable rtffr group, and let $n > 0$ be an integer. There is a bijection of finite sets, $\lambda_n : \Gamma(G^n) \longrightarrow \mathrm{Pic}(E(G))$.*

Proof. It follows from the Jordan–Zassenhaus lemma 1.9 that there are at most finitely many isomorphism classes of fractional ideals of $E(G)$. Thus $\mathrm{Pic}(E(G))$ is a finite multiplicative abelian group.

Let $n > 0$ be an integer. To define λ_n, let H be locally isomorphic to G^n. By Theorem 2.4, $A_G(H)$ is a finitely generated projective right $E(G)$-module of rank n over the integral domain $E(G)$. By uniqueness of rank over $E(G)$ and by Corollary 2.12.1, there are invertible fractional right ideals $U_1, \ldots, U_n$ such that $A_G(H) = U_1 \oplus \cdots \oplus U_n$. We define

$$\lambda_n((H)) = (U_1) \cdots (U_n) \in \mathrm{Pic}(E(G)).$$

We will show that λ_n is well defined. Let $(H) \in \Gamma(G^n)$ and suppose that

$$A_G(H) \cong U_1 \oplus \cdots \oplus U_r \cong W_1 \oplus \cdots \oplus W_s$$

for some invertible fractional right ideals U_i and W_j. The uniqueness of rank over the integral domain $E(G)$ implies that $r = s$. Then by Property 2.7,

$$(U_1) \cdots (U_r) = (U_1 \cdots U_r) = (W_1 \cdots W_r) = (W_1) \cdots (W_r),$$

and hence $\lambda_n((H))$ is well defined.

To see that λ_n is a surjection, let U be an invertible fractional ideal of $E(G)$. By Theorem 2.4.2 there is a group H that is locally isomorphic to G such that $A_G(H) \cong U$. Then $G^{n-1} \oplus H$ is locally isomorphic to G^n. Since $(A_G(G)) = (E(G))$ is the identity in $\mathrm{Pic}(E(G))$,

$$\lambda_n((G^{n-1} \oplus H)) = \underbrace{(E(G)) \cdots (E(G))}_{n-1} (A_G(H)) = (U).$$

Thus λ_n is a surjection.

Let $(H), (K) \in \Gamma(G^n)$ be such that $\lambda_n((H)) = \lambda_n((K))$. Using Corollary 2.12.1 we can write

$$A_G(H) \cong U_1 \oplus \cdots \oplus U_r \text{ and } A_G(K) \cong W_1 \oplus \cdots \oplus W_s$$

for some invertible fractional ideals U_i and W_j. By our choice of H and K and by Theorem 2.4.2, $A_G(H)$, $A_G(K)$, and $A_G(G^n) = E(G)^n$ are locally isomorphic. Uniqueness of rank over the integral domain $E(G)$ implies that $n = r = s$. Moreover, since then

$$(U_1 \cdots U_n) = (U_1) \cdots (U_n) = \lambda_n((H))$$
$$= \lambda_n((K)) = (W_1) \cdots (W_n) = (W_1 \cdots W_n),$$

Property 2.7 implies that

$$A_G(H) \cong U_1 \oplus \cdots \oplus U_n \cong W_1 \oplus \cdots \oplus W_n \cong A_G(K).$$

Then by Theorem 2.4.2, $H \cong K$, and the proof is complete.

Given an abelian group G, the *class number of G* is the number

$$\boxed{h(G)}$$

of isomorphism classes of groups H that are locally isomorphic to G.

Corollary 2.16. *Let G be a cocommutative strongly indecomposable rtffr group. Then*

$$h(G^n) = h(G) = h(E(G)) = \mathrm{card}(\mathrm{Pic}(E(G)))$$

is finite for each integer $n > 0$.

Our first example shows how one might use Lemma 2.15 to construct groups.

Example 2.17. For each square-free integer h there are integrally closed strongly indecomposable rank-two groups $G_h \ncong H_h$ such that $G_h^h \cong H_h^h$ while $G_h^k \ncong H_h^k$ for each integer $1 \le k < h$.

Proof: Let $h > 0$ be a square-free integer. There is a quadratic number field $\mathbf{k}_h$ such that $h(\mathbf{k}_h) = h$. Let $\overline{E}_h = \overline{E}$ be the ring of algebraic integers in $\mathbf{k}_h$. By Butler's theorem 1.10 we can choose a strongly indecomposable rank two group $\overline{G}_h = \overline{G}$ such that

$$\mathrm{End}(\overline{G}) = E(\overline{G}) = \overline{E}.$$

By Corollary 2.16 and our choice of h,

$$h(\overline{G}) = h(\overline{E}) = h(\mathbf{k}_h) = h \ne 1,$$

so $\mathrm{Pic}(\overline{E}) \ne 1$. Moreover, since the group $\mathrm{Pic}(\overline{E})$ has square-free order h, $\mathrm{Pic}(\overline{E})$ is cyclic. We choose a generator (U) for $\mathrm{Pic}(\overline{E})$ of order h, and then we use λ_1 in Lemma 2.15 to choose $(H) \in \Gamma(G)$ such that $\lambda_1((H)) = (U)$. For integers $1 \le k < h$,

$$\lambda_k(G^k) = (\overline{E})^k = (\overline{E}) \ne (U)^k = \lambda_k(H^k),$$

while

$$\lambda_h(G^h) = (\overline{E}) = (U)^h = \lambda_h(H^h).$$

Hence, by Lemma 2.15, $G^k \not\cong H^k$ for integers $1 \le k < h$, while $G^h \cong H^h$.

2.3 Internal cancellation

In this section we apply Lemma 2.15 to cancellation problems. The next example looks like failure of unique factorization in finite abelian groups. It is a generalization of an example due to D. M. Arnold [9, Example 5.4]. The *class number* $h(\mathbf{k})$ of the algebraic number field is the number of isomorphism classes of fractional ideals of the ring of algebraic integers $\overline{E}$ of $\mathbf{k}$.

Example 2.18. There are strongly indecomposable rtffr groups A, B, and C of rank two such that $A \oplus A \cong B \oplus C$ but $A \not\cong B$ and $A \not\cong C$.

Proof: Let $\mathbf{k}$ be a quadratic number field with class number $h(\mathbf{k}) = h \ge 2$, and let $\overline{E}$ be the ring of algebraic integers in $\mathbf{k}$. (See [103, page 181] for such a field $\mathbf{k}$.) Since $\overline{E}$ is a free abelian group of rank two, Butler's theorem 1.10 constructs a torsion-free group A of rank two such that $\text{End}(A) \cong \overline{E}$. Since $h \ge 2$ there is an invertible ideal $I \subset \overline{E}$ such that $I \not\cong \overline{E}$. Let $J = I^*$. Then one can show that

$$I \oplus J \cong \overline{E} \oplus IJ \cong \overline{E} \oplus \overline{E}.$$

By the Arnold–Lady theorem 1.13 there are groups $B, C \in \mathbf{P}_o(A)$ such that $\mathbf{H}_A(B) \cong I$ and $\mathbf{H}_A(C) = J$. Then

$$\mathbf{H}_A(A \oplus A) \cong \overline{E} \oplus \overline{E} \cong I \oplus J \cong \mathbf{H}_A(B \oplus C).$$

Another application of the Arnold–Lady theorem 1.13, $A \oplus A \cong B \oplus C$. But since $\overline{E} \not\cong I, J$ we have $A \not\cong B, C$.

Part 2 of the next result slightly extends Example 2.18.

Theorem 2.19. *Let G be a cocommutative strongly indecomposable rtffr group.*

1. *If H is a group such that $G \oplus G \cong G \oplus H$ then $G \cong H$.*
2. *If H is locally isomorphic to G then there is a group K unique to H such that $G \oplus G \cong H \oplus K$.*

Proof: 1. Given $G \oplus G \cong G \oplus H$ then by definition of λ_2,

$$(A_G(G)) = (E(G))(E(G)) = \lambda_2((G \oplus G))$$
$$= \lambda_2((G \oplus H)) = (E(G))(A_G(H)) = (A_G(H)).$$

Hence $A_G(G) \cong A_G(H)$, and by Theorem 2.4.2, $G \cong H$.

2. Suppose that H is locally isomorphic to G. Then $(H) \in \Gamma(G)$ so that $(A_G(H)) \in$ Pic$(E(G))$. The group Pic$(E(G))$ contains $(A_G(H))^{-1}$, and by Lemma 2.15 there is a unique group K that is locally isomorphic to G such that

$$(A_G(K)) = \lambda_1((K)) = (A_G(H))^{-1}.$$

Because $(E(G))$ is the identity in Pic$(E(G))$,

$$\lambda_2((G \oplus G)) = (E(G))(E(G)) = (E(G))$$
$$= (A_G(H))(A_G(H))^{-1} = (A_G(H))(A_G(K)) = \lambda_2((H \oplus K)),$$

so by Lemma 2.15, $G \oplus G \cong H \oplus K$. This completes the proof.

2.20. Jónsson's theorem [9, Corollary 7.9] can be stated as follows. Given an rtffr group G there are integers $t, e_1, \ldots, e_t > 0$ and strongly indecomposable pairwise non-quasi-isomorphic groups $G_1, \ldots, G_t$ such that G is quasi-isomorphic to $G_1^{e_1} \oplus \cdots \oplus G_t^{e_t}$ and this decomposition is unique to G up to the quasi-isomorphism classes of the G_i.

2.21. By [9, Theorem 9.10], if t, e_i, G_i are as in (2.20) and if $G = G_1^{e_1} \oplus \cdots \oplus G_t^{e_t}$ then $E(G) = E(G_1^{e_1}) \times \cdots \times E(G_t^{e_t})$. (You might try proving this result as an exercise.)

The rtffr group G has the *internal cancellation property* if given $H, K, L \in \mathbf{P}_o(G)$ such that $H \oplus K \cong H \oplus L$ then $K \cong L$.

Theorem 2.22. *A cocommutative strongly indecomposable rtffr group G has the internal cancellation property.*

Proof: Suppose that $H \oplus K \cong H \oplus L$ for some $H, K, L \in \mathbf{P}_o(G)$. By Corollary 2.12.2, write $H = H_1 \oplus \cdots \oplus H_r$, $K = K_1 \oplus \cdots \oplus K_s$, and $L = L_1 \oplus \cdots \oplus L_t$, where the groups H_i, K_i, and L_i are locally isomorphic to G. Because G is strongly indecomposable, the H_i, K_i, and L_i are strongly indecomposable. Jónsson's theorem implies that $s = t$. Then H is locally isomorphic to G^r, K and L are locally isomorphic to G^s, and $H \oplus K \cong H \oplus L$ is locally isomorphic to G^{r+s}. Apply λ_{r+s} to show that

$$\lambda_r((H))\lambda_s((K)) = \lambda_{r+s}((H \oplus K)) = \lambda_{r+s}((H \oplus L)) = \lambda_r((H))\lambda_s((L)).$$

Cancel the term $\lambda_r((H))$ from this equation of group elements. Then $\lambda_s((K)) = \lambda_s((L))$. Hence $K \cong L$ by Lemma 2.15.

Theorem 2.23. *Choose integers $t, e_1, \ldots, e_t > 0$ and let $G = G_1^{e_1} \oplus \cdots \oplus G_t^{e_t}$ be a direct sum of cocommutative strongly indecomposable pairwise non-quasi-isomorphic rtffr groups $G_1, \ldots, G_t$. Then G has the internal cancellation property.*

Proof: By Theorem 2.22, G_i, and so $G_i^{e_i}$, have internal cancellation for each $i = 1, \ldots, t$. By Theorem 2.4.2, $E(G_i^{e_i})$ has internal cancellation for each $i = 1, \ldots, t$, so by (2.21), $E(G)$ has internal cancellation. Hence by Theorem 2.4.2, G has internal cancellation.

Corollary 2.24. *Choose integers $t, e_1, \ldots, e_t > 0$ and let $G = G_1^{e_1} \oplus \cdots \oplus G_t^{e_t}$ be a direct sum of strongly indecomposable pairwise non-quasi-isomorphic groups $G_1, \ldots, G_t$. If each G_i has square-free rank then G has the internal cancellation property.*

2.4 Power cancellation

The group G has the *power cancellation property* if given a group H and an integer $n > 0$ such that $G^n \cong H^n$ then $G \cong H$. In this section we take advantage of the fact that $h(G)$ is the order of a finite abelian group to investigate the power cancellation property.

Lemma 2.25. [9, Corollary 7.17] *Let G be an rtffr group and let $n > 0$ be an integer. If $G^n \cong H^n$ then G is locally isomorphic to H.*

We will use Lemma 2.25 instead of Warfield's theorem 2.1 because we wish to slightly generalize Warfield's theorem.

Theorem 2.26. *Let G be a cocommutative strongly indecomposable rtffr group, and let H be a group. Then G is locally isomorphic to H iff $G^{h(G)} \cong H^{h(G)}$.*

Proof: Suppose that G is locally isomorphic to H. Since $A_G(\cdot)$ is an additive functor, $A_G(H)$ is locally isomorphic to $E(G)$. By Lemma 2.14, $A_G(H)$ is an invertible fractional ideal in $E(G)$, so $(A_G(H)) \in \text{Pic}(E(G))$. Furthermore, since $A_G(\cdot)$ is an additive functor,

$$A_G(H^{h(G)}) \cong A_G(H)^{h(G)}.$$

Since $\text{Pic}(E(G))$ is an abelian group of order $h(G)$, and by the definition of $\lambda_{h(G)}$, we have

$$\lambda_{h(G)}((H^{h(G)})) = (A_G(H))^{h(G)} = (E(G)) = \lambda_{h(G)}((G^{h(G)})),$$

where in this case $(A_G(H))^{h(G)}$ is a group product. By Lemma 2.15, $G^{h(G)} \cong H^{h(G)}$. The converse follows from Lemma 2.25, so that the proof is complete.

In a finite abelian group X of order n, and if $x \in X$, then $x^m = 1$ implies $x = 1$ if m is relatively prime to n. We can use this elementary fact to classify $h(G)$.

Theorem 2.27. *Let G be a cocommutative strongly indecomposable rtffr group. The following are equivalent for an integer $n > 0$.*

1. n is relatively prime to $h(G)$.
2. If H is a group such that $G^n \cong H^n$ then $G \cong H$.

Proof: $1 \Rightarrow 2$. Assume that n is relatively prime to $h(G)$, and suppose that $G^n \cong H^n$ for some group H. By Lemma 2.25, H is locally isomorphic to G so that $(H) \in \Gamma(G)$.

By Theorem 2.4.2 and Lemma 2.14, $A_G(H)$ is an invertible fractional ideal of $E(G)$, and $\lambda_n((H^n)) = (A_G(H))^n$ where $(A_G(H))^n$ is a product of group elements in $\text{Pic}(E(G))$. By the definition of λ_n we can write

$$(E(G)) = \lambda_n((G^n)) = \lambda_n((H^n)) = (A_G(H))^n.$$

Hence $(E(G)) = (A_G(H))^n$, so that by Corollary 2.16, the order of $(A_G(H))$ divides n and $h(E(G)) = h(G)$. Since n is relatively prime to $h(G)$, the order of $(A_G(H))$ is 1. It follows that

$$\lambda_1((G)) = (E(G)) = (A_G(H)) = \lambda_1((H))$$

so by Lemma 2.15, $G \cong H$. This proves part 2.

$2 \Rightarrow 1$. We prove the contrapositive. Assume that $p \in \mathbf{Z}$ is a prime divisor of n and of $h(G)$. Since $h(G) = h(E(G))$ is the order of the finite abelian group $\text{Pic}(E(G))$ there is a $(U) \in \text{Pic}(E(G))$ of order p. Then $(E(G)) \neq (U)$ and $(E(G)) = (U)^p$. By Lemma 2.15, there is a $(G) \neq (H) \in \Gamma(G)$ such that $\lambda_1((H)) = (U)$. Then

$$\lambda_p((G^p)) = (E(G)) = (U)^p = \lambda_p((H^p)).$$

Another application of Lemma 2.15 shows us that $G^p \cong H^p$. By our choice of p there is an integer b such that $pb = n$. Then $G^n = [G^p]^b \cong [H^p]^b = H^n$. This completes the proof.

The proof of the next corollary follows immediately from Theorem 2.27.

Corollary 2.28. *Let G be a cocommutative strongly indecomposable rtffr group. Let $p \in \mathbf{Z}$ be a prime. Then $h(G)$ is divisible by p iff there is a group H that is locally isomorphic to G such that $G \not\cong H$ and $G^p \cong H^p$.*

Corollary 2.29. *Let G be a cocommutative strongly indecomposable rtffr group. Then $h(G)$ is odd iff $G \oplus G \cong H \oplus H$ implies that $G \cong H$ for any group H.*

The next result shows why we consider the class number $h(G)$ of a cocommutative strongly indecomposable rtffr group G to be a measure of how far G is from satisfying the power cancellation property.

Theorem 2.30. *Let G be a cocommutative strongly indecomposable rtffr group. Then $h(G) = 1$ iff G has the power cancellation property.*

Proof. Assume that $h(G) = 1$, and let $n > 0$ be an integer. Then n is relatively prime to $h(G)$ so that, by Theorem 2.27, $G^n \cong H^n$ implies that $G \cong H$ for any group H. Thus, G has the power cancellation property.

Conversely, let p be a prime divisor of $h(G) \neq 1$. Then by Corollary 2.28, G fails to have the power cancellation property. This completes the proof.

Corollary 2.31. *Let G be a cocommutative strongly indecomposable rtffr group. If G has the power cancellation property then G^n has the power cancellation property for each integer $n > 0$.*

Proof: Suppose that G has the power cancellation property. By Theorem 2.30, $h(G) = 1$. Then Corollary 2.16 and Theorem 2.30 show us that $h(G^n) = h(G) = 1$. Hence by Theorem 2.30, G^n has the power cancellation property. This completes the proof.

2.5 Unique decomposition

Let G be a group. Then G has a *unique decomposition* if

1. There is an integer $t > 0$ and indecomposable groups $G_1, \ldots, G_t$ such that $G = G_1 \oplus \cdots \oplus G_t$, and
2. If $G = G_1' \oplus \cdots \oplus G_s'$ for some integer $s > 0$ and some indecomposable groups $G_1', \ldots, G_s'$ then $s = t$ and after a permutation of the subscripts, $G_i \cong G_i'$ for each $i = 1, \ldots, t$.

We say that G has a Σ-*unique decomposition* if G^n has a unique decomposition for each integer $n > 0$. We examine $h(G)$ for those groups G that have a Σ-unique decomposition.

Theorem 2.32. *Let G be a cocommutative strongly indecomposable rtffr group. Then $h(G) = 1$ iff G has a Σ-unique decomposition.*

Proof: Say $h(G) \neq 1$ and choose a prime divisor p of $h(G)$. Then by Corollary 2.28 there is a group H such that $G \not\cong H$ but such that $G^p \cong H^p$. Thus G does not have a Σ-unique decomposition.

Suppose, conversely, that $h(G) = 1$. Let $n > 0$ be an integer and write $G^n = H_1 \oplus \cdots \oplus H_m$ for some indecomposable groups $H_1, \cdots, H_m$. By Theorem 2.4, $A_G(H_i)$ is an indecomposable projective right $E(G)$-module. Since $E(G)$ is an rtffr integral domain, Corollary 2.12.1 implies that $A_G(H_i)$ is an invertible fractional ideal of $E(G)$. Thus by Lemma 2.14, $A_G(H_i)$ is locally isomorphic to $E(G) = A_G(G)$, and so by Theorem 2.4.2, H_i is locally isomorphic to G. Since $h(G) = 1$, $G \cong H_i$ for $i = 1, \ldots, m$. Thus G^n is quasi-isomorphic to G^m. Jónsson's theorem implies that $n = m$, and so G has a Σ-unique decomposition. This completes the proof.

Corollary 2.33. *Let G be a cocommutative strongly indecomposable rtffr group. Then G has the power cancellation property iff G has a Σ-unique decomposition.*

Proof: Apply Theorem 2.30 and Theorem 2.32.

Proposition 2.34. *Let G be a cocommutative strongly indecomposable rtffr group. Then G has a Σ-unique decomposition iff G^n has a Σ-unique decomposition for each integer $n > 0$.*

Proof: Suppose that G^n has a Σ-unique decomposition. Let $m > 0$ be an integer and let $G^m = H_1 \oplus \cdots \oplus H_t$ for some integer $t > 0$ and some indecomposable groups $H_1, \ldots, H_t$. By Corollary 2.12.2 each H_i is locally isomorphic to G, so by Jónsson's theorem, $m = t$. Then $(G^n)^m = H_1^n \oplus \cdots \oplus H_m^n$ so by hypothesis $H_i \cong G$ for each

$i = 1, \ldots, t$. Thus G has a Σ-unique decomposition. The converse is clear so the proof is complete.

Proposition 2.35. *Let G be a cocommutative strongly indecomposable rtffr group. Then G has a Σ-unique decomposition iff $G \oplus G$ has a unique decomposition.*

Proof: Assume that G does not have a Σ-unique decomposition. There is a direct sum $G^n \cong H_1 \oplus \cdots \oplus H_t$ for some indecomposable groups $H_1, \ldots, H_t$, that are locally isomorphic to G (Corollary 2.12.2), but such that $G \not\cong H_1$. By Theorem 2.19.2 there is a group K such that $G \oplus G \cong H_1 \oplus K$. By our choice of $G \not\cong H_1$, $G \oplus G$ does not have a unique decomposition. The converse is clear so the proof is complete.

We classify groups $G = G_1^{e_1} \oplus \cdots \oplus G_t^{e_t}$ that have a Σ-unique decomposition.

Theorem 2.36. *Let $G = G_1^{e_1} \oplus \cdots \oplus G_t^{e_t}$ for some integers $t, e_1, \ldots, e_t > 0$ and some cocommutative strongly indecomposable pairwise non-quasi-isomorphic groups $G_1, \ldots, G_t$. The following are equivalent for G:*

1. $h(G_1) \cdots h(G_t) = 1$.
2. *Each group $G_1, \ldots, G_t$ has a Σ-unique decomposition.*
3. G *has a Σ-unique decomposition.*

Proof: 1 $\Leftrightarrow$ 2 follows from Theorem 2.32.

2 $\Leftrightarrow$ 3. We prove both implications at once. By Theorem 2.4, G has a Σ-unique decomposition as a group iff $E(G)$ has a Σ-unique decomposition as an $E(G)$-module. By (2.21), $E(G) = E(G_1^{e_1}) \times \cdots \times E(G_t^{e_t})$, so that $E(G)$ has a Σ-unique decomposition iff $E(G_i^{e_i})$ has a Σ-unique decomposition for each $i = 1, \ldots, t$ iff (by Theorem 2.4), $G_i^{e_i}$ has a Σ-unique decomposition for each $i = 1, \ldots, t$ iff (by Proposition 2.35), G_i has a Σ-unique decomposition for each $i = 1, \ldots, t$. This completes the proof.

A theorem of J. D. Reid's [9, Theorem 3.3] states that if G is a strongly indecomposable rank two torsion-free group then $\text{End}(G)$ is commutative. Thus the strongly indecomposable rank two groups G are cocommutative . Moreover, the possible quasi-endomorphism rings are $\mathbf{Q}$, $\mathbf{Q}[\sqrt{d}]$ for some square-free integer d, and $\mathbf{Q}[x]/(x^2)$ for an indeterminant x. Thus $\mathbf{Q}E(G)$ is either $\mathbf{Q}$ or $\mathbf{Q}[\sqrt{d}]$ for some square-free integer d.

Let X be the set of square-free integers d for which $h(\mathbf{Q}[\sqrt{d}]) = 1$.

Theorem 2.37. *1. A strongly indecomposable rank two group has the internal cancellation property.*
2. If $\mathbf{Q}E(G) = \mathbf{Q}$ or, if G is integrally closed and $\mathbf{Q}E(G) = \mathbf{Q}[\sqrt{d}]$ for some $d \in X$ then G has the power cancellation property and G has a Σ-unique decomposition.

Proof: 1. Apply Theorem 2.22.

2. If $\mathbf{Q}E(G) = \mathbf{Q}$ then $E(G)$ is a *pid*, so by Corollary 2.16, $h(G) = h(E(G)) = 1$. Apply Theorem 2.30 and Corollary 2.33 to see that G has the power cancellation property and a Σ-unique decomposition.

Let $d \in X$ and let $\overline{E}$ be the ring of algebraic integers in $\mathbf{Q}[\sqrt{d}]$. Because $d \in X$, $h(\overline{E}) = h(\mathbf{Q}[\sqrt{d}]) = 1$. Since G is integrally closed, $G = \overline{G}$, and so $E(G) = E(\overline{G}) = \overline{E}[C^{-1}]$ for some multiplicatively closed $C \subset \mathbf{Q}[\sqrt{d}]$. Take, e.g., $C = \{c \in \mathbf{Q}[\sqrt{d}] \mid cE(\overline{G}) = E(\overline{G})\}$. Furthermore, if we let $U_1, \ldots, U_t$ be a representative list of the isomorphism classes of ideals of $E(G)$ then there are ideals $I_1, \ldots, I_t$ of $\overline{E}$ such that $U_i = I_i[C^{-1}]$ for each $i = 1, \ldots, t$. Since the U_i are pairwise nonisomorphic, the I_i are pairwise nonisomorphic. Hence $h(\overline{E}[C^{-1}]) \leq h(\overline{E})$, so that

$$h(G) = h(E(G)) \leq h(\overline{E}) = 1$$

by Corollary 2.16. Then by Theorem 2.30 and Corollary 2.33, G has the power cancellation property and a Σ-unique decomposition. This completes the proof.

2.6 Algebraic number fields

Let $\mathbf{k}$ be an algebraic number field with ring of algebraic integers $\overline{E}$. The *class number* of $h(\mathbf{k})$ is the number of isomorphism classes of fractional ideals of $\overline{E}$. Given an algebraic number field $\mathbf{k}$, it is a classic problem of algebra to find the class number $h(\mathbf{k})$ of $\mathbf{k}$. In particular, a problem that goes back to Gauss and Kummer is to determine the quadratic algebraic number fields $\mathbf{k}$ such that $h(\mathbf{k}) = 1$. This problem is related to Fermat's Last Theorem. See [103, page 193]. We will link the determination of $h(\mathbf{k})$ to the power cancellation property of cocommutative strongly indecomposable rtffr groups G.

Let $\mathbf{k}$ be an algebraic number field, let $\overline{E}$ denote the ring of algebraic integers in $\mathbf{k}$, and let $h(\mathbf{k})$ denote the *class number of* $\mathbf{k}$. That is, $h(\mathbf{k}) = h(\overline{E})$.

Let

$$\boxed{\Omega(\overline{E}) = \{\text{rtffr groups } G \mid E(G) = \overline{E}\}.}$$

Observe that if $G \in \Omega(\overline{E})$ then G is a cocommutative strongly indecomposable rtffr group.

Example 2.38. Let $\mathbf{k}$ be an algebraic number field and let $\overline{E}$ be the ring of algebraic integers in $\mathbf{k}$. The additive structure of $\overline{E}$ is a free abelian group, so by Butler's theorem 1.10, there is an rtffr group $\overline{E} \subset G \subset \mathbf{k}$ such that $\text{End}(G) = \overline{E}$. Thus $\Omega(\overline{E}) \neq \emptyset$.

Theorem 2.39. *Let* $\mathbf{k}$ *be an algebraic number field. The following are equivalent for the integer* $n > 0$.

1. *n is relatively prime to $h(\mathbf{k})$.*
2. *Given $G \in \Omega(\overline{E})$ then $G^n \cong H^n$ implies that $G \cong H$ for any group H.*
3. *There is a $G \in \Omega(\overline{E})$ such that $G^n \cong H^n$ implies that $G \cong H$ for any group H.*

Proof: $1 \Rightarrow 2$. Let n be relatively prime to $h(\mathbf{k})$ and let $G \in \Omega(\overline{E})$. By Corollary 2.16, n is relatively prime to $h(\mathbf{k}) = h(\overline{E}) = h(E(G)) = h(G)$. Then by Theorem 2.27, $G^n \cong H^n$ implies that $G \cong H$ for any group H.

$2 \Rightarrow 3$ follows from Example 2.38.

$3 \Rightarrow 1$. Suppose that there is a $G \in \Omega(\overline{E})$ such that if $G^n \cong H^n$ then $G \cong H$ for any group H. By Corollary 2.16 and Theorem 2.27, n is relatively prime to $h(G) = h(E(G)) = h(\overline{E}) = h(\mathbf{k})$. This proves part 1 and completes the proof.

Theorem 2.40. *Let $\mathbf{k}$ be an algebraic number field. The following are equivalent.*

1. $h(\mathbf{k}) = 1$.
2. *If $G \in \Omega(\overline{E})$ then each group that is locally isomorphic to G is isomorphic to G.*
3. *Each $G \in \Omega(\overline{E})$ has the power cancellation property.*
4. *Some $G \in \Omega(\overline{E})$ has the power cancellation property.*
5. *Each $G \in \Omega(\overline{E})$ has a Σ-unique decomposition.*
6. *Some $G \in \Omega(\overline{E})$ has a Σ-unique decomposition.*

Proof: $1 \Leftrightarrow 2$. Let $G \in \Omega(\overline{E})$. Then $h(\mathbf{k}) = h(\overline{E}) = 1$ iff (by Corollary 2.16), $h(G) = h(E(G)) = h(\overline{E}) = 1$ iff part 2 is true.

$1 \Rightarrow 3$. Let $G \in \Omega(\overline{E})$. By Corollary 2.16 and part 1, $h(G) = h(E(G)) = h(\overline{E}) = h(\mathbf{k}) = 1$. Then by Theorem 2.30, G has the power cancellation property. This is part 3.

$3 \Rightarrow 4$ follows from Example 2.38.

$4 \Rightarrow 1$. Suppose that $G \in \Omega(\overline{E})$ has the power cancellation property. Then by Corollary 2.16 and Theorem 2.30, $1 = h(G) = h(E(G)) = h(\overline{E}) = h(\mathbf{k})$. This proves part 1.

$3 \Leftrightarrow 5$ and $4 \Leftrightarrow 6$ follow from Corollary 2.33. This completes the proof.

Let

$$\Omega(\mathbf{k}) = \{\text{rtffr groups } G \mid QE(G) = \mathbf{k}\}.$$

Each $G \in \Omega(\mathbf{k})$ is a cocommutative strongly indecomposable rtffr group. An rtffr group $G \in \Omega(\mathbf{k})$ is integrally closed iff $E(G)$ is a Dedekind domain.

Theorem 2.41. *Let $\mathbf{k}$ be an algebraic number field. The following are equivalent.*

1. $h(\mathbf{k}) = 1$.
2. *Given an integrally closed group in $G \in \Omega(\mathbf{k})$ then each group locally isomorphic to G is isomorphic to G.*
3. *Every integrally closed group in $\Omega(\mathbf{k})$ has the power cancellation property.*
4. *Every integrally closed group in $\Omega(\mathbf{k})$ has a Σ-unique decomposition.*

Proof: $2 \Rightarrow 1$. Since $\Omega(\overline{E}) \subset \Omega(\mathbf{k})$, $2 \Rightarrow 1$ follows from Theorem 2.40.

$1 \Rightarrow 3$. Let $G \in \Omega(\mathbf{k})$ be such that $E(G)$ is a Dedekind domain, but such that G does not have the power cancellation property. By Corollary 2.16 and Theorem 2.30, $1 \neq h(G) = h(E(G))$, so that $E(G)$ is a Dedekind domain that is not a *pid*. Choose a multiplicatively closed set $C \subset \overline{E}$ such that $E(G) = \overline{E}[C^{-1}]$. Choose, e.g., $C = \{c \in \overline{E} \mid cE(G) = E(G)\}$. Each ideal of $E(G)$ is then of the form $I[C^{-1}]$ for some ideal $I \subset \overline{E}$. Hence some ideal $I \subset \overline{E}$ is not principal, and so $h(\mathbf{k}) \neq 1$.

$3 \Rightarrow 4$. See Corollary 2.33.

$4 \Rightarrow 2$. Let $G \in \Omega(\mathbf{k})$, and suppose that $G \not\cong H$ are locally isomorphic rtffr groups. By Theorem 2.19.2, there is a group K such that $G \oplus G \cong H \oplus K$. Our choice of $G \not\cong H$ implies that G does not have a Σ-unique decomposition. This proves the negation of part 4, which completes the proof.

2.7 Exercises

Let G and H denote rtffr groups. Let $E(G) = \operatorname{End}(G)/\mathcal{N}(\operatorname{End}(G))$. Let E be an rtffr ring (= a ring for which $(E, +)$ is a reduced torsion-free finite rank ring. Given a semi-prime rtffr ring, E let $\overline{E}$ be an integrally closed subring of $\mathbf{Q}E$ such that $E \subset \overline{E}$ and $\overline{E}/E$ is finite. Let τ be the largest ideal in the center, S of E such that $\tau \overline{E} \subset E$.

1. Without referring to Theorem 2.26, show that if $G^n \cong H^n$ for some integer $n > 0$ and group H then G is locally isomorphic to H.
2. Prove that if E is an rtffr ring and if M and N are finitely generated projective E-modules then M is locally isomorphic to N iff $M^n \cong N^n$ for some integer $n > 0$.
3. Prove Theorem 2.3 without using [9].
4. Let I be a fractional ideal in the commutative rtffr integral domain E. The following are equivalent:

   (a) I is invertible over E.
   (b) I is locally isomorphic to E.
   (c) $I \cong I'$ for some ideal I' of E such that $I' + \tau E = E$.

5. Let $h > 0$ be an integer. There are groups G and H such that $G^k \cong H^k$ iff $k \equiv 0 \pmod{h}$.
6. Prove Theorem 2.4.2.
7. Suppose that G and H are locally isomorphic. Show that $G \oplus G \cong H \oplus K$ for some group K.
8. If $G \oplus K \cong H \oplus K$, show that G is locally isomorphic to H.
9. Show that for any group H and integer $0 < k < h(G)$, $G^k \not\cong H^k$.
10. Show that $h(G)$ is an odd integer iff $G \oplus G \cong H \oplus H$ implies that $G \cong H$ for each group H.
11. Let G be a strongly indecomposable cocommutative rtffr group. Show that G has Σ-unique decomposition iff $E(G)$ is a *pid*.

2.8 Problems for future research

1. Prove the Jordan–Zassenhaus lemma 1.9 for rtffr rings E.
2. Let G be a strongly indecomposable rtffr group. Show that G has a Σ-unique decomposition iff $E(G)$ if each right ideal of $E(G)$ is principal.
3. Let D be a finite-dimensional $\mathbf{Q}$-division algebra. Define the class number $h(D)$ and then prove a version of Theorem 2.19 for D.

3

Mayer–Vietoris sequences

Let $\overline{G}$ be the *integral closure* of G (see page 13.)

Using the Mayer–Vietoris sequence [19] we factor the integer $h(G)$. Let **k** be a quadratic number field. We will connect unique factorization in **k** with the sequence of rational primes, and with direct sum properties of cocommutative strongly indecomposable torsion-free finite rank groups. From this evidence we see why the direct sum decomposition problems in abelian group theory have been considered to be hard. They imply hard problems in number theory.

3.1 The sequence of groups

We study a long exact sequence of multiplicative finite abelian groups and find certain equations of integers that relate $h(G)$ and $h(\overline{G})$. This reintroduces the unit group into the study of the power cancellation property.

Property 3.1. 1. Let G be a cocommutative strongly indecomposable rtffr group with integral closure $\overline{G}$.
2. $E = E(G)$ is an rtffr integral domain, its integral closure $\overline{E} = E(\overline{G})$ is a Dedekind domain, and $\overline{E}/E$ is finite. Hence $\overline{E}$ is the integral closure of E.
3. There is a nonzero ideal τ of $\overline{E}$ such that $\tau \subset E \subset \overline{E}$ and such that $\overline{E}/\tau$ is finite.

Given a ring R let $u(R)$ denote the *unit group* of R. Assume Property 3.1 and let

$$K = \{x \in u(\overline{E}) \mid 1 + x \in \tau\}.$$

Since $\tau \subset E, K \subset u(E)$.

Lemma 3.2. *Assume Property 3.1*

1. $u(\overline{E})/K$ and $u(E)/K$ are finite groups.
2. $L = \mathrm{card}(u(\overline{E})/u(E))$ is a finite cardinal.

Proof: 1. Since $\overline{E}$ is an rtffr integral domain and since $\tau \neq 0$ is an ideal in $\overline{E}, \overline{E}/\tau$ is finite. Since K is the kernel of the natural mapping $u(\overline{E}) \longrightarrow u(\overline{E}/\tau), u(\overline{E})/K$ embeds in the finite group $u(\overline{E}/\tau)$. Thus $u(E)/K \subset u(\overline{E})/K$ are finite.

   2. By part 1, $u(\overline{E})/u(E) \cong \dfrac{u(\overline{E})/K}{u(E)/K}$ is a finite group. This completes the proof.

By Property 3.1.3 there is a commutative square

$$
\begin{array}{ccc}
E & \overset{\iota}{\longrightarrow} & \overline{E} \\
{\scriptstyle \pi}\downarrow & & \downarrow{\scriptstyle \bar{\pi}} \\
E/\tau & \underset{\bar{\iota}}{\longrightarrow} & \overline{E}/\tau
\end{array}
$$

of rings, ring inclusions ι and $\bar{\iota}$, and natural projections π and $\bar{\pi}$. Thus there is a long exact sequence

$$
1 \longrightarrow u(E) \overset{\phi}{\longrightarrow} u(\overline{E}) \times u(E/\tau) \overset{\rho}{\longrightarrow} u(\overline{E}/\tau)
$$
$$
\longrightarrow \mathrm{Pic}(E) \longrightarrow \mathrm{Pic}(\overline{E}) \times \mathrm{Pic}(E/\tau) \longrightarrow \mathrm{Pic}(\overline{E}/\tau)
$$

of multiplicative groups called the *Mayer–Vietoris sequence* [19, page 482]. The map ρ is defined by

$$
\rho(x, y + \tau) = (x^{-1} + \tau)(y + \tau)
$$

for each $x \in u(\overline{E})$, and for each $y + \tau \in u(E/\tau)$. Furthermore,

$$
\phi(x) = (x, x + \tau) \text{ for each } x \in u(E).
$$

Because $\overline{E}/\tau$ and E/τ are finite (Artinian) commutative rings, [94, Theorem 37.22] states that $\mathrm{Pic}(\overline{E}/\tau) = \mathrm{Pic}(E/\tau) = 1$. Hence there is a long exact sequence

$$
1 \to u(E) \overset{\phi}{\longrightarrow} u(\overline{E}) \times u(E/\tau) \overset{\rho}{\longrightarrow} u(\overline{E}/\tau)
$$
$$
\to \mathrm{Pic}(E) \to \mathrm{Pic}(\overline{E}) \to 1 \qquad (3.1)
$$

of multiplicative groups. Since

$$
\phi(K) = \{(x, 1 + \tau) \,|\, x \in K\} \subset \ker \rho
$$

there is a natural isomorphism,

$$
\frac{u(\overline{E}) \times u(E/\tau)}{\phi(K)} \cong \frac{u(\overline{E})}{K} \times u(E/\tau).
$$

Hence there is a long exact sequence

$$
1 \to \frac{u(E)}{K} \overset{\bar{\phi}}{\longrightarrow} \frac{u(\overline{E})}{K} \times u(E/\tau) \overset{\bar{\rho}}{\longrightarrow} u(\overline{E}/\tau)
$$
$$
\to \mathrm{Pic}(E) \to \mathrm{Pic}(\overline{E}) \to 1 \qquad (3.2)
$$

of finite groups where $\overline{\rho}(xK, y+\tau) = \rho(x, y+\tau)$. One readily proves that image $\rho =$ image $\overline{\rho}$ so that

$$\text{coker } \rho = \text{coker } \overline{\rho}. \tag{3.3}$$

Lemma 3.3. *Assume Property 3.1, let* $\widehat{m} = \text{card}(u(\overline{E}/\tau))$, *let* $\widehat{n} = \text{card}(u(E/\tau))$, *and let* $L = \text{card}(u(\overline{E})/u(E))$. *Then*

$$\text{card}(\text{coker } \rho) = \frac{\widehat{m}}{L\widehat{n}}.$$

Proof: Let $M = \text{card}(u(\overline{E})/K)$ and let $N = \text{card}(u(E)/K)$. By Lemma 3.2.1, M and N are finite cardinals, so we have

$$L = \text{card}(u(\overline{E})/u(E)) = \text{card}\left(\frac{u(\overline{E})/K}{u(E)/K}\right) = \frac{M}{N}. \tag{3.4}$$

The exactness of (3.2) and the equation (3.4) reveals that

$$\text{card}(\text{image } \overline{\rho}) = \frac{\text{card}(u(\overline{E})/K) \cdot \text{card}(u(E/\tau))}{\text{card}(u(E)/K)} = \frac{M\widehat{n}}{N} = L\widehat{n}.$$

Hence (3.3) implies that

$$\text{card}(\text{coker } \rho) = \text{card}(\text{coker } \overline{\rho}) = \frac{\text{card}(u(\overline{E}/\tau))}{\text{card}(\text{image } \overline{\rho})} = \frac{\widehat{m}}{L\widehat{n}}.$$

This completes the proof.

Theorem 3.4. *Assume Property 3.1, let* $\widehat{m} = \text{card}(u(\overline{E}/\tau))$, *let* $\widehat{n} = \text{card}(u(E/\tau))$, *and let* $L = \text{card}\left(u(\overline{E})/u(E)\right)$. *Then*

$$h(G) = h(\overline{G})\frac{\widehat{m}}{L\widehat{n}}.$$

Proof: By Property 3.1.2, $E = E(G)$ and $\overline{E} = E(\overline{G})$, so by Corollary 2.16, $h(G) = \text{card}(\text{Pic}(E))$ and $h(\overline{G}) = \text{card}(\text{Pic}(\overline{E}))$. By the exactness of (3.1) and Lemma 3.3 we have

$$\text{card}(\text{Pic}(E)) = \text{card}(\text{Pic}(\overline{E})) \cdot \text{card}(\text{coker } \rho) = \text{card}(\text{Pic}(\overline{E})) \cdot \frac{\widehat{m}}{L\widehat{n}}.$$

Hence $h(G) = h(\overline{G}) \cdot \frac{\widehat{m}}{L\widehat{n}}$, which completes the proof.

Corollary 3.5. *Assume Property 3.1, let* $\widehat{m} = \text{card}(u(\overline{E}/\tau))$, *let* $\widehat{n} = \text{card}(u(E/\tau))$, *and let* $L = \text{card}\left(u(\overline{E})/u(E)\right)$. *Then*

1. $h(\overline{G})$ *divides* $h(G)$.
2. $L\widehat{n}$ *divides* $\widehat{m}$.

The question of when $h(G) = h(\overline{G})$ arises naturally from the previous result. The next result shows that $h(G) = h(\overline{G})$ when a weak sort of unit lifting property holds. Assuming Property 3.1.3, let

$$\frac{u(\overline{E}) + \tau}{\tau} = \{x + \tau \,|\, x \in u(\overline{E})\}.$$

Theorem 3.6. *Assume Property 3.1. Then*

$$u\big(\overline{E}/\tau\big) = \left[\frac{u(\overline{E}) + \tau}{\tau}\right] \cdot u(E/\tau)$$

iff $h(G) = h(\overline{G})$.

Proof: Assume that

$$u\big(\overline{E}/\tau\big) = \left[\frac{u(\overline{E}) + \tau}{\tau}\right] \cdot u(E/\tau)\,.$$

The expression

$$\left[\frac{u(\overline{E}) + \tau}{\tau}\right] \cdot u(E/\tau) = \text{image } \rho$$

in (3.1), so that ρ is a surjection. By the exactness of (3.1), $\text{Pic}(E) \cong \text{Pic}(\overline{E})$, so that by Property 3.1.2,

$$\text{card}(\text{Pic}(E(G))) = \text{card}(\text{Pic}(E))$$
$$= \text{card}(\text{Pic}(\overline{E})) = \text{card}(\text{Pic}(E(\overline{G}))). \qquad (3.5)$$

Hence Corollary 2.16 shows us that $h(G) = h(\overline{G})$.

Conversely, assume that $h(G) = h(\overline{G})$. Corollary 2.16 implies that

$$\text{card}(\text{Pic}(E)) = \text{card}(\text{Pic}(E(G)))$$
$$= \text{card}(\text{Pic}(E(\overline{G}))) = \text{card}(\text{Pic}(\overline{E})).$$

The surjection $\text{Pic}(E) \longrightarrow \text{Pic}(\overline{E})$ in (3.1) is then an isomorphism. Hence ρ is a surjection, whence

$$\text{image } \rho = u\big(\overline{E}/\tau\big) = \left[\frac{u(\overline{E}) + \tau}{\tau}\right] \cdot u(E/\tau)\,.$$

This completes the proof.

The group G has the *cancellation property* if given groups H and K such that $G \oplus H \cong G \oplus K$ then $H \cong K$. It has been known since the 1950s that the lifting of units modulo an integer n is connected with the cancellation property. The following result is due to L. Fuchs.

Lemma 3.7. [9, Lemma 8.10] *Let A be an rtffr group and let $n \neq 0 \in \mathbf{Z}$ be such that $nA \neq A$. Assume that there is a finite rank torsion-free group G such that*

1. *$\mathrm{End}(G) = \mathbf{Z}$,*
2. *$\mathrm{Hom}(A, G) = \mathrm{Hom}(G, A) = 0$, and*
3. *There is an epimorphism $\beta : G \to A/nA$.*

If A has the cancellation property then every unit of $\mathrm{End}(A)/n\mathrm{End}(A)$ lifts to a unit of $\mathrm{End}(A)$.

Corollary 3.8. *Assume Property 3.1. If each unit of $\overline{E}/\tau$ lifts to a unit of $\overline{E}$ then $h(G) = h(\overline{G})$.*

Proof: Suppose that units of $\overline{E}/\tau$ lift to units of $\overline{E}$. Then the map ρ in (3.1) is a surjection, so that $\mathrm{Pic}(E) \cong \mathrm{Pic}(\overline{E})$. By Corollary 2.16 and Property 3.1.2, $h(G) = \mathrm{card}(\mathrm{Pic}(E)) = \mathrm{card}(\mathrm{Pic}(\overline{E})) = h(\overline{G})$.

Example 3.9. Let $\tau \subset E \subset \overline{E}$ be rtffr integral domains such that $\overline{E}/\tau = \prod_{i=1}^{n} \mathbf{Z}/2\mathbf{Z}$ for some integer $n > 1$ and

$$E/\tau = \{(x, \cdots, x) \,|\, x \in \mathbf{Z}/2\mathbf{Z}\}.$$

Then $u(\overline{E}/\tau) = u(E/\tau) = \{1+\tau\}$. Hence each unit of $\overline{E}/\tau$ lifts to a unit of $\overline{E}$, and each unit of E/τ lifts to a unit of E. By Butler's theorem 1.10 there is a group $E \subset G \subset \mathbf{Q}E$ such that $E = \mathrm{End}(G)$. Let $\overline{G}$ be the integral closure of G. (See Lemma 2.9.) Then by Corollary 3.8, $h(G) = h(\overline{G})$. A ring $\overline{E}$ with this type of factorization of $2\overline{E}$ is the *pid* $\overline{E} = \mathbf{Q}[\sqrt{3}]$ in which $2 = (-5+3\sqrt{3})(5+3\sqrt{3})$. Then $h(G) = h(E) = h(\overline{E}) = 1 = h(\overline{G})$.

The following result shows that if we are interested in investigating the power cancellation property in the group G then we need only consider those groups G such that $E(\overline{G})$ is a *pid*.

Theorem 3.10. *Assume Property 3.1.*

1. *If $h(\overline{G}) \neq 1$ then $h(G) \neq 1$.*
2. *Suppose that $h(\overline{G}) = 1$. Then $h(G) = 1$ iff*

$$u(\overline{E}/\tau) = \left[\frac{u(\overline{E}) + \tau}{\tau} \right] \cdot u(E/\tau).$$

Proof: Part 1 follows from Corollary 3.5.1.

2. Suppose that $h(\overline{G}) = 1$. Then by Theorem 3.6, $h(G) = 1$ iff

$$u(\overline{E}/\tau) = \left[\frac{u(\overline{E}) + \tau}{\tau} \right] \cdot u(E/\tau).$$

This completes the proof.

Theorem 3.11. *Assume Property 3.1, and let τ be a prime ideal of $\overline{E}$ that contains a rational prime p. Suppose that E/τ is a subfield of $\overline{E}/\tau$, let*

$$e = [E/\tau : \mathbf{Z}/p\mathbf{Z}], \quad f = [\overline{E}/\tau : E/\tau],$$

and let $L = \text{card}(u(\overline{E})/u(E))$. Then

$$h(G) = \frac{h(\overline{G})}{L} \sum_{i=0}^{f-1} p^{ei}.$$

Proof. Since $[E/\tau : \mathbf{Z}/p\mathbf{Z}] = e$, $\text{card}(E/\tau) = p^e$, and since $[\overline{E}/\tau : E/\tau] = f$, $\text{card}(\overline{E}/\tau) = p^{ef}$. Consequently,

$$\frac{\widehat{m}}{\widehat{n}} = \frac{\text{card}(u(\overline{E}/\tau))}{\text{card}(u(E/\tau))} = \frac{p^{ef} - 1}{p^e - 1} = \sum_{i=0}^{f-1} p^{ei}.$$

Apply this to Theorem 3.4 with complete the proof.

We calculate $h(G)$ in the case where $u(\overline{E})$ is a finite group. For instance, the ring $\overline{E}$ of algebraic integers in the field $\mathbf{Q}[\sqrt{-d}]$ has finite unit group when $d > 0$ is a square-free integer. See [103, Proposition 4.2]. Compare the next theorem with Theorem 3.4.

Theorem 3.12. *Assume Property 3.1 and assume that $\overline{E}$ has a finite unit group. Let*

$$m = \text{card}(u(\overline{E})),$$
$$n = \text{card}(u(E)),$$
$$\widehat{m} = \text{card}(u(\overline{E}/I)), \text{ and}$$
$$\widehat{n} = \text{card}(u(E/I)).$$

Then

1. $h(G) = h(\overline{G}) \dfrac{\widehat{m}n}{m\widehat{n}}$, and

2. $h(G) = 1$ iff $h(\overline{G})\widehat{m}n = m\widehat{n}$.

Proof. Since $u(\overline{E})$ and $u(E)$ are finite groups, m and n are integers such that

$$L = \text{card}(u(\overline{E})/u(E)) = \frac{m}{n}.$$

Then by Theorem 3.4, $h(G) = h(\overline{G}) \dfrac{\widehat{m}n}{m\widehat{n}}$. This proves the theorem.

Corollary 3.13. *Assume Property 3.1 and assume that* $\overline{E}$ *is a* pid *with finite unit group. Let*

$$m = \text{card}(u(\overline{E})),$$

$$n = \text{card}(u(E)),$$

$$\widehat{m} = \text{card}(u(\overline{E}/I)), \text{ and}$$

$$\widehat{n} = \text{card}(u(E/I)).$$

Then $h(G) = \dfrac{\widehat{m}n}{m\widehat{n}}.$

3.2 Analytic methods

We link the power cancellation property to the sequence of rational primes. Specialize the notation of Property 3.1 as follows.

Let **k** be an algebraic number field with $\overline{E}$ its ring of algebraic integers and with *class number* $h(\mathbf{k})$. Let G be an *rtffr* abelian group with class number $h(G)$.

Property 3.14. 1. Let **k** be an algebraic number field, let $[\mathbf{k} : \mathbf{Q}] = f$, and let $h(\mathbf{k})$ denote the class number of **k**. Let $\overline{E}$ be the ring of algebraic integers in **k**. Then $\overline{E}$ is a ring whose additive group $(\overline{E}, +)$ is a free abelian group of finite rank f. Then $\mathbf{Z}/p\mathbf{Z}\text{-dim}(\overline{E}/p\overline{E}) = f$ for each rational prime p.

2. For each rational prime p let $E(p) = \mathbf{Z} + p\overline{E}$. Let $G(p)$ be a reduced torsion-free rank f abelian group such that $\text{End}(G(p)) \cong E(p)$. These groups exist by Butler's theorem 1.10. There is a torsion-free reduced group $\overline{G}(p)$ of rank f such that $\overline{G}(p)/G(p)$ is finite, and $\text{End}(\overline{G}(p)) = \overline{E}$. Then $E(\overline{G}(p)) = \overline{E}$ is the integral closure of the rtffr integral domain $E(p)$, and $G(p)$ is a cocommutative strongly indecomposable rtffr group.

3. $u(R)$ is the unit group in R. Let

$$\widehat{m}_p = \text{card}(u(\overline{E}/p\overline{E})),$$

$$\widehat{n}_p = \text{card}(u(E(p)/p\overline{E})), \text{ and}$$

$$L(p) = \text{card}\left(\frac{u(\overline{E})}{u(E(p))}\right).$$

We say that sequences $s(n)$ and $t(n)$ for positive $n \in \mathbf{Z}$ are *asymptotically equal* if $\lim_{n>0} s(n)/t(n) = 1$.

Theorem 3.15. *Assume Property 3.14. Let* $h(\mathbf{k}) = h$. *Then* $\{L(p)h(G(p)) \mid$ *rational primes* $p\}$ *is asymptotically equal to the sequence* $\{hp^{f-1} \mid$ *rational primes* $p\}$.

Proof: There are at most finitely many rational primes that ramify in **k**, so let us avoid those primes. By Theorem 3.4,

$$L(p)h(G(p))\frac{\widehat{n}_p}{\widehat{m}_p} = h(\overline{G}(p)). \qquad (3.6)$$

Because $\text{End}(\overline{G}(p)) = \overline{E}$, Corollary 2.16, implies that $h(\overline{G}(p)) = h(\overline{E}) = h(\mathbf{k}) = h$. Hence

$$L(p)h(G(p))\frac{\widehat{n}_p}{\widehat{m}_p} = h. \tag{3.7}$$

Since p does not ramify in $\mathbf{k}$, there are distinct prime ideals $I_1, \ldots, I_g$ in $\overline{E}$ and integers $f_1, \ldots, f_g$ such that $\Sigma_{i=1}^g f_i = f$, such that

$$p\overline{E} = I_1 \cap \cdots \cap I_g,$$

and $[\overline{E}/I_i : \mathbf{Z}/p\mathbf{Z}] = f_i$ for each $i = 1, \ldots, g$. Then

$$\overline{E}/p\overline{E} = \frac{\overline{E}}{I_1} \times \cdots \times \frac{\overline{E}}{I_g}$$

so that

$$u(\overline{E}/p\overline{E}) = u\left(\frac{\overline{E}}{I_1}\right) \times \cdots \times u\left(\frac{\overline{E}}{I_g}\right).$$

Since $\overline{E}/I_i$ is a finite field of characteristic p,

$$\widehat{m}_p = (p^{f_1} - 1) \cdots (p^{f_g} - 1). \tag{3.8}$$

Since $E(p)/p\overline{E} \cong \mathbf{Z}/p\mathbf{Z}, \widehat{n}_p = p - 1$.

Form the polynomial of degree $f - 1$,

$$x^{f-1} + Q_p(x) = \frac{(x^{f_1} - 1) \cdots (x^{f_g} - 1)}{x - 1}. \tag{3.9}$$

The coefficients of $(x^{f_2} - 1) \cdots (x^{f_g} - 1)$ are multinomial coefficients $\binom{f-1}{r_1 \cdots r_t}$ for some partitions $r_1, \ldots, r_t$ of $f - 1$. These coefficients are bounded above by $(f - 1)!$. The coefficients of $x^{f-1} + Q_p(x)$ in (3.9) are then bounded above by $f!$. Consequently, $Q_p(x)$ has degree $\leq f - 2$, and the coefficients of $Q_p(x)$ are bounded above by $f!$. Hence

$$\lim_p \frac{p^{f-1} + Q_p(p)}{p^{f-1}} = 1 + \lim_p \frac{Q_p(p)}{p^{f-1}} = 1. \tag{3.10}$$

Now, $p^{f-1} + Q_p(p) = \frac{\widehat{m}_p}{\widehat{n}_p}$ when p replaces x in (3.9), so by (3.7),

$$\frac{L(p)h(G(p))}{p^{f-1} + Q_p(p)} = L(p)h(G(p))\frac{\widehat{n}_p}{\widehat{m}_p} = h. \tag{3.11}$$

Furthermore,

$$\frac{L(p)h(G(p))}{p^{f-1}} = \frac{\left(\dfrac{L(p)h(G(p))}{p^{f-1}}\right)}{\left(\dfrac{L(p)h(G(p))}{p^{f-1}+Q_p(p)}\right)} \cdot \frac{L(p)h(G(p))}{p^{f-1}+Q_p(p)}$$

$$= \frac{p^{f-1}+Q_p(p)}{p^{f-1}} \cdot h$$

by (3.11). Using the limit in (3.10) we see that

$$\lim_p \frac{L(p)h(G(p))}{hp^{f-1}} = 1.$$

Therefore, $\{L(p)h(G(p)) \mid$ rational primes $p\}$ is asymptotically equal to $\{hp^{f-1} \mid$ rational primes $p\}$.

Corollary 3.16. *Let* **k** *be a quadratic number field, and let* $h(\mathbf{k}) = h$. *Then* $\{L(p)h(G(p)) \mid$ *rational primes* $p\}$ *is asymptotically equal to the sequence* $\{hp \mid$ *rational primes* $p\}$.

Theorem 3.17. *Let* **k** *be an algebraic number field and let* $h > 0$ *be an integer. The following are equivalent.*

1. $h(\mathbf{k}) = h$.
2. *The sequence* $\{L(p)h(G(p)) \mid$ *rational primes* $p\}$ *is asymptotically equal to the sequence* $\{hp^{f-1} \mid$ *rational primes* $p\}$.

Proof: $1 \Rightarrow 2$. This is Theorem 3.15.

$2 \Rightarrow 1$. The sequence $\{L(p)h(G(p)) \mid$ rational primes $p\}$ is asymptotically equal to the sequence $\{hp^{f-1} \mid$ rational primes $p\}$ for some integer $h > 0$. Then by Theorem 3.15 and part 2,

$$\lim_p \frac{L(p)h(G(p))}{h(\mathbf{k})p^{f-1}} = 1 = \lim_p \frac{L(p)h(G(p))}{hp^{f-1}}.$$

Hence $h(\mathbf{k}) = h$ which completes the proof.

Corollary 3.18. *Let* **k** *be a quadratic number field and let* $h > 0$ *be an integer. The following are equivalent.*

1. $h(\mathbf{k}) = h$.
2. *The sequence* $\{L(p)h(G(p)) \mid$ *rational primes* $p\}$ *is asymptotically equal to the sequence* $\{hp \mid$ *rational primes* $p\}$.

Theorem 3.19. *Assume Property 3.14. Suppose that* **k** *is a quadratic number field. Then* $h(\mathbf{k}) = 1$ *iff the sequence* $\{h(G(p))L(p) \mid$ *primes* $p\}$ *is asymptotically equal to the sequence of rational primes.*

Proof: Apply Theorem 3.17.

Theorem 3.20. *Assume Property 3.14. Specifically,* $\mathbf{k}$ *is a quadratic number field,* $E(G(p)) = \mathbf{Z} + p\overline{E}$ *for rational primes* p, *and* $\overline{G}(p)$ *is the integral closure of* $G(p)$. *Then* $\overline{G}(p)$ *has the power cancellation property iff the sequence* $\{h(G(p))L(p) \mid$ *primes* $p\}$ *is asymptotically equal to the sequence of rational primes.*

Proof: Apply Theorems 3.19 and 2.30.

If $\overline{E}$ is the ring of algebraic integers in $\mathbf{k} = \mathbf{Q}[\sqrt{-d}]$ for some square-free integer $d = 2$ or $d > 3$ then $u(\overline{E}) = \{1, -1\}$, [103, Proposition 4.2]. Then also $u(E(p)) = \{1, -1\}$ for each rational prime p, so that $\operatorname{card}(u(\overline{E})/u(E(p))) = L(p) = 1$. We have stumbled upon the hypothesis for the next result.

Corollary 3.21. *Assume Property 3.14. Suppose that there is an integer* ℓ *such that* $L(p) = \ell$ *for each prime* p, *and let* $h = h(\mathbf{k})$. *Then the sequence* $\{\frac{\ell}{h}h(G(p)) \mid$ *primes* $p\}$ *is asymptotically equal to the sequence of rational primes.*

Corollary 3.22. *Assume Property 3.14 and say that* $\mathbf{k} = \mathbf{Q}[\sqrt{-1}]$. *Then the sequence* $\{2 \cdot h(G(p)) \mid$ *primes* $p\}$ *is asymptotically equal to the sequence of rational primes.*

Proof: It is known that $u(\overline{E}) = \{\pm 1, \pm i\}$ and that $u(E(p)) = \{1, -1\}$ for each prime p. Then $L(p) = \operatorname{card}(u(\overline{E})/u(E(p))) = 2$ for each rational prime p. Since $\overline{E}$ is a *pid*, Corollary 2.16 implies that $h(\overline{G}(p)) = h(\overline{E}) = 1$ for each rational prime p. Now apply Corollary 3.21. This completes the proof.

Corollary 3.23. *Assume Property 3.14 and say that* $\mathbf{k} = \mathbf{Q}[\sqrt{-3}]$. *Then the sequence* $\{3 \cdot h(G(p)) \mid$ *primes* $p\}$ *is asymptotically equal to the sequence of rational primes.*

Proof: It is known that $u(\overline{E}) = \{\pm 1, \pm \omega, \pm \overline{\omega}\}$ where $\omega = \frac{1}{2} + \frac{\sqrt{-3}}{2}$, and that $\overline{E} = \mathbf{Z}[\omega]$ is a *pid*, [103, Theorem 3.2 and Proposition 4.2]. Then $u(E(p)) = \{1, -1\}$ for each rational prime p since $E(p) \neq \overline{E}$. Thus $L(p) = \operatorname{card}(u(\overline{E})/u(E(p))) = 3$ for each rational prime p. Now apply Corollary 3.21. This completes the proof.

Let $d \neq 0, 1$ be a square-free integer and let $\mathbf{k} = \mathbf{Q}[\sqrt{-d}]$. For $d = 2$ or $d > 3$, $\{1, -1\}$ is the set of units in $\overline{E}$ = the ring of algebraic integers in $\mathbf{k}$. Then $u(E(p)) = u(\overline{E})$ for each prime p so that $L(p) = 1$.

Corollary 3.24. *Assume Property 3.14 and say that* $\mathbf{k} = \mathbf{Q}[\sqrt{-d}]$ *for some square-free* $d = 2$ *or* $d > 3$. *Then* $\{\frac{1}{h(\mathbf{k})} \cdot h(G(p)) \mid$ *primes* $p\}$ *is asymptotically equal to the sequence of rational primes.*

Corollary 3.25. *Assume Property 3.14 and say that* $\mathbf{k} = \mathbf{Q}[\sqrt{-d}]$ *for some square-free* $d = 2$ *or* $d > 3$. *Then* $h(\mathbf{k}) = 1$ *iff the sequence* $\{h(G(p)) \mid$ *primes* $p\}$ *is asymptotically equal to the sequence of rational primes.*

Corollary 3.26. *Assume Property 3.14 and say that* $\mathbf{k} = \mathbf{Q}[\sqrt{-d}]$ *for some square-free* $d = 2$ *or* $d > 3$. *Then* $\{h(G(p)) \mid \text{primes } p\}$ *is asymptotically equal to the sequence of rational primes for at most finitely many* d.

Proof: There are at most finitely many d such that $h(\mathbf{k}) = 1$. See [103, Theorem 10.5]. $\quad\blacksquare$

3.3 Exercises

Let G and H denote rtffr groups. Let $E(G) = \text{End}(G)/\mathcal{N}(\text{End}(G))$. Let E be an rtffr ring ($=$ a ring for which $(E, +)$ is a reduced torsion-free finite rank ring). Given a semi-prime rtffr ring E, let $\overline{E}$ be an integrally closed subring of $\mathbf{Q}E$ such that $E \subset \overline{E}$ and $\overline{E}/E$ is finite. Let τ be the largest ideal in the center, S of E such that $\tau\overline{E} \subset E$. Let p be a rational prime, let $\overline{E}$ be a prime rtffr integrally closed ring, and let $E(p) = \mathbf{Z} + p\overline{E}$.

1. Show that $\text{Pic}(A) = 1$ if A is a commutative Artinian ring.
2. Let $\mathbf{Q}\overline{E} = \mathbf{Q}[\sqrt{-1}]$. Find $\text{card}(u(\overline{E})/u(E(p)))$ for rational primes p.
3. In the notation of Theorem 3.1.6, assume that $p^k \in \tau$ is a primary ideal. That is, assume that $p^k \in \tau$ for some rational prpime p and integer $k > 0$, assume that $\overline{E}/\sqrt{\tau}$ is a field, and assume that $\sqrt{\tau}/\tau$ has finite $\mathbf{Z}/p\mathbf{Z}$-dimension. Calculate $\text{card}(u(\overline{E}/\tau))$ and $\text{card}(u(E(p)/\tau))$.

3.4 Problems for future research

1. Let $\mathbf{k}$ be an algebraic number field with ring of algebraic integers $\overline{E}$. Determine the rational primes p such that units of $\overline{E}/p\overline{E}$ lift to units of $\overline{E}$.
2. Let $\mathbf{k}$ be an algebraic number field with ring of algebraic integers $\overline{E}$. Determine the rational primes p such that units of $E(p)/p\overline{E}$ lift to units of $\overline{E}$.
3. This is a hard one. Let $\mathbf{k} = \mathbf{Q}[\sqrt{-d}]$ for $d \neq 2$, $d > 3$. Give an algorithm with output the integers $\{h(G(p)) \mid \text{rational primes } p\}$, thus giving an algorithm whose output could be used to study $\pi(n)$.

4

Lifting units

In this chapter we examine the *unit lifting problem* for rtffr rings E. The problem is to determine the ideals I in E such that each unit of E/I lifts to a unit of E. We are led to find $h(G(p))$ for quadratic number fields $\mathbf{k}$ and groups $G(p)$ in

$$\Omega(\mathbf{k}) = \{\text{integrally closed rtffr } G \mid QE(G) = \mathbf{k}\}.$$

4.1 Units and sequences

As in Chapter 3, assume the following.

Property 4.1. 1. Let G be a cocommutative strongly indecomposable rtffr group with integral closure $\overline{G}$. See Lemma 2.9.

2. Then $E = E(G)$ is an rtffr integral domain, its integral closure $\overline{E} = E(\overline{G})$ is a Dedekind domain, and $\overline{E}/E$ is finite. Thus $\overline{E}$ is the integral closure of E.

3. There is a nonzero ideal τ of $\overline{E}$ such that $\tau \subset E \subset \overline{E}$ and such that $\overline{E}/\tau$ is finite.

The results in Chapter 3, especially Theorem 3.10, shows us that the power cancellation property is closely connected with the lifting of units from E/I for any ideal $I \subset E$ to E. Thus, we examine the unit lifting property for rings $E(p)$ and $\overline{E}$. We show that unit lifting in $E(p)$ can occur at only finitely many rational primes p. We begin with a pair of rings and ideals for which units lift.

Proposition 4.2. *Assume Property 4.1, and let $\mathbf{Z} + p\overline{E} = E(p) \subset \overline{E}$ for $p = 2, 3$. Then each unit of $E(p)/p\overline{E}$ lifts to a unit of $E(p)$.*

Proof: For $p = 2$, $E(2)/2\overline{E} \cong \mathbf{Z}/2\mathbf{Z}$ has exactly one unit, so each unit of $E(2)/2\overline{E}$ lifts to a unit of $E(2)$.

For $p = 3$, $E(3)/3\overline{E} \cong \mathbf{Z}/3\mathbf{Z}$ has exactly two units, so $1 + 3\overline{E}$ and $-1 + 3\overline{E}$. Thus, each unit of $E(3)/3\overline{E}$ lifts to a unit of $E(3)$. This completes the proof.

Theorem 4.3. *Assume Property 4.1. Suppose that we have subrings $\tau \subset E \subset E' \subset \overline{E}$. If each unit of E'/τ lifts to a unit of E' then each unit of E/τ lifts to a unit of E. Furthermore, if $E' = E(G')$ then $h(G) = h(G')$.*

Proof: By hypothesis we have $\tau \subset E \subset E' \subset \overline{E}$, so there is a Mayer–Vietoris sequence

$$1 \to u(E) \overset{\phi}{\longrightarrow} u(E') \times u(E/\tau) \overset{\rho}{\longrightarrow} u(E'/\tau)$$

$$\to \text{Pic}(E) \to \text{Pic}(E') \to 1 \qquad (4.1)$$

of multiplicative groups. By hypothesis, each unit of E'/τ lifts to a unit of E', so (4.1) contains a short exact sequence

$$1 \longrightarrow u(E) \longrightarrow u(E') \times u(E/\tau) \overset{\rho}{\longrightarrow} u\left(E'/\tau\right) \longrightarrow 1.$$

The lifting property ensures that $u(E/\tau) \subset u(E'/\tau) = \rho(u(E'))$, and that a unit $\bar{x} \in E/\tau$ lifts to a unit $x \in E'$. Because $\bar{x}^{-1} \in E/\tau, x^{-1} \in E$. Thus, each unit in E/τ lifts to a unit of E.

Furthermore, the exactness of (4.1) and an application of Corollary 2.16 shows us that

$$h(G) = h(E) = \text{card}(\text{Pic}(E)) = \text{card}(\text{Pic}(E')) = h(E') = h(G').$$

This completes the proof.

The next result is interesting since it shows that in one important instance the power cancellation property is inherited by subgroups of finite index. This kind of inheritance is rare in torsion-free finite-rank groups.

Corollary 4.4. *Assume Property 4.1. Suppose that each unit of $\overline{E}/\tau$ lifts to a unit of $\overline{E}$. Then the following are equivalent.*

1. Each rtffr group G' such that $\tau \subset E(G') \subset \overline{E}$ has the power cancellation property.
2. $\overline{G}$ has the power cancellation property.

Proof: $1 \Rightarrow 2$ is clear.

$2 \Rightarrow 1$. Let G' be an rtffr group such that $\tau \subset E(G') \subset \overline{E}$. There is a long exact sequence (3.1) temporarily letting $E = E(G')$. By hypothesis, each unit of $\overline{E}/\tau$ lifts to a unit of $\overline{E}$, so ρ is onto, and by the exactness of (3.1),

$$h(E(G')) = \text{Pic}(E(G')) \cong \text{Pic}(\overline{E}) = h(\overline{E}).$$

By Corollary 2.16, $h(\overline{G}) = h(\overline{E})$, and by part 2 and Theorem 2.30, $h(\overline{G}) = 1$. Another application of Corollary 2.16 shows us that

$$h(G') = \text{card}(\text{Pic}(E(G'))) = \text{card}(\text{Pic}(\overline{E})) = 1,$$

so by Theorem 2.30, G' has the power cancellation property. This proves part 1 and completes the proof.

The following is a useful lemma.

Lemma 4.5. *Let A be a finite $\mathbf{Z}/p\mathbf{Z}$-algebra. There is a semi-simple $\mathbf{Z}/p\mathbf{Z}$-subalgebra $B \subset A$ such that $B \cong A/\mathcal{N}(A)$, such that $A = B \oplus \mathcal{N}(A)$ and*

$$\mathrm{card}(u(A)) = \mathrm{card}(u(B)) \cdot \mathrm{card}(\mathcal{N}(A)).$$

Proof: By the Wedderburn theorem 1.7 there is a semi-simple $\mathbf{Z}/p\mathbf{Z}$-subalgebra $B \subset A$ such that $B \cong A/\mathcal{N}(A)$ and

$$A = B \oplus \mathcal{N}(A) \tag{4.2}$$

as B-modules.

Let $x \oplus y \in u(A)$ for some $x \in B$ and $y \in \mathcal{N}(A)$. Then $x \oplus y + \mathcal{N}(A) = x + \mathcal{N}(A)$ is a unit in $A/\mathcal{N}(A) \cong B$ so that $x \in u(B)$. Conversely, let $x \oplus y \in A$ for some $x \in u(B)$ and $y \in \mathcal{N}(A)$. Then $x \oplus y + \mathcal{N}(A) = x + \mathcal{N}(A)$ is a unit of $A/\mathcal{N}(A) \cong B$. Since units lift modulo the nil radical of A, $x \oplus y$ is a unit of A. Then

$$u(A) = \{x \oplus y \,\big|\, x \in u(B) \text{ and } y \in \mathcal{N}(A)\},$$

and thus $\mathrm{card}(u(A)) = \mathrm{card}(u(B)) \cdot \mathrm{card}(\mathcal{N}(A))$. This completes the proof.

We further specialize Property 3.14.

Property 4.6. 1. Let $\overline{E}$ be the ring of algebraic integers in the algebraic number field $\mathbf{k} = \mathbf{Q}\overline{E}$.
2. For each rational prime p let $E(p) = \mathbf{Z} + p\overline{E}$, let $G(p)$ be an rtffr group such that $E(G(p)) = E(p)$, and let $\overline{G}(p)$ be the integral closure of $G(p)$. Then $E(\overline{G}(p)) = \overline{E}$ is the integral closure of the rtffr integral domain $E(p)$, and $G(p)$ is a cocommutative strongly indecomposable rtffr group.
3. Let $\widehat{m}_p = \mathrm{card}(u(\overline{E}/p\overline{E}))$, let $\widehat{n}_p = \mathrm{card}(u(E(p)/p\overline{E}))$, and let $L(p) = \mathrm{card}\left(\dfrac{u(\overline{E})}{u(E(p))}\right)$.
4. Let $p \in \mathbf{Z}$ be a prime, let $k > 0$ be an integer, and let τ be a nonzero prime ideal of $\overline{E}$ such that $p \in \tau^k \subset E(p) \subset \overline{E}$, and such that $E(p)/\tau^k$ is a field.
5. Let $[E(p)/\tau^k : \mathbf{Z}/p\mathbf{Z}] = e \geq 1$ and $[\overline{E}/\tau : E(p)/\tau^k] = f \geq 1$. Then $[\overline{E}/\tau : \mathbf{Z}/p\mathbf{Z}] = ef \geq 1$, and $\mathbf{Z}/p\mathbf{Z}\text{-dim}(\tau/\tau^k) = ef(k-1) \geq 0$.

Example 4.7. The Property 4.6.4 for τ, $E(p)$, and $\overline{E}$ in (4.6) occurs in the following natural way. Let p be a rational prime, let $p \in \tau \subset \overline{E}$ be a prime ideal, and let $k \geq 1$ an integer such that $p \in \tau^k$. Then $\overline{E}/\tau$ is a field and $\overline{E}/\tau^k$ is a finite $\mathbf{Z}/p\mathbf{Z}$-algebra with Jacobson radical τ/τ^k. By the Wedderburn theorem 1.7, there is a subalgebra $B \subset \overline{E}/\tau^k$ such that $B \cong \overline{E}/\tau$ is a field and $\overline{E}/\tau^k = B \oplus \tau/\tau^k$ as B-modules. Choose a subring $\tau^k \subset E(p) \subset \overline{E}$ such that $E(p)/\tau^k \subset B$. Then Property 4.6.4 is satisfied by $p, k, \tau, E(p)$, and $\overline{E}$. Since at most finitely many rational primes p ramify in $\mathbf{k}$, there are at most finitely many rational primes p for which $k \neq 1$.

The next three results bound the primes $p \in \mathbf{Z}$ at which we can lift units from $\overline{E}/p\overline{E}$ to $\overline{E}$.

Lemma 4.8. *Assume Property 4.6. By Property 4.6.4 there is a prime p, an integer $k > 0$, and a prime ideal $\tau \subset \overline{E}$ such that $p \in \tau^k$. Then $\mathrm{card}(u(\overline{E}/\tau^k)) = p^{ef(k-1)}(p^{ef} - 1)$.*

Proof: Since τ is a prime ideal in the Dedekind domain $\overline{E}$, $\mathcal{N}(\overline{E}/\tau^k) = \tau/\tau^k$, so that by Lemma 4.5,

$$\mathrm{card}(u(\overline{E}/\tau^k)) = \mathrm{card}(u(\overline{E}/\tau)) \cdot \mathrm{card}(\tau/\tau^k).$$

By Property 4.6.5,

$$[\overline{E}/\tau : \mathbf{Z}/p\mathbf{Z}] = ef \text{ and } \mathbf{Z}/p\mathbf{Z}\text{-dim}(\tau/\tau^k) = ef(k-1),$$

so we have

$$\mathrm{card}(u(\overline{E}/\tau^k)) = p^{ef(k-1)}(p^{ef} - 1).$$

This proves the lemma.

Lemma 4.9. *Assume Property 4.6. Suppose that $\mathrm{card}(u(\overline{E})) = m$ is finite. By Property 4.6.4 there is a rational prime p, an integer $k > 0$, and a prime ideal $\tau \subset \overline{E}$ such that $p \in \tau^k$. If each unit of $\overline{E}/\tau^k$ lifts to a unit of $\overline{E}$ then*

1. *$p^{ef(k-1)}(p^{ef} - 1) \leq m$, and*
2. *if $p > 2$ then $p^{ef(k-1)}(p^{ef} - 1)$ is not relatively prime to m.*

Proof: 1. By Lemma 4.8, $\mathrm{card}(u(\overline{E}/\tau^k)) = p^{ef(k-1)}(p^{ef}-1)$. If $p^{ef(k-1)}(p^{ef}-1) > m$ then we cannot lift the $p^{ef(k-1)}(p^{ef} - 1)$ different units of $\overline{E}/\tau^k$ to the m units of $\overline{E}$.

2. If $p > 2$ and if $p^{ef(k-1)}(p^{ef} - 1)$ is relatively prime to m then by Lemma 4.8, $(u(\overline{E}) + \tau)/\tau = 1$. Since $p > 2$, $u(\overline{E}/\tau) \neq 1$, so some element of $u(\overline{E}/\tau)$ does not lift to an element of $u(\overline{E})$. This completes the proof.

Theorem 4.10. *Let $\overline{E}$ be an rtffr Dedekind domain with finite unit group. There are at most finitely many prime ideals $\tau \subset \overline{E}$ and finitely many integers $k > 0$ such that $\tau^k \cap \mathbf{Z}$ is a prime ideal in $\mathbf{Z}$, and such that each unit of $\overline{E}/\tau^k$ lifts to a unit of $\overline{E}$.*

Proof: Let $\mathrm{card}(u(\overline{E})) = m$ be finite. The set $\mathcal{P} = \{\text{primes } p \in \mathbf{Z} \mid \text{ there are integers } s > 0 \text{ and } t \geq 0 \text{ such that } p^{st}(p^s - 1) \leq m\}$ is finite. Let $\mathcal{Q} = \{\text{primes } p \in \mathbf{Z} \mid \text{ there is a prime ideal } \tau \subset \overline{E} \text{ and an integer } k > 0 \text{ such that } p \in \tau^k \text{ and such that each unit of } \overline{E}/\tau^k \text{ lifts to a unit of } \overline{E}\}$. By Lemma 4.9.1, $\mathcal{Q} \subset \mathcal{P}$ so $\mathcal{Q}$ is finite as required.

4.2 Calculations with primary ideals

Review (4.6). Consider the fixed rational prime p and the ring $\overline{E}$ of algebraic integers in the algebraic number field **k**. Let $k > 0 \in \mathbf{Z}$ and $E(p)$ be such that

$$\boxed{p\overline{E} \subset \tau^k \subset E(p) \subset \overline{E}}$$

for some prime ideal $\tau \subset \overline{E}$, and such that $E(p)/\tau^k$ is a field. Let

$$
L(p) = \text{card}\left(\frac{u(\overline{E})}{u(E(p))}\right).
$$

Let $G(p)$ be an rtffr group such that $G(p)$ is in

$$
\Omega(E(p)) = \{\text{rtffr groups } G \mid E(G) \cong E(p)\}.
$$

In the next several sections we calculate $L(p)$ and $h(G(p))$ over quadratic number fields **k**.

Lemma 4.11. *Assume Property 4.6. Then*

$$
\frac{\text{card}(u(\overline{E}/\tau^k))}{\text{card}(u(E(p)/\tau^k))} = p^{ef(k-1)}\sum_{j=0}^{f-1}p^{ej}.
$$

Proof. By Property 4.6.4, $E(p)/\tau^k$ is a field and $[E(p)/\tau^k : \mathbf{Z}/p\mathbf{Z}] = e > 0$. Then $\text{card}(u(E(p)/\tau^k)) = p^e - 1$. By Property 4.6.5, $\text{card}(u(\overline{E}/\tau)) = p^{ef} - 1$ and $\text{card}(\tau/\tau^k) = p^{ef(k-1)}$, so by Lemma 4.5,

$$
\text{card}(u(\overline{E}/\tau^k)) = \text{card}(u(\overline{E}/\tau)) \cdot \text{card}(\tau/\tau^k) = p^{ef(k-1)}(p^{ef} - 1).
$$

Thus

$$
\frac{\text{card}(u(\overline{E}/\tau^k))}{\text{card}(u(E(p)/\tau^k))} = \frac{p^{ef(k-1)}(p^{ef} - 1)}{p^e - 1} = p^{ef(k-1)}\sum_{j=0}^{f-1}p^{ej},
$$

which completes the proof.

Theorem 4.12. *Assume Property 4.6. If G has the power cancellation property then*

$$
L(p) = \frac{\text{card}(u(\overline{E}/\tau^k))}{\text{card}(u(E(p)/\tau^k))} = p^{ef(k-1)}\sum_{j=0}^{f-1}p^{ej}.
$$

Proof. By Corollary 3.5.2 and Lemma 4.11, $L(p)$ divides

$$
\frac{\text{card}(u(\overline{E}/\tau^k))}{\text{card}(u(E(p)/\tau^k))} = p^{ef(k-1)}\sum_{i=0}^{f-1}p^{ei}.
$$

Conversely, recall the integers e, f, k from Property 4.6.5, and assume that G has the power cancellation property. Corollary 2.16 and Theorem 2.30 imply that $\text{card}(\text{Pic}(E(p))) = h(G) = 1$, so that the map ρ in (3.1) is a surjection. The map ρ

induces a map $\rho^* : u(\overline{E}) \longrightarrow u(\overline{E}/\tau^k)$ defined by $\rho^*(x) = \rho(x, 1 + \tau^k) = x^{-1} + \tau^k$ for each $x \in u(\overline{E})$. Then

$$u(\overline{E}/\tau^k) = \rho(u(\overline{E}) \times u(E(p)/\tau^k)) = \rho^*(u(\overline{E})) \cdot u(E(p)/\tau^k). \qquad (4.3)$$

The inclusion $\rho^*(u(E(p))) \subset u(E(p)/\tau^k)$ implies that

$$\rho^*(u(E(p))) \subset \rho^*(u(\overline{E})) \cap u(E(p)/\tau^k). \qquad (4.4)$$

Consequently, $u(E(p))$ is contained in the kernel of the canonical induced surjection

$$\bar{\rho}^* : u(\overline{E}) \longrightarrow \frac{\rho^*(u(\overline{E}))}{\rho^*(u(\overline{E})) \cap u(E(p)/\tau^k)} : x \longmapsto \overline{\rho^*(x)}.$$

Hence by (4.3) and (4.4),

$$\frac{u(\overline{E}/\tau^k)}{u(E(p)/\tau^k)} \simeq \frac{\rho^*(u(\overline{E})) \cdot u(E(p)/\tau^k)}{u(E(p)/\tau^k)} \simeq \frac{\rho^*(u(\overline{E}))}{\rho^*(u(\overline{E})) \cap u(E(p)/\tau^k)}$$

is a quotient of $u(\overline{E})/u(E(p))$. Thus $L(p)$ is divisible by $\dfrac{\operatorname{card}(u(\overline{E}/\tau^k))}{\operatorname{card}(u(E(p)/\tau^k))}$, and

therefore $L(p) = \dfrac{\operatorname{card}(u(\overline{E}/\tau^k))}{\operatorname{card}(u(E(p)/\tau^k))}$. This completes the proof.

Corollary 4.13. *Assume Property 4.6, recall that $p \in \tau^k$ from Property 4.6.4, and suppose that $k \neq 1$. If $p > L(p)$ then G does not have the power cancellation property. Specifically, if $u(\overline{E}) = u(E(p))$ then G does not have the power cancellation property.*

Proof: Suppose that $p > L(p)$. If $f = 1$ then because $k - 1 \neq 0$,

$$p^{ef(k-1)} \sum_{j=0}^{f-1} p^{ej} \geq p > L(p).$$

Otherwise $f \geq 2$, so that

$$p^{ef(k-1)} \sum_{j=0}^{f-1} p^{ej} \geq 1 + p > L(p).$$

In either case,

$$p^{ef(k-1)} \sum_{j=0}^{f-1} p^{ej} \neq L(p),$$

so by Theorem 4.12, G does not have the power cancellation property.

Specifically, if $u(\overline{E}) = u(E(p)))$ then $p > L(p) = 1$ for all ramified rational primes p. In this case, G does not have the power cancellation property. This completes the proof.

Corollary 4.14. *Assume Property 4.6, and assume that the prime p in Property 4.6.4 is a ramified odd prime. If $L(p) \le 2$ then G does not have the power cancellation property.*

Proof: By hypothesis, $p > 2 \ge L(p)$, so by Corollary 4.13, G does not have the power cancellation property.

Corollary 4.15. *Assume Property 4.6, recall that $p \in \tau^k$ from Property 4.6.4, and suppose that $k \ne 1$. If G has the power cancellation property then $p \le L(p)$.*

Theorem 4.16. *Assume Property 4.6, and suppose that $\mathrm{card}(u(\overline{E}))$ is finite. There are at most finitely many rational primes p and nonzero primary ideals J such that $p \in J \subset \overline{E}$ and such that $h(\mathbf{Z} + J) = 1$.*

Proof: Let $\mathcal{P} = \{(k,f,p,\tau) \mid (i) \, k,f > 0$ are integers, p is a rational prime, and $\tau \subset \overline{E}$ is a prime ideal such that (ii) $p \in \tau^k$, (iii) $[\overline{E}/\tau : \mathbf{Z}/p\mathbf{Z}] = f$, and (iv) $h(\mathbf{Z} + \tau^k) = 1\}$. We claim that $\mathcal{P}$ is a finite set.

Let $p_1, \dots, p_t \in \mathbf{Z}$ be the finite list of the primes p_i that ramify in the algebraic number field $\mathbf{k}$. Since $\overline{E}/p_i\overline{E}$ is finite for each prime p_i on our list, there are at most finitely many prime ideals $\tau_1, \dots, \tau_s$ in $\overline{E}$ and finitely many integers $k(i,j) > 1$ such that $p_i \in \tau_j^{k(i,j)}$. Let $f_j = [\overline{E}/\tau_j : \mathbf{Z}/p\mathbf{Z}]$. Then there are at most finitely many quadruples $(k(i,j), f_j, p_i, \tau_j) \in \mathcal{P}$ with $k(i,j) > 1$.

Otherwise, we are given $(1, f, p, \tau) \in \mathcal{P}$. Let $Q(a,x) = \sum_{j=0}^{a-1} x^j$ and fix a residue number f. Some calculus shows us that $Q(f,x)$ is eventually a one-to-one function. Then as p ranges over the infinite set of primes, the sequence $\{Q(f,p) \mid \text{primes } p\}$ of integers is unbounded.

By hypothesis,

$$h(\mathbf{Z} + \tau^1) = 1 = [(\mathbf{Z} + p\overline{E})/p\overline{E} : \mathbf{Z}/p\mathbf{Z}],$$

so by Theorems 2.30 and 4.12, $Q(f,p)$ divides $\mathrm{card}(u(\overline{E})) = m$ for each rational prime p such that $(1, f, p, \tau) \in \mathcal{P}$. ($L(p)$ is a quotient of m.) There are then at most finitely many rational primes $p_1, \dots, p_t$ in the p component of quadruples in $\mathcal{P}$ for each residue number f. Consequently, there are at most finitely many possible components τ in $(1, f, p, \tau) \in \mathcal{P}$.

Now fix a rational prime $p = p_j$ and let f vary. We have observed that $Q(f,p)$ divides m. However, $Q(f,p) \ge Q(f,2) > m$ for sufficiently large f. Then there is a finite list $f_1, \dots, f_t$ of possible residue numbers f in quadruples $(1, f, p, \tau) \in \mathcal{P}$. As claimed, $\mathcal{P}$ is finite.

Let $\mathcal{Q} = \{(p, \tau^k) \mid (i) \, p$ is a rational prime and $k > 0$ is an integer, (ii) τ is a prime ideal $p \in \tau \subset \overline{E}$, and (iii) $h(\mathbf{Z} + \tau^k) = 1\}$. There is a natural surjection $\mathcal{P} \to \mathcal{Q}$, so $\mathcal{Q}$ is finite. Hence there are at most finitely many rational primes p and nonzero primary ideals $p \in \tau^k \subset \overline{E}$ such that $h(\mathbf{Z} + \tau^k) = 1$. This completes the proof.

Example 4.17. Given a square-free integer $d = 2$ or $d > 3$, the ring of algebraic integers $\overline{E}$ in $\mathbf{Q}[\sqrt{-d}]$ has unit group $\{1, -1\}$, so any subring $E(p) \ne \overline{E}$ has unit group $\{1, -1\}$. Thus $L(p) = 1$ for any rational prime p. Given an rtffr group $G(p)$

such that $\mathbf{Z} + p\overline{E} = E(G(p)) \subset \overline{E}$, then card $\left(\dfrac{u(\overline{E})}{u(E(G(p)))} \right) = 1$. Hence $G(p)$ fails

to have the power cancellation property. However, there are only finitely many $d > 0$ such that $h(\overline{E}) = h(\mathbf{Q}[\sqrt{-d}]) = 1$, and for these d, $\overline{G}(p)$ has the power cancellation property.

Example 4.18. Assume that $\overline{E} = \mathbf{Z}[\sqrt{-1}]$. Evidently, $5 = (2 + i)(2 - i)$ is semi-prime in $\overline{E}$. Thus there is a prime ideal $5 \in \tau \subset \overline{E}$ such that $\overline{E}/\tau \cong \mathbf{Z}/5\mathbf{Z}$. The 4 units $\{\pm 1, \pm i\}$ of $u(\overline{E})$ map onto the 4 units of $u(\overline{E}/\tau)$ since $1 + \tau \neq i + \tau$. Thus each unit of $\overline{E}/\tau$ lifts to a unit of $\overline{E}$.

We show that some unit of $\overline{E}/5\overline{E}$ does not lift to a unit of $\overline{E}$. Given a prime $p \geq 5$, then $p\overline{E}$ is prime or $p\overline{E}$ is semi-prime in $\overline{E}$. Inasmuch as card$(u(\overline{E}/5\overline{E})) = 16 > 4 = $ card$(\overline{E})$, some unit of $\overline{E}/p\overline{E}$ does not lift to a unit of $\overline{E}$.

4.3 Quadratic number fields

We will examine $h(G)$ for groups G such that $\mathbf{QE}(G) = \mathbf{k} = \mathbf{Q}[\sqrt{d}]$ for square-free integer d.

Example 4.19. Assume Property 4.6. Fix a rational prime p, and let $\overline{E}$ be the ring of algebraic integers in the *quadratic number field* $\mathbf{k}$. By Property 4.6.4, $k > 0$ is an integer, p is a rational prime, and $\tau \subset \overline{E}$ is a prime ideal such that $p \in \tau^k \subset E(p) \subset \overline{E}$ and $E(p)/\tau^k \cong \mathbf{Z}/p\mathbf{Z}$. Let $L(p) = $ card$(u(\overline{E})/u(E(p)))$.

Since $\overline{E}/\tau$ is a field extension of $\mathbf{Z}/p\mathbf{Z}$, we let

$$[\overline{E}/\tau : \mathbf{Z}/p\mathbf{Z}] = g \geq 1,$$

and then

$$\mathbf{Z}/p\mathbf{Z}\text{-dim}(\tau/\tau^k) = g(k - 1) \geq 0.$$

Evidently, $gk = 2$.

CASE 1: For $g = 1$, $k = 2$ and

$$\text{card}(u(\overline{E}/\tau)) = \text{card}(u(\mathbf{Z}/p\mathbf{Z})) = p - 1.$$

By Lemma 4.5,

$$\text{card}(u(\overline{E}/\tau^2)) = \text{card}(u(\overline{E}/\tau)) \cdot \text{card}(\tau/\tau^2) = p(p - 1),$$

so by Lemma 3.3,

$$\text{card}(\text{coker } \rho) = \frac{p(p - 1)}{L(p) \cdot (p - 1)} = \frac{p}{L(p)}. \tag{4.5}$$

Hence $L(p) = 1$ or $L(p) = p$.

If $L(p) = p$ then card(coker ρ) $= 1$, thus ρ is a surjection, and so $h(G(p)) = h(\overline{G}(p))$ for each rational prime p with $g = 1$. Hence, by Corollary 2.16 and

Theorem 2.30, $G(p)$ has the power cancellation property iff $\overline{G}(p)$ has the power cancellation property iff $\overline{E}$ is a *pid*. Otherwise, $L(p) = 1$. In this case, (4.5) and Theorem 3.4 imply that $h(G(p)) = h(\overline{G}(p))p > 1$ for each rational prime p such that $g = 1$. By Theorem 2.30, such a group $G(p)$ does not have the power cancellation property.

CASE 2: Say $g = 2$. Then $k = 1$, and hence

$$\text{card}(u(\overline{E}/\tau)) = p^2 - 1 \text{ and card}(u(E(p)/\tau)) = p - 1.$$

By Lemma 3.3,

$$\text{card}(\text{coker } \rho) = \frac{p^2 - 1}{L(p) \cdot (p - 1)} = \frac{p + 1}{L(p)}, \tag{4.6}$$

so $L(p)$ divides $p + 1$.

If $L(p) = p + 1$ then the map ρ in (3.1) is a surjection, so by Corollary 2.16, $h(G(p)) = h(\overline{G}(p))$. Thus, by Corollary 2.16 and Theorem 2.30, $G(p)$ has the power cancellation property iff $\overline{G}(p)$ has the power cancellation property iff $\overline{E}$ is a *pid*. Otherwise, $L(p) < p + 1$. Then by (4.6),

$$h(G(p)) = h(\overline{G}(p))\frac{p + 1}{L(p)} > 1,$$

so that by Theorem 2.30, $G(p)$ does not have the power cancellation property.

CASE 3: Say $g = 2$ and $p = 2$. Then by (4.6), $L(p) = 1$ or $L(p) = 3$. If $L(p) = 3$ then $h(G(p)) = h(\overline{G}(p))$. By Corollary 2.16 and Theorem 2.30, $G(p)$ has the power cancellation property iff $\overline{E}$ is a *pid*. Otherwise, $L(p) = 1$. Then by (4.6),

$$h(G(p)) = h(\overline{G}(p))3 > 1,$$

so that $G(p)$ does not have the power cancellation property.

4.4 The Gaussian integers

We specialize notation.

Property 4.20. Assume Property 4.6.

1. Further assume that $\overline{E} = \mathbf{Z}[i]$ where $i = \sqrt{-1}$. By Property 4.6, p is a rational prime and $E(p) = \mathbf{Z} + p\overline{E} = \{a + pbi \,|\, a, b \in \mathbf{Z}\}$.

   Then $\overline{E}$ is a *pid*, $u(\overline{E}) = \{\pm 1, \pm i\}$, $u(E(p)) = \{1, -1\}$, and $\text{card}(u(E/p\overline{E})) = p - 1$.

Example 4.21. Assume Property 4.20. By Property 4.20.1, $\overline{E}$ is a *pid*, so by Corollary 2.16, $h(\overline{G}(p)) = h(\overline{E}) = 1$. By Theorem 2.30, $\overline{G}(p)$ has the power cancellation property. Since $u(\overline{E}) = \{\pm 1, \pm i\}$ and $u(E(p)) = \{\pm 1\}$,

$$L(p) = \text{card}(u(\overline{E})/u(E(p))) = 2.$$

If $p > 2 = \text{card}(u(E(p)))$ then Corollary 4.13 implies that $G(p)$ does not have the power cancellation property.

We improve on the previous example.

Example 4.22. Assume Property 4.20. Observe that $L(p) = 2$ for all rational primes p. Compare this example with Corollary 3.22.

CASE 1: $p = 2$. Since $2\overline{E} = \tau^2$ for some prime ideal $\tau \subset \overline{E}, \overline{E}/2\overline{E} \cong \mathbf{Z}/2\mathbf{Z} \oplus \tau/\tau^2$ as groups, so that $\text{card}(u(\overline{E}/2\overline{E})) = 1 \cdot 2$ by Lemma 4.5. By Corollary 2.16 and Property 4.20.1,

$$h(\overline{G}(2)) = h(\overline{E}) = 1,$$

so that $\text{Pic}(\overline{E}) = 1$. Since $1 - i \notin 2\overline{E}$, the units $1 + 2\overline{E}$ and $i + 2\overline{E}$ are different units in $\overline{E}/2\overline{E}$. Thus the 2 units of $\overline{E}/2\overline{E}$ lift to units of $\overline{E}$. Hence the map ρ in (3.1) is a surjection, so that $\text{Pic}(E(G(2))) = 1$. By Corollary 2.16, $h(G(2)) = 1$, and by Theorem 2.30, $G(2)$ has the power cancellation property.

CASE 2: $p \geq 3$ is a prime in $\overline{E}$. Then $\overline{E}/p\overline{E}$ is a field extension of $\mathbf{Z}/p\mathbf{Z}$ of degree 2, so that $\text{card}(u(\overline{E}/p\overline{E})) = p^2 - 1$. By Corollary 3.13,

$$h(G(p)) = \frac{(p^2 - 1)}{2 \cdot (p - 1)} = \frac{p + 1}{2} \neq 1,$$

so by Theorem 2.30, $G(p)$ does not have the power cancellation property.

CASE 3: $p \geq 5$ in $\overline{E}$ and $p\overline{E} = \tau \cap J$ for some prime ideals $\tau \neq J \subset \overline{E}$. Hence $\overline{E}/p\overline{E} \cong \mathbf{Z}/p\mathbf{Z} \times \mathbf{Z}/p\mathbf{Z}$, and thus $\text{card}(u(\overline{E}/p\overline{E})) = (p - 1)^2$. It follows from Corollary 3.13 that

$$h(G(p)) = \frac{(p - 1)^2}{2 \cdot (p - 1)} = \frac{p - 1}{2} \neq 1,$$

so by Theorem 2.30, $G(p)$ does not have the power cancellation property.

4.5 Imaginary quadratic number fields

Suppose that $G(p)$ is an rtffr group such that $QE(G(p)) = \mathbf{Q}[\sqrt{-d}]$ for some square-free integer $d = 2$ or $d > 3$. We determine those rational primes p such that $G(p)$ has the power cancellation property.

Property 4.23. Assume Property 4.6. Let $d = 2$ or $d > 3$ be a square-free integer.

Example 4.24. Let $\overline{E}$ be the ring of algebraic integers in $\mathbf{Q}[\sqrt{-3}]$. Then $u(\overline{E}) = \{\pm 1, \pm \omega, \pm \omega^2\}$ where $\omega = \frac{1}{2} + \frac{\sqrt{3}}{2}i$. Since 3 is a perfect square in $\overline{E}$, $3\overline{E} = \tau^2$ for some ideal $3\overline{E} \subset \tau \subset \overline{E}$. Then $\overline{E}/\tau \cong \mathbf{Z}/3\mathbf{Z}$ so that by Lemma 4.5,

$$\widehat{m}_3 = \text{card}(u(\overline{E}/\tau^2)) = \text{card}(u(\overline{E}/\tau)) \cdot \text{card}(\tau/\tau^2) = 2 \cdot 3.$$

Let $E(3) = \mathbf{Z} + \tau^2$. Then $u(E(3)/\tau^2) = u(\mathbf{Z}/3\mathbf{Z}) = \{\pm 1\}$, so that $\widehat{n}_3 = 2$. Hence each unit of $E(3)/\tau^2$ lifts to a unit of $E(3)$. Observe that

$$L(3) = \frac{\widehat{m}_3}{\widehat{n}_3} = 3.$$

The units 1 and $\frac{1}{2} + \frac{\sqrt{-3}}{2}$ in $\overline{E}$ map to different units of $\overline{E}/\tau^2$ since the differences $\frac{1}{2} \pm \frac{\sqrt{-3}}{2} - 1 = -\frac{1}{2} \pm \frac{\sqrt{-3}}{2}$ are units of $\overline{E}$. Hence the $m_3 = 6$ units of $\overline{E}$ map onto the $\widehat{m}_3 = 6$ units of $\overline{E}/\tau^2$. That is, each unit of $\overline{E}/\tau^2$ lifts to a unit of $\overline{E}$. Moreover, because $\overline{E}$ is a *pid*, Corollary 2.16 and Theorem 3.12 imply that

$$h(G(3)) = h(\overline{G}(3)) \frac{\widehat{m}_3}{L(3) \cdot \widehat{n}_3} = h(\overline{E}) \frac{\widehat{m}_3}{L(3) \cdot \widehat{n}_3} = 1.$$

Hence $G(3)$ has the power cancellation property.

By [103, Proposition 4.2], the ring of algebraic integers $\overline{E}$ in $\mathbf{Q}[\sqrt{-d}]$ has exactly two units for square-free integers $d = 2$ or $d > 3$. Thus $u(\overline{E}) = u(E(p)) = \{\pm 1\}$. That is, $L(p) = 1$ for each rational prime p. Furthermore, $E(p)/p\overline{E} \cong \mathbf{Z}/p\mathbf{Z}$, so that $\mathrm{card}(u(E(p)/p\overline{E})) = p - 1$. By Corollary 4.14, G does not have the power cancellation property for $p > 2$. We will calculate $h(G(p))$.

Example 4.25. Assume Property 4.23. Let $\overline{E}$ be the ring of algebraic integers in $\mathbf{Q}[\sqrt{-d}]$.

CASE 1: Suppose that $p\overline{E} = \tau^2$ for some prime ideal $\tau \subset \overline{E}$. By Lemma 4.5, $\mathrm{card}(u(\overline{E}/p\overline{E})) = p(p - 1)$, and hence

$$h(G(p)) = h(\overline{G}(p)) \frac{p(p - 1)}{1 \cdot (p - 1)} = h(\overline{G}(p))p$$

by Theorem 3.4. Then $h(G(p)) \geq p$.

CASE 2: Suppose that $p\overline{E}$ is prime in $\overline{E}$. Then $\overline{E}/p\overline{E}$ is a field of degree 2 over $\mathbf{Z}/p\mathbf{Z}$, so that $\mathrm{card}(u(\overline{E}/p\overline{E})) = p^2 - 1$. Thus

$$h(G(p)) = h(\overline{G}(p)) \frac{(p^2 - 1)}{1 \cdot (p - 1)} = h(\overline{G}(p))(p + 1),$$

and so $h(G(p)) \geq p + 1$.

CASE 3: Suppose that $p > 2$ is a rational prime such that $p\overline{E} = \tau \cap J$ for some prime ideals $I \neq J \subset \overline{E}$. By the Chinese Remainder Theorem,

$$\overline{E}/p\overline{E} \cong \overline{E}/I \times \overline{E}/J \cong \mathbf{Z}/p\mathbf{Z} \times \mathbf{Z}/p\mathbf{Z}.$$

Thus $\mathrm{card}(u(E(p)/p\overline{E})) = p - 1$ and $\mathrm{card}(u(\overline{E}/p\overline{E})) = (p - 1)^2$. Hence

$$h(G(p)) = h(\overline{G}(p)) \frac{(p - 1)^2}{1 \cdot (p - 1)} = h(\overline{G}(p))(p - 1).$$

Since $p > 2$, $h(G(p)) \geq p - 1 > 1$.

CASE 4: If $p = 2$ and if $2\overline{E} = I \cap J$ for some prime ideals $I \neq J \subset \overline{E}$, then

$$u(\overline{E}/2\overline{E}) \cong u(\overline{E}/I) \times u(\overline{E}/J) \cong u(\mathbf{Z}/2\mathbf{Z}) \times u(\mathbf{Z}/2\mathbf{Z}) = \{(1, 1)\}.$$

The one unit of $\overline{E}/2\overline{E}$ lifts to a unit of $\overline{E}$. Hence by Corollary 2.16 and Theorem 3.4,

$$h(G(p)) = h(\overline{G}(p)) \frac{1}{1 \cdot 1} = h(\overline{E}) = h(\mathbf{Q}[\sqrt{-d}]),$$

which by [103, Theorem 10.5] equals 1 only for $d \in \{163, 67, 43, 19, 11, 7, 3, 2, 1\}$.

CASE 5: If $p = 2$ and if $2\overline{E} = I^2$ for some prime ideal $I\overline{E}$, then $\overline{E}/I \cong \mathbf{Z}/2\mathbf{Z}$. Then

$$E(2)/I^2 = \mathbf{Z} + I^2/I^2 \cong \mathbf{Z}/2\mathbf{Z}$$

and

$$\overline{E}(2)/I^2 = E(2) \oplus I/I^2,$$

so that

$$\mathrm{card}(u(E(2)) = 1 \text{ and } \mathrm{card}(u(\overline{E}(2))) = 2.$$

Then

$$h(G(2)) = h(\overline{E}) \frac{\widehat{m_3}}{L(3)\widehat{n_3}} = 2,$$

so that $G(2)$ does not have the power cancellation property.

4.6 Exercises

1. Examine the units of $\mathbf{Q}[\sqrt{d}]$ for $d = -1, 2, 3$ and $d > 3$.
2. Fill in the details of the proof of Theorem 4.16.
3. Fill in the details to Examples 4.19 and 4.22.

4.7 Problems for future research

1. Let $E(p) \subset E \subset \overline{E}$ be a subring and let G be a group such that $E(G) = E$. Follow as in Examples 4.19 and 4.22 to show that G does or does not have the power cancellation property.

5

The conductor

The general professional opinion is that groups that possess a unique decomposition or that possess the refinement property are rare. While the Baer–Kulikov–Kaplansky theorem implies that a direct sum $G = \oplus_i G_i$ of rank one groups G_i has the refinement property, examples by Corner and Fuchs–Loonstra (see [59]) show that under some mild hypotheses there are subgroups of finite index in G that do not possess a (locally) unique decomposition. This juxtaposition of uniqueness with nonuniqueness invites us to study further. We can make some sense of this seemingly counterintuitive coincidence by examining direct sum decompositions of *semi-primary groups*.

As in the previous chapters, G is an rtffr group and

$$\boxed{E(G) = \mathrm{End}(G)/\mathcal{N}(\mathrm{End}(G)).}$$

Our goal in this chapter is to study direct sum decompositions of certain *reduced torsion-free finite-rank rtffr* abelian groups by introducing an ideal τ of $E(G)$ called a *conductor of G*. This ideal induces a natural ring decomposition $E(G) = E(G)(\tau) \times E(G)^\tau$ and a natural direct sum decomposition $G = G(\tau) \oplus G^\tau$. A close examination of τ will uncover conditions under which G has a direct sum decomposition that is unique up to the local isomorphism classes of the indecomposable direct summands of G. The uncovered conditions are general enough to include most of the groups occurring in modern abelian group literature.

5.1 Introduction

Throughout this chapter, E denotes a semi-prime rtffr ring with center $\mathrm{center}(E) = S$. Then E is a *Noetherian* (on both sides) ring and $\mathbf{Q}E$ is a semi-simple Artinian ring. The ring E is a prime ring iff $\mathbf{Q}E$ is a simple Artinian ring.

An rtffr ring $\overline{E}$ is a *maximal order* if $\overline{E}$ is a prime ring, and if given a ring $\overline{E} \subset E' \subset \mathbf{Q}\overline{E}$ such that $E'/\overline{E}$ is finite then $\overline{E} = E'$. In this case, $\overline{S} = \mathrm{center}(\overline{E})$ is a Dedekind domain and $\overline{E}$ is a finitely generated projective $\overline{S}$-module. These definitions are in step with the definitions of a classical maximal order as given in [9, 94].

The semi-prime rtffr ring $\overline{E}$ is *integrally closed* if given a ring $\overline{E} \subset E' \subset \mathbf{Q}\overline{E}$ such that $E'/\overline{E}$ is finite then $\overline{E} = E'$. Given a semi-prime rtffr ring $\overline{E}$ then there is a product

$E' = \overline{E}_1 \times \cdots \times \overline{E}_t$ of maximal orders such that $E'/\overline{E}$ is finite, [9, Corollary 10.14(c)]. Since $\overline{E}$ is integrally closed, $\overline{E} = E'$. It follows that $\overline{E}$ is integrally closed iff

$$\overline{E} = \overline{E}_1 \times \cdots \times \overline{E}_t$$

for some maximal orders $\overline{E}_1, \ldots, \overline{E}_t$. For this reason, several of the structural results for classic maximal orders in [9, 94] easily lift to structural results for integrally closed rtffr rings.

At all times, $\overline{E}$ denotes an integrally closed, semi-prime, rtffr ring. Then rtffr ring E is integrally closed iff the localization E_p is integrally closed for each rational prime p, [9, Theorem 11.7]. $\mathrm{Mat}_n(\overline{E})$ is integrally closed for each integer $n > 0$, [9, Corollary 10.3]. Integrally closed rtffr rings are hereditary (both sides) rings, [9, Theorem 11.5].

An *E-lattice* is a finitely generated right E-submodule of the right E-module $\mathbf{Q}E^{(n)}$ for some integer $n > 0$. If U and V are $\overline{E}$-lattices then by [9, Corollary 12.3], U is locally isomorphic to V iff $\mathbf{Q}U \cong \mathbf{Q}V$ as right $\mathbf{Q}\overline{E}$-modules. Given a semi-prime rtffr ring E, there is an integrally closed ring $\overline{E} \subset \mathbf{Q}E$ and an integer $n \neq 0$ such that $n\overline{E} \subset E \subset \overline{E}$, [9, page 115]. Specifically, $\overline{E}/E$ is finite.

If R is an rtffr ring then by the Beaumont–Pierce theorem 1.8 there is a semi-prime subring $T \subset R$ such that

$$R \doteq T \oplus \mathcal{N}(R).$$

Let $\mathbf{Q}T = B$, a semi-simple finite-dimensional $\mathbf{Q}$-algebra. Then $\mathbf{Q}R = B \oplus \mathcal{N}(\mathbf{Q}R)$. Let

$$T' = \{b \in B \mid b \oplus c \in \mathrm{End}(G) \text{ for some } c \in \mathcal{N}(\mathbf{Q}\mathrm{End}(G))\}.$$

Evidently $E(G) \cong T'$. We will identify

$$E(G) = T' \subset B.$$

Then by the Beaumont–Pierce theorem 1.8,

$$\mathrm{End}(G) \doteq E(G) \oplus \mathcal{N}(\mathrm{End}(G))$$

as groups.

Property 5.1. Let G be an rtffr group and let

$$E(G) \subset \overline{E} \subset B \subset \mathbf{Q}\mathrm{End}(G)$$

be an integrally closed ring such that $\overline{E}/E(G)$ is finite. By Lemma 2.9 and its proof, the rtffr group $\overline{G} = \overline{E}G \subset \mathbf{Q}G$ satisfies

1. $G \subset \overline{G} \subset \mathbf{Q}G$,
2. $\overline{G}/G$ is finite, and
3. $E(\overline{G}) = \overline{E}$.

Our discussion requires more detailed information about the conductor τ of E. Recall $E, \overline{E}, S$, and $\overline{S}$ from Property 3.1.

Property 5.2. 1. There are maximal orders $\overline{E}_1, \ldots, \overline{E}_t$ such that

$$\boxed{\overline{E} = \overline{E}_1 \times \cdots \times \overline{E}_t.}$$

Let $\overline{S}_i = \text{center}(\overline{E}_i)$ for each $i = 1, \ldots, t$. Then

$$\boxed{\overline{S} = \overline{S}_1 \times \cdots \times \overline{S}_t.}$$

2. The largest ideal $\tau \subset S$ such that

$$\boxed{\tau \overline{E} \subset E \subset \overline{E}}$$

is called the *conductor* of E (relative to $\overline{E}$). Since $\overline{E}/E$ is finite, S/τ, $E/\tau E$, and $\overline{E}/\tau \overline{E}$ are finite.

3. Since $\tau \overline{S} \subset \mathbf{Q}S \cap E = S$ and since τ is maximal in S with respect to property (2) above, $\tau = \tau \overline{S}$ is an ideal of $\overline{S}$. We write

$$\boxed{\tau = \tau_1 \oplus \cdots \oplus \tau_t}$$

where $\tau_i = \tau \overline{S}_i$.

There can be many integrally closed rings $\overline{E} \subset \mathbf{Q}E$ such that $\overline{E}/E$ is finite, and for each containment $E \subset \overline{E}$ there is a conductor $\tau = \tau(E, \overline{E})$.

5.2 Some functors

Let $\mathbf{P}_o(G) = \{\text{groups } H \mid H \oplus H' \cong G^n$ for some integer $n > 0$ and group $H'\}$. Let $\mathbf{P}_o(R)$ denote the category of finitely generated right modules over the ring R.

To study local isomorphism we will need a small collection of functors and categories. For example, the Arnold–Lady theorem 1.13 gives the category equivalence in the following theorem. Let (X) denote the isomorphism class of X and let $[X]$ denote the local isomorphism class of X.

Theorem 5.3. *Let G be an rtffr group and let $\mathbf{H}_G(\cdot) = \text{Hom}(G, \cdot)$. There is a category equivalence*

$$\mathbf{H}_G(\cdot) : \mathbf{P}_o(G) \longrightarrow \mathbf{P}_o(\text{End}(G))$$

that induces bijections

1. $\{(H) \mid H \in \mathbf{P}_o(G)\} \longrightarrow \{(W) \mid W \in \mathbf{P}_o(\text{End}(G))\}$, *and*
2. $\eta_G : \{[H] \mid H \in \mathbf{P}_o(G)\} \longrightarrow \{[W] \mid W \in \mathbf{P}_o(\text{End}(G))\}$.

A simpler setting for studying finitely generated projective right R-modules is achieved by considering $R/\mathcal{N}(R)$-modules.

From Theorems 2.3 and 2.4, recall the full functors

$$A_R(\cdot) : \mathbf{P}_o(R) \longrightarrow \mathbf{P}_o(R/\mathcal{N}(R))$$
$$A_G(\cdot) : \mathbf{P}_o(G) \longrightarrow \mathbf{P}_o(E(G)).$$

The second functor $A_G(\cdot)$ is defined by

$$A_G(\cdot) = (A_{\mathrm{End}(G)} \circ \mathbf{H}_G)(\cdot)$$

and $A_{\mathrm{End}(G)}(\cdot)$ induces a bijection

$$\alpha_{\mathrm{End}(G)} : \{[U] \,\big|\, U \in \mathbf{P}_o(\mathrm{End}(G))\} \longrightarrow \{[W] \,\big|\, W \in \mathbf{P}_o(E(G))\}.$$

Then $A_G(\cdot)$ induces bijections

1. $\{(H) \,\big|\, H \in \mathbf{P}_o(G)\} \longrightarrow \{(W) \,\big|\, W \in \mathbf{P}_o(E(G))\}$, and
2. $(\alpha_{\mathrm{End}(G)} \circ \eta_G) : \{[H] \,\big|\, H \in \mathbf{P}_o(G)\} \longrightarrow \{[W] \,\big|\, W \in \mathbf{P}_o(E(G))\}$.

Recall that the ring R is *semi-perfect* if

1. $R/\mathcal{J}(R)$ is semi-simple Artinian and
2. Given an $\bar{e}^2 = \bar{e} \in R/\mathcal{J}(R)$ there is an $e^2 = e \in R$ such that $\bar{e} = e + \mathcal{J}(R)$. That is, *idempotents lift modulo $\mathcal{J}(R)$*.

Theorem 5.4. *Let R be a ring, let $\mathbf{S}_o(R)$ denote the category of finitely generated semi-simple right R-modules, and let*

$$B_R(\cdot) = \cdot \otimes_R R/\mathcal{J}(R).$$

If R is a semi-perfect ring then the full functor

$$B_R(\cdot) : \mathbf{P}_o(R) \longrightarrow \mathbf{S}_o(R)$$

induces a bijection

$$\beta_R : \{(P) \,\big|\, P \in \mathbf{P}_o(R)\} \longrightarrow \{(W) \,\big|\, W \in \mathbf{S}_o(R)\}.$$

Proof: Over a semi-perfect ring R, semi-simple right modules are of the form $P/P\mathcal{J}(R)$ for some finitely generated projective R-module P [7], so the assignment $(P) \mapsto (B_R(P))$ is a surjection $\mathbf{P}_o(R)/\cong \longrightarrow \mathbf{S}_o(R)/\cong$. Let $P, Q \in \mathbf{P}_o(R)$ be such that $P/P\mathcal{J}(R) = B_R(P) \cong B_R(Q) = Q/Q\mathcal{J}(R)$. By Lemma 2.2, $P \cong Q$. Then $(P) \mapsto (B_R(P))$ defines an injective assignment, and hence $B_R(\cdot) : \mathbf{P}_o(R) \longrightarrow \mathbf{S}_o(R)$ induces the bijection $\beta_G : \mathbf{P}_o(R)/\cong \longrightarrow \mathbf{S}_o(R)/\cong$. The functor $B_R(\cdot)$ is full because objects in $\mathbf{P}_o(R)$ are projective right R-modules.

5.3 Conductor of an *rtffr* ring

We study the conductor τ of the semi-prime rtffr ring E.

Let

$$C_\tau = \{x \in S \mid x + \tau \text{ is a unit of } S/\tau\}.$$

Then C_τ is a multiplicatively closed subset of S containing 1, so we can localize S at C_τ. To press the importance of τ in this localization we let

$$X_\tau = X[C_\tau^{-1}].$$

Let $(\cdot)_\tau$ denote the localization functor. Given an S-module X let

$$X(\tau) = \{x \in X \mid xc = 0 \text{ for some } c \in C_\tau\}.$$

We let

$$X^\tau = X/X(\tau)$$

Then $X_\tau = (X/X(\tau))_\tau = (X^\tau)_\tau$.

Lemma 5.5. [46, Lemma 3.2.1] *Let E be a semi-prime rtffr ring with conductor τ, and let M be a maximal ideal of S. Then $\tau \not\subset M$ iff E_M is integrally closed.*

The conductor of the semi-prime rtffr ring E gives a local condition that shows us when an E-lattice is a projective E-module.

Lemma 5.6. [46, Lemma 3.2.3] *Let E be a semi-prime rtffr ring with conductor τ and let U be an E-lattice. Then U is a projective right E-module iff the localization U_τ is a projective right E_τ-module.*

Theorem 5.7. *Let E be a semi-prime rtffr ring with conductor τ. There are integrally closed semi-prime Noetherian rings $S(\tau)$ and $E(\tau)$, and semi-prime rtffr rings S^τ and E^τ such that*

1. *$S = S(\tau) \times S^\tau$ and $E = E(\tau) \times E^\tau$,*
2. *$S(\tau)_\tau = E(\tau)_\tau = 0$, and*
3. *$S_\tau = (S^\tau)_\tau$ and $E_\tau = (E^\tau)_\tau$.*

Proof: One shows that $S(\tau)$ is an ideal of S such that $S/S(\tau)$ is C_τ-torsion-free. Thus $S/S(\tau)$ is an S-lattice. Furthermore, $S_\tau \cong (S/S(\tau))_\tau$ is a projective S_τ-module. By Lemma 5.6, $S/S(\tau)$ is then a projective S-module. Hence $S = S(\tau) \oplus S^\tau$ for some ideal $S^\tau \cong S/S(\tau)$ of S. Because S is commutative, $S = S(\tau) \times S^\tau$. It follows that $S_\tau \cong (S/S(\tau))_\tau \cong (S^\tau)_\tau$.

In a similar manner $E = E(\tau) \oplus E^\tau$ where $E^\tau \cong E/E(\tau)$. Since E is a semi-prime rtffr ring and since $E(\tau)$ is an ideal, $E = E(\tau) \times E^\tau$. Since $S(\tau)$ and $E(\tau)$ are $\mathcal{C}_\tau$-torsion S-modules, $S(\tau)_\tau = E(\tau)_\tau = 0$.

Furthermore, let M be a maximal ideal in S.

If $\tau \not\subset M$ then by Lemma 5.5, E_M is an integrally closed ring so that $E(\tau)_M$ is integrally closed.

On the other hand, if $\tau \subset M$ then $\mathcal{C}_\tau \subset \mathcal{C}_M$. Thus

$$E(\tau)_M = (E(\tau)_M)_\tau = (E(\tau)_\tau)_M = 0.$$

Thus $E(\tau)_M$ is integrally closed for each maximal ideal $M \subset S$, and hence $E(\tau)$ is integrally closed. This completes the proof.

Corollary 5.8. *Let E be a semi-prime rtffr ring and let U be an E-lattice. Then $U(\tau) = UE(\tau)$, $U^\tau = UE^\tau$, and $U = U(\tau) \oplus U^\tau$. Moreover, $U^\tau \subset (U^\tau)_\tau \cong U_\tau$ as right E_τ-modules.*

Proof: Since U is an E-lattice there is an integer $n > 0$ such that $U \subset E^n$. By Theorem 5.7, $E = E(\tau) \times E^\tau$, so that

$$UE(\tau) \subset U(\tau) \subset E(\tau)^n, \quad UE^\tau \subset (E^\tau)^n$$

and

$$U \cong U \otimes_E E \cong U \otimes_E (E(\tau) \times E^\tau) \cong UE(\tau) \oplus UE^\tau.$$

Furthermore,

$$(UE^\tau)(\tau) \subset (E^\tau)^n \cap E(\tau)^n = 0,$$

so that by the Modular Law,

$$U(\tau) = U \cap U(\tau) = (UE(\tau) \oplus UE^\tau) \cap U(\tau)$$
$$= UE(\tau) \oplus (UE^\tau \cap U(\tau)) = UE(\tau)$$

Then $U(\tau) = UE(\tau)$ and $U^\tau = UE^\tau$, whence $U = U(\tau) \oplus U^\tau$.

Moreover, because $U^\tau(\tau) = 0$, $U^\tau \subset (U^\tau)_\tau = U_\tau$. This completes the proof.

5.4 Local correspondence

The main result of this section shows how localization at the conductor can be used to translate local isomorphism into isomorphism. We invoke the Properties 5.2 without reference. Given a right E-module U, let $U = U(\tau) \oplus U^\tau$ as in Corollary 5.8.

Let E be a semi-prime rtffr ring. Define a functor $L_\tau(\cdot)$ as follows.

$$L_\tau(\cdot) : \mathbf{P}_o(E) \longrightarrow \mathbf{P}_o(\mathbf{Q}E(\tau) \times E_\tau)$$
$$L_\tau(U) = \mathbf{Q}U(\tau) \oplus U_\tau.$$

Since localization functors are additive and exact, $L_\tau(\cdot)$ is a well-defined additive exact functor. Let

$$\lambda_\tau[U] = (L_\tau(U)) = (\mathbf{Q}U(\tau) \oplus U_\tau)$$

for each $U \in \mathbf{P}_o(E)$.

Theorem 5.9. *Let E be a semi-prime rtffr ring with conductor τ. Then*

$$\lambda_\tau : \{[U] \,\big|\, U \in \mathbf{P}_o(E)\} \longrightarrow \{(W) \,\big|\, W \in \mathbf{P}_o(\mathbf{Q}E(\tau) \times E_\tau)\}$$

is a functorial bijection.

Proof: The proof is a series of lemmas.

Lemma 5.10. [9, Corollary 12.3] *Let $\overline{E}$ be an integrally closed semi-prime rtffr ring. The $\overline{E}$-lattices U and V are locally isomorphic iff $\mathbf{Q}U \cong \mathbf{Q}V$ as $\mathbf{Q}\overline{E}$-modules.*

Lemma 5.11. *Let E be a semi-prime rtffr ring with conductor τ, and let U, V be E-lattices. Then U is locally isomorphic to V iff $\mathbf{Q}U(\tau) \cong \mathbf{Q}V(\tau)$ and U^τ is locally isomorphic to V^τ.*

Proof: By Theorem 5.7, $E = E(\tau) \times E^\tau$, where $E(\tau)$ is an integrally closed Noetherian semi-prime rtffr ring, and E^τ is a Noetherian semi-prime rtffr ring. It is clear that U is locally isomorphic to V iff $U(\tau)$ is locally isomorphic to $V(\tau)$ as $E(\tau)$-lattices, and U^τ is locally isomorphic to V^τ as E^τ-lattices. Since $E(\tau)$ is integrally closed, Lemma 5.10 states that $U(\tau)$ is locally isomorphic to $V(\tau)$ iff $\mathbf{Q}U(\tau) \cong \mathbf{Q}V(\tau)$. This proves the lemma.

Corollary 5.12. *Let E be a semi-prime rtffr ring with conductor τ. Then λ_τ is a well-defined function.*

Proof: Suppose that $U, V \in \mathbf{P}_o(E)$ are locally isomorphic. By Lemma 5.11, $\mathbf{Q}U(\tau) \cong \mathbf{Q}V(\tau)$ and U^τ is locally isomorphic to V^τ. One proves that because E_τ is a semi-local ring, $(U^\tau)_\tau \cong (V^\tau)_\tau$. Then by Corollary 5.8,

$$U_\tau \cong (U^\tau)_\tau \cong (V^\tau)_\tau \cong V_\tau.$$

Hence

$$L_\tau(U) = \mathbf{Q}U(\tau) \oplus U_\tau \cong \mathbf{Q}V(\tau) \oplus V_\tau = L_\tau(V),$$

and so λ_τ is well defined.

Lemma 5.13. *Let E be a semi-prime rtffr ring with conductor τ.*

1. *For each finitely generated projective $\mathbf{Q}E(\tau)$-module W there is a finitely generated projective projective right $E(\tau)$-module U such that $\mathbf{Q}U \cong W$.*
2. *For each finitely generated projective E_τ-module W there is a finitely generated projective E^τ-module U such that $U_\tau \cong W$.*

Proof: 1. Write $W = w_1\mathbf{Q}E(\tau) + \cdots + w_s\mathbf{Q}E(\tau)$ for some finite set $\{w_1, \ldots, w_s\} \subset W$. Let $U = w_1E(\tau) + \cdots + w_sE(\tau)$ be the $E(\tau)$-lattice generated by $w_1, \ldots, w_s$. Let $\oplus_{i=1}^s x_iE(\tau)$ be a free right $E(\tau)$-module on $\{x_1, \ldots, x_s\}$. There is an $E(\tau)$-module surjection $\pi : \oplus_{i=1}^s x_iE(\tau) \to U$ such that $\pi(x_i) = w_i$ for each i. Because localization is exact, the image of $1 \otimes \pi : \oplus_{i=1}^s x_i\mathbf{Q}E(\tau) \to \mathbf{Q}U \subset W$ is

$$\mathbf{Q}U = w_1\mathbf{Q}E(\tau) + \cdots + w_s\mathbf{Q}E(\tau) = W.$$

Because $E(\tau)$ is integrally closed, U is a finitely generated projective right $E(\tau)$-module. This proves part 1.

2. Let $W \in \mathbf{P}_o(E_\tau)$ and write $W = w_1E_\tau + \cdots + w_sE_\tau$ for some finite set $\{w_1, \ldots, w_s\} \subset W$. Let $U = w_1E^\tau + \cdots + w_sE^\tau$ be the E^τ-lattice generated by $w_1, \ldots, w_s$. Let $\oplus_{i=1}^s x_iE^\tau$ be a free right E^τ-module on $\{x_1, \ldots, x_s\}$. There is an E^τ-module surjection $\pi : \oplus_{i=1}^s x_iE^\tau \to U$ such that $\pi(x_i) = w_i$ for each i. Because $E_\tau \cong (E^\tau)_\tau$, and because localization is exact, the image of $\pi_\tau : \oplus_{i=1}^s x_iE_\tau \to U_\tau \subset W$ is

$$U_\tau = w_1E_\tau + \cdots + w_sE_\tau = W.$$

Furthermore, $U_\tau = W$ is a finitely generated projective E_τ-module, so by Lemma 5.6, U is a finitely generated projective E-module. The lemma is proved.

Corollary 5.14. *Let E be a semi-prime rtffr ring with conductor τ. Then λ_τ is a surjection.*

Proof: Given $W \in \mathbf{P}_o(\mathbf{Q}E(\tau) \times E_\tau)$, $W = W(\tau) \oplus W^\tau$ for some finitely generated projective modules $W(\tau)$ over $\mathbf{Q}E(\tau)$ and W^τ over E_τ. By Lemma 5.13 there are $U', U'' \in \mathbf{P}_o(E)$ such that $\mathbf{Q}U' = W(\tau)$ and $(U'')_\tau = W_\tau$. Since $L_\tau(U') = \mathbf{Q}U' = W(\tau)$ and since $L_\tau(U'') = U''_\tau = W_\tau$, $\lambda_\tau[U' \oplus U''] = (W(\tau) \oplus W_\tau) = (W)$. Hence λ_τ is a surjection.

It remains to show that λ_τ is an injection. Given an integer n notice that S_n is formed by inverting integers that are relatively prime to n, while S_{nS} is localization of S at elements that map to units in S/S_{nS}. These localizations agree in our setting.

Lemma 5.15. *Let S and τ be as in (5.2). Suppose that $S = S^\tau$, and let $n \neq 0 \in \tau$ be an integer. Then $S_n = S_{nS}$.*

Proof: Let τ be the conductor of E, let $n \neq 0 \in \tau$, let $\{T_1, \ldots, T_s\}$ be a complete set of the maximal ideals of S that contain τ, and let $\{M_1, \ldots, M_r, T_1, \ldots, T_s\}$ be a complete set of the maximal ideals of S that contain nS. Let

$$C_n = \{c \in S \mid c + nS \text{ is a unit in } S/nS\}.$$

We claim that $S_n \subset S_{nS} \subset QS$ naturally. Since S/nS is a $\mathbf{Z}/n\mathbf{Z}$-algebra, and since $nS \subset \tau$, we have

$$\{m \in \mathbf{Z} \mid m + n\mathbf{Z} \text{ is a unit in } \mathbf{Z}/n\mathbf{Z}\} \subset C_n \subset C_\tau.$$

Since $S = S^\tau$ is a C_τ-torsion-free S-module, S is a C_n-torsion-free S-module. That is, $yc \neq 0$ for each element $c \in C_n$ and $y \neq 0 \in S$. Now the semi-simple ring QS is the localization of S at the regular elements of S. Furthermore, since $S = S^\tau$, S is C_τ-torsion-free, so the elements of C_n are regular. Hence, as claimed, there is an inclusion $S \subset S_{nS} \subset QS$ that lifts to one $S_n \subset S_{nS} \subset QS$.

We show that $S_n = S_{nS}$. Let $N \in \{M_1, \ldots, M_r, T_1, \ldots, T_s\}$. Since $n \in nS \subset N$, $C_n \subset C_N$. Thus

$$(S_n)_N = (S_N)_n = S_N = (S_N)_{nS} = (S_{nS})_N.$$

Say, on the other hand, that $N \notin \{M_1, \ldots, M_r, T_1, \ldots, T_s\}$. Then $n \notin N$, so that

$$QS_N = (S_N)_n = (S_n)_N.$$

By the above claim and the exactness of localization,

$$(S_n)_N \subset (S_{nS})_N \subset QS_N.$$

Thus $QS_N = (S_n)_N = (S_{nS})_N$. By the Local-Global Theorem 1.5, $S_n = S_{nS}$, which completes the proof of the lemma.

Lemma 5.16. *Let E be a semi-prime rtffr ring with conductor τ. Let $n \neq 0 \in \tau$ be an integer. Let $E = E^\tau$ and let U and $V \in \mathbf{P}_o(E)$. If $U_\tau \cong V_\tau$ then $U_n \cong V_n$.*

Proof: Suppose that $U_\tau \cong V_\tau$. Let $M_1, \ldots, M_r, T_1, \ldots, T_s$ be a complete list of the distinct maximal ideals of S that contain nS, where $T_1, \ldots, T_s$ is the list of maximal ideals of S that contain τ.

Let $T \in \{T_1, \ldots, T_s\}$ and let

$$C_T = \{c \in S \mid c + T \text{ is a unit in } S/T\}.$$

Since T is maximal, S/T is a field, a C_T-torsion-free divisible S-module. Moreover, $C_\tau \subset C_T$. Then for each right E-module X,

$$X/XT \cong (X/XT)_\tau \cong X_\tau/X_\tau T_\tau.$$

Hence, the isomorphism $U_\tau \cong V_\tau$ of E_τ-modules induces an isomorphism

$$\frac{U}{UT} \cong \frac{U_\tau}{U_\tau T_\tau} \cong \frac{V_\tau}{V_\tau T_\tau} \cong \frac{V}{VT}$$

so that

$$\bigoplus_{i=1}^{s} \frac{U}{UT_i} \cong \bigoplus_{i=1}^{s} \frac{V}{VT_i}$$

as right E-modules.

Next, let $M \in \{M_1, \ldots, M_r\}$. By hypothesis, $E = E^\tau$, so that E is $\mathcal{C}_\tau$-torsion-free. Hence $E \subset E_\tau \subset \mathbf{Q}E$, so that $\mathbf{Q}E \cong \mathbf{Q}E_\tau$. Then, since $U, V \in \mathbf{P}_o(E)$ and since $U_\tau \cong V_\tau$,

$$\mathbf{Q}U = \mathbf{Q}U_\tau = \mathbf{Q}V_\tau = \mathbf{Q}V.$$

Localizing at M yields $\mathbf{Q}U_M \cong \mathbf{Q}V_M$. Since $M \notin \{T_1, \ldots, T_s\}$, Lemma 5.5 states that E_M is integrally closed. Then Lemma 5.11 implies that U_M is locally isomorphic to V_M. Since E_M is a semi-local ring, $U_M \cong V_M$ as right E_M-modules. Thus

$$U/UM \cong U_M/U_M M_M \cong V_M/V_M M_M \cong V/VM,$$

so that

$$\bigoplus_{i=1}^{r} \frac{U}{UM_i} \oplus \bigoplus_{j=1}^{s} \frac{U}{UT_j} \cong \bigoplus_{i=1}^{r} \frac{V}{VM_i} \oplus \bigoplus_{j=1}^{s} \frac{V}{VT_j}. \tag{5.1}$$

Let

$$J = M_1 \cap \cdots \cap M_r \cap T_1 \cap \cdots \cap T_s.$$

Then

$$J_{nS} = \mathcal{J}(S_{nS}) \subset \mathcal{J}(E_{nS})$$

by our choice of maximal ideals $\{M_1, \ldots, M_r, T_1, \ldots, T_s\}$ in S. By applying the Chinese Remainder Theorem to (5.1), we have

$$\frac{U}{UJ} \cong \bigoplus_{i=1}^{r} \frac{U}{UM_i} \oplus \bigoplus_{j=1}^{s} \frac{U}{UT_j}.$$

Then by the isomorphism (5.1) we can write that

$$U_{nS}/U_{nS}J_{nS} \cong U/UJ \cong V/VJ \cong V_{nS}/V_{nS}J_{nS}.$$

Since U and V are projective E-modules, U_{nS} and V_{nS} are projective E_{nS}-modules. Then by Lemma 2.2, $U_{nS} \cong V_{nS}$ as left E_{nS}-modules. Putting this together with Lemma 5.15 yields the equation

$$U_n \cong U \otimes_S S_n \cong U \otimes_S S_{nS} \cong U_{nS} \cong V_{nS} \cong V \otimes_S S_{nS} \cong V \otimes_S S_n \cong V_n$$

which proves the lemma.

Lemma 5.17. *Let E be a semi-prime rtffr ring with conductor τ. If $U, V \in \mathbf{P}_o(E)$ then U and V are locally isomorphic right E-modules iff $\mathbf{Q}U(\tau) \cong \mathbf{Q}V(\tau)$ as right $\mathbf{Q}E(\tau)$-modules and $U_\tau \cong V_\tau$ as right E_τ-modules.*

Proof: For $X \in \mathbf{P}_o(E)$, write $X = X(\tau) \oplus X^\tau$ as in Corollary 5.8. Suppose that $\mathbf{Q}U(\tau) \cong \mathbf{Q}V(\tau)$ and that $U_\tau \cong V_\tau$.

Since $U(\tau)$ and $V(\tau)$ are $E(\tau)$-lattices, Lemma 5.11 and $\mathbf{Q}U(\tau) \cong \mathbf{Q}V(\tau)$ show us that $U(\tau)$ is locally isomorphic to $V(\tau)$.

Let $n > 0 \in \mathbf{Z}$ be given. Since S/τ is finite, we may assume without loss of generality that $n \in \tau$. By Corollary 5.8,

$$(U^\tau)_\tau \cong U_\tau \cong V_\tau \cong (V^\tau)_\tau,$$

so by Lemma 5.16,

$$(U^\tau)_n \cong (V^\tau)_n$$

as right E^τ-modules. There are then maps

$$f_n : (U^\tau)_n \to (V^\tau)_n \text{ and } g_n : (V^\tau)_n \to (U^\tau)_n$$

such that $f_n g_n = 1 = g_n f_n$. Since U^τ and V^τ are finitely generated there is an integer m relatively prime to n such that

$$mf_n : U^\tau \to V^\tau \text{ and } mg_n : V^\tau \to U^\tau$$

and $(mf_n)(mg_n) = m^2 1 = (mg_n)(mf_n)$. Since n was arbitrarily taken, U^τ is locally isomorphic to V^τ. Thus $U = U(\tau) \oplus U^\tau$ is locally isomorphic to $V = V(\tau) \oplus V^\tau$. The converse is clear, so the proof of the lemma is complete.

Corollary 5.18. *Let E be a semi-prime rtffr ring with conductor τ. Then λ_τ is an injection.*

Proof: follows from Lemma 5.17.

Proof of Theorem 5.9 completed: By Corollaries 5.14 and 5.18, λ_τ is a bijection. This completes the proof of the theorem.

5.5 Exercises

Let G and H denote rtffr groups. Let $E(G) = \text{End}(G)/\mathcal{N}(\text{End}(G))$. Let E be an rtffr ring. Let $S = \text{center}(E)$. Given a semi-prime rtffr ring E let $\overline{E}$ be an integrally closed subring of $\mathbf{Q}E$ such that $E \subset \overline{E}$ and $\overline{E}/E$ is finite. Let τ be the conductor of E.

1. Let E be semi-prime and let $I \subset E$ be a fractional right ideal. Then I is locally isomorphic to E iff $I \cong I'$ such that $I' + \tau E = E$.

2. Let M and N be finitely generated projective right E-modules. The following are equivalent.

   (a) M is locally isomorphic to N.
   (b) $M/M\mathcal{N}(E)$ is locally isomorphic to $N/N\mathcal{N}(E)$
   (c) $M^n \cong N^n$ for some integer $n > 0$.

3. Let E be a semi-perfect ring. Let M and N be finitely generated projective right E-modules.

   (a) $M \cong N$ iff $M/M\mathcal{J}(E) \cong N/N\mathcal{J}(E)$.
   (b) If K is a simple right E-module then $K = M/M\mathcal{J}(E)$ for some indecomposable E-module M.
   (c) $M/M\mathcal{J}(E)$ is indecomposable iff M is indecomposable.

4. Prove Lemma 5.10.

5. Prove Lemma 5.5.

6. Let E be a prime rtffr ring and let U be an E-lattice. Prove that U is a projective E-module iff U_τ is a projective E_τ-module.

5.6 Problems for future research

1. Show that the conductor τ is unique to the prime *rtffr* ring E.

2. Characterize the rtffr groups G such that $\text{End}(G)$ is semi-perfect.

3. Find a better proof of Lemma 5.16.

6

Conductors and groups

Fix an rtffr group G. A *conductor of G* is a conductor τ of $E(G)$. Thus there is an integrally closed ring $\overline{E}$ such that $E(G) \subset \overline{E} \subset QE(G)$, $\overline{E}/E(G)$ is finite, and such that τ is the largest ideal in $S = \text{center}(E(G))$ such that $\tau\overline{E} \subset E(G)$. Let

$$E_\tau(G) = QE(G)(\tau) \times E(G)_\tau.$$

Notice the difference between $E(G)_\tau$, a localization of $E(G)$, and the ring product $E_\tau(G)$. Define an additive functor $L_G(\cdot)$ as the composition of functors,

$$L_G(\cdot) = L_\tau(\cdot) \circ A_{\text{End}(G)}(\cdot) \circ H_G(\cdot) : \mathbf{P}_o(G) \longrightarrow \mathbf{P}_o(E_\tau(G))$$

and let

$$\lambda_G[H] = (L_G(H))$$

for each $H \in \mathbf{P}_o(G)$, where $[X]$ is the local isomorphism class of X and (X) is the isomorphism class of X. In this chapter we will study direct sum decompositions of G in light of the conductor τ of G and the functor $L_G(\cdot)$.

6.1 *Rtffr* groups

Theorem 6.1. *Let G be an rtffr ring with conductor τ. There is a functorial bijection*

$$\lambda_G : \{[H] \mid H \in \mathbf{P}_o(G)\} \longrightarrow \{(W) \mid W \in \mathbf{P}_o(E_\tau(G))\}.$$

Proof: Let G be an rtffr group with a conductor τ. By Theorem 5.3, $H_G(\cdot)$ induces a bijection

$$\eta_G : \{[H] \mid H \in \mathbf{P}_o(G)\} \longrightarrow \{[W] \mid W \in \mathbf{P}_o(\text{End}(G))\}$$

such that $\eta_G[H] = [\mathbf{H}_G(H)]$. Theorem 2.3 shows us that the functor $A_{\mathrm{End}(G)}(\cdot) :$ $\mathbf{P}_o(\mathrm{End}(G)) \longrightarrow \mathbf{P}_o(E(G))$ induces a bijection

$$\alpha_G : \{[P]\,|\,P \in \mathbf{P}_o(\mathrm{End}(G))\} \longrightarrow \{[Q]\,|\,Q \in \mathbf{P}_o(E(G))\}$$

such that $\alpha_G[P] = [A_{\mathrm{End}(G)}(P)]$. By Theorem 5.9, $L_\tau(\cdot)$ induces a bijection

$$\lambda_\tau : \{[Q]\,|\,Q \in \mathbf{P}_o(E(G))\} \longrightarrow \{(W)\,|\,W \in \mathbf{P}_o(E_\tau(G))\}$$

such that $\lambda_\tau[Q] = (QQ(\tau) \oplus Q_\tau)$. Then

$$\lambda_G[H] = (\lambda_\tau \circ \alpha_G \circ \eta_G)[H] = (L_G(H))$$

defines a bijection

$$\lambda_G : \{[H]\,|\,H \in \mathbf{P}_o(G)\} \longrightarrow \{(W)\,|\,W \in \mathbf{P}_o(E_\tau(G))\}.$$

This completes the proof.

Corollary 6.2. *Let G be an rtffr group and assume that $E(G)$ is an integrally closed ring. There is a bijection*

$$\boxed{\lambda_G(\cdot) : \{[H]\,|\,H \in \mathbf{P}_o(G)\} \longrightarrow \{(W)\,|\,W \in \mathbf{P}_o(\mathbf{Q}E(G))\}}$$

such that

$$\boxed{\lambda_G[H] = (L_G(H))}$$

Proof: Proceed as in Theorem 6.1 using Lemma 5.10 to show that H is locally isomorphic to K iff $\alpha_G \circ \eta_G(H)$ is locally isomorphic to $\alpha_G \circ \eta_G(K)$ over the integrally closed ring $E(G)$, iff

$$\mathbf{Q}(\alpha_G \circ \eta_G(H)) \cong \mathbf{Q}(\alpha_G \circ \eta_G(K))$$

as $\mathbf{Q}E(G)$-modules.

Corollary 6.3. *Let G be an rtffr group with conductor $\tau \neq S$, and assume that $E(G)$ has no nonzero integrally closed ring factor E'. There is a bijection*

$$\lambda_G(\cdot) : \{[H]\,|\,H \in \mathbf{P}_o(G)\} \longrightarrow \{(W)\,|\,W \in \mathbf{P}_o(E(G)_\tau)\}.$$

Proof: By Theorem 5.7 and by the hypothesis on G, $E(G)(\tau) = 0$ and $E_\tau(G) = E(G)_\tau$. Apply Theorem 6.1 to complete the proof of the corollary.

In the next result, we will show that if $E_\tau(G)$ is semi-perfect then local direct summands of G^n behave like finitely generated semi-simple modules over $E_\tau(G)$.

Theorem 6.4. *Let G be an rtffr group with conductor* τ. *Let* $\mathbf{S}_o(E_\tau(G))$ *be the category of finitely generated semi-simple right* $E_\tau(G)$*-modules, and suppose that* $E_\tau(G)$ *is semi-perfect.*

1. There is a functor

$$C_G(\cdot) = (B_{E_\tau(G)} \circ L_\tau \circ A_{\text{End}(G)} \circ \mathbf{H}_G)(\cdot) : \mathbf{P}_o(G) \longrightarrow \mathbf{S}_o(E_\tau(G)).$$

2. $C_G(\cdot)$ *induces a bijection*

$$\gamma_G : \{[H] \, \big| \, H \in \mathbf{P}_o(G)\} \longrightarrow \{(M) \, \big| \, M \in \mathbf{S}_o(E_\tau(G))\}.$$

Proof: Part 1 follows from Theorems 2.3, 5.3, 5.4, and 5.9. For part 2, let

$$\gamma_G = \beta_{E_\tau(G)} \circ \lambda_\tau \circ \alpha_{\text{End}(G)} \circ \eta_G.$$

This completes the proof.

6.2 Direct sum decompositions

The power of Theorem 6.1 is that it translates local isomorphism classes of an rtffr group G into isomorphism classes of finitely generated projective right modules over the semi-local semi-prime ring $E_\tau(G)$. For this reason, when discussing direct sum decompositions of G we can instead study these decompositions over the pleasantly structured ring $E_\tau(G)$.

Arnold's theorem [9, Theorem 12.6] states that H is locally isomorphic to $G_1 \oplus G_2$ as rtffr groups iff $H = H_1 \oplus H_2$ for some rtffr groups H_1, H_2 such that H_j and G_j are locally isomorphic for $j = 1, 2$.

Theorem 6.5. *Let G be an rtffr group with conductor* τ, *let* $L_G(\cdot)$ *be the functor defined above, and let* $H \in \mathbf{P}_o(G)$. *Then* $L_G(H) \cong W_1 \oplus W_2$ *as right* $E_\tau(G)$*-modules iff* $H \cong H_1 \oplus H_2$ *for some rtffr groups* H_1, H_2 *such that* $L_G(H_j) \cong W_j$ *for* $j = 1, 2$.

Proof: Suppose that $L_G(H) = W \cong W_1 \oplus W_2$ for some $H \in \mathbf{P}_o(G)$ and some W, W_1, and $W_2 \in \mathbf{P}_o(E_\tau(G))$. By Theorem 6.1, the functor

$$L_G(\cdot) = L_\tau \circ A_{\text{End}(G)} \circ \mathbf{H}_G$$

induces a bijection

$$\lambda_G : \{[H] \, \big| \, H \in \mathbf{P}_o(G)\} \longrightarrow \{(W) \, \big| \, W \in \mathbf{P}_o(E(G))\}.$$

By Theorem 2.3, $(A_{\text{End}(G)} \circ \mathbf{H}_G)(H) = U$ is a finitely generated projective right $E(G)$-module such that $L_\tau(U) \cong W$, and similarly because λ_τ is a bijection there are $U_1, U_2 \in \mathbf{P}_o(E_\tau(G))$ such that $L_\tau(U_i) \cong W_i$ for $i = 1, 2$. By Theorems 5.3 and 2.3 there are $H_1, H_2 \in \mathbf{P}_o(G)$ such that

$$(A_{\text{End}(G)} \circ \mathbf{H}_G)(H_i) \cong U_i \text{ for } i = 1, 2.$$

Thus

$$L_G(H) = W \cong W_1 \oplus W_2 \cong L_G(H_1) \oplus L_G(H_2) = L_G(H_1 \oplus H_2).$$

Since λ_G is a bijection, Theorem 6.1 states that H is locally isomorphic to $H_1 \oplus H_2$. Arnold's theorem [9, Theorem 12.6] shows us that $H \cong H'_1 \oplus H'_2$ for some H'_i that is locally isomorphic to H_i for $i = 1, 2$. This completes the proof of the first implication. The converse is clear, so the proof is complete.

Theorem 6.6. *Let G be an rtffr group with conductor τ. Then*

$$G = G(\tau) \oplus G^\tau$$

where $L_G(G(\tau)) = \mathbf{Q}E(\tau)$ and $L_G(G^\tau) = E(G)_\tau$.

Proof: We have

$$L_G(G) = E_\tau(G) = \mathbf{Q}E(\tau) \oplus E(G)_\tau.$$

By Theorem 6.5, $G = G(\tau) \oplus G^\tau$ for some groups $G(\tau)$ and G^τ such that $L_G(G(\tau)) = \mathbf{Q}E(\tau)$, and such that $L_G(G^\tau) = E(G)_\tau$. This completes the proof.

Since each $P \in \mathbf{P}_o(E_\tau(G))$ can be written as $P(\tau) \oplus P^\tau$ where $P(\tau)$ is the $\mathcal{C}_\tau$-torsion $E(G)$-submodule of P, an application of Theorem 6.5 proves the following theorem.

Theorem 6.7. *Let G be an rtffr group with conductor τ. Then*

1. *Each $H \in \mathbf{P}_o(G)$ can be written uniquely as $H \cong H(\tau) \oplus H^\tau$ where $H(\tau) \in \mathbf{P}_o(G(\tau))$ and $H^\tau \in \mathbf{P}_o(G^\tau)$.*
2. *There is a functorial bijection*

$$\mathbf{Q}\lambda_G : \{(H) \,\big|\, H \in \mathbf{P}_o(G(\tau))\} \longrightarrow \{(W) \,\big|\, W \in \mathbf{P}_o(\mathbf{Q}E(G)(\tau))\}.$$

3. *There is a functorial bijection*

$$\lambda_G^\tau : \{(H) \,\big|\, H \in \mathbf{P}_o(G^\tau)\} \longrightarrow \{(W) \,\big|\, W \in \mathbf{P}_o(E(G)^\tau)\}.$$

Proof: 1. Let $H \in \mathbf{P}_o(G)$ and let $(A_{\mathrm{End}(G)} \circ \mathbf{H}_G)(H) = U \in \mathbf{P}_o(E(G))$. By Corollary 5.8, $U = U(\tau) \oplus U^\tau$ where $U(\tau)$ is a finitely generated projective $E(G)(\tau)$-module, and U^τ is a finitely generated projective $E(G)^\tau$. By Theorems 5.3 and 2.3 there are groups $H(\tau)$ and H^τ such that $(A_{\mathrm{End}(G)} \circ \mathbf{H}_G)(H(\tau)) = U(\tau)$ and $(A_{\mathrm{End}(G)} \circ \mathbf{H}_G)(H^\tau) = U^\tau$. Furthermore,

$$(A_{\mathrm{End}(G)} \circ \mathbf{H}_G)(H) \cong U(\tau) \oplus U^\tau$$

$$\cong (A_{\mathrm{End}(G)} \circ \mathbf{H}_G)(H(\tau) \oplus H^\tau).$$

Hence $H \cong H(\tau) \oplus H^\tau$ by Theorems 2.3 and 5.3.

The proof of parts 2 and 3 is just a matter of showing that λ_G restricts to maps $\mathbf{Q}\lambda_G$ and λ_G^τ. The details are left to the reader. This completes the proof.

Remark 6.8. We conclude from Theorem 6.7 that

1. Direct sum decompositions of $G(\tau)$ are functorially equivalent to direct sum decompositions of finitely generated semi-simple modules over the semi-simple Artinian ring $\mathbf{Q}E(\tau)$.
2. Direct sum decompositions of G^τ are functorially equivalent to direct sum decompositions of finitely generated projective modules over the semi-local semi-prime rtffr ring $E(G)_\tau$.

Corollary 6.9. *Let G be an rtffr group with conductor τ and let $H \in \mathbf{P}_o(G)$. Then H is an indecomposable rtffr group iff $L_G(H)$ is an indecomposable finitely generated projective right $E_\tau(G)$-module.*

Corollary 6.10. *Let G be an rtffr group with conductor τ. The following are equivalent.*

1. *G is indecomposable.*
2. *If G is locally isomorphic to H then H is indecomposable.*
3. *$E(G)$ is indecomposable as a right $E(G)$-module.*
4. *$E_\tau(G)$ is indecomposable as a right $E_\tau(G)$-module.*

Proof: $1 \Rightarrow 2$. Apply Arnold's theorem.

$2 \Rightarrow 3$. Say that G is locally isomorphic to H and that

$$(A_{\text{End}(G)} \circ \mathbf{H}_G)(H) \cong U_1 \oplus U_2.$$

By Theorem 2.3, $H \cong H_1 \oplus H_2$ where

$$(A_{\text{End}(G)} \circ \mathbf{H}_G)(H_i) \cong U_i \text{ for } i = 1, 2.$$

$3 \Rightarrow 4$. Say $L_G(G) = E_\tau(G) \cong U_1 \oplus U_2$. Then by Theorem 6.1, $G = G_1 \oplus G_2$ where $L_G(G_i) \cong U_i$ for $i = 1, 2$. Thus

$$E(G) = (A_{\text{End}(G)} \circ \mathbf{H}_G)(G_1) \oplus (A_{\text{End}(G)} \circ \mathbf{H}_G)(G_2),$$

which proves that $E(G)$ has a nontrival decomposition.

$4 \Rightarrow 1$. Say $G = G_1 \oplus G_2$. Then

$$E_\tau(G) = L_\tau(G) = L_\tau(G_1) \oplus L_\tau(G_2)$$

is decomposable. This completes the proof.

The rtffr group G has a *(locally) unique decomposition* if

1. $G = G_1 \oplus \cdots \oplus G_t$ for some indecomposable rtffr groups $G_1, \ldots, G_t$, and
2. If $G \cong H_1 \oplus \cdots \oplus H_s$ for some indecomposable rtffr groups $H_1, \ldots, H_s$ then $s = t$ and G_i is (locally) isomorphic to H_i after a permutation of the subscripts.

We say that the rtffr group G has the *(local) refinement property* if

1. $G = G_1 \oplus \cdots \oplus G_t$ for some indecomposable rtffr groups $G_1, \ldots, G_t$, and
2. Each $H \in \mathbf{P}_o(G)$ is (locally) isomorphic to a direct sum of rtffr groups in $\{G_1, \ldots, G_t\}$.

Theorem 6.11. *Let G be an rtffr group with conductor τ. The following are equivalent.*

1. *G has a locally unique decomposition.*
2. *The localization $E(G)_\tau$ has a unique decomposition as a right $E(G)_\tau$-module.*

Proof: $2 \Leftrightarrow 1$. By Theorem 6.1, there is a functorial bijection

$$\lambda_G : \{[H] \,\big|\, H \in \mathbf{P}_o(G)\} \longrightarrow \{(W) \,\big|\, W \in \mathbf{P}_o(E_\tau(G))\}.$$

Then G has a locally unique decomposition iff $E_\tau(G)$ has a unique direct sum decomposition. Recall that

$$\boxed{E_\tau(G) = \mathbf{Q}E(G)(\tau) \times E(G)_\tau,}$$

where $\mathbf{Q}E(G)(\tau)$ is a semi-simple Artinian ring. Direct sum decompositions of $\mathbf{Q}E(G)$ are unique. Then G has a locally unique decomposition iff $E(G)_\tau$ has a unique decomposition. This completes the proof.

Theorem 6.12. *Let G be an rtffr group with conductor τ. The following are equivalent.*

1. *G has the local refinement property.*
2. *The localization $E(G)_\tau$ has the refinement property as a right $E(G)_\tau$-module.*

There is at least one nontrivial class of rtffr groups G that satisfies the conditions in Theorems 6.11 and 6.12.

Theorem 6.13. *Let G be an rtffr group with conductor τ. Then G has a locally unique decomposition and the local refinement property if the localization $E(G)_\tau$ is a semi-perfect ring.*

Proof: Since $E(G)_\tau$ is a semi-perfect ring it satisfies Theorems 6.11.2 and 6.12.2, so G satisfies Theorems 6.11.1 and 6.12.1.

Corollary 6.14. *Let G be an rtffr group with conductor τ. Then G is an indecomposable group and it has the local refinement property if $E_\tau(G)$ is a local ring.*

Proof: A local ring is an indecomposable semi-perfect ring. Now apply Theorem 6.13.

6.3 Locally semi-perfect rings

Given Theorem 6.13, we are motivated to find semi-prime rtffr rings E with conductor τ such that the localization E_τ is a semi-perfect ring. In this section, we give some conditions on τ and E such that the localization E_τ is semi-perfect.

An ideal $I \neq 0$ in a commutative ring S is called a *primary ideal* if S/I is a local ring with nilpotent Jacobson radical.

Property 6.15. Recall that τ, τ_i, E, S, $\overline{E}$, $\overline{S}$, $\overline{E}_i$, and $\overline{S}_i$ satisfy (5.2).

1. Assume that the ideal τ_i is primary in $\overline{S}_i$ for each $i = 1, \ldots, t$.
2. Assume that there are integers $n_1, \ldots, n_t > 0$ and algebraic number fields $\mathbf{k}_1, \ldots, \mathbf{k}_t$ such that

$$\mathbf{Q}\overline{E}_i = \mathrm{Mat}_{n_i}(\mathbf{k}_i).$$

Lemma 6.16. *Let E be a semi-perfect ring, and let $e^2 = e$, $f^2 = f \in E$ be such that $eE \cong fE$. There is a unit $u \in E$ such that $ueu^{-1} = f$.*

Proof: We have $eE \cong fE$. Since $e^2 = e$ and $f^2 = f$ we can write

$$E = eE \oplus (1-e)E = fE \oplus (1-f)E,$$

and since E is semi-perfect $(1-e)E \cong (1-f)E$ by the Azumaya–Krull–Schmidt theorem 1.4. Thus there is an isomorphism $u : E \to E$ such that

$$u(eE) = fE$$

and

$$u((1-e)E) = (1-f)E$$

so that $ue \in fE$ and $u(1-e) \in (1-f)E$. Evidently u is multiplication by a unit u of E, so $ueu^{-1} \in ueE = fE$ and $u(1-e)u^{-1} \in u(1-e)E = (1-f)E$. Hence

$$
\begin{aligned}
1 &= u1u^{-1} \\
&= u(e + (1-e))u^{-1} \\
&= ueu^{-1} \oplus u(1-e)u^{-1} \\
&\in fE \oplus (1-f)E.
\end{aligned}
$$

Since $1 = f \oplus (1-f)$ can be written in exactly one way as an element in a direct sum we have $ueu^{-1} = f$. This completes the proof of the lemma.

Lemma 6.17. *Let E be a semi-perfect ring, and let $J \subset \mathcal{J}(E)$ be an ideal. Then idempotents lift modulo J in E.*

Proof: Let $J \subset \mathcal{J}(E)$ and let $e \in E$ be such that $e^2 - e \in J \subset \mathcal{J}(E)$. Given $x \in E$ let $\bar{x} = x + J \in E/J$.

Because E is semi-perfect there is an $f^2 = f \in E$ such that $e - f \in \mathcal{J}(E)$. Then $\bar{e}, \bar{f} \in E/J$ are idempotents such that $\bar{e} - \bar{f} \in \mathcal{J}(E/J) = \mathcal{J}(E)/J$. We will show that there is a unit $\bar{u} \in E/J$ such that $\bar{u}\bar{f}\bar{u}^{-1} = \bar{e}$.

We have $\bar{e} + \mathcal{J}(E/J) = \bar{f} + \mathcal{J}(E/J)$ and $\bar{e}^2 = \bar{e}, \bar{f}^2 = \bar{f} \in E/J$ so that

$$\frac{\bar{e}E/J}{\bar{e}\mathcal{J}(E/J)} \cong (\bar{e} + \mathcal{J}(E/J))\frac{E/J}{\mathcal{J}(E/J)}$$

$$= (\bar{f} + \mathcal{J}(E/J))\frac{E/J}{\mathcal{J}(E/J)} \cong \frac{\bar{f}E/J}{\bar{f}\mathcal{J}(E/J)}.$$

By Lemma 6.16 there is a unit $\bar{u} \in E/J$ such that $\bar{e} = \bar{u}\bar{f}\bar{u}^{-1}$.

Since units lift modulo $J \subset \mathcal{J}(E)$, $\bar{u} = u + J$ for some unit $u \in E$ and hence $ufu^{-1} \equiv e \pmod{J}$. Thus idempotents lift modulo J in E.

We need to know the following.

Lemma 6.18. [97, Theorem 3.84] *If U is a finitely presented S-module and if $\mathcal{C}$ is multiplicatively closed subset of S that contains 1, then*

$$(\mathrm{End}_S(U))[\mathcal{C}^{-1}] \cong \mathrm{End}_{S[\mathcal{C}^{-1}]}(U[\mathcal{C}^{-1}])$$

as rings.

Lemma 6.19. *Let $\overline{E}$ be an integrally closed semi-prime rtffr ring with center $\overline{S}$, and assume Property 6.15. The localization $\overline{E}_\tau$ is a semi-perfect ring.*

Proof: As in (5.2), write $\overline{E} = \overline{E}_1 \times \cdots \times \overline{E}_t$ and $\overline{S} = \overline{S}_1 \times \cdots \times \overline{S}_t$, where the $\mathbf{Q}\overline{E}_i$ are the simple Artinian factors of $\mathbf{Q}\overline{E}$ and where $\overline{S}_i$ is the center of $\overline{E}_i$. Then $\overline{E}_i$ is a finitely generated projective $\overline{S}_i$-module for each $i = 1, \ldots, t$. (See [9, Theorem 9.10(d)].) By Property 6.15.2, $\mathrm{Mat}_{n_1}(\mathbf{k}_1), \ldots, \mathrm{Mat}_{n_t}(\mathbf{k}_t)$ are the simple factors of $\mathbf{Q}\overline{E}$. Thus, after a permutation of subscripts,

$$\mathbf{Q}\overline{E}_i \cong \mathrm{Mat}_{n_i}(\mathbf{k}_i) \text{ for each } i = 1, \ldots, t.$$

Fix $i \in \{1, \ldots, t\}$. Then

$$\mathbf{Q}\overline{S}_i = \mathrm{center}(\mathbf{Q}\overline{E}_i) = \mathrm{center}(\mathrm{Mat}_{n_i}(\mathbf{k}_i)) = \mathbf{k}_i.$$

Let V_i be a simple left module over the simple Artinian ring $\mathbf{Q}\overline{E}_i$ and let $U_i = \overline{E}_i v_i \neq 0$ be a nonzero cyclic left $\overline{E}_i$-submodule of V_i. Then

$$U_i \subset \mathbf{Q}U_i = \mathbf{Q}\overline{E}_i v_i = V_i \cong \mathbf{k}_i^{(n_i)}.$$

Since $\overline{S}_i$ is a Dedekind domain, and since $\overline{E}_i$ is a finitely generated projective $\overline{S}_i$-module, we see that U_i is a finitely generated projective $\overline{S}_i$-module. Thus $\mathrm{End}_{\overline{S}_i}(U_i)$ is a finitely generated torsion-free $\overline{S}_i$-module. Furthermore,

$$\overline{E}_i \subset \mathrm{End}_{\overline{S}_i}(U_i) \subset \mathrm{End}_{\mathbf{k}_i}(\mathbf{Q}U_i) = \mathrm{End}_{\mathbf{k}_i}(\mathbf{k}_i^{(n_i)}) = \mathrm{Mat}_{n_i}(\mathbf{k}_i) \cong \mathbf{Q}\overline{E}_i.$$

Thus $\mathrm{End}_{\overline{S}_i}(U_i)/\overline{E}_i$ is a finitely generated torsion $\overline{S}_i$-module. By hypothesis, $\overline{E}_i$ is integrally closed, so $\overline{E}_i = \mathrm{End}_{\overline{S}_i}(U_i)$.

Since U_i is a finitely generated projective $\overline{S}_i$-module, Lemma 6.18 implies that

$$(\overline{E}_i)_\tau = (\overline{E}_i)_{\tau_i} \cong \mathrm{End}_{(\overline{S}_i)_{\tau_i}}((U_i)_{\tau_i}).$$

Inasmuch as the Property 6.15.1 states that τ_i is a primary ideal in $\overline{S}_i$, we see that

$$\frac{\overline{S}_i}{\tau_i} \cong \left(\frac{\overline{S}_i}{\tau_i}\right)_{\tau_i} \cong \frac{(\overline{S}_i)_{\tau_i}}{(\tau_i)_{\tau_i}}$$

is a local ring. The maximal ideals of $(\overline{S}_i)_{\tau_i}$ are exactly the localizations N_{τ_i} of maximal ideals N of $\overline{S}_i$ that contain τ_i. Thus $(\overline{S}_i)_{\tau_i}$ has only one maximal ideal, whence $(\overline{S}_i)_{\tau_i}$ is a local ring.

Moreover, since $(U_i)_{\tau_i}$ is then a projective module over the local ring $(\overline{S}_i)_{\tau_i}$, $(U_i)_{\tau_i} \cong (\overline{S}_i)_{\tau_i}^{(m_i)}$ for some integer $m_i > 0$. Then

$$(\overline{E}_i)_\tau \cong \mathrm{End}_{(\overline{S}_i)_{\tau_i}}((U_i)_{\tau_i}) \cong \mathrm{Mat}_{m_i}((\overline{S}_i)_{\tau_i}).$$

Since $(\overline{S}_i)_{\tau_i}$ is a local ring, $\mathrm{Mat}_{m_i}((\overline{S}_i)_{\tau_i})$ is a semi-perfect ring. Thus $(\overline{E}_i)_{\tau_i}$ is semi-perfect, and hence $\overline{E}_\tau$ is semi-perfect. This completes the proof of the lemma.

Lemma 6.20. *Assume Property 6.15. Idempotents of $E_\tau / E_\tau \tau_\tau$ lift to idempotents of E_τ.*

Proof: Let $e \in E_\tau$ be such that $e^2 - e \in \tau_\tau E_\tau$. Inasmuch as $\tau \overline{E} \subset E \subset \overline{E}$, setting $I = \tau_\tau^2 \overline{E}_\tau$ gives us

$$I = \tau_\tau(\tau_\tau \overline{E}_\tau) \subset \tau_\tau E_\tau \subset \mathcal{J}(S_\tau)E_\tau \subset \mathcal{J}(E_\tau).$$

Since idempotents lift modulo the nilpotent ideal $\tau_\tau E_\tau / I \subset E_\tau / I$, there is an $(f + I)^2 = f + I \in E_\tau / I \subset \overline{E}_\tau / I$ such that $e - f \in \tau_\tau E_\tau$. Now, by Lemma 6.19, $\overline{E}_\tau$ is semi-perfect, and since $I = \tau_\tau^2 \overline{E}_\tau \subset \mathcal{J}(\overline{E}_\tau)$, Lemma 6.17 states that there is a $g^2 = g \in \overline{E}_\tau$ such that $f + I = g + I$. Thus $g \in f + I \subset E_\tau$, and hence idempotents in $E_\tau / \tau_\tau E_\tau$ lift to idempotents in E_τ.

Theorem 6.21. *Let E be a semi-prime rtffr ring with conductor τ that satisfies (6.15). Then $\mathbf{Q}E(\tau) \times E_\tau$ is a semi-perfect ring.*

Proof: Since E is semi-prime, $\mathbf{Q}E(\tau)$ is semi-simple, hence semi-perfect. Thus it suffices to assume that $E(\tau) = 0$.

Let $e^2 - e \in \mathcal{J}(E_\tau)$. Since $\mathcal{J}(E_\tau)/\tau_\tau E_\tau$ is a nilpotent ideal in the finite ring $E_\tau / \tau_\tau E_\tau$, there is an

$$(f + \tau_\tau E_\tau)^2 = (f + \tau_\tau E_\tau) \in E_\tau / \tau_\tau E_\tau$$

such that

$$(f + \tau_\tau E_\tau) - (e + \tau_\tau E_\tau) \in \mathcal{J}(E_\tau)/\tau_\tau E_\tau.$$

By Lemma 6.20, the idempotent $f + \tau_\tau E_\tau$ lifts to an idempotent of E_τ, so there is an idempotent $g \in E_\tau$ such that $f - g \in \tau_\tau E_\tau$. Then

$$e - g = (e - f) + (f - g) \in \mathcal{J}(E_\tau),$$

and hence idempotents lift modulo $\mathcal{J}(E_\tau)$.

Moreover, τ has finite index in S. Thus $E/\tau E \cong E_\tau/\tau_\tau E_\tau$ is finite. Since $\tau_\tau E_\tau \subset \mathcal{J}(\tau_\tau E_\tau)$, $E_\tau/\mathcal{J}(E_\tau)$ is a (finite) semi-simple Artinian ring. Thus E_τ is a semi-perfect ring. This completes the proof of the theorem.

Corollary 6.22. *Suppose that E is a prime rtffr ring with center S and conductor τ such that $\mathbf{Q}E \cong \mathrm{Mat}_n(\mathbf{k})$ for some integer $n > 0$ and some algebraic number field field $\mathbf{k}$. Let $\overline{S}$ be the integral closure of S in $\mathbf{k}$. If τ is a primary ideal in $\overline{S}$ then E_τ is a semi-perfect ring.*

Proof. Since $\mathbf{Q}E = \mathrm{Mat}_n(\mathbf{k})$ for some algebraic number field $\mathbf{k}$ then $S = \mathrm{center}(E)$ is an integral domain with field of quotients $\mathbf{k}$. There is a unique Dedekind domain $S \subset \overline{S}$ such that $\overline{S}/S$ is finite. By assumption τ is primary in $\overline{S}$. Thus E and S satisfy (6.15). Now apply Theorem 6.21.

6.4 Balanced semi-primary groups

Let G be an rtffr group, and let $\overline{G}$ be the integral closure of G. Then $E(G) \subset E(\overline{G})$, $E(\overline{G})$ is integrally closed, and $E(\overline{G})/E(G)$ is finite.

The results of this section will give conditions on the containment $G \subset \overline{G}$ under which G has a locally unique decomposition and the local refinement property.

The rtffr group G is a *Dedekind group* if $E(G)$ is a Dedekind domain. If G is a Dedekind group then G is integrally closed and strongly indecomposable.

The set $\{G_1, \ldots, G_t\}$ of strongly indecomposable rtffr groups is said to be *nilpotent* if $G_i \not\cong G_j$ for integers $i \neq j \in \{1, \ldots, t\}$. (See section 18.4 of this text or [46, page 64] for more details on nilpotent sets of groups). Suppose there is a nilpotent set $\{G_1, \ldots, G_t\}$ of Dedekind rtffr groups and integers $n_1, \ldots, n_t > 0$ such that

$$\overline{G} = G_1^{n_1} \oplus \cdots \oplus G_t^{n_t}. \tag{6.1}$$

We say that $\overline{G}$ is a *semi-Dedekind rtffr group*. For each $i = 1, \ldots, t$ let

$$\mathcal{N}(\mathrm{End}(G_i)) \subset P_i \subset \mathrm{End}(G_i)$$

be an ideal of finite index in $\mathrm{End}(G_i)$, let

$$P_o = P_1 \oplus \cdots \oplus P_t \subset \mathrm{End}(\overline{G}), \tag{6.2}$$

and let G be any group such that

$$P_o\overline{G} \subset G \subset \overline{G}.$$

We have an unambiguous definition for the action

$$P_o(G_i^{n_i}) = (P_iG_i)^{n_i}$$

so that

$$P_o\overline{G} = P_1G_1^{n_1} \oplus \cdots \oplus P_tG_t^{n_t}$$

is well-defined. We will use the notations $\overline{G}$, G_i, P_o, and P_i in the way that they are used in (6.1) and (6.2).

We say that P_o is an index of G in $\overline{G}$. We say that P_o is a *semi-primary index in* $\text{End}(\overline{G})$ if for each $i = 1, \ldots, t$, P_i is a primary ideal in $\text{End}(G_i)$. We say that G *has semi-primary index in* $\overline{G}$ if there is a semi-primary index $P_o \subset \text{End}(\overline{G})$ such that $P_o\overline{G} \subset G \subset \overline{G}$. We say that G *is a balanced subgroup of* $\overline{G}$ if $G \subset \overline{G}$ and if $E(G) \subset E(\overline{G})$. We say that G is a *balanced semi-primary group* if there is a balanced embedding $G \subset \overline{G}$ in which G has semi-primary index in $\overline{G}$.

The next result explains our interest in balanced semi-primary rtffr groups.

Theorem 6.23. *Let G be a balanced semi-primary subgroup of a semi-Dedekind rtffr group $\overline{G}$. Let τ be the conductor of G relative to the inclusion $E(G) \subset E(\overline{G})$. Then*

1. *$E(G)_\tau$ is a semi-perfect ring,*
2. *G has a locally unique decomposition, and*
3. *G has the local refinement property.*

Proof: The proof depends upon the next lemma.

Lemma 6.24. *Let G be a balanced semi-primary rtffr subgroup of a semi-Dedekind rtffr group. Then G satisfies the conditions in Property 6.15.*

Proof: Let G be a balanced semi-primary rtffr group. There is a semi-Dedekind rtffr group $\overline{G}$ as in (6.1) and a semi-primary index P_o as in (6.2) such that $P_o\overline{G} \subset G \subset \overline{G}$ and $E(G) \subset E(\overline{G})$. Since P_i has finite index in $\text{End}(G_i)$ for each $i = 1, \ldots, t$, $G \doteq \overline{G}$.

For each $i = 1, \ldots, t$, $E(G_i)$ is a Dedekind domain whose field of fractions $\mathbf{Q}E(G_i) = \mathbf{k}_i$ is an algebraic number field. Since $\{G_1, \ldots, G_t\}$ is a nilpotent set, (2.21) implies that

$$\mathbf{Q}E(\overline{G}) = \text{Mat}_{n_1}(\mathbf{k}_1) \times \cdots \times \text{Mat}_{n_t}(\mathbf{k}_t).$$

Since $G \doteq \overline{G}$, $\text{End}(G) \doteq \text{End}(\overline{G})$, so that $\mathbf{Q}E(G) \cong \mathbf{Q}E(\overline{G})$ as rings. Thus $\mathbf{Q}E(G)$ satisfies Property 6.15.2.

As for Property 6.15.1, for each $i = 1, \ldots, t$, P_i maps onto a nonzero proper primary ideal T_i in the Dedekind domain $E(G_i)$. Let $T_o = T_1 \oplus \cdots \oplus T_t$. Then T_o is an ideal of finite index in center$(E(G)) = \overline{S} = E(G_1) \times \cdots \times E(G_t)$. Since $P_o\overline{G} \subset G$ and since $E(G) \subset E(\overline{G})$,

$$T_o E(\overline{G}) = P_o E(\overline{G}) \subset E(G) \subset E(\overline{G}).$$

Let $\tau = \tau_1 \oplus \cdots \oplus \tau_t$ be the conductor of $E(G)$ relative to $E(\overline{G})$. By assumption, τ is the largest ideal in $\overline{S}$ such that $\tau E(\overline{G}) \subset E(G)$ so we have $T_o \subset \tau$. Thus $T_i \subset \tau_i \subset E(G_i)$ for each $i = 1, \ldots, t$. Since T_i is a primary ideal of finite index in $E(G_i)$, τ_i is a primary ideal in $E(G_i)$. Thus $E(G)$ satisfies Property 6.15.1, which completes the proof of the lemma.

Proof of Theorem 6.23. Lemma 6.24 shows us that G satisfies the conditions in Property 6.15.1. Theorem 6.21 states that then the localization $E(G)_\tau$ is a semi-perfect ring. Hence Theorem 6.13 states that G has the local refinement property and a locally unique decomposition. This completes the proof of the theorem.

Theorem 6.25. *Let G be a balanced semi-primary subgroup of a semi-Dedekind rtffr group $\overline{G}$. Let τ be the conductor of G relative to the inclusion $E(G) \subset E(\overline{G})$, and let $S_o(E_\tau(G))$ be the category of finitely generated semi-simple $E_\tau(G)$-modules. Then*

1. *There is a functor*

$$C_G(\cdot) = (B \circ L_\tau \circ A \circ H_G)(\cdot) : P_o(G) \longrightarrow S_o(E_\tau(G)).$$

2. *$C_G(\cdot)$ induces a bijection*

$$\gamma_G : \{[H] \,\big|\, H \in P_o(G)\} \longrightarrow \{(M) \,\big|\, M \in S_o(E_\tau(G))\}.$$

Proof: Apply Theorem 6.4 and Theorem 6.23.

6.5 Examples

Let

$$\boxed{\mathcal{D} = \{\text{rtffr groups } G \,\big|\, E(G) \text{ is a Dedekind domain.}\}}$$

Then $\mathcal{D}$ is a set of integrally closed, strongly indecomposable, rtffr groups such that $E(G)$ is commutative. By Lemma 2.9, there is a group $G \subset \overline{G}$ such that $G \doteq \overline{G}$, such that $E(G) \subset E(\overline{G})$, and such that $E(\overline{G})$ is an rtffr Dedekind domain. Then G is balanced in $\overline{G}$. Thus $\mathcal{D}$ contains a rich collection of groups.

Lemma 6.26. *If $E(G)$ is a pid then $G \in \mathcal{D}$.*

Direct sums of groups G such that $E(G)$ is a *pid* are studied in [13].

Lemma 6.27. *Let G be an integrally closed, strongly indecomposable rtffr group of square-free rank. Then $G \in \mathcal{D}$.*

Proof: By Lemma 2.6, $E(G)$ is a commutative integral domain if G is a strongly indecomposable rtffr group of square-free rank. Thus $E(G)$ is a Dedekind domain if G is an integrally closed strongly indecomposable of square-free rank.

Lemma 6.28. *If G is an integrally closed, strongly indecomposable, rtffr group of rank one, two or three then $G \in \mathcal{D}$.*

For the purposes of this paragraph, let

$$H = H_1 \oplus \cdots \oplus H_s,$$

where $H_i \subset \mathbf{Q}$ for each $i = 1, \ldots, s$. A *kernel group* is a group G that fits into an exact sequence

$$0 \to G \longrightarrow H \overset{\sigma}{\longrightarrow} \mathbf{Q},$$

where σ is the summation map defined by $\sigma(x_1, \ldots, x_s) = x_1 + \cdots + x_s$ where $x_i \in H_i$ for each $i = 1, \ldots, s$. Dually a *cokernel group* is a group G that fits into an exact sequence

$$0 \to X \longrightarrow H \overset{\pi}{\longrightarrow} G \to 0,$$

where X is a pure rank one subgroup of H. We call G a *bracket group* if G is either a kernel group or a cokernel group. If G is a strongly indecomposable bracket group, then it is known that $\mathrm{End}(G) \subset \mathbf{Q}$ is a *pid*. See [9, 85, 95].

Corollary 6.29. *Each strongly indecomposable bracket group is in $\mathcal{D}$.*

A *strongly homogeneous group* is a group G such that for any pure rank one subgroups $X, Y \subset G$ there is an automorphism $\alpha : G \to G$ such that $\alpha(X) = Y$. It is known that if G is a strongly indecomposable strongly homogeneous rtffr group then $\mathrm{End}(G)$ is a *pid* in which the primes $p \in \mathbf{Z}$ such that $pG \neq G$ are primes in $\mathrm{End}(G)$. See [9].

Corollary 6.30. *Each strongly indecomposable strongly homogeneous rtffr group is in $\mathcal{D}$.*

The group G is a *Murley group* if $\mathbf{Z}/p\mathbf{Z}\text{-dim}(G/pG) \leq 1$ for each prime $p \in \mathbf{Z}$. It is known that if G is a strongly indecomposable Murley group then $\mathrm{End}(G)$ is a *pid* in which $p \in \mathbf{Z}$ is a prime in $\mathrm{End}(G)$ if $pG \neq G$. See [9].

Corollary 6.31. *Each indecomposable rtffr Murley group is in $\mathcal{D}$.*

Summarizing, the class $\mathcal{D}$ contains

1. Integrally closed strongly indecomposable rtffr groups G of square-free rank,
2. Rank one groups G,
3. Integrally closed, strongly indecomposable, rank two groups G,
4. Integrally closed, strongly indecomposable, rank three groups G,
5. Groups G such that $E(G)$ is a *pid*,

6. Strongly indecomposable bracket groups G,
7. Strongly indecomposable strongly homogeneous groups G, and
8. Indecomposable Murley groups G.

Our goal now is to use the objects in $\mathcal{D}$ to construct balanced semi-primary rtffr groups G.

Theorem 6.32. *Let $\{G_1, \ldots, G_t\}$ be a nilpotent subset of $\mathcal{D}$. Let $\overline{G}$ be the direct sum (6.1) for some integers $n_1, \ldots, n_t > 0$. If $G \subset \overline{G}$ is a balanced semi-primary subgroup of $\overline{G}$, then G has a locally unique decomposition and the local refinement property.*

Proof: Since G is a balanced semi-primary subgroup of the semi-Dedekind group $\overline{G}$, an application of Theorem 6.23 shows us that G has a locally unique decomposition and the local refinement property. This completes the proof.

Corollary 6.33. *Let $\{G_1, \ldots, G_t\}$ be a nilpotent set such that $\text{End}(G_i)$ is a pid for each $i = 1, \ldots, t$. Let $\overline{G}$ be the direct sum (6.1) for some integers $n_1, \ldots, n_t > 0$. Suppose that G is a balanced semi-primary subgroup of $\overline{G}$. Then G has a locally unique decomposition and the local refinement property.*

Corollary 6.34. *Let $\{G_1, \ldots, G_t\}$ be a nilpotent subset of $\mathcal{D}$ such that $E(G_1) \cong E(G_i)$ for each $i = 1, \ldots, t$, and let $\overline{G}$ be a direct sum as in (6.1). If $P \subset E(G_1)$ is a primary ideal, then each balanced subgroup $G \subset \overline{G}$ such that*

$$P\overline{G} \subset G \subset \overline{G}$$

has a locally unique decomposition and the local refinement property.

Corollary 6.35. *Let $\overline{G} \in \mathcal{D}$, let $n > 1$ be an integer, and let $P \subset E(\overline{G})$ be a primary ideal. Then any balanced subgroup $P\overline{G}^n \subset G \subset \overline{G}^n$ has a locally unique decomposition and the local refinement property.*

Corollary 6.36. *Let $\overline{G} \in \mathcal{D}$, and let $P \subset E(\overline{G})$ be a primary ideal. Then any subgroup $P\overline{G} \subset G \subset \overline{G}$ is a balanced semi-primary rtffr group. Hence G has the local refinement property.*

Proof: In this case $E(G)$ is an integral domain, $E(G) \doteq E(\overline{G})$, and $E(\overline{G})$ is the unique integrally closed ring in the quasi-equality class of $E(G)$ [9]. Hence $E(\overline{G})$ is the integral closure of $E(G)$. Thus $E(G) \subset E(\overline{G})$ and $G \subset \overline{G}$ is a balanced embedding.

If H is a rank one group then $\text{End}(H) \subset \mathbf{Q}$ is a *pid* and a given prime $p \in \mathbf{Z}$ is either a unit in $\text{End}(H)$ or a prime. Thus the rank one groups are Dedekind groups whose primary ideals are generated by powers of primes in $\mathbf{Z}$.

We say that $\overline{G}$ is *homogeneous completely decomposable of type σ* if $\overline{G} \cong H^{(n)}$ for some rank one group H of type σ. If $\overline{G}$ is a completely decomposable rtffr group then

$$\overline{G} = G[\sigma_1] \oplus \cdots \oplus G[\sigma_t]$$

where $G[\sigma_i]$ is a homogeneous completely decomposable group of type σ_i, and the types $\sigma_1, \ldots, \sigma_t$ are distinct.

Theorem 6.37. *Let* $\overline{G} = G[\sigma_1] \oplus \cdots \oplus G[\sigma_t]$ *be a completely decomposable rtffr group. Let* $G \subset \overline{G}$ *be a balanced subgroup such that for each* $i = 1, \ldots, t$ *there is a prime* $p_i \in \mathbf{Z}$ *such that* $(G + G[\sigma_i])/G$ *is a finite* p_i-*group. Then* G *has a locally unique decomposition and the local refinement property.*

Proof: By hypothesis, $G[\sigma_i] = G_i^{(n_i)}$ for some rank one group G_i of type σ_i, for each $i = 1, \ldots, t$. The reader can show that $\{G_1, \ldots, G_t\}$ is a nilpotent set. Suppose that $G \subset \overline{G}$ is a balanced subgroup, and that for each $i = 1, \ldots, t$, there is a rational prime p_i such that $(G + G[\sigma_i])/G$ is a finite p_i-group. There are integers $m_1, \ldots, m_t > 0$ such that

$$p_1^{m_1} G[\sigma_1] \oplus \cdots \oplus p_t^{m_t} G[\sigma_t] \subset G \subset G[\sigma_1] \oplus \cdots \oplus G[\sigma_t].$$

Then G is a balanced semi-primary rtffr group. Hence Theorem 6.23 states that G has a locally unique decomposition and the local refinement property.

Let $G_1, \ldots, G_t \subset \mathbf{Q}$ be rank one groups such that $G_i \ncong G_j$ for each $i \neq j$, and let $\overline{G}$ be as in (6.1). Let $q \in \mathbf{Z}$ be a prime power and let G be a group such that

$$q\overline{G} \subset G \subset \overline{G}.$$

We may then assume that $\mathrm{End}(G) \subset \mathrm{End}(\overline{G})$, [85]. Then G is a balanced semi-primary subgroup of $\overline{G}$. It is traditional to call G an *acd group with primary regulating quotient.*

Corollary 6.38. [58, Theorem 3.5]. *Let* $\overline{G}$ *be a completely decomposable rtffr group and let* $q \in \mathbf{Z}$ *be a prime power. If* G *is a group such that* $q\overline{G} \subset G \subset \overline{G}$ *then* G *has a locally unique decomposition and the local refinement property.*

6.6 Exercises

Let G and H denote rtffr groups. Let $E(G) = \mathrm{End}(G)/\mathcal{N}(\mathrm{End}(G))$. Let E be an *rtffr* ring. Let $S = \mathrm{center}(E)$. Given a semi-prime rtffr ring E, let $\overline{E}$ be an integrally closed subring of $\mathbf{Q}E$ such that $E \subset \overline{E}$ and $\overline{E}/E$ is finite. Let τ be the conductor of E.

1. Show that if τ and σ are conductor s of G then $\sqrt{\tau} = \sqrt{\sigma}$.
2. If τ and σ are conductor s of G then $E(G)_\tau \cong E(G)_\sigma$ naturally.
3. Say G is locally isomorphic to $H = H_1 \oplus H_2$. Then $G \cong G_1 \oplus G_2$ where G_i is locally isomorphic to H_i for $i = 1, 2$.
4. If $E(G)$ is prime then $G(\tau) = G$ or $G(\tau) = 0$.
5. As in Lemma 6.3.5, show that $(\overline{E}_i)_\tau = (\overline{E}_i)_{\tau_i} = \overline{E}_{\tau_i}$.
6. Let G be a balanced semi-primary group with conductor τ. Make a list of (direct sum) properties of G that are characterized in terms of right ideals and modules in $E(G)_\tau$.
7. Show that the endomorphism ring of a bracket group G is a *pid*.

6.7 Problems for future research

1. Extend Lemma 6.19.
2. Show that if there is a semi-primary embedding $G \to \overline{G}$ then the canonical inclusion $G \subset \overline{G}$ is a balanced semi-primary embedding.
3. Study direct sum decomposition properties of rtffr groups G such that $P^k \overline{E} \subset E(G) \subset \overline{E}$ for some product of a single prime ideal $P \subset S$.

7

Invertible fractional ideals

Fix an rtffr group G. Our purpose in this chapter is to study the groups that are locally isomorphic to G by reducing the problem to studying the fractional right ideals of a prime rtffr ring E that are locally isomorphic to E. As we showed in Theorem 2.39, the local isomorphism property is related to the determination of the class number $h(\mathbf{k})$ of an algebraic number field $\mathbf{k}$. We further reduce the study of local isomorphism to the commutative case by extending a theorem of Swan's that shows that for maximal orders $\overline{E}$ with center $\overline{S}$ in a simple finite-dimensional $\mathbf{Q}$-algebra, the reduced norm $\overline{nr}$ on fractional right ideals of $\overline{E}$ induces an isomorphism $\overline{\nu} : \mathrm{Cl}(\overline{E}) \to \mathrm{Pic}(\overline{S})$, where $\mathrm{Cl}(\overline{E})$ is the group of isomorphism classes of fractional right ideals of $\overline{E}$, and where $\mathrm{Pic}(\overline{S})$ is the group of isomorphism classes of fractional ideals of the Dedekind domain $\overline{S}$. There are also applications to the local isomorphism problem for rtffr groups G and to the class number of an algebraic number field.

7.1 Introduction

Let G be a *rtffr* group. Then $E(G)$ is a semi-prime Noetherian rtffr ring. We say that G has the *power cancellation property* if for each integer $n > 0$ and group H, $G^n \cong H^n$ implies that $G \cong H$. Let E be an rtffr ring and let (X, Y) be a pair of groups or a pair of E-modules. Then X is *locally isomorphic to* Y if for each integer $n > 0$ there exist an integer m relatively prime to n and maps $f : X \to Y$ and $g : Y \to X$ such that $gf = m1_X$ and $fg = m1_Y$. There is a deep connection between local isomorphism and power cancellation of rtffr groups.

Recall that Warfield's theorem 2.1 states that rtffr groups G and H are locally isomorphic iff there exists an integer $n > 0$ such that $G^n \cong H^n$. Thus we will further reduce our study of the power cancellation property by studying locally isomorphic (rtffr) groups.

Theorem 7.4 below shows that the set of isomorphism classes of groups locally isomorphic to G is bijective with the set of isomorphism classes of fractional right ideals that are locally isomorphic to $E(G)$. Thus we can study the groups H that are locally isomorphic to G by studying the fractional right ideals I that are locally isomorphic to $E(G)$.

We are led to study the power cancellation property for rtffr abelian groups G by studying the fractional right ideals I of a Noetherian prime rtffr ring E that are locally isomorphic to E.

For the remainder of this chapter, E is a prime rtffr ring with center S. Then S is an integral domain whose additive structure is an rtffr abelian group. There is a maximal S-order $\overline{E}$ in the simple $\mathbf{Q}$-algebra $\mathbf{Q}E$ such that $E \subset \overline{E}$ and $\overline{E}/E$ is finite. Let τ be the largest ideal of S such that $\tau\overline{E} \subset E$. We say that τ is a *conductor of E*. See Chapter 5. The right E-submodule I of $\mathbf{Q}E$ is a *fractional right ideal of E* if $\mathbf{Q}I = \mathbf{Q}E$, or equivalently if I contains a unit of E. Let

$$\mathrm{id}_E(E, \tau) = \{\text{fractional right ideals } I \subset E \mid I + \tau E = E\}.$$

Then $\mathrm{id}_E(E, \tau)$ is a moniod but unless E is commutative, $\mathrm{id}_E(E, \tau)$ need not be a group.

See [94] for a complete discussion of the reduced norm. Let $x \in \mathbf{Q}E$ be a unit and let x_L be left multiplication on $\mathbf{Q}E$ by x. Then x_L has a characteristic polynomial

$$\mathrm{char.poly.}_x(t)$$

as a $\mathbf{Q}S$-linear transformation of $\mathbf{Q}E$. Furthermore, there is an integer $m > 0$ such that $\mathbf{Q}S\text{-dim}(\mathbf{Q}E) = cm^2$ for some integer c. Then $\mathrm{char.poly.}_x(t)$ is an m-th power of a polynomial called the *reduced characteristic polynomial*, $\mathrm{red.char.poly.}_x(t)$. By [94, Theorem 9.14], we have

$$\mathrm{char.poly.}_x(t) = (\mathrm{red.char.poly.}_x(t))^m.$$

Now let $x \in \mathbf{Q}\overline{E}$ be a unit. Then $\mathrm{red.char.poly.}_x(t)$ is a polynomial over $\mathbf{Q}\overline{S}$, the center of $\mathbf{Q}\overline{E}$. The classic reduced norm of x over $\mathbf{Q}\overline{S}$ is

$$\overline{\mathrm{nr}}(x) = \mathrm{red.char.poly.}_x(0).$$

This norm is multiplicative $\overline{\mathrm{nr}}(xy) = \overline{\mathrm{nr}}(x)\overline{\mathrm{nr}}(y)$ and

$$x \in \overline{E} \implies \overline{\mathrm{nr}}(x) \in \overline{S}.$$

Thus $\overline{\mathrm{nr}}(x)$ is a unit in $\overline{S}$ if x is a unit in $\overline{E}$.

The classic reduced norm of a fractional right idedal $\overline{I}$ of $\overline{E}$ is

$$\overline{\mathrm{nr}}(\overline{I}) = \text{the fractional ideal of } \overline{S} \text{ generated by } \overline{\mathrm{nr}}(x) \text{ for } x \in \overline{I}.$$

We extend Swan's theorem 7.1. Recall that if $\overline{E}$ is a prime rtffr ring with center $\overline{S}$, then $\mathbf{Q}\overline{E}$ is a simple finite-dimensional $\mathbf{Q}$-algebra, $\mathbf{Q}\overline{S}$ is an algebraic number field. If $\overline{E}$ is a maximal $\overline{S}$-order in $\mathbf{Q}\overline{E}$ then $\mathrm{Cl}(\overline{E})$ is the group of stable isomorphism classes of fractional right ideals of $\overline{E}$, and $\mathrm{Pic}(\overline{S})$ is the group of isomorphism classes of fractional ideals of $\overline{S}$.

Theorem 7.1. [94, Theorem 35.14] *Let $\overline{E}$ be a maximal $\overline{S}$-order in the simple* **Q**-*algebra* **Q**$\overline{E}$. *Then the classic reduced norm* $\overline{\mathrm{nr}}$ *induces an isomorphism of groups* $\overline{\nu} : \mathrm{Cl}(\overline{E}) \to \mathrm{Pic}(S)$.

Furthermore, let E be a prime rtffr ring with center S. The fractional right ideal I of E is *invertible* if $I^*I = E$ where $I^* = \{q \in \mathbf{Q}E \,|\, qI \subset E\}$. Then $\mathrm{Pic}(S) \cong \mathrm{Inv}(S)/\mathrm{Pr}(S)$ where

$$\boxed{\mathrm{Inv}(S) = \text{the group of invertible fractional ideals of } S,}$$

where $\mathrm{Pr}(S)$ is the set of *principal fractional ideals* of S. The multiplication in $\mathrm{Pic}(S)$ is $(I)(J) = (IJ)$ and $(I)^{-1} = (I^{-1})$, where $I^{-1} = \{q \in \mathbf{Q}S \,|\, qI \subset S\}$.

7.2 Functors and bijections

We begin by reviewing the functorial connection between G and $E(G)$. Let U be a strongly indecomposable *rtffr* group, let $k > 0$ be an integer, and let G have finite index in U^k. Since U is strongly indecomposable, $\mathbf{Q}E(U)$ is a division **Q**-algebra, and then $\mathbf{Q}E(G) \cong \mathrm{Mat}_{k \times k}(\mathbf{Q}E(U))$, [9, Corollary 7.8 and Theorem 9.10]. Then $E(G)$ is a Noetherian prime ring whose additive structure is a *rtffr* group. See also [46].

Let

$$\boxed{\begin{array}{c} \Gamma(G) = \text{the set of isomorphism classes of groups } H \\ \text{that are locally isomorphic to } G, \end{array}}$$

and let

$$\boxed{\begin{array}{c} \Gamma(E) = \text{the set of isomorphism classes of fractional right} \\ \text{ideals } I \text{ of } E \text{ that are locally isomorphic to } E. \end{array}}$$

We require some elementary results about local isomorphism.

Lemma 7.2. *Let G and H be rtffr groups. Let E be an rtffr ring and let I be a fractional ideal of E.*

1. *If H is locally isomorphic to G then $G \oplus G \cong H \oplus K$ for some group K.*
2. *If I is locally isomorphic to E then $E \oplus E \cong I \oplus J$ for some fractional right ideal J.*

Proof: Suppose that G and H are locally isomorphic, and let $k > 0$ be an integer. There are maps $f_k : G \to H$ and $g_k : H \to G$ and an integer $n > 0$ that is relatively prime to k such that $f_k g_k = n1_H$ and $g_k f_k = n1_G$. There are maps $f_n : G \to H$ and $g_n : H \to G$ and an integer $m > 0$ that is relatively prime to $n > 0$ such that $f_n g_n = m1_H$ and $g_n f_n = m1_G$. There are then integers a and b such that

$$an + bm = 1.$$

Define maps

$$f_k \oplus f_n : G \oplus G \to H$$

$$(ag_k, bg_n) : H \to G \oplus G$$

defined by

$$(f_k \oplus f_n)(x, y) = f_k(x) + f_n(y) \text{ for each } x, y \in G$$

$$(ag_k, bg_n)(x) = (ag_k(x), bg_n(x)) \text{ for each } x \in G.$$

Then for each $x \in H$ we have

$$(f_k \oplus f_n)(ag_k, bg_n)(x) = af_k g_k(x) + bf_n g_n(x) = (an + bm)x = x.$$

Hence $(f_k \oplus f_n)$ splits, and consequently $G \oplus G \cong H \oplus \ker(f_k \oplus f_n)$, as required by the lemma.

2. Proceed as in part 1.

It follows from Lemma 7.2 that $\Gamma(G) \subset \mathbf{P}_o(G)$, and that $\Gamma(E) \subset \mathbf{P}_o(E)$.

Lemma 7.3. *Let E be a semi-prime rtffr ring. If P and Q are finitely generated projective right E-modules then P is locally isomorphic to Q iff $P^n \cong Q^n$ for some integer $n > 0$.*

Proof: Let P and Q be finitely generated projective right E-modules such that P is locally isomorphic to Q. Since $(E, +)$ is a *rtffr* group, Corner's theorem 1.11 states that $E \cong \mathrm{End}(G)$ for some rtffr group G. By The Arnold–Lady theorem 1.13 there are groups $H, K \in \mathbf{P}(G)$ such that $\mathrm{Hom}(G, H) \cong P$ and $\mathrm{Hom}(G, K) \cong Q$. Since the additive functors Hom and $\otimes_{\mathrm{End}(G)} G$ take multiplication by an integer to multiplication by the same integer, H and K are locally isomorphic groups. By Warfield's theorem 2.1, $H^n \cong K^n$ for some integer $n > 0$, and thus

$$P^n \cong \mathrm{Hom}(G, H)^n \cong \mathrm{Hom}(G, H^n) \cong \mathrm{Hom}(G, K^n) \cong \mathrm{Hom}(G, K)^n \cong Q^n.$$

Conversely, suppose that $P^n \cong Q^n$. Choose groups $H, K \in \mathbf{P}_o(G)$ such that $\mathrm{Hom}(G, H) \cong P$ and $\mathrm{Hom}(G, K) \cong Q$. Then by the Arnold–Lady theorem 1.13, $H^n \cong K^n$. By Warfield's theorem 2.1, H and K are locally isomorphic as groups so that $P \cong \mathrm{Hom}(G, H)$ is locally isomorphic to $Q \cong \mathrm{Hom}(G, K)$. This concludes the proof.

Let (X) be the isomorphism class of the group or module X.

Theorem 7.4. *Let G be a rtffr group. The functor $A_G : \mathbf{P}_o(G) \to \mathbf{P}_o(E(G))$ such that*

$$A_G(\cdot) = \mathrm{Hom}(G, \cdot) \otimes_{\mathrm{End}(G)} E(G)$$

induces a bijection $\gamma : \Gamma(G) \to \Gamma(E(G))$ such that $\gamma((H)) = (A_G(H))$.

Proof: The functor $A_G(\cdot)$ is given in Theorem 2.4, where it is shown that $A_G(\cdot)$ induces a bijection between the set of isomorphism classes of $\mathbf{P}_o(G)$ and the set of isomorphism classes of $\mathbf{P}_o(E(G))$.

Suppose that H is locally isomorphic to K and choose an integer $m > 0$ such that $H \oplus H' \cong K \oplus K' \cong G^m$. Since the additive functor preserves composition and multiplication by an integer, it is readily seen that $A_G(H)$ and $A_G(K)$ are locally isomorphic finitely generated projective right $E(G)$-modules. Then γ is well defined. Furthermore, by theorem 2.4, each $P \in \mathbf{P}_o(E(G))$ is isomorphic to a right $E(G)$-module $A_G(H)$. Thus γ is a surjection. Next, let H and K be such that $A_G(H)$ is locally isomorphic to $A_G(K)$. By Lemma 7.3, there is an integer $n > 0$ such that

$$A_G(H^n) \cong A_G(H)^n \cong A_G(K)^n \cong A_G(K^n).$$

Then by Theorem 2.4, $H^n \cong K^n$ which by Warfield's theorem 2.1 implies that H is locally isomorphic to K. Therefore $\gamma : \Gamma(G) \to \Gamma(E(G))$ is a bijection.

Thus we can study the local isomorphism property in G by using γ and by studying the local isomorphism property in $E(G)$.

7.3 The square

We begin by writing a notation that remains in effect for the length of the chapter.

Property 7.5. 1. Let E be a prime *rtffr* ring with center S,
2. Let $\overline{E}$ be a maximal S-order such that E is a subring of $\overline{E}$ and $\overline{E}/E$ is a finite group,
3. Let $\overline{S}$ be the center of $\overline{E}$.
4. Let τ be a conductor of E. That is, τ is the largest ideal in S such that $\tau\overline{E} \subset E$.

Then E and $\overline{E}$ are Noetherian prime rings that are finitely generated torsion-free S-modules, S and $\overline{S}$ are Noetherian commutative integral domains, and $\overline{S}$ is a Dedekind domain. These facts are found in [9, Theorems 9.9, 9.10, Corollary 10.13, and Theorem 11.3]. We will use these facts without reference.

We will make extensive use of the following localization in S. Let X be an S-module and let $I \neq 0$ be an ideal in S. Then define

$$C_I = \{c \in S \,|\, c + I \text{ is a unit in the ring } S/I\}$$

and let

$$X_I = X[C_I^{-1}].$$

Given a maximal ideal M of S, S/M is a field so that in this case, $C_I = S \setminus M$. It is easily seen that $(X_I)_J = (X_J)_I$ if I, J are nonzero ideals of S. Moreover, if $I \subset J$ are nonzero ideals of S then $C_I \subset C_J$ so that $(X_J)_I = (X_I)_J = X_J$.

We will refer to the following lemma often. The E-module M is an *E-lattice* if it is a finitely generated E-submodule of $\mathbf{Q}E^n$ for some integer $n > 0$.

Lemma 7.6. *Let E be a semi-prime rtffr ring, and let M and N be E-lattices. The following are equivalent.*

1. *M is locally isomorphic to N.*
2. *$M_I \cong N_I$ as right E_I-modules for each ideal $I \subset S$ that has finite index in S.*
3. *$M_P \cong N_P$ as right E_P-modules for each maximal ideal $P \subset S$.*

Proof: Proof: $1 \Rightarrow 2$ is an exercise.

$2 \Rightarrow 3$ is clear since by [46, Lemma 4.2.3], S/P is finite for each maximal ideal P of S.

$3 \Rightarrow 1$. Suppose that $M_P \cong N_P$ for each maximal ideal $P \subset S$. Let $n > 0$ be an integer. Let $p_1, \ldots, p_t$ be the prime divisors of n, and let $P_1, \ldots, P_t$ be the maximal ideals of S containing $p_1, \ldots, p_t$, respectively. Since $M_{P_i} \cong N_{P_i}$ for each $i = 1, \ldots, t$, we choose maps $f_i : M_{P_i} \to N_{P_i}$ and $g_i : N_{P_i} \to M_{P_i}$ and integers $m_i > 0$ that are relatively prime to p_i such that $f_i g_i = m_i 1_{N_{P_i}}$ and $g_i f_i = m_i 1_{M_{P_i}}$ for each $i = 1, \ldots, t$. Since M and N are finitely generated E-modules, there are integers a_i relatively prime to p_i for each $i = 1, \ldots, t$ such that $(a_i f_i) : M \to N$ and $(a_i g_i) : N \to M$, and such that $a_i^2 f_i g_i = a_i^2 m_i$. Notice that $a_i^2 m_i$ is relatively prime to p_i. We can assume without loss of generality that $a_i^2 m_i$, $a_i f_i$, and $a_i g_i$ were chosen to begin with.

By the Chinese Remainder Theorem, there are maps $f : M \to N$ and $g : N \to M$ such that $f \equiv f_i (\text{mod } P_i)$ and $g \equiv g_i (\text{mod } P_i)$ for each $i = 1, \ldots, t$. Then

$$fg \equiv f_i g_i \equiv m_i (\text{mod } P_i)$$

for each $i = 1, \ldots, t$. Since m_i is relatively prime to the rational prime p_i for each $i = 1, \ldots, t$, there is by the Chinese Remainder Theorem an integer m that is relatively prime to p_i, and such that $m \equiv m_i (\text{mod } p_i)$ for each $i = 1, \ldots, t$. That is, $fg \equiv m (\text{mod } P_i)$ for each $i = 1, \ldots t$.

Since m is relatively prime to p_i for each prime p_i, m is a unit in $\text{End}_E(M)_{P_i}$ for each $i = 1, \ldots, t$. Then f is an isomorphism $f : M_{P_i} \to N_{P_i}$ for each $i = 1, \ldots, t$. Let $J = P_1 \cap \cdots \cap P_t$. By the Local-Global Theorem 1.5, $X_J = \bigcap_{i=1}^{t} X_{P_i}$, so that f induces an isomorphism $f_J : M_J \to N_J$. Let $h_J : M_J \to N_J$ be the inverse of f_J. Since M and N are finitely generated, there is an integer $k > 0$ relatively prime to $p_1 \cdots p_t \in J$ such that $kh_J(N) \subset M$. Hence $(f_J)(kh_J) = k1_M$ and $(kh_J)(f_J) = k1_M$. Since k is not divisible by the prime divisors $p_1, \ldots, p_t$ of n, k is relatively prime to n. This proves part 1 and thus completes the logical cycle.

Lemma 7.7. *Assume Property 7.5 and let $I \subset E$ be a fractional right ideal of E.*

1. *If $\sigma \subset S$ is an ideal and if $I + \sigma E = E$, then $I_M = E_M$ for each maximal ideal $\sigma \subset M \subset S$.*
2. *$E_M = \overline{E}_M$ and $S_M = \overline{S}_M$ for each maximal ideal $\tau \not\subset M \subset S$.*
3. *I is an invertible right ideal of E if $I \cong I'$ for some fractional right ideal $I' \subset E$ such that $I' + \tau E = E$.*
4. *I is locally isomorphic to E iff $I \cong I'$ for some fractional right ideal $I' \subset E$ such that $I' + \tau E = E$.*
5. *If I is locally isomorphic to E then I is invertible.*

Proof: 1. Let I be a fractional right ideal of E such that $I + \sigma E = E$. Then for each maximal ideal M of S, $I_M + \sigma_M E_M = E_M$. If $\sigma \subset M \subset S$ then $\sigma_M \subset M_M = \mathcal{J}(S_M)$. Since E_M is a finitely generated S_M-module, Nakayama's theorem 1.1 implies that $I_M = I_M + \sigma_M E_M = E_M$.

2. If $\tau \not\subset M \subset S$, then $\tau_M = S_M$. Thus the inclusion $\tau \overline{E} \subset E \subset \overline{E}$ leads us to the chain of inclusions

$$\overline{E}_M = \tau_M \overline{E}_M \subset E_M \subset \overline{E}_M,$$

so that $E_M = \overline{E}_M$. Furthermore, $S_M = \text{center}(E_M) = \text{center}(\overline{E}_M) = \overline{S}_M$.

3. Let I be a fractional right ideal of E such that $I + \tau E = E$. If $\tau \subset M \subset S$ then $I_M = E_M$ by part 1. If, otherwise, $\tau \not\subset M \subset S$ then $E_M = \overline{E}_M$ by part 2. It follows that I_M is a fractional right ideal over the maximal S-order E_M. Such right ideals are invertible. Since E is Noetherian, the Local-Global Theorem 1.5 implies that I is invertible.

4. Let $I + \tau E = E$. If $\tau \subset M \subset S$ is a maximal ideal then $I_M = E_M$ by part 1. If, $\tau \not\subset M \subset S$ is a maximal ideal then by part 2, $E_M = \overline{E}_M$ and $S_M = \overline{S}_M$. Since $\overline{E}_M$ is a maximal order over the discrete valuation domain $\overline{S}_M$, $I_M \cong \overline{E}_M = E_M$, [94, Theorem 18.7(ii)]. Then by Lemma 7.6 I is locally isomorphic to E.

Conversely, suppose that I is locally isomorphic to E. Then I is a projective right E-module and

$$I/\tau I \cong I_\tau/\tau_\tau I_\tau \cong E_\tau/\tau_\tau E_\tau = E/\tau E.$$

Since I is projective, this isomorphism lifts to a mapping $f : I \to E$ such that $f(I) + \tau E = E$. Then $f(I)_\tau + \tau_\tau E_\tau = E_\tau$. Since $\tau_\tau \subset \mathcal{J}(S_\tau)$ and since E is a finitely generated S-module [9], Nakayama's theorem 1.1 shows us that $f(I)_\tau = E_\tau$. Thus $f(I)$ is a fractional ideal of E, so that $I' = f(I) \cong I$ satisfies $I' + \tau E = E$.

5. Apply parts 3 and 4. This completes the proof of the lemma.

Lemma 7.8. *Assume Property 7.5 and let I be a fractional ideal of S.*

1. *I is an invertible ideal of S iff $I \cong I'$ for some invertible fractional ideal $I' \subset S$ such that $I' + \tau = S$.*
2. *I is invertible over S iff I is locally isomorphic to S.*

Proof: 1. Let I be invertible over S. Then I is a finitely generated projective generator over S, and so $I/\tau I$ is a finitely generated projective generator over S/τ. Since S/τ is finite, S/τ is a product of local rings $S/\tau = A_1 \times \cdots \times A_t$, and each A_i is generated by $I/\tau I$. We write $I/\tau I = B_1 \oplus \cdots \oplus B_t$, where $B_i = (I/\tau I)A_i$. Since A_i is a local direct summand of S/τ, there is a surjection $f_i : B_i \to A_i$, and hence there is a surjection

$$f = \oplus_i f_i : I/\tau I = B_1 \oplus \cdots \oplus B_t \to A_1 \times \cdots \times A_t = S/\tau.$$

Since I is projective, f lifts to a map $g : I \to S$ such that $g(I) + \tau = S$. Since S is an integral domain, g is an injection, so that $I' \cong g(I)$ is an ideal of S such that $I' + \tau = S$. To prove the converse proceed as in Lemma 7.7.3.

2. Suppose that I is invertible over S. Then I is a projective S-module. Let $M \subset S$ be a maximal ideal. Because S is a Noetherian integral domain, I_M is a projective S_M-module of torsion-free rank one over the local ring S_M. Hence $I_M \cong S_M$, and because M was arbitrarily taken, Lemma 7.6 states that I is locally isomorphic to S. The proof there easily extends to localizations I_M at maximal ideals M of S.

Conversely, apply Lemma 7.7.5. This proves the lemma.

Lemma 7.9. *Assume Property 7.5, let $I \subset E$, and let $\bar{I} \subset \bar{E}$ be fractional right ideals.*

1. *If $\bar{I} + \tau \bar{E} = \bar{E}$ then $\bar{I} = (\bar{I} \cap E)\bar{E}$.*
2. *If $I + \tau E = E$ then $I = I\bar{E} \cap E$.*
3. *If $\bar{I} + \tau \bar{E} = \bar{E}$ then $(\bar{I} \cap E) + \tau E = E$.*

Proof: 1. Let $\bar{I} + \tau \bar{E} = \bar{E}$, and choose a maximal ideal $M \subset S$.

If $\tau \subset M$ then by Lemma 7.7.1, $\bar{I}_M = \bar{E}_M$ so that

$$[(\bar{I} \cap E)\bar{E}]_M = (\bar{I}_M \cap E_M)\bar{E}_M = (\bar{E}_M \cap E_M)\bar{E}_M = \bar{E}_M = \bar{I}_M.$$

If, otherwise, $\tau \not\subset M$ then by Lemma 7.7.2, $E_M = \bar{E}_M$ so that

$$[(\bar{I} \cap E)\bar{E}]_M = (\bar{I}_M \cap E_M)\bar{E}_M = (\bar{I}_M \cap \bar{E}_M)\bar{E}_M = \bar{I}_M$$

because $\bar{I} \subset \bar{E}$. By the Local-Global Theorem 1.5, $(\bar{I} \cap E)\bar{E} = \bar{I}$.

2. Given a maximal ideal M of S such that $\tau \subset M$ then by Lemma 7.7.1, $I_M = E_M$, so that

$$(I\bar{E} \cap E)_M = I_M\bar{E}_M \cap E_M = E_M\bar{E}_M \cap E_M = E_M = I_M.$$

Given a maximal ideal M of S such that $\tau \not\subset M$, then by Lemma 7.7.2, $E_M = \bar{E}_M$, so that

$$(I\bar{E} \cap E)_M = I_M\bar{E}_M \cap E_M = I_M E_M \cap E_M = I_M$$

because $I \subset E$. Then by the Local-Global Theorem 1.5, $I\bar{E} \cap E = I$.

3. Assume $\bar{I} + \tau \bar{E} = \bar{E}$ and let $M \subset S$ be a maximal ideal. Then $[(\bar{I} \cap E) + \tau E]_M = (\bar{I}_M \cap E_M) + \tau_M E_M$. If $\tau \subset M$ then by Lemma 7.7.1, $\bar{I}_M = \bar{E}_M$, so that

$$(\bar{I}_M \cap E_M) + \tau_M E_M = (\bar{E}_M \cap E_M) + \tau_M E_M = E_M + \tau_M E_M = E_M.$$

If, otherwise, $\tau \not\subset M$ then by Lemma 7.7.2, $E_M = \bar{E}_M$. Hence

$$(\bar{I}_M \cap E_M) + \tau_M E_M = (\bar{I}_M \cap \bar{E}_M) + \tau_M \bar{E}_M = \bar{I}_M + \tau_M \bar{E}_M = \bar{E}_M$$

by our choice of $\bar{I}$. This completes the proof.

Notice that in the next result, τ is the conductor of E and not necessarily of S.

Corollary 7.10. *Assume Property 7.5, let $I \subset S$, and let $\bar{I} \subset \bar{S}$ be fractional right ideals.*

1. *If $\bar{I} + \tau \bar{S} = \bar{S}$ then $\bar{I} = (\bar{I} \cap S)\bar{S}$.*
2. *If $I + \tau S = S$ then $I = \bar{I}\bar{S} \cap S$.*
3. *If $\bar{I} + \tau \bar{S} = \bar{S}$ then $(\bar{I} \cap S) + \tau S = S$.*

Proof: 1. Let $\bar{I} + \tau \bar{S} = \bar{S}$, and choose a maximal ideal $M \subset S$. If $\tau \subset M$ then by Lemma 7.7.1, $\bar{I}_M = \bar{S}_M$ so that

$$[(\bar{I} \cap S)\bar{S}]_M = (\bar{I}_M \cap S_M)\bar{S}_M = (\bar{S}_M \cap S_M)\bar{S}_M = \bar{S}_M = \bar{I}_M.$$

If, otherwise, $\tau \not\subset M$ then by Lemma 7.7.2, $S_M = \bar{S}_M$ so that

$$[(\bar{I} \cap S)\bar{S}]_M = (\bar{I}_M \cap S_M)\bar{S}_M = (\bar{I}_M \cap \bar{S}_M)\bar{S}_M = \bar{I}_M$$

because $\bar{I} \subset \bar{S}$. By the Local-Global Theorem 1.5, $(\bar{I} \cap S)\bar{S} = \bar{I}$.

Parts 2 and 3 can be proved in manner similar to the corresponding proofs in Lemma 7.9. This completes the proof.

Assume Property 7.5, and let

$$\boxed{\text{id}_E(E, \tau) = \{\text{fractional right ideals } I \subset E \mid I + \tau E = E\}.}$$

Similarly define $\text{id}_E(\bar{E}, \tau)$, $\text{id}_E(S, \tau)$, and $\text{id}_E(\bar{S}, \tau)$. By Lemma 7.7.4, I is locally isomorphic to E iff $I \cong I'$ for some $I' \in \text{id}_E(E, \tau)$. Specifically, by Lemma 7.7.5 each $I \in \text{id}_E(E, \tau)$ is invertible.

Recall the classic reduced norm $\overline{\text{nr}}$ from page 78. For each $\bar{I} \subset \bar{E}$, $\overline{\text{nr}}(\bar{I}) \subset \bar{S}$, and $\overline{\text{nr}}(\bar{E}) = \bar{S}$. Given $I \in \text{id}_E(E, \tau)$ define

$$\boxed{\text{nr}(I) = S_\tau \cap \overline{\text{nr}}(I\bar{E}) = S_\tau \cap \bigcap_{\tau \not\subset N \subset S} \overline{\text{nr}}(I\bar{E})_N.}$$

This definition of nr cannot be made in the method we used to define $\overline{\text{nr}}$ on page 78. Given an ideal $\bar{I} \subset \bar{E}$, $\overline{\text{nr}}(\bar{I})$ is the fractional right ideal of $\bar{S}$ generated by $\overline{\text{nr}}(x)$ for each $x \in \bar{I}$. Since $x \in \bar{E}$ is integral over $\bar{S}$ we have $\overline{\text{nr}}(x) \in \bar{S}$, so that $\overline{\text{nr}}(\bar{I}) \subset \bar{S}$. However, for a fractional right ideal $I \subset E$, each $x \in I$ will be integral over $\bar{S}$, but not necessarily over S. Hence, if defined element by element, it would happen that $\text{nr}(I) \not\subset S$ for some fractional right ideal I of E. This would impede our discussion of fractional right ideals I such that $I + \tau E = E$. See diagram (7.1) and Lemma 7.12.

Theorem 7.11. *Assume Property 7.5. There is a commutative square*

$$\begin{array}{ccc} \mathrm{id}_E(E,\tau) & \xrightarrow{\sigma \;\cong} & \mathrm{id}_E(\overline{E},\tau) \\[2mm] \Big\downarrow{\scriptstyle \mathrm{nr}} & & \Big\downarrow{\scriptstyle \overline{\mathrm{nr}}} \\[2mm] \mathrm{id}_E(S,\tau) & \xrightarrow[\overline{\sigma} \;\cong]{} & \mathrm{id}_E(\overline{S},\tau) \end{array} \qquad (7.1)$$

of monoids and epimorphisms of monoids. Moreover, the maps σ and $\overline{\sigma}$ are isomorphisms of monoids.

Proof: The set $\mathrm{id}_E(E,\tau)$ is a monoid since for each $I, J \in \mathrm{id}_E(E,\tau)$ we have

$$IJ + \tau E = IJ + \tau IE + \tau E = I(J + \tau E) + \tau E = IE + \tau E = E.$$

This gives us the corners of (7.1).

We define and examine the maps in (7.1) in a series of results. Define the map $\sigma : \mathrm{id}_E(E,\tau) \to \mathrm{id}_E(\overline{E},\tau)$ by $\sigma(I) = I\overline{E}$ for each $I \in \mathrm{id}_E(E,\tau)$. Define the map $\overline{\sigma} : \mathrm{id}_E(S,\tau) \to \mathrm{id}_E(\overline{S},\tau)$ by $\overline{\sigma}(I) = I\overline{S}$ for each $I \in \mathrm{id}_E(S,\tau)$.

Let $I \in \mathrm{id}_E(E,\tau)$. Since $I + \tau E = E$ we have $I\overline{E} + \tau\overline{E} = \overline{E}$, or equivalently that, $\sigma(I) = I\overline{E} \in \mathrm{id}_E(\overline{E},\tau)$. Hence σ is well-defined. Similarly, $\overline{\sigma}$ is well-defined. The next lemma shows that $\mathrm{nr} : \mathrm{id}_E(E,\tau) \to \mathrm{id}_E(S,\tau)$ and $\overline{\mathrm{nr}} : \mathrm{id}_E(\overline{E},\tau) \to \mathrm{id}_E(\overline{S},\tau)$ are well-defined maps.

Lemma 7.12. *Assume Property 7.5. Given $I \in \mathrm{id}_E(E,\tau)$ then $\mathrm{nr}(I) + \tau = S$.*

Proof: The ideal $K \subset S$ satisfies $K + \tau = S$ if $K \not\subset M$ for each maximal ideal $\tau \subset M \subset S$. Let $I \in \mathrm{id}_E(E,\tau)$ and let $\tau \subset M \subset S$ be a maximal ideal. Because $S_N \otimes_S S_M = (S_N)_M = \mathbf{Q}S$ for maximal ideals $N \neq M$, we have

$$\mathrm{nr}(I)_M = \left(S_\tau \cap \bigcap_{\tau \not\subset N \subset S} \overline{\mathrm{nr}}(I\overline{E})_N \right)_M$$

$$= (S_\tau)_M \cap \bigcap_{\tau \not\subset N \subset S} (\overline{\mathrm{nr}}(I\overline{E})_N)_M$$

$$= S_M \cap \bigcap_{\tau \not\subset N \subset S} \mathbf{Q}\overline{\mathrm{nr}}(I\overline{E})$$

$$= S_M.$$

Then $\mathrm{nr}(I) \not\subset M$, so that $\mathrm{nr}(I) + \tau = S$.

Lemma 7.13. *Assume Property 7.5. Let $I \in \mathrm{id}_E(E,\tau)$. Then $\overline{\sigma}(\mathrm{nr}(I)) = \overline{\mathrm{nr}}(\sigma(I))$.*

Proof: Let $I \in \mathrm{id}_E(E,\tau)$ and let $\tau \subset M \subset S$ be a maximal ideal. Because $\mathrm{nr}(I) + \tau = S$ (Lemma 7.12), we have

$$\overline{\sigma}(\mathrm{nr}(I))_M = (\mathrm{nr}(I)\overline{S})_M = \mathrm{nr}(I)_M \overline{S}_M = S_M \overline{S}_M = \overline{S}_M$$

by Lemma 7.7.1. Since $I + \tau E = E$, $I\overline{E} + \tau\overline{E} = \overline{E}$, so Lemma 7.12 shows us that $\overline{\mathrm{nr}}(I\overline{E}) + \tau = \overline{S}$. Then by Lemma 7.7.2,

$$\overline{\mathrm{nr}}(\sigma(I))_M = \overline{\mathrm{nr}}(I\overline{E})_M = \overline{S}_M.$$

Hence $\overline{\sigma}(\mathrm{nr}(I))_M = \overline{\mathrm{nr}}(\sigma(I))_M$ for each maximal ideal $\tau \subset M \subset S$.

Next suppose that $\tau \not\subset M \subset S$ is a maximal ideal. Then by the definition of nr,

$$\overline{\sigma}(\mathrm{nr}(I))_M = (\mathrm{nr}(I)\overline{S})_M = \mathrm{nr}(I)_M \overline{S}_M = \overline{\mathrm{nr}}(I\overline{E})_M = \overline{\mathrm{nr}}(\sigma(I))_M.$$

Then by the Local-Global Theorem 1.5, $\overline{\sigma}(\mathrm{nr}(I)) = \overline{\mathrm{nr}}(\sigma(I))$, as was required by the lemma.

Thus the diagram (7.1) exists and is commutative by Lemmas 7.9, 7.12, and 7.13, although we have yet to show that the maps are multiplicative. The next two lemmas show that $\sigma, \overline{\sigma}$, nr, and $\overline{\mathrm{nr}}$ are multiplicative maps.

Lemma 7.14. *Assume Property 7.5. Then $\sigma(IJ) = \sigma(I)\sigma(J)$ for each $I, J \in \mathrm{id}_E(E, \tau)$ and $\overline{\sigma}(IJ) = \overline{\sigma}(I)\overline{\sigma}(J)$ for each $I, J \in \mathrm{id}_E(S, \tau)$.*

Proof: Let $I, J \in \mathrm{id}_E(E, \tau)$. Given a maximal ideal $\tau \subset M \subset S$ then by Lemma 7.7.1, $I_M = J_M = E_M = (IJ)_M$. Thus

$$[(I\overline{E})(J\overline{E})]_M = (I_M\overline{E}_M)(J_M\overline{E}_M) = (E_M\overline{E}_M)(E_M\overline{E}_M)$$
$$= E_M\overline{E}_M = (IJ)_M\overline{E}_M = [(IJ)\overline{E}]_M.$$

Given a maximal ideal $\tau \not\subset M \subset S$ then by Lemma 7.7.2, $E_M = \overline{E}_M$. Thus

$$[(I\overline{E})(J\overline{E})]_M = (I_M\overline{E}_M)(J_M\overline{E}_M)$$
$$= (I_M E_M)(J_M\overline{E}_M) = (I_M J_M)\overline{E}_M = [(IJ)\overline{E}]_M.$$

By the Local-Global Theorem 1.5 $\sigma(I)\sigma(J) = (I\overline{E})(J\overline{E}) = (IJ)\overline{E} = \sigma(IJ)$.

Lemma 7.15. $\mathrm{nr}(IJ) = \mathrm{nr}(I)\mathrm{nr}(J)$ *for each $I, J \in \mathrm{id}_E(E, \tau)$.*

Proof: Let $I, J \in \mathrm{id}_E(E, \tau)$. Given a maximal ideal $\tau \subset M \subset S$, then

$$\mathrm{nr}(IJ)_M = \left(S_\tau \cap \bigcap_{\tau \not\subset N \subset S} \overline{\mathrm{nr}}(IJ\overline{E})_N \right)_M$$
$$= (S_\tau)_M \cap \bigcap_{\tau \not\subset N \subset S} (\overline{\mathrm{nr}}(IJ\overline{E})_N)_M$$

$$= S_M \cap \bigcap_{\tau \not\subset N \subset S} \mathbf{Q}\overline{\mathrm{nr}}(IJ\overline{E})$$

$$= S_M.$$

Furthermore,

$$[\mathrm{nr}(I)\mathrm{nr}(J)]_M = \left(S_\tau \cap \bigcap_{\tau \not\subset N \subset S} \overline{\mathrm{nr}}(I\overline{E})_N \right)_M \left(S_\tau \cap \bigcap_{\tau \not\subset N \subset S} \overline{\mathrm{nr}}(J\overline{E})_N \right)_M$$

$$= \left(S_M \cap \bigcap_{\tau \not\subset N \subset S} \mathbf{Q}\overline{\mathrm{nr}}(I\overline{E}) \right) \left(S_M \cap \bigcap_{\tau \not\subset N \subset S} \mathbf{Q}\overline{\mathrm{nr}}(J\overline{E}) \right)$$

$$= S_M.$$

Hence $\mathrm{nr}(IJ)_M = [\mathrm{nr}(I)\mathrm{nr}(J)]_M$ for each maximal ideal $\tau \subset M \subset S$.

Next, choose a maximal ideal $\tau \not\subset M \subset S$. The reader can show that $(S_\tau)_M = \mathbf{Q}S$. Thus we have

$$\mathrm{nr}(IJ)_M = \left(S_\tau \cap \bigcap_{\tau \not\subset N \subset S} \overline{\mathrm{nr}}(IJ\overline{E})_N \right)_M$$

$$= (S_\tau)_M \cap (\overline{\mathrm{nr}}(IJ\overline{E})_M)_M \cap \bigcap_{\tau \not\subset N \neq M \subset S} (\overline{\mathrm{nr}}(IJ\overline{E})_N)_M$$

$$= \mathbf{Q}S \cap \overline{\mathrm{nr}}(IJ\overline{E})_M \cap \bigcap_{\tau \not\subset N \neq M \subset S} \mathbf{Q}\overline{\mathrm{nr}}(IJ\overline{E})$$

$$= \overline{\mathrm{nr}}(IJ\overline{E})_M \tag{7.2}$$

Also,

$$[\mathrm{nr}(I)\mathrm{nr}(J)]_M = \left(S_\tau \cap \bigcap_{\tau \not\subset N \subset S} \overline{\mathrm{nr}}(I\overline{E})_N \right)_M \left(S_\tau \cap \bigcap_{\tau \not\subset N \subset S} \overline{\mathrm{nr}}(J\overline{E})_N \right)_M$$

$$= \left(\mathbf{Q}S \cap \overline{\mathrm{nr}}(I\overline{E})_M \cap \bigcap_{\tau \not\subset N \neq M \subset S} \mathbf{Q}\overline{\mathrm{nr}}(I\overline{E}) \right) \tag{7.3}$$

$$\times \left(\mathbf{Q}S \cap \overline{\mathrm{nr}}(J\overline{E})_M \cap \bigcap_{\tau \not\subset N \neq M \subset S} \mathbf{Q}\overline{\mathrm{nr}}(J\overline{E}) \right)$$

$$= \overline{\mathrm{nr}}(I\overline{E})_M \overline{\mathrm{nr}}(J\overline{E})_M. \tag{7.4}$$

Finally, by Lemma 7.14, $IJ\overline{E} = (I\overline{E})(J\overline{E})$, so that by (7.2) and (7.4),

$$\mathrm{nr}(IJ)_M = \overline{\mathrm{nr}}(IJ\overline{E})_M = \overline{\mathrm{nr}}((I\overline{E})(J\overline{E}))_M$$

$$= \overline{\mathrm{nr}}(I\overline{E})_M \overline{\mathrm{nr}}(J\overline{E})_M = [\mathrm{nr}(I)\mathrm{nr}(J)]_M$$

for each maximal ideal $\tau \not\subset M \subset S$. Thus by the Local-Global Theorem 1.5, $\mathrm{nr}(IJ) = \mathrm{nr}(I)\mathrm{nr}(J)$. This completes the proof.

Theorem 7.16. *Assume Property 7.5. There are isomorphisms of monoids* σ : $\mathrm{id}_E(E, \tau) \rightarrow \mathrm{id}_E(\overline{E}, \tau)$ *and* $\overline{\sigma}$: $\mathrm{id}_E(S, \tau) \rightarrow \mathrm{id}_E(\overline{S}, \tau)$.

Proof: By Lemma 7.14, it suffices to show that σ is a bijection. Choose $\overline{I} \in \mathrm{id}_E(\overline{E}, \tau)$. By Lemma 7.9.3, $\overline{I} \cap E \in \mathrm{id}_E(E, \tau)$, and by Lemma 7.9.1, $\sigma(\overline{I} \cap E) = (\overline{I} \cap E)\overline{E} = \overline{I}$. Thus σ is a surjection.

Suppose that $\sigma(I) = \sigma(J)$ for some $I, J \in \mathrm{id}_E(E, \tau)$. Then $I\overline{E} = J\overline{E}$, and by Lemma 7.9.2, $I = I\overline{E} \cap E = J\overline{E} \cap E = J$. Thus σ is an injection, and hence σ is a bijection. Similarly, $\overline{\sigma}$: $\mathrm{id}_E(S, \tau) \rightarrow \mathrm{id}_E(\overline{S}, \tau)$ is an isomorphism. This completes the proof of Theorem 7.16.

Theorem 7.17. *Assume Property 7.5. Then* nr : $\mathrm{id}_E(E, \tau) \rightarrow \mathrm{id}_E(S, \tau)$ *and* $\overline{\mathrm{nr}}$: $\mathrm{id}_E(\overline{E}, \tau) \rightarrow \mathrm{id}_E(\overline{S}, \tau)$ *are epimorphisms of monoids.*

Proof: Let $Q \in \mathrm{id}_E(S, \tau)$. Then $Q_M = S_M$ for each maximal ideal $\tau \subset M \subset S$. One can prove that Q is a finite product of maximal ideals $\tau \not\subset M_i \subset S$. Then by Lemma 7.15, it suffices to show that if Q is a maximal ideal such that $Q + \tau = S$ then $Q = \mathrm{nr}(L)$ for some $L \in \mathrm{id}_E(E, \tau)$.

Let Q be a maximal ideal of S such that $\tau \not\subset Q \subset S$. By Lemma 7.7.2, $S_Q = \overline{S}_Q$, so that Q_Q is the maximal ideal in $\overline{S}_Q$. Since then $Q_Q = (Q\overline{S})_Q$, $(Q\overline{S})_Q$ is the maximal ideal in $\overline{S}_Q$. For each maximal ideal $\tau \subset M \neq Q \subset S$, $Q_M = S_M$, so $(Q\overline{S})_M = \overline{S}_M$. Furthermore, for each maximal ideal $\tau \not\subset M \neq Q \subset S$, $Q_M = S_M = \overline{S}_M = (Q\overline{S})_M$, so by the Local-Global Theorem 1.5, $Q\overline{S}$ is a maximal ideal in $\overline{S}$. Furthermore, since $\overline{E}$ is a finitely generated $\overline{S}$-module, there is a maximal right ideal $\overline{L} \subset \overline{E}$ such that

$$Q\overline{E} \subset (Q\overline{S})\overline{E} \subset \overline{L} \subset \overline{E}.$$

Since $Q + \tau = S$, $Q\overline{E} + \tau\overline{E} = \overline{E}$, so $\tau\overline{E} \not\subset \overline{L}$. That is, $\overline{L} + \tau\overline{E} = \overline{E}$. Also, since $Q\overline{S}$ is a maximal ideal in $\overline{S}$, we can appeal to [94, Theorem 24.13] to show that $Q\overline{S} = \overline{\mathrm{nr}}(\overline{L})$.

By Lemma 7.9, $\overline{L} \cap E \in \mathrm{id}_E(E, \tau)$ and $(\overline{L} \cap E)\overline{E} = \overline{L}$. By our choice of the maximal ideal $\tau \not\subset Q \subset S$, $Q_M = S_M$ for each $\tau \subset M \subset S$, the Local-Global Theorem 1.5 shows us that $Q_\tau = S_\tau$. Furthermore, because $S_M = \overline{S}_M$ for each maximal ideal $\tau \not\subset M \subset S$, (Lemma 7.7.2), we have

$$\mathrm{nr}(\overline{L} \cap E) = S_\tau \cap \bigcap_{\tau \not\subset M \subset S} \overline{\mathrm{nr}}((\overline{L} \cap E)\overline{E})_M$$

$$= S_\tau \cap \bigcap_{\tau \not\subset M \subset S} \overline{\mathrm{nr}}(\overline{L})_M$$

$$= S_\tau \cap \bigcap_{\tau \not\subset M \subset S} (Q\overline{S})_M$$

$$= Q_\tau \cap \bigcap_{\tau \not\subset M \subset S} Q_M$$

$$= Q.$$

Hence nr is a surjection. Similarly, $\overline{\mathrm{nr}}$ is a surjection.

This completes the proof of Theorem 7.11.

7.4 Isomorphism classes

Continue to assume the Properties 7.5. We will define $\mathrm{Cl}(E, \tau)$ presently. Modules M and N are *stably isomorphic* if there is a free module F such that $M \oplus F \cong N \oplus F$. Isomorphic modules are stably isomorphic. Let $\mathrm{Cl}(\overline{E})$ denote the group of stable isomorphism classes of fractional right ideals of the maximal $\overline{S}$-order $\overline{E}$ in $\mathbf{Q}\overline{E}$. Since $\overline{E}$ is a maximal $\overline{S}$-order and since $\mathbf{Q}\overline{S}$ is an algebraic number field, Swan's theorem 7.1 states that the reduced norm $\overline{\mathrm{nr}}$ induces an isomorphism $\overline{\nu} : \mathrm{Cl}(\overline{E}) \to \mathrm{Pic}(\overline{S})$. As an application of Theorem 7.11, we will extend Swan's theorem to the nonmaximal order E and its non-Dedekind center S.

Theorem 7.18. *Assume Property 7.5. The commutative square (7.1) induces a commutative square*

$$
\begin{array}{ccc}
\mathrm{Cl}(E, \tau) & \xrightarrow{\ \delta\ } & \mathrm{Cl}(\overline{E}) \\[4pt]
\cong \Big\downarrow \nu & & \overline{\nu} \Big\downarrow \cong \\[4pt]
\mathrm{Pic}(S) & \xrightarrow[\ \overline{\delta}\ ]{} & \mathrm{Pic}(\overline{S})
\end{array}
\qquad (7.5)
$$

of finite groups and group epimorphisms. Moreover, ν and $\overline{\nu}$ are group isomorphisms.

Proof: Let (X) denote the isomorphism class of X. There is an equivalence relation $\sim$ on $\mathrm{id}_E(E, \tau)$ defined by

$$I \sim J \text{ if } \mathrm{nr}(I) \cong \mathrm{nr}(J).$$

Write

$$\mathrm{id}_E(E, \tau)/{\sim} = \mathrm{Cl}(E, \tau).$$

Let ϕ be the composition of maps

$$\mathrm{id}_E(E, \tau) \xrightarrow{\ \mathrm{nr}\ } \mathrm{id}_E(S, \tau) \xrightarrow{\ \pi\ } \mathrm{Pic}(S),$$

where $\pi(I) = (I)$ for ideals $I \subset S$. Let

$$\{I\} = \phi^{-1}(I) = \{J \in \mathrm{id}_E(E, \tau) \,\big|\, \mathrm{nr}(I) \cong \mathrm{nr}(J)\}.$$

Then ϕ induces a map $\nu : \mathrm{Cl}(E, \tau) \to \mathrm{Pic}(S)$ such that $\phi(\{I\}) = (\mathrm{nr}(I))$ for $I \in \mathrm{Cl}(E, \tau)$.

By the definition of the equivalence class $\{I\}$, ν is a well-defined injection. Next, let $(I) \in \mathrm{Pic}(S)$. Then I is an invertible ideal of S. By Lemma 7.8.1, there is an ideal $I' \cong I$ such that $I' + \tau = S$. That is, $I' \in \mathrm{id}_E(S, \tau)$. By Theorem 7.17, there is a $J \in \mathrm{id}_E(E, \tau)$ such that $\mathrm{nr}(J) = I'$. Hence $\nu(\{J\}) = (\mathrm{nr}(J)) = (I') = (I)$, so that ν is a surjection. Furthermore, by Lemma 7.15, ν is multiplicative. Therefore, $\nu : \mathrm{Cl}(E, \tau) \to \mathrm{Pic}(S)$ is an isomorphism. By Swan's theorem 7.1, $\overline{\nu}$ is a group isomorphism.

The map $\overline{\delta} : \mathrm{Pic}(S) \to \mathrm{Pic}(\overline{S})$ is defined by $\overline{\delta}((I)) = (I\overline{S})$. Let $(\overline{I}) \in \mathrm{Pic}(\overline{S})$. Since $\overline{S}$ is a Dedekind domain, $\overline{I}$ is invertible, so by Lemma 7.8.1, there is an ideal $\overline{I} \cong \overline{I}' \subset \overline{S}$ such that $\overline{I}' + \tau = \overline{S}$. Since $\overline{\delta}$ is induced by the epimorphism $\overline{\sigma}$ of monoids in (7.1), there is a $J \in \mathrm{id}_E(S, \tau)$ such that $\overline{\sigma}(J) = J\overline{S} = \overline{I}'$, so that $\overline{\delta}(J) = (\overline{I}') = (\overline{I})$. Thus $\overline{\delta}$ is an epimorphism of groups.

Define $\delta : \mathrm{Cl}(E, \tau) \to \mathrm{Cl}(\overline{E})$ in the only way possible to make the square commute. Since ν, $\overline{\delta}$, and $\overline{\nu}$ are epimorphisms, δ is an epimorphism of groups. Specifically, by the commutativity of (7.1),

$$\delta(\{I\}) = \overline{\nu}^{-1}\overline{\delta}(\nu(\{I\})) = \overline{\nu}^{-1}(\mathrm{nr}(I)\overline{S}) = \overline{\nu}^{-1}(\overline{\mathrm{nr}}(I\overline{E})) = \overline{\nu}^{-1}\overline{\nu}(I\overline{E}) = (I\overline{E}).$$

Thus,

$$\delta(\{I\}) = (I\overline{E}) \text{ for each } \{I\} \in \mathrm{Cl}(E, \tau)$$

and hence (7.5) is a commutative square induced by (7.1).

Finally, since S and $\overline{S}$ are *rtffr* integral domains, [9, Theorem 13.13] readily implies that $\mathrm{Pic}(S)$, and $\mathrm{Pic}(\overline{S})$ are finite groups. Inasmuch as ν and $\overline{\nu}$ are isomorphisms, $\mathrm{Cl}(E, \tau)$ and $\mathrm{Cl}(\overline{E})$ are finite groups.

Theorem 7.19. *Let E be a prime* rtffr *ring with center S and a conductor τ. There is an isomorphism $\mathrm{Cl}(E, \tau) \cong \mathrm{Pic}(S)$ of finite abelian groups.*

7.5 The equivalence class {*I*}

We give several partial classifications of the equivalence class $\{I\}$ for $I \in \mathrm{id}_E(E, \tau)$.

Theorem 7.20. *Assume Property 7.5 and let $I, J \in \mathrm{id}_E(E, \tau)$.*

1. *If $\mathrm{nr}(I) \cong \mathrm{nr}(J)$ then $I\overline{E} \cong J\overline{E}$.*
2. *If $I\overline{E} \cong J\overline{E}$ then $\mathrm{nr}(I)\overline{S} \cong \mathrm{nr}(J)\overline{S}$.*
3. *If $I\overline{E} = J\overline{E}$ then $\mathrm{nr}(I) = \mathrm{nr}(J)$.*

Proof: Let $I, J \in \mathrm{id}_E(E, \tau)$.

1. Suppose that $\mathrm{nr}(I) \cong \mathrm{nr}(J)$. By the commutativity of (7.1),

$$\overline{\mathrm{nr}}(I\overline{E}) = \mathrm{nr}(I)\overline{S} \cong \mathrm{nr}(J)\overline{S} = \overline{\mathrm{nr}}(J\overline{E})$$

so by Swan's theorem [94, Theorem 35.14], $I\overline{E} \cong J\overline{E}$.

2. We have $I\overline{E} \cong J\overline{E}$, so

$$\mathrm{nr}(I)\overline{S} = \overline{\mathrm{nr}}(I\overline{E}) \cong \overline{\mathrm{nr}}(J\overline{E}) = \mathrm{nr}(J)\overline{S}.$$

by the commutativity of (7.1).

3. Since $I, J \in \mathrm{id}_E(E, \tau)$, $\mathrm{nr}(I), \mathrm{nr}(J) \in \mathrm{id}_E(S, \tau)$ by Lemma 7.12. Suppose that $I\overline{E} = J\overline{E}$. Then $\overline{\mathrm{nr}}(I\overline{E}) = \overline{\mathrm{nr}}(J\overline{E})$ so that

$$\mathrm{nr}(I)\overline{S} = \overline{\mathrm{nr}}(I\overline{E}) = \overline{\mathrm{nr}}(J\overline{E}) = \mathrm{nr}(J)\overline{S}$$

by the commutativity of (7.1). Then by Lemma 7.9.2,

$$\mathrm{nr}(I) = \mathrm{nr}(I)\overline{S} \cap S = \mathrm{nr}(J)\overline{S} \cap S = \mathrm{nr}(J).$$

This completes the proof.

The next result shows us that the equivalence relation $\sim$ on $\mathrm{id}_E(E, \tau)$ is different from the equivalence relations isomorphism or stable isomorphism on $\mathrm{id}_E(E, \tau)$.

Theorem 7.21. *Assume Property 7.5 and assume that $S = \overline{S}$. Let $I, J \in \mathrm{id}_E(E, \tau)$. Then $I\overline{E} \cong J\overline{E}$ iff $\mathrm{nr}(I) \cong \mathrm{nr}(J)$ iff $\{I\} = \{J\}$.*

Proof: Suppose that $\mathrm{nr}(I) \cong \mathrm{nr}(J)$. By Theorem 7.20.1, $I\overline{E} \cong J\overline{E}$. Conversely, suppose that $I\overline{E} \cong J\overline{E}$. By Theorem 7.20.2 and because $S = \overline{S}$, $\mathrm{nr}(I) = \mathrm{nr}(J)\overline{S} \cong \mathrm{nr}(J)\overline{S} = \mathrm{nr}(J)$. The rest follows from the definition of $\{I\}$. This completes the proof.

An example is in order.

Example 7.22. Let $\overline{E}$ be the ring of algebraic integers in $\mathbf{Q}[\sqrt{-d}]$ for some integer $d = 2$ or $d > 3$ such that $\overline{E}$ is a *pid*. Choose a rational prime p such that $p\overline{E}$ is a perfect square in $\overline{E}$, and let $E(p) = \mathbf{Z} + p\overline{E}$. By [39, Example 14.3, Case 1], there are at least p ideals I of $E(p)$ that are locally isomorphic to $E(p)$. Take two nonisomorphic ideals I and J that are locally isomorphic to E. Since $\overline{E}$ is a *pid*, $I\overline{E} \cong \overline{E} \cong J\overline{E}$. Thus $\mathrm{nr}(I) \cong \mathrm{nr}(J)$ but $I \not\cong J$.

Remark 7.23. The next result needs some motivation. Assume Property 7.5. Suppose that $I, J \in \mathrm{id}_E(E, \tau)$ are fractional right ideals of E such that $I \cong J$. Since $\mathbf{Q}E$ is a simple ring, there is an $x \in \mathbf{Q}E$ such that $I = xJ$. Given a maximal ideal $\tau \subset M \subset S$, Lemma 7.7.1 states that $E_M = I_M = xJ_M = xE_M$. Thus x is a unit of E_M hence x is a unit of $\overline{E}_M$, whence $\overline{\mathrm{nr}}(x)$ is a unit in $\overline{S}_M$. At present we need additional hypotheses to see that $\overline{\mathrm{nr}}(x) \in S_M$.

Lemma 7.24. *Assume Property 7.5, and let $I, J \in \mathrm{id}_E(E, \tau)$. If $I = xJ$ for some $x \in \mathbf{Q}E$ such that $\overline{\mathrm{nr}}(x) \in S_\tau$ then $\mathrm{nr}(I) \cong \mathrm{nr}(J)$.*

Proof: Let $I = xJ \in \mathrm{id}_E(E, \tau)$ where $x \in \mathbf{Q}E$ is some element such that $\overline{\mathrm{nr}}(x) \in S_\tau$. Then $\overline{\mathrm{nr}}(x) \in S_M$ for each maximal ideal $\tau \subset M \subset S$. Choose a maximal ideal

$\tau \subset M \subset S$. Then $\overline{\mathrm{nr}}(x) \in S_M$. As in Remark 7.23, $\overline{\mathrm{nr}}(x)$ is a unit in E_M and in $\overline{S}_M$, so $\overline{\mathrm{nr}}(x)$ is a unit in $E_M \cap \overline{S}_M = S_M$. Then by the commutativity of (7.1),

$$\mathrm{nr}(xJ)_M \overline{S}_M = \overline{\mathrm{nr}}(xJ\overline{E})_M = \overline{\mathrm{nr}}(x)\overline{\mathrm{nr}}(J\overline{E})_M = \overline{\mathrm{nr}}(x)\mathrm{nr}(J)_M \overline{S}_M.$$

Since $\overline{\mathrm{nr}}(x)$ is a unit of S_M, $\overline{\mathrm{nr}}(x)\mathrm{nr}(J)_M = \mathrm{nr}(J)_M$, and since J and $I = xJ \in \mathrm{id}_E(E,\tau)$, an application of Lemma 7.12 implies that $\mathrm{nr}(xJ)_M, \mathrm{nr}(J)_M \in \mathrm{id}_E(S_M, \tau_M)$. Thus by Corollary 7.10.2,

$$\mathrm{nr}(I)_M = \mathrm{nr}(xJ)_M = \mathrm{nr}(xJ)_M \overline{S}_M \cap S_M$$
$$= [\overline{\mathrm{nr}}(x)\mathrm{nr}(J)_M]\overline{S}_M \cap S_M = \overline{\mathrm{nr}}(x)\mathrm{nr}(J)_M.$$

Hence $\mathrm{nr}(I)_M = \overline{\mathrm{nr}}(x)\mathrm{nr}(J)_M$.

Next, choose a maximal ideal $\tau \not\subset M \subset S$. Then $S_M = \overline{S}_M$ by Lemma 7.7.2. As above, the commutativity of (7.1) is used to show that

$$\mathrm{nr}(xJ)_M = \mathrm{nr}(J)_M \overline{S}_M = \overline{\mathrm{nr}}(xJ\overline{E})_M$$
$$= \overline{\mathrm{nr}}(x)\overline{\mathrm{nr}}(J\overline{E})_M = \overline{\mathrm{nr}}(x)\mathrm{nr}(J)_M \overline{S}_M = \overline{\mathrm{nr}}(x)\mathrm{nr}(J)_M.$$

Since $I = xJ \in \mathrm{id}_E(E,\tau)$, $\mathrm{nr}(I)_M = \mathrm{nr}(xJ)_M = \overline{\mathrm{nr}}(x)\mathrm{nr}(J)_M$.

Hence $\mathrm{nr}(I)_M = \overline{\mathrm{nr}}(x)\mathrm{nr}(J)_M$ for each maximal ideal $M \subset S$. An appeal to the Local-Global Theorem 1.5 then shows us that $\mathrm{nr}(I) = \overline{\mathrm{nr}}(x)\mathrm{nr}(J) \cong \mathrm{nr}(J)$. This completes the proof of the lemma.

It is well known that $\overline{\mathrm{nr}}(u(\overline{E})) \subset u(\overline{S})$. This is the inspiration for the hypotheses in the next result.

Theorem 7.25. *Assume Property 7.5 and assume that* $\overline{\mathrm{nr}}(u(E_\tau)) \subset S_\tau$. *Given* $I, J \in \mathrm{id}_E(E,\tau)$, *then* $I \cong J$ *implies that* $\mathrm{nr}(I) \cong \mathrm{nr}(J)$.

Proof: Apply the previous remark and lemma.

Remark 7.26. To this point, our Local-Global definition of nr and our techniques yield only partial results for the implication $I \cong J \implies \mathrm{nr}(I) \cong \mathrm{nr}(J)$. See the results in this section. However, ν is the natural way to complete the commutative diagram (7.5), and the isomorphism ν extends Swan's theorem 7.1. Thus we feel that nr has been appropriately defined. It would be helpful if there were a pointwise definition of nr, but at present we do not see a natural one.

7.6 Commutative domains

Given the isomorphism $\mathrm{Cl}(E, \tau) \cong \mathrm{Pic}(S)$ in Theorem 7.19, we have reduced our study of the power cancellation property in *rtffr* groups G to the study of the invertible fractional ideals of the commutative ring S.

Let us give the notation for the next few sections.

Property 7.27. 1. S denotes an integral domain whose additive structure is a *rtffr* group.
2. $\overline{S}$ denotes the *integral closure of S* in the algebraic number field $\mathbf{Q}S$.
3. τ is the largest ideal in S such that $\tau \overline{S} \subset S \subset \overline{S}$.

Let $\mathrm{id}_E(\overline{S})$ be the group of nonzero fractional ideals of $\overline{S}$. Since S has finite torsion-free rank, $\overline{S}$ is a Dedekind domain, and $\overline{S}/S$ is finite, [9, Corollary 10.12]. Let $\mathrm{Inv}(S)$ denote the group of invertible ideals of S, and let $\mathrm{Pr}(S) \subset \mathrm{Inv}(S)$ denote the principal fractional ideals of S.

Using a slightly different notation from the previous sections, $\mathrm{id}_E(S, \tau)$ denotes the *multiplicative abelian group generated* by ideals I of S such that $I + \tau = S$. Also, $\mathrm{Pr}(S, \tau)$ is the *multiplicative abelian group* of principal ideals in $\mathrm{id}_E(S, \tau)$. As a consequence of the Jordan–Zaussenhaus theorem 1.9, $\mathrm{Pic}(S)$ is a finite group.

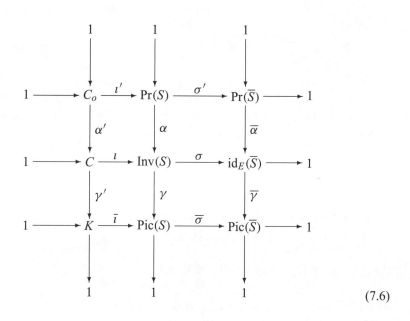

$$(7.6)$$

We will investigate $\mathrm{Inv}(S)$ by studying the 3×3 commutative diagram (7.6) that has exact rows and columns. (See below.)

Construction of diagram (7.6): The sets $\mathrm{Inv}(S)$ and $\mathrm{Pr}(S)$ are abelian groups. Since $\mathrm{Pic}(S) = \mathrm{Inv}(S)/\mathrm{Pr}(S)$ the middle and right hand columns of (7.6) are exact.

Define $\sigma : \mathrm{Inv}(S) \to \mathrm{id}_E(\overline{S})$ by $\sigma(I) = I\overline{S}$. Then σ is a group homomorphism that restricts to a group homomorphism $\sigma' : \mathrm{Pr}(S) \to \mathrm{Pr}(\overline{S})$. Furthermore, σ induces a well defined group homomorphism $\overline{\sigma} : \mathrm{Pic}(S) \to \mathrm{Pic}(\overline{S})$ such that $\overline{\sigma}(I) = (I\overline{S})$ for each isomorphism class $(I) \in \mathrm{Pic}(S)$. The rows containing σ, σ', and $\overline{\sigma}$ are extended to include their kernels C, C_o, and K and the inclusion maps ι, ι', and $\overline{\iota}$.

The Snake Lemma, (see [97]), produces the exact left hand column. That is, (7.6) is a commutative diagram of abelian groups with exact columns. It remains to show that $\sigma' : \text{Pr}(S) \to \text{Pr}(\overline{S})$, $\sigma : \text{Inv}(S) \to \text{id}_E(\overline{S})$, and $\overline{\sigma} : \text{Pic}(S) \to \text{Pic}(\overline{S})$ are epimorphisms.

Lemma 7.28. *Assume Property 7.27. The maps σ', σ and $\overline{\sigma}$ in (7.6) are epimorphisms of abelian groups.*

Proof: If $x\overline{S} \in \text{Pr}(\overline{S})$ then $\sigma'(xS) = x\overline{S}$, so σ' is a surjection.

Let $(\overline{I}) \in \text{Pic}(\overline{S})$. Without loss of generality we may assume that $\overline{I} \subset \overline{S}$. By Lemma 7.8.1, there is an ideal $\overline{I} \cong \overline{I}'$ such that $\overline{I}' + \tau = \overline{S}$. Because $\tau \subset S$, the modular law implies that $(\overline{I}' \cap S) + \tau = S$, so by Lemma 7.8.1, $\overline{I}' \cap S$ is an invertible ideal of S. By Lemma 7.9.1, $(\overline{I}' \cap S)\overline{S} = \overline{I}'$, and hence $\overline{\sigma}(\overline{I}' \cap S) = (\overline{I}') = (\overline{I})$. Thus $\overline{\sigma}$ is an epimorphism.

By the Snake Lemma applied to the middle and rightmost columns, the map σ is an epimorphism. This completes the proof of the lemma.

Construction of diagram (7.6): By Lemma 7.28, σ', σ, and $\overline{\sigma}$ are surjections, so the rows of (7.6) are exact. Thus (7.6) is a commutative diagram with exact rows and columns. This completes the construction of (7.6).

7.7 Cardinality of the kernels

Given a ring R, let $u(R)$ denote the group of units of R. We have a commutative diagram (7.6) with exact rows and columns. Observe that $C_o = \{xS \subset \mathbf{Q}S \mid x\overline{S} = \overline{S}\}$ and that $C = \{\text{fractional } S\text{-ideals } I \mid I\overline{S} = \overline{S}\}$.

Theorem 7.29. *Assume Property 7.27 and consider the kernel C_o in (7.6). Then $C_o \cong u(\overline{S})/u(S)$ and $u(\overline{S})/u(S)$ is a finite group.*

Proof: Since $\sigma'(xS) = x\overline{S} = \overline{S}$ for each $xS \in C_o$, there is an epimorphism $\phi : u(\overline{S}) \to C_o$ such that $\phi(x) = xS$ for each $x \in u(\overline{S})$. Evidently, $x \in \ker \phi$ iff $xS = \phi(x) = S$ iff $x \in u(S)$, so $\ker \phi = u(S)$. That is, $C_o \cong u(\overline{S})/u(S)$.

We must show that $u(\overline{S})/u(S)$ is a finite group. Consider the mapping

$$\rho : u(\overline{S}) \to \prod_{\tau \subset M} u(\overline{S}_M)/u(S_M)$$

defined by $\rho(x) = (xu(S_M) \mid \text{maximal ideals } \tau \subset M \subset S)$. Given $x \in u(\overline{S})$, then $xu(S_M) \in u(\overline{S}_M)/u(S_M)$ for each maximal ideal $\tau \subset M \subset S$. Hence ρ is well defined.

Let $x \in \ker \nu \subset u(\overline{S})$. Given a maximal ideal $\tau \subset M \subset S$, then $x \in u(S_M)$ by our choice of x. Furthermore, given a maximal ideal $\tau \not\subset M \subset S$, then by Lemma 7.7.2, $S_M = \overline{S}_M$, so that $x \in u(\overline{S}_M) = u(S_M)$. Hence x is a unit in S_M for each maximal ideal M of S. The Local-Global Theorem 1.5 then implies that $x \in u(S)$, and hence $\ker \rho = u(S)$. That is,

$$u(\overline{S})/u(S) \subset \prod_{\tau \subset M} u(\overline{S}_M)/u(S_M). \tag{7.7}$$

It suffices to show that $u(\overline{S}_M)/u(S_M)$ is finite for each maximal ideal $\tau \subset M \subset S$. There is a natural group homomorphism

$$u(\overline{S}_M) \to \frac{u(\overline{S}_M) + \tau_M}{\tau_M} \subset u(\overline{S}_M/\tau_M)$$

defined by $x \mapsto x + \tau_M$. The kernel of this map is $1 + \tau_M =$ the set of units in $\overline{S}_M$ that are congruent to 1 modulo τ_M. So $u(\overline{S}_M)/(1 + \tau_M) \subset u(\overline{S}_M/\tau_M)$. Since $\overline{S}_M$ is a *rtffr* integral domain, $\overline{S}_M/\tau_M$ is a bounded hence finite ring, whence $u(\overline{S}_M)/1 + \tau_M$ is a finite group. Furthermore,

$$\frac{u(\overline{S}_M)}{u(S_M)} \cong \frac{u(\overline{S}_M)/(1 + \tau_M)}{u(S_M)/(1 + \tau_M)}$$

so that $u(\overline{S}_M)/u(S_M)$ is a finite group for each maximal ideal $\tau \subset M \subset S$. Finally, because $\overline{S}$ is a *rtffr* integral domain, there are only finitely many maximal ideals $\tau \subset M \subset S$. Thus the inclusion (7.7) shows us that $u(\overline{S})/u(S)$ is (contained in) a finite group. This completes the proof.

Corollary 7.30. *Assume Property 7.27. The kernels C_o, C, and K in (7.6) are finite groups.*

Proof: By the Jordan–Zassenhaus lemma 1.9, $\mathrm{Pic}(S)$ is finite, so K is finite. Appeal to Theorem 7.29 and the exact left-hand column of (7.6) to show that C_o and C are finite.

Theorem 7.31. *Assume Property 7.27.*

1. $\mathrm{Inv}(S) \cong F \times C$ where F is a free abelian group and C is a finite abelian group.
2. $\mathrm{Pr}(S) \cong E \times C_o$ where E is a free abelian group and C_o is a finite abelian group.

Proof: 1 and 2. Because $\overline{S}$ is a Dedekind domain, $F = \mathrm{id}_E(\overline{S})$ and $E = \mathrm{Pr}(\overline{S}) \subset \mathrm{id}_E(\overline{S})$ are free abelian groups. Because the rows of (7.6) are exact, the multiplicative groups decompose as $\mathrm{Inv}(S) = F \times C$ and $\mathrm{Pr}(S) = F' \times C_o$. By Corollary 7.30, C and C_o are finite abelian groups. This completes the proof of the theorem.

7.8 Relatively prime to τ

Let $\mathrm{Pr}(S, \tau) = \mathrm{id}_E(S, \tau) \cap \mathrm{Pr}(S)$. Similarly define $\mathrm{Pr}(\overline{S}, \tau)$. Assuming (7.27), we construct diagram (7.8).

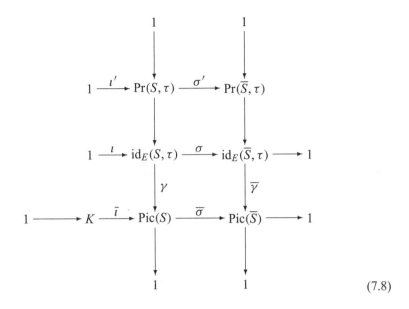

$$(7.8)$$

Construction of diagram (7.8): Assume Property 7.27.

Let $(I) \in \text{Pic}(S)$. By Lemma 7.8.1, there is an $I' \in \text{id}_E(S, \tau)$ such that $I \cong I'$. Then $\gamma(I') = (I)$ so that γ is an epimorphism. Because the map $\gamma : \text{Inv}(S) \to \text{Pic}(S)$ in (7.6) is an epimorphism, the First Isomorphism Theorem implies that

$$\text{Pic}(S) \cong \frac{\text{Inv}(S)}{\text{Pr}(S)} = \frac{\text{id}_E(S, \tau)\text{Pr}(S)}{\text{Pr}(S)} \cong \frac{\text{id}_E(S, \tau)}{\text{id}_E(S, \tau) \cap \text{Pr}(S)} = \frac{\text{id}_E(S, \tau)}{\text{Pr}(S, \tau)}.$$

Thus the center column in (7.8) is exact. Similarly the right-hand column of (7.8) is exact.

Next, restrict the maps σ', σ, and $\bar{\sigma}$ in (7.6) to define the maps σ', σ, and $\bar{\sigma}$ in (7.8). Since (7.6) is commutative, (7.8) is commutative. The bottom row of (7.8) is the exact bottom row of (7.6). Therefore, (7.8) is a commutative diagram with exact bottom row and exact columns.

Theorem 7.32. *Assume Property 7.27. The map $\sigma : \text{id}_E(S, \tau) \to \text{id}_E(\bar{S}, \tau)$ in (7.8) is an isomorphism of abelian groups.*

Proof: Evidently $\sigma : \text{id}_E(S, \tau) \to \text{id}_E(\bar{S}, \tau)$ is a group homomorphism.

Let (S, τ) be the set of ideals I in S such that $I + \tau = S$. Similarly define $(\bar{S}, \tau)$.

Let $\bar{I} \subset \bar{S}$ be such that $\bar{I} + \tau = \bar{S}$. By the Modular Law $(\bar{I} \cap S) + \tau = S$, so that there is a set function $\jmath : (\bar{S}, \tau) \to (S, \tau)$ given by $\jmath(\bar{I}) = \bar{I} \cap S$.

For $\bar{I} \subset (\bar{S}, \tau)$ $\sigma \jmath(\bar{I}) = (\bar{I} \cap S)\bar{S} = \bar{I}$ by Lemma 7.9.1. For $I \in (S, \tau)$, $\jmath\sigma(I) = I\bar{S} \cap S = I$ by Lemma 7.9.2. Hence $\sigma : (S, \tau) \to (\bar{S}, \tau)$ is a bijection.

Now, the subgroup $\text{id}_E(\bar{S}, \tau) \subset \text{id}_E(\bar{S})$ is an abelian group generated by $(\bar{S}, \tau)$. Since $\sigma : (S, \tau) \to (\bar{S}, \tau)$ is a surjection, σ lifts to a surjection $\sigma : \text{id}_E(S, \tau) \to \text{id}_E(\bar{S}, \tau)$.

Let $I \in \mathrm{id}_E(S, \tau)$ be such that $\sigma(I) = \overline{S}$. Because (S, τ) generates $\mathrm{id}_E(S, \tau)$, there are $J, L \in (S, \tau)$ such that $I = JL^{-1}$. Then $\sigma(JL^{-1}) = \overline{S}$ so that $\sigma(J) = \sigma(L)$. Since σ is a bijection on (S, τ), $J = L$, whence $I = JL^{-1} = S$. That is, σ is a monomorphism on $\mathrm{id}_E(\overline{S}, \tau)$. Hence $\sigma : \mathrm{id}_E(S, \tau) \to \mathrm{id}_E(\overline{S}, \tau)$ is an isomorphism of abelian groups. This completes the proof of the theorem.

Corollary 7.33. *Assume Property 7.27. In diagram (7.8), $\ker \sigma = \ker \sigma' = 1$.*

We have constructed the commutative diagram (7.8) having exact rows and columns.

Corollary 7.34. *Assume Property 7.27. Then $\mathrm{Pr}(S, \tau)$ and $\mathrm{id}_E(S, \tau)$ are free abelian groups.*

Proof: Because $\overline{S}$ is a Dedekind domain, $\mathrm{id}_E(\overline{S})$ is a free abelian group. Then by Theorem 7.32,

$$\mathrm{Pr}(S, \tau) \subset \mathrm{id}_E(S, \tau) \cong \mathrm{id}_E(\overline{S}, \tau) \subset \mathrm{id}_E(\overline{S})$$

are free abelian groups.

Theorem 7.35. *Assume Property 7.27. There is a long exact sequence*

$$1 \to \mathrm{Pr}(S, \tau) \xrightarrow{\sigma'} \mathrm{Pr}(\overline{S}, \tau) \longrightarrow \mathrm{Pic}(S) \xrightarrow{\overline{\sigma}} \mathrm{Pic}(\overline{S}) \to 1 \qquad (7.9)$$

of abelian groups.

Proof: Consider the kernels in (7.8). There is an exact sequence $1 \to 1 \to K \to K \to 1$. Then by the Snake Lemma applied to the bottom two rows of (7.8) there is a short exact sequence

$$1 \to \mathrm{Pr}(S, \tau) \xrightarrow{\sigma'} \mathrm{Pr}(\overline{S}, \tau) \longrightarrow K \longrightarrow 1.$$

Since $K = \ker \overline{\sigma}$ there is an exact sequence (7.9).

7.9 Power cancellation

We specify our setting.

Property 7.36. 1. Fix a reduced strongly indecomposable *rtffr* group U and an integer $k > 0$. Let G be a subgroup of finite index in U^k.
2. $E(G) = E$ is a prime *rtffr* ring whose center S is an integral domain, [9, Theorem 9.10].
3. As in previous sections, $E(G) = E$ is a subring of finite index in a maximal S-order $\overline{E} \subset \mathbf{Q}E$. We denote the center of $\overline{E}$ by $\overline{S}$.
4. Let τ be the largest ideal in S such that $\tau\overline{E} \subset E \subset \overline{E}$.

We give several applications to direct sum decompositions of *rtffr* groups. The group G has the *power cancellation property* if for each integer $n > 0$ and group H,

$G^n \cong H^n$ implies that $G \cong H$. We remind the reader that Theorem 2.40 shows that the power cancellation property is intimately connected with the problem of determining those algebraic number fields with unique factorization.

Theorem 7.37. *Assume Property 7.36. The following are equivalent for G.*

1. *G has the power cancellation property.*
2. *Each right ideal of $E(G)$ that is locally isomorphic to $E(G)$ is principal.*
3. *If I is a fractional right ideal of $E(G)$ such that $I + \tau E(G) = E(G)$ then I is principal.*

Proof: Recall that $E = E(G)$.

$3 \Leftrightarrow 2$. Assume part 3, and let I be locally isomorphic to E. By Lemma 7.9.3, there is an $I' \cong I$ such that $I' + \tau E = E$. By part 3, $I' \cong E$. This proves part 2.

Assume part 2, and let I be a fractional right ideal of E such that $I + \tau E = E$. By Lemma 7.9.3, I is locally isomorphic to E, so by part 2, $I \cong E$. This proves part 3.

$2 \Rightarrow 1$. Suppose that each right ideal of E that is locally isomorphic to E is principal, and suppose that H is an abelian group such that $G^n \cong H^n$ for some integer $n > 0$. By Warfield's theorem 7.3, G is locally isomorphic to H, so that by Theorem 7.4, $A_G(G)$ is locally isomorphic to $A_G(H)$. By hypothesis, $A_G(G) \cong E \cong A_G(H)$, so by Theorem 7.4, $G \cong H$. Thus G has the power cancellation property.

$1 \Rightarrow 2$. Suppose that G has the power cancellation property, and let I be a right ideal of E that is locally isomorphic to E. Then I is a fractional right ideal of E such that $I^n \cong E^n$ for some integer $n > 0$ (Lemma 7.3). By Theorem 7.4 there is an $H \in \mathbf{P}_o(G)$ such that $A_G(H) = I$. Then we have

$$A_G(H^n) \cong A_G(H)^n \cong I^n \cong E^n \cong A_G(G)^n \cong A_G(G^n).$$

Then by Theorem 7.4, $H^n \cong G^n$. Since G has the power cancellation property, $H \cong G$, and hence

$$I \cong A_G(H) \cong A_G(G) \cong E.$$

This completes the logical cycle.

The following result has a stronger conclusion if we assume that E is a commutative integral domain. See Corollary 2.16 and Theorem 2.30.

Theorem 7.38. *Let U be a strongly indecomposable rtffr group, let $k > 0$ be an integer, and let G be quasi-equal to U^k.*

1. *G has the power cancellation property if each invertible fractional right ideal of E is principal.*
2. *If G has the power cancellation property then every fractional right ideal of $\overline{E}$ is principal.*

Proof: 1. Suppose that each invertible fractional right ideal of E is principal. Let I be a fractional right ideal of E that is locally isomorphic to E. By Lemma 7.7.5, I is invertible, so I is principal. Then by Theorem 7.37, G has the power cancellation property.

2. Suppose that G has the power cancellation property. Let $\bar{I}$ be a fractional right ideal of $\bar{E}$. It is known that $\bar{I}$ is locally isomorphic to the maximal $\bar{S}$-order $\bar{E}$. (In [94], we would read that $\bar{I}$ and $\bar{E}$ have the same genus class.) Then by Lemma 7.7.4, $\bar{I} \in \mathrm{id}_E(\bar{E}, \tau)$, so by Theorem 7.16, there is an $I \in \mathrm{id}_E(E, \tau)$ such that $I\bar{E} = \bar{I}$. Since $I + \tau E = E$, Lemma 7.7.4 states that I is locally isomorphic to E. Thus, by Theorem 7.4, there is a group H such that $A_G(H) = I$ and H is locally isomorphic to G. By Warfield's theorem 7.3, $G^n \cong H^n$ for some integer $n > 0$. Since G has the power cancellation property, $G \cong H$ so that

$$\bar{E} = E\bar{E} \cong A_G(G)\bar{E} \cong A_G(H)\bar{E} = I\bar{E} = \bar{I}.$$

This completes the proof of the theorem.

The group G has Σ-*unique decomposition* if for each integer $n > 0$ and for each direct summand P of G^n (i) there is a finite direct sum $P = P_1 \oplus \cdots \oplus P_t$ of indecomposable groups $P_1, \ldots, P_t$, and (ii) if $P \cong Q_1 \oplus \cdots \oplus Q_s$ for some indecomposable groups $Q_1, \ldots, Q_s$ then $s = t$ and after a permutation of the subscripts $P_i \cong Q_i$ for each $i = 1, \ldots, t$. The problem of describing the abelian groups with a Σ-unique decomposition has a long history. See [9, 46, 59].

The following result is proved in [39] with a stronger conclusion under the assumption that E is a commutative integral domain. See Theorem 2.32.

Theorem 7.39. *Let U be a strongly indecomposable* rtffr *group, let $k > 0$ be an integer, and let G be an indecomposable group that is quasi-equal to U^k. If G has Σ-unique decomposition then G has the power cancellation property.*

Proof: Suppose that G has Σ-unique decomposition. Let H be a group and $n > 0$ be an integer such that $G^n \cong H^n$. Write $H = H_1 \oplus \cdots \oplus H_s$ for some indecomposable groups $H_1, \ldots, H_s$. Then $G^n \cong H_1^n \oplus \cdots \oplus H_s^n$. Since G has Σ-unique decomposition, and since G is indecomposable, the indecomposable direct summands $H_1, \ldots, H_s$ are isomorphic to G. By comparing ranks, $s = 1$, and hence $H = H_1 \cong G$. Thus G has power cancellation. This completes the proof.

Theorem 7.40. *Let E be a prime* rtffr *ring. If some $G \in \Omega(E)$ has Σ-unique decomposition then each right ideal of E that is locally isomorphic to E is principal.*

Proof: Suppose that $G \in \Omega(E)$ has Σ-unique decomposition. By Theorem 7.39, G has the power cancellation property, so by Theorem 7.41, each right ideal of E that is locally isomorphic to E is principal. This completes the proof.

7.10 Algebraic number fields

The following theorem and Theorem 2.30 seem to indicate that a more realistic generalization of unique factorization to noncommutative rings is condition 1 of the next theorem, and not the principal right ideal condition.

Given a semi-prime ring R let

$$\Omega(R) = \{\text{abelian groups } G \,|\, E(G) \cong R\}.$$

Theorem 7.41. *Let E be a prime* rtffr *ring. The following are equivalent.*

1. *Each right ideal of E that is locally isomorphic to E is principal.*
2. *Each $G \in \Omega(E)$ has the power cancellation property.*
3. *Some $G \in \Omega(E)$ has the power cancellation property.*
4. *Some* rtffr *group $G \in \Omega(E)$ has the power cancellation property.*

Proof: $1 \Rightarrow 2$. Let $G \in \Omega(E)$. Then $E(G) = E$ satisfies the condition in part 1. By Theorem 7.37, G has the power cancellation property, so part 2 is satisfied.

$2 \Rightarrow 4$. Assume part 2. By Corner's theorem 1.11, there is a *rtffr* group G such that $E \cong \text{End}(G)$ as rings. Since E is prime, $E(G) = \text{End}(G)$, so $G \in \Omega(E) \neq \emptyset$. By part 2, G has the power cancellation property, so that part 4 is true.

$4 \Rightarrow 3$ is clear.

$3 \Rightarrow 1$. Choose $G \in \Omega(E)$ that has the power cancellation property. By Theorem 7.37, $E(G) = E$ satisfies part 1. This completes the logical cycle.

Theorem 7.42. *Let* **k** *be an algebraic number field, let $\overline{E}$ be the ring of algebraic numbers in* **k**. *Then* **k** *has unique factorization if there is a subring E of finite index in $\overline{E}$ and a strongly indecomposable* rtffr *group $G \in \Omega(E)$ such that G has the power cancellation property.*

Proof: Suppose that E is a subring of finite index in $\overline{E}$, and suppose that some strongly indecomposable *rtffr* group $G \in \Omega(E)$ has the power cancellation property. Observe that $\overline{E}$ is a maximal order containing E. Let $\overline{I}$ be an ideal of $\overline{E}$. Since $\overline{E}$ is a Dedekind domain, $\overline{I}$ is invertible, and so $\overline{I}$ is locally isomorphic to $\overline{E}$ (Lemma 7.8.2). By Theorem 7.38.2, $\overline{I} \cong \overline{E}$, so **k** has unique factorization. This completes the proof.

Theorem 7.43. *Let* **k** *be an algebraic number field, and let $\overline{E}$ be the ring of algebraic numbers in* **k**.

1. **k** *has unique factorization if there is a subring E of finite index in $\overline{E}$ such that each ideal of E that is locally isomorphic to E is principal.*
2. **k** *has unique factorization if there is a subring E of finite index in $\overline{E}$ with conductor τ such that given a ideal I of E such that $I + \tau = E$ then I is principal.*

Proof: 1. Suppose that there is a subring E of finite index in $\overline{E}$ such that each ideal of E that is locally isomorphic to E is principal. By Corner's theorem 1.11 there is a *rtffr* group G such that $E(G) \cong E$. By Theorem 7.41, G has the power cancellation property. Then by Theorem 7.42, **k** has unique factorization.

2. Suppose that there is a subring E of finite index in $\overline{E}$ with conductor τ such that given a ideal I of E that satisfies $I + \tau = E$, then I is principal. By Corner's theorem 1.11, there is a *rtffr* group G such that $E(G) \cong E$. By Theorem 7.37, G has the power cancellation property, so by Theorem 7.42, **k** has unique factorization.

Theorem 7.44. *Let* **k** *be an algebraic number field, and let* $\overline{E}$ *be the ring of algebraic integers in* **k**. *Suppose there is a subring* E *of finite index in* $\overline{E}$ *such that some* $G \in \Omega(E)$ *that has* Σ-*unique decomposition. Then* **k** *has unique factorization.*

Proof: Apply Theorem 7.40 and Theorem 7.43.

7.11 Exercises

Let G and H denote rtffr groups. Let $E(G) = \text{End}(G)/\mathcal{N}(\text{End}(G))$. Let E be an rtffr ring (= a ring for which $(E, +)$ is a *rtffr* ring. Let $S = \text{center}(E)$. Given a semi-prime *rtffr* ring E let $\overline{E}$ be an integrally closed subring of $\mathbf{Q}E$ such that $E \subset \overline{E}$ and $\overline{E}/E$ is finite. Let τ be the conductor of E.

1. Show that $E(G)$ is a Noetherian semi-prime ring.
2. Prove that $\text{id}_E(E, \tau)$ is a monoid.
3. Prove that $\overline{\text{nr}}(xy) = \overline{\text{nr}}(x)\overline{\text{nr}}(y)$ for $x, y \in \mathbf{Q}\overline{E}$.
4. Prove that E-lattices M and N are locally isomorphic iff $M \oplus M \cong N \oplus K$ for some E-lattice K.
5. Let G and H be rtffr groups. Show that G is locally isomorphic to H iff for each integer $n > 0$ there are maps $f : G \to H$ and $g : H \to G$, and an integer m that is relatively prime to n such that $fg = m1_H$ and $gf = m1_G$.
6. Prove parts 2 and 3 of Corollary 7.10.

7.12 Problems for future research

1. Give a group structure for $\Gamma(G)$.
2. Show that $I \cong J$ implies that $\text{nr}(I) \cong \text{nr}(J)$.
3. See (7.6). Describe the ideals in the finite groups C and C_o.
4. Extend Theorem 7.9.3 to groups G such that $E(G)$ is a semi-prime ring.

8

$\mathcal{L}$-groups

The rtffr group G is an *Eichler* group if none of the simple ring factors of $\mathbf{Q}\mathrm{End}(G)$ is a totally definite quaternion algebra. The rtffr group G is called a $\mathcal{J}$-*group* if G is isomorphic to each subgroup of finite index in G, and G is called an $\mathcal{L}$-*group* if G is locally isomorphic to each subgroup of finite index in G. We show that Eichler $\mathcal{L}$-groups are $\mathcal{J}$-groups, and that $\mathrm{End}(G)$ is a domain if G is an indecomposable $\mathcal{L}$-group.

8.1 $\mathcal{J}$-groups, $\mathcal{L}$-groups, and $\mathcal{S}$-groups

The rtffr group G is an *Eichler* group if none of the simple ring factors of $\mathbf{Q}\mathrm{End}(G)$ is a totally definite quaternion algebra. The group G is called a $\mathcal{J}$-*group* if G is isomorphic to each subgroup of finite index in G, and G is called an $\mathcal{L}$-*group* if G is locally isomorphic to each subgroup of finite index in G.

Evidently $\mathcal{J}$-groups $\Rightarrow \mathcal{L}$-groups. C. Murley [87] proves that pure subgroups of the $\mathbf{Z}$-adic completion $\widehat{\mathbf{Z}}$ are $\mathcal{J}$-groups. The reader can show that $\mathbf{Z} \oplus K$ is a $\mathcal{J}$-group if K is a $\mathcal{J}$-group. $\mathcal{J}$-groups are introduced by D. M. Arnold in [10]. $\mathcal{L}$-groups are introduced in [46]. Further reseach on $\mathcal{J}$-groups can be found in [57] where we show that if $\mathbf{Q}\mathrm{End}(G)$ is commutative then G is a Murley group iff G is a $\mathcal{J}$-group.

Theorem 8.1. [57, Proposition III.6] *If G is an Eichler group then G is a finitely faithful $\mathcal{S}$-group iff G is a $\mathcal{J}$-group.*

Example 8.2. There is a strongly indecomposable $\mathcal{J}$-group that is not a Murley group. See exercise 1. and [10, 57].

8.2 Eichler groups

8.3. Given a group G, The Wedderburn theorem 1.7 states that $\mathbf{Q}\mathrm{End}(G) = B \oplus \mathcal{N}(\mathbf{Q}\mathrm{End}(G))$ for some semi-simple $\mathbf{Q}$-algebra B. Let

$$E(G) = \{b \mid b \oplus c \in \mathrm{End}(G) \text{ for some } c \in \mathcal{N}(\mathbf{Q}\mathrm{End}(G))\}.$$

Then $E(G)$ is a semi-prime subring of $\mathbf{Q}\mathrm{End}(G)$. By the Beaumont–Pierce theorem 1.8, $\mathrm{End}(G) \doteq T \oplus \mathcal{N}(\mathrm{End}(G))$ for some semi-prime ring T such that $\mathbf{Q}T = B$. Thus $T \subset E(G)$ and $\mathrm{End}(G) \doteq E(G) \oplus \mathcal{N}(\mathrm{End}(G))$.

Let i, j, k be such that $-1 = i^2 = j^2 = k^2 = ijk$ and let **k** be an algebraic number field. The **k**-algebra $D = \mathbf{k}1 \oplus \mathbf{k}i \oplus \mathbf{k}j \oplus \mathbf{k}k$ is called a *totally definite quaternion algebra* if D is a division algebra, and each embedding of **k** into the complex numbers is an embedding of **k** into the reals. If we let $\mathbf{k} = \mathbf{Q}$ then D is the **Q**-algebra of Hamiltonian quaternions. Thus the Hamiltonian quaternions form a totally definite quaternion algebra.

Write

$$\mathbf{Q}E(G) = A_1 \times \cdots \times A_t$$

for some simple nonzero **Q**-algebras $A_1, \ldots, A_t$. We say that G is an *Eichler group* if G is rtffr and if none of the simple factors $A_1, \ldots, A_t$ is a totally definite quaternion algebra.

Since $E(G)$ is an rtffr semi-prime ring there are classical maximal orders $\overline{E}_1, \ldots, \overline{E}_t$ in the simple Artinian **Q**-algebras $A_1, \ldots, A_t$, respectively, such that

$$E(G) \subset \overline{E}_1 \times \cdots \times \overline{E}_t$$

and such that $(\overline{E}_1 \times \cdots \times \overline{E}_t)/E(G)$ is finite, [9, Theorem 9.10]. So $E(G)$ is integrally closed iff $E(G) = \overline{E}_1 \times \cdots \times \overline{E}_t$.

Have you proved this Lemma yet, reader?

Lemma 8.4. *Let* $G \doteq G_1^{e_1} \oplus \cdots \oplus G_t^{e_t}$ *for some strongly indecomposable, pairwise non-quasi-isomorphic, rtffr groups* $G_1, \ldots, G_t$ *and some integers* $e_1, \ldots, e_t > 0$. *Then*

$$\mathbf{Q}E(G) = \mathbf{Q}E(G_1^{e_1}) \times \cdots \times \mathbf{Q}E(G_t^{e_t})$$

and

$$\mathbf{Q}E(G_i^{e_i}) = \mathrm{Mat}_{e_i \times e_i} \mathbf{Q}E(G_i)$$

for each $i = 1, \ldots, t$. *In particular,* $\mathbf{Q}E(G_1^{e_1}), \ldots, \mathbf{Q}E(G_t^{e_t})$ *are the simple ring factors of* $\mathbf{Q}E(G)$.

Proof: See [9, Theorem 9.10].

Theorem 8.5. [94, Corollary 35.12(iv)] *Let* $r \geq 2$ *and let* $U_1, \ldots, U_r, V_1, \ldots, V_r$ *be left fractional right ideals in* $\overline{E}$. *Then*

$$U_1 \oplus \cdots \oplus U_t \cong V_1 \oplus \cdots \oplus V_t$$

iff

$$(\overline{\mathrm{nr}}(U_1)) \cdots (\overline{\mathrm{nr}}(U_t)) = (\overline{\mathrm{nr}}(V_1)) \cdots (\overline{\mathrm{nr}}(V_t))$$

where the product takes place in $\mathrm{Pic}(\overline{S})$.

Lemma 8.6. *Let $G \doteq G_1^{e_1} \oplus \cdots \oplus G_t^{e_t}$ for some strongly indecomposable, pairwise non-quasi-isomorphic, rtffr groups $G_1, \ldots, G_t$ and some integers $e_1, \ldots, e_t > 0$. The following are equivalent.*

1. *G is an Eichler group.*
2. *If $\mathbf{QE}(G_i)$ is a totally definite quaternion algebra for some integer $i \in \{1, \ldots, t\}$ then $e_i > 1$.*

Proof: $2 \Rightarrow 1$. Suppose that $\mathbf{QE}(G_j)$ is a totally definite quaternion algebra for some $j \in \mathcal{J} \subset \{1, \ldots, t\}$. Then $e_j > 1$ so that $\mathbf{QE}(G_j^{e_j})$ is not a domain for each $j \in \mathcal{J}$. It follows that $\mathbf{QE}(G_i^{e_i})$ is not a totally definite quaternion algebra for each $i \in \{1, \ldots, t\}$. Lemma 8.4 shows us that the rings $\mathbf{QE}(G_i^{e_i})$ are the simple factors of the ring $\mathbf{QE}(G)$, so that G is an Eichler group.

$1 \Rightarrow 2$. We prove the contrapositive. Suppose that $\mathbf{QE}(G_1)$ is a totally definite quaternion algebra and that $e_1 = 1$. Then $\mathbf{QE}(G_1)$ is a simple factor of the ring $\mathbf{QE}(G)$, so that G is not an Eichler group. This concludes the proof.

Compare the next result with Example 8.8. The group G constructed there is not an Eichler group since $\mathbf{QEnd}(G) = $ the Hamiltonian quaternions.

Theorem 8.7. *Let G be an Eichler group, suppose that $E(G)$ is an integrally closed ring, and let H be an rtffr group. If $G \oplus G \cong G \oplus H$ then $G \cong H$.*

Proof: Suppose that $E(G)$ is integrally closed, and assume that $G \oplus G \cong G \oplus H$. Write $\mathbf{QE}(G) = A_1 \times \cdots \times A_t$ where for each $i = 1, \ldots, t$, $A_i = \mathbf{QE}(G_i^{e_i})$ is a simple ring factor of $\mathbf{QE}(G)$. Since $E(G)$ is integrally closed, $E(G) = \overline{E}_1 \times \cdots \times \overline{E}_t$ where $\overline{E}_i$ is a classical maximal order in A_i.

Because G and H are rtffr groups, Jónsson's theorem and $G \oplus G \cong G \oplus H$ implies that, $G \cong H$, so that $E(G) = A(G) \doteq A(H)$. That is, $A(H)$ is a fractional right ideal of $E(G)$ such that

$$E(G) \oplus E(G) \cong A(G) \oplus A(G) \cong A(G) \oplus A(H) \cong E(G) \oplus A(H).$$

Let $A(H)\overline{E}_i = I_i$. We then have $\overline{E}_i \oplus \overline{E}_i \cong \overline{E}_i \oplus I_i$ for each $i = 1, \ldots, t$.

Let S_i be the center of $\overline{E}_i$ for each $i = 1, \ldots, t$. Then $\mathrm{nr}(\overline{E}_i) = S_i$. From $\overline{E}_i \oplus \overline{E}_i \cong \overline{E}_i \oplus I_i$ and Theorem 8.5 we see that

$$\mathrm{nr}(\overline{E}_i) = \mathrm{nr}(\overline{E}_i) \cdot \mathrm{nr}(\overline{E}_i) \cong \mathrm{nr}(\overline{E}_i) \cdot \mathrm{nr}(I_i) \cong \mathrm{nr}(I_i).$$

Inasmuch as G is an Eichler group, the simple ring factor $\mathbf{QE}(G_i^{e_i})$ of $\mathbf{QE}(G)$ is not a totally definite quaternion algebra for any $i = 1, \ldots, t$. Then by Theorem 8.5, $\overline{E}_i \cong I_i$. Consequently, $A(G) = E(G) \cong A(H)$, and so $G \cong H$ by Theorem 2.4.

In the last line of the above theorem, we conclude from $\mathrm{nr}(E_i) = \mathrm{nr}(I_i)$ that $E_i \cong I_i$. This matter is peculiar to Eichler groups, as the next example shows.

Example 8.8. Let D denote a totally definite quaternion $\mathbf{Q}$-algebra. By [94, midpage 279] there is a maximal $\mathbf{Z}$-order $\mathcal{O} \subset D$ and a nonprincipal right ideal $I \subset \mathcal{O}$ such that

$\text{nr}(\mathcal{O}) = \text{nr}(I)$. The reduced norm is taken in $S = \text{center}(\mathcal{O})$. Obviously, $\mathcal{O} \not\cong I$ but $\text{nr}(\mathcal{O}) \cdot \text{nr}(\mathcal{O}) \cong \text{nr}(\mathcal{O}) \cdot \text{nr}(I)$. Since there are two direct summands here, and since d is a finite-dimensional $\mathbf{Q}$-algebra, then Theorem 8.5, states that $\mathcal{O} \oplus \mathcal{O} \cong \mathcal{O} \oplus I$.

Use Corner's theorem to construct an rtffr group G such that $\mathcal{O} = \text{End}(G)$ and let $H = IG$. Repeated uses of the Arnold–Lady theorem 1.13 proves the following implications. Since I is projective, $\text{Hom}(G, H) = I$. Since $\mathcal{O}$ is locally isomorphic to I, G is locally isomorphic to H. Since $\mathcal{O} \oplus \mathcal{O} \cong \mathcal{O} \oplus I$, $G \oplus G \cong G \oplus H$. Since $\mathcal{O}$ and I are not isomorphic, G and H are not isomorphic. Thus G is locally isomorphic to H but not isomorphic to H.

8.3 Direct sums of $\mathcal{L}$-groups

We introduce a bit of power.

Theorem 8.9. [9, E. L. Lady, Corollary 7.17] *Let H, K, and L be rtffr groups. If $H \oplus L$ is locally isomorphic to $K \oplus L$ then H is locally isomorphic to K.*

Theorem 8.10. [9, D.M. Arnold, Corollary 12.9(b)] *Let G, H, and K be rtffr groups. If G is locally isomorphic to $H \oplus K$ then $G \cong H' \oplus K'$ for some groups H' and K' which are locally isomorphic to H and K, respectively.*

Lemma 8.11. *Let G be an $\mathcal{L}$-group. Then $E(G)$ is an integrally closed ring.*

Proof: Let $E = E(G)$. By Lemma 2.9 there is a group $\overline{G} \doteq G$ such that $E(\overline{G})$ is an integrally closed ring. Because G is an $\mathcal{L}$-group, G is locally isomorphic to $\overline{G}$ and then Warfield's theorem 2.1 states that $G^{(e)} \cong \overline{G}^{(e)}$ for some integer $e > 0$. Then

$$\text{Mat}_{e \times e}(\text{End}(G)) \cong \text{End}(G^{(e)}) \cong \text{End}(\overline{G}^{(e)}) \cong \text{Mat}_{e \times e}(\text{End}(\overline{G}))$$

so that

$$\text{Mat}_{e \times e}(E(G)) \cong \text{Mat}_{e \times e}(E(\overline{G}))$$

is an integrally closed ring. Since *integrally closed* is a Morita invariant property [94], $E(G)$ is integrally closed. This completes the proof.

Lemma 8.12. *Let G be an rtffr $\mathcal{L}$-group. If G is locally isomorphic to $H \oplus K$ then H is an $\mathcal{L}$-group.*

Proof: Let G be locally isomorphic to $H \oplus K$ and suppose that $H' \doteq H$. Then G, $H \oplus K$, and $H' \oplus K$ are quasi-isomorphic. Since G is assumed to be an $\mathcal{L}$-group $H \oplus K$ and $H' \oplus K$ are locally isomorphic groups. Then Theorem 8.9 states that H is locally isomorphic to H', whence H is an $\mathcal{L}$-group. This completes the proof.

Lemma 8.13. *Let G be an rtffr $\mathcal{L}$-group. Then $G = G_1 \oplus \cdots \oplus G_t$ for some strongly indecomposable $\mathcal{L}$-groups $G_1, \ldots, G_t$.*

Proof: By Jónsson's theorem there are strongly indecomposable rtffr groups $G'_1, \cdots, G'_t$ such that $G \doteq G'_1 \oplus \cdots \oplus G'_t$. Since G is an $\mathcal{L}$-group, G is locally isomorphic to $G'_1 \oplus \cdots \oplus G'_t$. Then by an induction on the number of summands in Theorem 8.10 there are rtffr strongly indecomposable $G_1, \ldots, G_t$ such that G_i is locally isomorphic to G'_i for each $i = 1, \ldots, t$, and $G = G_1 \oplus \cdots \oplus G_t$. By Lemma 8.12 each G_i is an $\mathcal{L}$-group. This completes the proof.

Theorem 8.14. *Let $e > 0$ be an integer, let G be an indecomposable rtffr group, and assume that G^e is an $\mathcal{L}$-group. Then $\mathrm{End}(G^e)$ is a prime ring.*

Proof: Let G be an indecomposable rtffr group, and assume that G^e is an $\mathcal{L}$-group. By Lemma 8.12, G is an $\mathcal{L}$-group, so by Lemma 8.13 we have a strongly indecomposable rtffr $\mathcal{L}$-group G. Hence $E(G)$ is a domain, so that

$$E(G^e) = \mathrm{Mat}_{e \times e}(E(G))$$

is a prime ring. It then suffices to show that $\mathcal{N}(\mathrm{End}(G)) = 0$. We assume for the sake of contradiction that $\mathcal{N}(\mathrm{End}(G)) \neq 0$. The proof is a series of numbered implications and their proofs.

Let $E = \mathrm{End}(G)$, let $\mathcal{N} = \mathcal{N}(E)$, let $E(G) = E/\mathcal{N}$, and let $\mathcal{M} = \{r \in E \mid \mathcal{N} r = 0\}$. Then $E(G)$ is a *rtffr* ring, and $\mathcal{M}$ is an ideal in E that is pure as a subgroup of E.

Property 8.15. Since G is reduced and of finite rank the reader will prove that there is an integer $m > 0$ such that $\cap_{k>0} m^k G_m = 0$. Let n be some power of m.

Property 8.16. By hypothesis, G is strongly indecomposable. In particular, $\mathbf{Q}E(G)$ is a division algebra and $\mathbf{Q}\mathcal{N} \neq 0$ is the unique largest right ideal in $\mathbf{Q}E$. Specifically, because $\mathbf{Q}\mathcal{N}$ is nilpotent, $0 \neq \mathbf{Q}\mathcal{M} \subset \mathbf{Q}\mathcal{N}$. Thus $0 \neq E \cap \mathbf{Q}\mathcal{M} = \mathcal{M} \subset \mathcal{N}$.

Property 8.17. Let $I = nE + \mathcal{N}$ and let $J = \mathrm{Hom}(G, IG)$. We will show that $J_n = xE_n$ for some regular ($=$ nonzero divisor) element $x \in E_n$. Since I is an ideal of E, J is an ideal of E. Since G is an $\mathcal{L}$-group and since $nG \subset IG \subset G$, IG is locally isomorphic to G. By Warfield's theorem 2.1, there is an integer $f \neq 0$ such that $(IG)^f \cong G^f$, so $J^f \cong E^f$ as right E-modules. By the Arnold–Lady theorem 1.13 and Warfield's theorem 2.1, J is a projective right E-module that is locally isomorphic to E as a right module. Thus, there is an integer k relatively prime to n, and there are E-module maps $f_n : E \longrightarrow J$ and $g_n : J \longrightarrow E$ such that $f_n g_n = k1_{IG}$ and $g_n f_n = k1_G$. Localizing f_n and g_n at n shows that $E_n \cong J_n$ as right E_n-modules, so that $J_n = xE_n$ for some regular $x \in E_n$.

Property 8.18. Because G is strongly indecomposable and because $\mathcal{N} \subset I$,

$$\mathcal{N} = \mathrm{Hom}(G, \mathcal{N}G) \subset \mathrm{Hom}(G, IG) = J.$$

The element x was found in Property 8.17. The reader will show that because each $u \in \mathbf{Q}E \setminus \mathbf{Q}\mathcal{N}$ is a unit in $\mathbf{Q}E$, $y \in \mathcal{N}_n$ iff $xy \in \mathcal{N}_n$. By Property 8.17, $J_n = xE_n$ so that $\mathcal{N}_n = xK_n$ for some right ideal $K_n \subset E_n$. Hence $\mathcal{N}_n = K_n$ so that $\mathcal{N}_n = x\mathcal{N}_n$. By Property 8.16, $\mathcal{M} \subset \mathcal{N}$, so $\mathcal{M}_n = x\mathcal{M}_n$ in an analogous manner.

Property 8.19. We claim that $xE(G)_n = E(G)_n x$. By Property 8.17 and our choice of I there are quasi-equal ideals $nE_n \subset I_n \subset J_n = xE_n \subset E_n$, so

$$E_n \subset \operatorname{End}_{E_n}(J_n) = xE_n x^{-1}$$

are quasi-equal rings. Viewing these rings as subrings of $\mathbf{Q}E$ we have quasi-equal rings

$$E(G)_n \cong \frac{E_n + \mathbf{Q}\mathcal{N}}{\mathbf{Q}\mathcal{N}} \subset \frac{xE_n x^{-1} + \mathbf{Q}\mathcal{N}}{\mathbf{Q}\mathcal{N}} \cong xE(G)_n x^{-1}.$$

By Lemma 8.11, $E(G)$, and hence $E(G)_n$, are integrally closed rings. Thus $E(G)_n = xE(G)_n x^{-1}$ and hence $xE(G)_n = E(G)_n x$, as claimed. In other words, $xE_n + \mathcal{N}_n = E_n x + \mathcal{N}_n$.

Property 8.20. We claim that $\operatorname{Hom}_{E_n}(E_n, \mathcal{M}_n) = \operatorname{Hom}_{E_n}(J_n, \mathcal{M}_n)$. The inclusion

$$\operatorname{Hom}_{E_n}(E_n, \mathcal{M}_n) \subset \operatorname{Hom}_{E_n}(J_n, \mathcal{M}_n)$$

is the restriction map. Conversely, let $f \in \operatorname{Hom}_{E_n}(J_n, \mathcal{M}_n)$. By Property 8.17, $J_n = xE_n$, and by Property 8.18, $\mathcal{M}_n = x\mathcal{M}_n$, so

$$f \in \operatorname{Hom}_E(xE, x\mathcal{M}_n) = x\mathcal{M}_n x^{-1} \subset \mathbf{Q}\mathcal{M}.$$

By Property 8.16 and because $E_n \doteq J_n$ we have

$$f(\mathbf{Q}\mathcal{N}_n) \subset \mathbf{Q}\mathcal{N}_n \cdot f(\mathbf{Q}E_n) \subset \mathbf{Q}\mathcal{N}_n \cdot \mathbf{Q}\mathcal{M}_n = 0.$$

Hence $f(\mathcal{N}_n) = 0$. Then by Properties 8.17 and 8.19,

$$\begin{aligned}
\operatorname{Hom}_{E_n}(J_n, \mathcal{M}_n) &= \operatorname{Hom}_{E_n}(J_n/\mathcal{N}_n, \mathcal{M}_n) \\
&\overset{(8.17)}{=} \operatorname{Hom}_{E_n}((xE_n + \mathcal{N}_n)/\mathcal{N}_n, \mathcal{M}_n) \\
&\overset{(8.19)}{=} \operatorname{Hom}_{E_n}(xE(G)_n, \mathcal{M}_n) \\
&= \operatorname{Hom}_{E_n}(E(G)_n x, \mathcal{M}_n).
\end{aligned}$$

Thus there is a $u \in \mathcal{M}_n$ such that $f(rx) = ru$ for each $r \in E(G)_n$. By Property 8.18, $\mathcal{M}_n = x\mathcal{M}_n$, so the product $x^{-1}u$ exists in $\mathcal{M}_n$. The map f then lifts to a map $\bar{f} \in \operatorname{Hom}_{E_n}(E_n, \mathcal{M}_n)$ such that $\bar{f}(r) = r(x^{-1}u)$ for each $r \in E_n$. As claimed, $\operatorname{Hom}_{E_n}(E_n, \mathcal{M}_n) = \operatorname{Hom}_{E_n}(J_n, \mathcal{M}_n)$.

Property 8.21. By Properties 8.17 and 8.18, $xE_n = J_n \subset E_n$ and $\mathcal{M} = x\mathcal{M}$. Thus

$$\begin{aligned}
\mathcal{M}_n = \operatorname{Hom}_{E_n}(E_n, \mathcal{M}_n) &\overset{(8.20)}{=} \operatorname{Hom}_{E_n}(J_n, \mathcal{M}_n) \\
&= \operatorname{Hom}_{E_n}(xE_n, x\mathcal{M}_n) = x\mathcal{M}_n x^{-1}.
\end{aligned}$$

Then $\mathcal{M}_n x = x\mathcal{M}_n = \mathcal{M}_n$.

Property 8.22. Finally,

$$\mathcal{M}_n J_n G \overset{(8.17)}{=\!=} \mathcal{M}_n(xG) \overset{(8.21)}{=\!=} \mathcal{M}_n G$$

and $\mathcal{M}_n$ is nilpotent, Property 8.16, so $\mathcal{M}_n G$ is superfluous in $J_n G$. Then

$$J_n G = I_n G = n E_n G + \mathcal{M}_n G = n E_n G = n G_n.$$

Hence $\mathcal{M}_n G \subset n G_n$.

By Property 8.15, $n = m^k$ for some integer $k > 0$, so that $X_n = X_m$ for any torsion-free group X. Then by Property 8.22, $\mathcal{M}_m G_m = \mathcal{M}_n G_n \subset n G_n = m^k G_m$. Furthermore, $\bigcap_{k>0} m^k G_m = 0$. Thus $\mathcal{M} G \subset \mathcal{M}_m G_m = 0$, whence $\mathcal{M} = 0$. This contradiction to Property 8.16 shows us that $\mathcal{N} = 0$. I.e. E is a semi-prime ring, which completes the proof.

Theorem 8.23. *Let G be an indecomposable rtffr L-group. Then* $\mathrm{End}(G)$ *is a classical maximal order in the division ring* $\mathbf{Q}\mathrm{End}(G)$.

Proof: Since G is indecomposable, Lemma 8.13 implies that G is strongly indecomposable. Then $\mathbf{Q}\mathrm{End}(G)$ is a local ring, and so $\mathbf{Q}E(G)$ is a division algebra. Since G is an indecomposable rtffr L-group, Theorem 8.14 states that $\mathcal{N}(\mathrm{End}(G)) = 0$. Then $E(G) = \mathrm{End}(G)/\mathcal{N}(\mathrm{End}(G)) = \mathrm{End}(G)$, and by Lemma 8.11, $e(G)$ is integrally closed. Hence $\mathrm{End}(G)$ is a classical maximal order in the division algebra $\mathbf{Q}\mathrm{End}(G)$. This completes the proof.

8.4 Eichler L-groups are J-groups

Evidently, a $\mathcal{J}$-group, is an $\mathcal{L}$-group, is an $\mathcal{S}$-group. If $\mathrm{End}(G)$ is semi-prime then Lemma 8.24 shows us that an $\mathcal{L}$-group is a finitely faithful $\mathcal{S}$-group.

Lemma 8.24. *Let G be an rtffr L-group. If* $\mathrm{End}(G)$ *is semi-prime then G is a finitely faithful E-flat group.*

Proof: Suppose that $\mathrm{End}(G)$ is semi-prime and let I be a right ideal of $\mathrm{End}(G)$. By Lemma 8.11, $E(G) = \mathrm{End}(G)$ is integrally closed, hence an hereditary ring, whence I is a projective ($=$ flat) right $\mathrm{End}(G)$-module. An application of $I \otimes_{\mathrm{End}(G)} \cdot$ to the exact sequence

$$0 \to G \longrightarrow \mathbf{Q}G \longrightarrow \mathbf{Q}G/G \to 0$$

produces the exact sequence

$$0 = \mathrm{Tor}^1_{\mathrm{End}(G)}(I, \mathbf{Q}G/G) \longrightarrow I \otimes_{\mathrm{End}(G)} G \longrightarrow I \otimes_{\mathrm{End}(G)} \mathbf{Q}G$$

of groups. Consequently $I \otimes_{\mathrm{End}(G)} G \longrightarrow I \otimes_{\mathrm{End}(G)} \mathbf{Q}G : r \otimes x \longmapsto r \otimes x$ is an injection. Moreover, $\mathbf{Q}\mathrm{End}(G)$ is a semi-simple Artinian algebra, so that $\mathbf{Q}G$ is a projective left $\mathbf{Q}\mathrm{End}(G)$-module. Since $\mathbf{Q}\mathrm{End}(G)$ is a flat left $\mathrm{End}(G)$-module $\mathbf{Q}G$ is a flat left $\mathrm{End}(G)$-module. Thus the canonical map

$$I \otimes_{\mathrm{End}(G)} \mathbf{Q}G \longrightarrow \mathbf{Q}G : r \otimes x \longmapsto rx$$

is an injection, so that the composite map $I \otimes_{\mathrm{End}(G)} G \to G : r \otimes x \longmapsto rx$ is an injection. Hence G is a flat left $\mathrm{End}(G)$-module.

Next suppose that I is a maximal right ideal of finite index in $\mathrm{End}(G)$ such that $IG = G$. Since I is a finitely generated projective right $\mathrm{End}_R(G)$-module the Arnold Lady Theorem 1.13 implies that $I \cong \mathrm{End}(G)$, say $I = x\mathrm{End}(G)$ for some $x \in \mathrm{End}(G)$. Then

$$G = IG = x\mathrm{End}(G)G = xG.$$

Since G has finite rank x is an automorphism of G. That is $I = \mathrm{End}(G)$ so that G is finitely faithful. This completes the proof.

The next three results examine the implications $\mathcal{S}$-group $\Rightarrow$ $\mathcal{L}$-group $\Rightarrow$ $\mathcal{J}$-group.

Theorem 8.25. *Let $e > 0$ be an integer, let G be an indecomposable group, and let G^e be an Eichler group. The following are equivalent.*

1. *G^e is a $\mathcal{J}$-group.*
2. *G^e is an $\mathcal{L}$-group.*
3. *G^e is a finitely faithful $\mathcal{S}$-group.*

Proof: $\mathcal{J}$-group $\Rightarrow$ $\mathcal{L}$-group $\Rightarrow$ $\mathcal{S}$-group. Furthermore, the $\mathcal{L}$-group G^e has semi-prime endomorphism ring by Theorem 8.14. Thus $1 \Rightarrow 2 \Rightarrow 3$ follows from Lemma 8.24.

$3 \Rightarrow 1$. Since G^e is an Eichler group, this implication follows from Theorem 8.1. This completes the proof.

Theorem 8.26. *Eichler $\mathcal{L}$-groups are $\mathcal{J}$-groups.*

Proof: Let G be an rtffr $\mathcal{L}$-group and let $H \doteq G$. By Jónsson's theorem $G \doteq G' = G_1^{e_1} \oplus \cdots \oplus G_t^{e_t}$ for some nonzero, pairwise non-quasi-isomorphic, strongly indecomposable groups $G_1, \ldots, G_t$ and some integers $e_1, \ldots, e_t > 0$. Since G is an $\mathcal{L}$-group, G' is an $\mathcal{L}$-group, so that H is locally isomorphic to G'. By Arnold's theorem 8.10, $H = H_1 \oplus \cdots \oplus H_t$ where $H_1, \ldots, H_t$ are rtffr groups such that $G_i^{e_i}$ and H_i are locally isomorphic for $i = 1, \ldots, t$. By Lemma 8.12, $G_i^{e_i}$ is an $\mathcal{L}$-group, and since $e_i > 1$ if $\mathbf{Q}E(G_i)$ is a totally definite quaternion algebra (Lemma 8.6), we see that $\mathbf{Q}E(G_i^{e_i})$ is not a totally definite quaternion algebra. Then $G_i^{e_i}$ is an Eichler $\mathcal{L}$-group and G_i is indecomposable. An appeal to Theorem 8.25 shows us that $G_i^{e_i}$ is a $\mathcal{J}$-group. Hence $G_i^{e_i} \cong H_i$ for each $i = 1, \ldots, t$, so that $G' \cong H$. Therefore G' is a $\mathcal{J}$-group. But then $G \cong G'$ is a $\mathcal{J}$-group. This completes the proof.

Corollary 8.27. *An Eichler group G is a $\mathcal{J}$-group iff it is an $\mathcal{L}$-group.*

We will require the following result. We present only as much of it as we need to proceed.

Theorem 8.28. [46, Theorem 17.10] *The following are equivalent for an rtffr group G.*

1. $\text{Ext}(G, G)$ *is a torsion-free group.*
2. G *is a finitely faithful S-group.*
3. $r_p(G)^2 = r_p(\text{End}(G))$ *for each prime p, where $r_p(X)$ is the $\mathbf{Z}/p\mathbf{Z}$-dimension of X/pX.*
4. $\text{End}(G)/p\text{End}(G) \cong \text{Mat}_{r_p(G)}(\mathbf{Z}/p\mathbf{Z})$ *for each prime p.*

Theorem 8.29. *The following are equivalent for the rtffr group G.*

1. $\text{End}(G)$ *is semi-prime and G is an $\mathcal{L}$-group.*
2. G *is a finitely faithful S-group.*

Proof: Assume part 1: G is an $\mathcal{L}$-group with semi-prime endomorphism ring. By Lemma 8.24, G is finitely faithful. $\mathcal{L}$-groups are S-groups, so we have proved part 2.

Assume part 2. Let G be a finitely faithful S-group. By Theorem 8.28, $\text{Ext}^1_{\mathbf{Z}}(G, G)$ is a torsion-free group, hence $\text{Ext}^1_{\mathbf{Z}}(G^2, G^2)$ is a torsion-free group, whence G^2 is a finitely faithful S-group. By Lemma 8.6, G^2 is an Eichler group, so by Theorem 8.1, G^2 is a $\mathcal{J}$-group. Then by Lemma 8.12, G is an $\mathcal{L}$-group.

Because G is a finitely faithful S-group, Theorem 8.28 states that $\text{End}(G)/p\text{End}(G)$ is a simple Artinian algebra for each prime $p \in \mathbf{Z}$. Thus $\mathcal{N}(\text{End}(G)) \subset p\text{End}(G)$ for each prime $p \in \mathbf{Z}$. Consequently, $\mathcal{N}(\text{End}(G)) = p\mathcal{N}(\text{End}(G))$ for each prime $p \in \mathbf{Z}$. Since G is reduced, $\mathcal{N}(\text{End}(G)) = 0$, which completes the proof.

Example 8.30. Here is an example of an $\mathcal{L}$-group with nonzero nilradical. Let H be an indecomposable $\mathcal{J}$-group as in Example 8.2. Then $\text{Hom}(H, \mathbf{Z}) = 0$ so that $G = \mathbf{Z} \oplus H$ is a $\mathcal{J}$-group with nonzero nilradical $\text{Hom}(\mathbf{Z}, H)$.

8.5 Exercises

Let G and H denote rtffr groups. Let $E(G) = \text{End}(G)/\mathcal{N}(\text{End}(G))$. Let E be an rtffr ring. Let $S = \text{center}(E)$. Given a semi-prime rtffr ring E let $\overline{E}$ be an integrally closed subring of $\mathbf{Q}E$ such that $E \subset \overline{E}$ and $\overline{E}/E$ is finite. Let τ be the conductor of E.

1. Construct an example of a strongly indecomposable $\mathcal{L}$-group that is not a Murley group. Hint: Use Theorem 1.12.
2. Let $D = \mathbf{Q}1 \oplus \mathbf{Q}i \oplus \mathbf{Q}j \oplus \mathbf{Q}k$ where i, j, k are the *Hamiltonian quaternions*. $i^2 = j^2 = k^2 = ijk = -1$.
   (a) Show that D is a division $\mathbf{Q}$-algebra of dimension 4.
   (b) Let H be the *Hurwitz quaternions*. That is, H is the subring of D generated by $\frac{1}{2}(1 + i + j + k), 1, i, j, k$. Show that H is a maximal $\mathbf{Z}$-order in D.
   (c) Let E be the subring of H generated by $1, i, j, k$. Show that E is a Noetherian hereditary prime ring.

3. Suppose that G has finite index in $G_1^{e_1} \oplus \cdots \oplus G_t^{e_t}$ for some integers $t, e_1, \ldots, e_t$ and some strongly indecomposable rtffr groups $G_1, \ldots, G_t$. Then

   (a) $\mathbf{Q}E(G_1) = D_1$ is a division $\mathbf{Q}$-algebra,
   (b) $\mathbf{Q}E(G_1^{e_1}) \cong \mathrm{Mat}_{e_1 \times e_1}(D_1)$, and
   (c) $\mathbf{Q}E(G) \cong \mathrm{Mat}_{e_1 \times e_1}(D_1) \times \cdots \times \mathrm{Mat}_{e_t \times e_t}(D_t)$.

4. Let H be the ring of Hurwitz quaternions given in Exercise 2.

   (a) Show that $H = E(G)$ for some strongly indecomposable rtffr group G.
   (b) Show that G is not an Eichler group.
   (c) Find an example of a right iedal $I \subset H$ such that $I \ncong H$.
   (d) Show that $G \oplus G$ is an Eichler group.
   (e) Show that $G \oplus G \cong G \oplus H$ implies that $G \cong H$.

5. Fill in the details of the proof of Theorem 8.14.

8.6 Problems for future research

1. Please find a shorter proof of Theorem 8.26.
2. Prove that an $\mathcal{L}$-group is a finitely faithful $\mathcal{S}$-group.

9

Modules and homotopy classes

9.1 Right endomorphism modules

We fix for the duration of this text an associative ring R and a right R-module G. The results of this chapter are found in [50, 48, 42].

The existence in Theorem 1.14 of a category equivalence $\mathbf{H}_G(\cdot) : \mathbf{P}_o(G) \longrightarrow \mathbf{P}_o(\mathrm{End}_R(G))$ allows us to translate questions on direct summands of $G^{(n)}$ for integers $n > 0$ into questions on finitely generated projective modules. This translation often results in a significant reduction in difficulty since $P \in \mathbf{P}_o(\mathrm{End}_R(G))$ is finitely generated while $H \in \mathbf{P}_o(G)$ is in general not finitely generated over any ring. In this chapter we will extend the category equivalence $\mathbf{H}_G(\cdot) : \mathbf{P}_o(G) \longrightarrow \mathbf{P}_o(\mathrm{End}_R(G))$ to larger classes of right $\mathrm{End}_R(G)$-modules. Our rationale for seeking such a theorem is as follows. Since $\mathbf{H}_G \mathbf{T}_G(P) \cong P$ for each $P \in \mathbf{P}_o(\mathrm{End}_R(G))$ and since each finitely generated right $\mathrm{End}_R(G)$-module M possesses a projective resolution some of whose terms are in $\mathbf{P}_o(\mathrm{End}_R(G))$ then it ought to be true that some category containing $\mathbf{P}_o(G)$ is equivalent to $\mathbf{Mod}\text{-}\mathrm{End}_R(G)$.

In this chapter we characterize functorially the category of right $\mathrm{End}_R(G)$-modules in terms of G. This characterization allows us to characterize ring-theoretic properties of $\mathrm{End}_R(G)$ in terms of G and group-theoretic properties of G in terms of $\mathrm{End}_R(G)$. In doing so we give partial solutions to the question *Which properties of G are induced by, (or conversely, induce), properties of $\mathrm{End}_R(G)$?* The literature on this question is extensive. See the references.

9.1.1 The homotopy of G-plexes

We discuss the groups and functors surrounding complexes.

The right H-module is *(finitely) G-generated* if H is a homomorphic image of a direct sum of (finitely many) copies of G, or equivalently if there is a $Q \in \mathbf{P}(G)$ (or $Q \in \mathbf{P}_o(G)$) and an R-submodule $K \subset Q$ such that $H \cong Q/K$.

Given a right R-module H *the G-socle of H* is

$$\mathbf{S}_G(H) = \sum \{f(G) \mid f \in \mathrm{Hom}_R(G, H)\}.$$

The G-socle of H is the largest G-generated R-submodule of H.

Given integers $k > 0$, the complex

$$C = \cdots C_{k+1} \xrightarrow{\gamma_{k+1}} C_k \xrightarrow{\gamma_k} C_{k-1} \cdots \xrightarrow{\gamma_1} C_0$$

is said to be *exact at k* if image $\gamma_{k+1} = \ker \gamma_k$. We assume that C is exact at 0 by attaching the cokernel map

$$\cdots C_0 \xrightarrow{\gamma_0} C_0/\text{image } \gamma_1 \longrightarrow 0.$$

We say that C is *exact* if C is exact at all integers $k > 0$.

The complex

$$P = \cdots \xrightarrow{\partial_3} P_2 \xrightarrow{\partial_2} P_1 \xrightarrow{\partial_1} P_0$$

is called a *projective resolution* if P is exact (i.e., image $\partial_{k+1} = \ker \partial_k$), and P_k is a projective right $\text{End}_R(G)$-module for each $k \geq 0$. P_k is called the *k-th term* of P. If $M \cong \text{coker } \partial_1 = P_0/\text{image } \partial_1$ then P is called a *projective resolution for M*. In this text a projective resolution P is usually one of $\text{End}_R(G)$-modules.

A complex

$$Q = \cdots \xrightarrow{\delta_3} Q_2 \xrightarrow{\delta_2} Q_1 \xrightarrow{\delta_1} Q_0$$

of right R-modules is called a *G-plex* if

1. $Q_k \in \mathbf{P}(G)$ for each $k \geq 0$, and
2. G has the following lifting property for each $k \geq 1$. Given a map $\phi : G \longrightarrow Q_k$ such that $\delta_k \phi = 0$ there is a map $\psi : G \longrightarrow Q_{k+1}$ such that $\phi = \delta_{k+1}\psi$ as in the commutative triangle below.

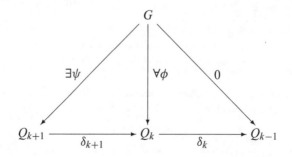

If $H = \text{coker } \delta_1 = Q_0/\text{image } \delta_1$ then Q *is a G-plex for H*. Q_k is called the *k-th term* of Q.

Let Q be a G-plex. The *homology groups* for Q are

$$H_k(Q) = \ker \delta_k / \text{image } \delta_{k+1} \text{ for integers } k > 0$$
$$H_0(Q) = \text{coker } \delta_1 = Q_0 / \text{image } \delta_1.$$

Notice that $H_k(Q)$ is calculated at Q_k for integers $k \geq 0$.

Remark 9.1. Unless otherwise stated

$$\boxed{\mathcal{P}, \, \mathcal{Q}, H_k(\mathcal{Q}), \partial_k, \delta_k, P_k, \text{ and } Q_k}$$

refer to *the complexes, maps,* and *modules defined above. In particular,* $\mathcal{P}$ and $\mathcal{Q}$ have infinitely many (possibly trivial) terms.

Given a projective resolution $\mathcal{P}$, we can apply the right exact additive functor $\mathbf{T}_G(\cdot) = \cdot \otimes_{\text{End}_R(G)} G$ to produce the complex

$$\boxed{\mathbf{T}_G(\mathcal{P}) = \cdots \xrightarrow{\partial_2 \otimes 1} \mathbf{T}_G(P_1) \xrightarrow{\partial_1 \otimes 1} \mathbf{T}_G(P_0).}$$

Given a G-plex $\mathcal{Q}$, we can apply the left exact additive functor $\mathbf{H}_G(\cdot)$ to produce the complex

$$\boxed{\mathbf{H}_G(\mathcal{Q}) = \cdots \xrightarrow{\delta_2^*} \mathbf{H}_G(Q_1) \xrightarrow{\delta_1^*} \mathbf{H}_G(Q_0)}$$

where $\delta^* = \text{Hom}_R(G, \delta)$.

Let G be a right R-module, let Q be a G-plex. For each integer $k \geq 1$ there is a short exact sequence

$$0 \longrightarrow \ker \delta_k \longrightarrow Q_k \xrightarrow{\delta_k} \text{image } \delta_k \longrightarrow 0 \tag{9.1}$$

We will say that a short exact sequence

$$0 \longrightarrow K \longrightarrow Q \xrightarrow{\delta} H \longrightarrow 0 \tag{9.2}$$

of right R-modules is *G-balanced* if G is projective relative to (9.2).

Lemma 9.2. *Let Q be a G-plex. Given an integer $k \geq 2$, each $Q \in \mathbf{P}(G)$ is projective relative to (9.1).*

Proof: Let $\phi : Q \to H_k$ be an R-module map and write $Q \oplus Q' = G^{(c)}$ for some cardinal c. The usual properties of direct sums shows us that we can lift ϕ if we can lift the restrictions $\phi|G$ for each copy of G in $G^{(c)}$. We have reduced our goal to proving that $Q = G$ is projective relative to (9.1). But G is projective relative to (9.1) by item 3 of the definition of G-plex.

The exactness and inexactness of the complexes $\mathbf{T}_G(\mathcal{P})$ and $\mathbf{H}_G(\mathcal{Q})$ are often interesting.

Lemma 9.3. *Let $\mathcal{Q}$ be a complex in which $Q_k \in \mathbf{P}(G)$ for each $k \geq 0$. Then $\mathcal{Q}$ is a G-plex iff $\mathbf{H}_G(\mathcal{Q})$ is exact.*

Proof: Let $\mathcal{Q}$ be a G-plex and assume that $\mathcal{Q}$ is a G-plex. The reader can show as an exercise that the lifting property in item 2 of the definition of G-plex implies that image $\delta_{k+1}^* = \ker \delta_k^*$.

Conversely if $\mathbf{H}_G(\mathcal{Q})$ is exact then the exactness of $\mathbf{H}_G(\mathcal{Q})$ is exactly the lifting property in item 2 of the definition of G-plex. This completes the proof. ∎

The next result guarantees that there are enough G-plexes.

Lemma 9.4. *Let G be a right R-module.*

1. *Let H be a G-generated R-module. There is a G-balanced short exact sequence (9.2) in which $Q \cong G^{(\mathcal{I})}$ for some index $\mathcal{I}$.*
2. *Let $\delta : Q \to H$ be a surjection in which $Q \in \mathbf{P}(G)$. There is a G-plex $\mathcal{Q}$ such that $Q_0 = Q$ and image $\delta_1 \cong \mathbf{S}_G(\ker \delta)$.*
3. *Let $Q_1, Q_0 \in \mathbf{P}(G)$, and let $\delta : Q_1 \to Q_0$. There is a G-plex $\mathcal{Q}$ such that $\delta_1 = \delta$.*

Proof: 1. Suppose that H is G-generated. For each $f \in \mathbf{H}_G(H)$ let $\alpha_f : G_f \longrightarrow G$ be an isomorphism, let

$$Q = \bigoplus_{f \in \mathrm{Hom}_R(G,H)} G_f,$$

let $\jmath_f : G_f \longrightarrow Q$ be the canonical injection, and using the universal property of direct sums choose a unique map $\delta : Q \longrightarrow H$ such that $\delta \jmath_f = f\alpha_f$ for each $f \in \mathbf{H}_G(H)$. Since H is G-generated, δ is a surjection. Let $g : G \longrightarrow H$ be an R-module map. Then $g = \delta(\jmath_g \alpha_g^{-1})$, so that δ is G-balanced.

2. Since H is G-presented, there is a short exact sequence (9.2) in which $Q = Q_0 \in \mathbf{P}(G)$ and $K = \ker \delta$ is G-generated. By part 1 we construct a G-balanced surjection $\delta_1 : Q_1 \longrightarrow K$ in which $Q_1 \cong G^{(\mathcal{I})}$ for some index set $\mathcal{I}$. Using induction, for each $k \geq 1$ we construct a G-balanced surjection $\delta_{k+1} : Q_{k+1} \longrightarrow \mathbf{S}_G(\ker \delta_k)$. Then

$$\cdots \xrightarrow{\delta_3} Q_2 \xrightarrow{\delta_2} Q_1 \xrightarrow{\delta_1} Q_0$$

is a G-plex such that $H \cong \mathrm{coker}\, \delta_1$.

3. Say $\delta : Q_1 \longrightarrow Q_0$ is as given in part 3 and let $\ker \delta = K_1$. By part 1 there is a G-balanced short exact sequence

$$0 \longrightarrow K_2 \longrightarrow Q_2 \xrightarrow{\delta_2} \mathbf{S}_G(K_1) \longrightarrow 0$$

of right R-modules where $Q_2 \in \mathbf{P}(G)$. Inductively, using part 1 construct for each $k \geq 1$ a G-balanced short exact sequence

$$0 \longrightarrow K_{k+1} \longrightarrow Q_{k+1} \overset{\delta_{k+1}}{\longrightarrow} \mathbf{S}_G(K_k) \longrightarrow 0$$

of right R-modules where $Q_{k+1} \in \mathbf{P}(G)$. Then

$$\cdots \overset{\delta_3}{\longrightarrow} Q_2 \overset{\delta_2}{\longrightarrow} Q_1 \overset{\delta}{\longrightarrow} Q_0$$

is a G-plex whose first map is δ. This completes the proof.

Lemma 9.5. *Let G be a self-small right R-module.*

1. *Given a projective resolution $\mathcal{P}$ then $\mathbf{T}_G(\mathcal{P})$ is a G-plex.*
2. *Given a right $\mathrm{End}_R(G)$-module M there is a G-plex $\mathcal{Q}$ such that $\mathbf{H}_G(\mathcal{Q})$ is a projective resolution for M.*

Proof: 1. Let $\mathcal{P}$ be a projective resolution, let $M \cong \mathrm{coker}\,\partial_1$, and form $\mathbf{H}_G\mathbf{T}_G(\mathcal{P})$. There is a commutative diagram

of right $\mathrm{End}_R(G)$-modules whose top row is exact. Since P_k is a direct summand of $\oplus_{\mathcal{I}}\mathrm{End}_R(G)$ for some index set $\mathcal{I}$, $\mathbf{T}_G(P_k)$ is a direct summand of $\mathbf{T}_G(\oplus_{\mathcal{I}}\mathrm{End}_R(G)) \cong \oplus_{\mathcal{I}}G$. Then $\mathbf{T}_G(P_k) \in \mathbf{P}(G)$. Furthermore, Theorem 1.13 states that Ψ_{P_k} is an isomorphism for each integer $k \geq 0$. Thus the bottom row of (9.3) is exact and hence by Lemma 9.3, $\mathbf{T}_G(\mathcal{P})$ is a G-plex.

2. Let M be a right $\mathrm{End}_R(G)$-module and let $\mathcal{P}$ be a projective resolution for M. By part 1, $\mathbf{T}_G(\mathcal{P})$ is a G-plex and, by Theorem 1.13, Ψ_{P_k} is an isomorphism for each $k \geq 0$. Then by diagram (9.3), $\mathcal{P} \cong \mathbf{H}_G\mathbf{T}_G(\mathcal{P})$. Hence by Lemma 9.3, $\mathcal{Q} = \mathbf{T}_G(\mathcal{P})$ is a G-plex such that $\mathbf{H}_G(\mathcal{Q})$ is a projective resolution for M. This completes the proof.

We will often refer to the following technical result.

Lemma 9.6. *Let G be a right R-module and let $\mathcal{Q} \in G$-**Plex**.*

1. *image $\delta_{k+1} = \mathbf{S}_G(\ker \delta_k)$ for each integer $k > 0$.*
2. *$H_k(\mathcal{Q}) = \ker \delta_k / \mathbf{S}_G(\ker \delta_k)$ for each integer $k > 0$.*

Proof: 1. Inasmuch as image δ_{k+1} is G-generated, image $\delta_{k+1} \subset \mathbf{S}_G(\ker \delta_k)$. On the other hand given a map $\phi : G \longrightarrow \ker \delta_k$ there is, by the definition of G-plex, a commutative diagram

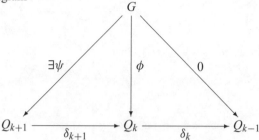

of right R-modules. Then $\phi(G) \subset$ image δ_{k+1} so that $\mathbf{S}_G(\ker \delta_k) \subset$ image δ_{k+1}. Hence image $\delta_{k+1} = \mathbf{S}_G(\ker \delta_k)$.

Part 2 follows immediately from part 1.

9.1.2 Homotopy and homology

We will refer to the next two diagrams often.

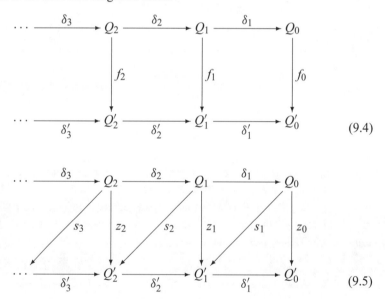

A *chain map* $f : Q \longrightarrow Q'$ is a sequence $f = (\cdots, f_2, f_1, f_0)$ of maps such that the diagram (9.4) commutes.

The chain map $z = (\cdots, z_2, z_1, z_0) : Q \to Q'$ is *null homotopic* if there is a diagram (9.5) in which the square boundaries commute (i.e., $z_{k-1}\delta_k = \delta'_k z_k$ for each integer $k > 0$), and

$$z_k = \delta'_{k+1} s_{k+1} + s_k \delta_k \text{ for each } k \geq 0 \text{ where } s_0 = \delta_0 = 0.$$

The sequence $s = (\cdots, s_3, s_2, s_1)$ is called a *homotopy for z*. We say that the chain maps f and g are *homotopic*, or that f *is homotopic to g* if $f - g$ is a null homotopic chain map. Evidently *is homotopic to* is an additive equivalence relation on the set of chain maps $f : Q \longrightarrow Q'$. Let $[f]$ denote *the homotopy (equivalence) class* of f. We say that Q and Q' have *the same homotopy type*, or equivalently that Q *has the homotopy type of* Q', if there are chain maps $f : Q \longrightarrow Q'$ and $g : Q' \longrightarrow Q$ such that $[1_Q] = [gf]$ and $[1_{Q'}] = [fg]$.

Let

> G-**Plex** = the category whose objects are G-plexes Q and whose morphisms $[f] : Q \longrightarrow Q'$ are the homotopy classes $[f]$ of chain maps $f : Q \longrightarrow Q'$.

Lemma 9.7. *Let G be a right R-module.*

1. G-**Plex** *is an additive category.*
2. *The G-plexes Q and Q' have the same homotopy type iff $Q \cong Q'$ in G-**Plex**.*

Proof: The zero object in G-**Plex** is $\cdots \longrightarrow 0 \longrightarrow 0$ and the zero map is $[\cdots, 0, 0]$. The rest of the proof is a routine matter of verifying that G-**Plex** satisfies the properties of an additive category.

Lemma 9.8. *Let P and P' be two projective resolutions of right R-modules M and M', respectively.*

1. $M \cong M'$ *iff P and P' have the same homotopy type.*
2. *If $\phi : M \longrightarrow M'$ then ϕ lifts to a chain map $f : P \longrightarrow P'$. If g is another lifting of ϕ then $[f] = [g]$.*

Proof: See [97].

For a proof of the next result see [97], or try to prove it without peeking.

Lemma 9.9. *Let C and C' be complexes of modules.*

1. *If $(\cdots, f_1, f_0) : C \longrightarrow C'$ is a chain map then for each $k \geq 0$, f induces a module map $H_k(f) : H_k(C) \longrightarrow H_k(C')$.*
2. *If f and g : C \longrightarrow C' are homotopic chain maps then for each $k \geq 0$, $H_k(f) = H_k(g)$.*
3. *If C and C' have the same homotopy type then for each $k \geq 0$, $H_k(C) \cong H_k(C')$.*

The following result will be used when we have occasion to induce chain maps between G-plexes.

Lemma 9.10. *Let Q, $Q' \in G\text{-}\mathbf{Plex}$, and let*

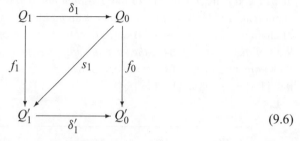

$$(9.6)$$

be a diagram of right R-modules such that $f_0\delta_1 = \delta_1'f_1$. Then

1. *If the square boundary commutes then (f_1, f_0) lifts to a chain map $f : Q \longrightarrow Q'$.*
2. *If $f_0 = \delta_1' s_1$ then f is a null homotopic map.*
3. *f is unique up to homotopy equivalence.*

Proof: 1. Given the square (9.6) with commutative boundary construct a commutative diagram (9.4) of right R-modules as follows. Because the first (= right most) square in (9.4) is commutative and because the rows in (9.4) are complexes, we have $\delta_1'f_1\delta_2 = f_0\delta_1\delta_2 = 0$. Thus by Lemma 9.6,

$$\text{image } f_1\delta_2 \subset \mathbf{S}_G(\ker \delta_1') = \text{image } \delta_2'.$$

Because Q is G-balanced, and because $Q_2 \in \mathbf{P}(G)$, there is a map $f_2 : Q_2 \longrightarrow Q_2'$ such that

$$f_2\delta_2' = f_1\delta_2$$

(see Lemma 9.2). Proceed inductively, constructing maps $f_k : Q_k \longrightarrow Q_k'$ such that the k-th square in (9.4) commutes. Then $f = (\cdots, f_2, f_1, f_0)$ is a chain map $f : Q \longrightarrow Q'$.

    2. Proceed as in part 1 lifting s_1 to a sequence of maps $(\cdots, s_2, s_1)$ as in the diagram (9.5) such that $f_k - s_k\delta_k = \delta_{k+1}s_{k+1}$. Hence f is a null homotopic chain map.

    3. If $f = (\cdots, f_2, f_1, f_0)$, $g = (\cdots, g_2, f_1, f_0) : Q \longrightarrow Q'$ are chain map liftings of (f_1, f_0) then $f - g$ is the chain map $(\cdots, f_2 - g_2, 0, 0)$ so that $s_1 = 0$ exists as in (9.6). By part 2 $f - g$ is null homotopic and thus $[f] = [g]$. This completes the proof.

9.1.3 Endomorphism modules as G-plexes

The theorems in this section do nothing less than characterize $\mathbf{Mod}\text{-}\text{End}_R(G)$. The additive functor $F(\cdot) : \mathbf{C} \longrightarrow \mathbf{D}$ is a *category equivalence* if

1. $F(\cdot)$ induces an isomorphism

$$\text{Hom}_{\mathbf{C}}(X, Y) \cong \text{Hom}_{\mathbf{D}}(F(X), F(Y))$$

    for all objects X, Y in $\mathbf{C}$, and
2. To each object $Z \in \mathbf{D}$ there is an object $X \in \mathbf{C}$ such that $F(X) \cong Z$ in $\mathbf{D}$.

$F(\cdot)$ is *full* if the induced map $\phi \mapsto F(\phi)$ in part 1 above is a surjection. $F(\cdot)$ is *faithful* if this map is an injection.

Theorem 9.11. [48, T. G. Faticoni] *Let G be a right R-module. If G is self-small then $\mathbf{H}_G(\cdot)$ lifts to a category equivalence*

$$\mathrm{h}_G(\cdot) : G\text{-}\mathbf{Plex} \longrightarrow \mathbf{Mod}\text{-}\mathrm{End}_R(G).$$

Proof: To begin with we will define $\mathrm{h}_G(\cdot)$. Given a map x let $x^* = \mathbf{H}_G(x)$.

Let $f = (\cdots, f_1, f_0) : \mathcal{Q} \longrightarrow \mathcal{Q}'$ be a chain map in G-**Plex**. An application of $\mathbf{H}_G(\cdot)$ produces the unique chain map

$$f^* = (\cdots, f_1^*, f_0^*) : \mathbf{H}_G(\mathcal{Q}) \longrightarrow \mathbf{H}_G(\mathcal{Q}').$$

By part 1 of Lemma 9.9 there is a unique map

$$H_0(f^*) : H_0(\mathbf{H}_G(\mathcal{Q})) \longrightarrow H_0(\mathbf{H}_G(\mathcal{Q}'))$$

of right $\mathrm{End}_R(G)$-modules. The functor $\mathrm{h}_G(\cdot) : G\text{-}\mathbf{Plex} \longrightarrow \mathbf{Mod}\text{-}\mathrm{End}_R(G)$ is then defined by

$$\mathrm{h}_G(\mathcal{Q}) = H_0(\mathbf{H}_G(\mathcal{Q})) \text{ for each } \mathcal{Q} \in G\text{-}\mathbf{Plex}$$

$$\mathrm{h}_G([f]) = H_0(f^*) \text{ for each chain map } f : \mathcal{Q} \longrightarrow \mathcal{Q}'.$$

We claim that $\mathrm{h}_G(\cdot) : G\text{-}\mathbf{Plex} \longrightarrow \mathbf{Mod}\text{-}\mathrm{End}_R(G)$ is well-defined. Obviously given a G-plex $\mathcal{Q}$ there is exactly one right $\mathrm{End}_R(G)$-module $\mathrm{h}_G(\mathcal{Q}) = H_0(\mathbf{H}_G(\mathcal{Q}))$. Let $f, g : \mathcal{Q} \longrightarrow \mathcal{Q}'$ be homotopic chain maps, and let $f = (\cdots, f_1, f_0)$. Since $\mathcal{Q}$ is a G-plex and since G is self-small, Theorem 1.13 and Lemma 9.3 show us that $\mathbf{H}_G(\mathcal{Q})$ is a projective resolution of right $\mathrm{End}_R(G)$-modules. Since $\mathbf{H}_G(\cdot)$ is an additive functor $f^*, g^* : \mathbf{H}_G(\mathcal{Q}) \longrightarrow \mathbf{H}_G(\mathcal{Q}')$ are homotopic chain maps of projective resolutions over $\mathrm{End}_R(G)$. Then by Lemma 9.9, f^* and g^* induce the same map $H_0(f^*) : H_0(\mathbf{H}_G(\mathcal{Q})) \longrightarrow H_0(\mathbf{H}_G(\mathcal{Q}'))$. Hence $\mathrm{h}_G([f]) : \mathrm{h}_G(\mathcal{Q}) \longrightarrow \mathrm{h}_G(\mathcal{Q}')$ is well defined. As claimed $\mathrm{h}_G(\cdot) : G\text{-}\mathbf{Plex} \longrightarrow \mathbf{Mod}\text{-}\mathrm{End}_R(G)$ is a well-defined functor.

To prove that $\mathrm{h}_G(\cdot) : G\text{-}\mathbf{Plex} \longrightarrow \mathbf{Mod}\text{-}\mathrm{End}_R(G)$ is a category equivalence we will show that to each $M \in \mathbf{Mod}\text{-}\mathrm{End}_R(G)$ there is a $\mathcal{Q} \in G\text{-}\mathbf{Plex}$ such that $\mathrm{h}_G(\mathcal{Q}) \cong M$ and that $\mathrm{h}_G(\cdot) : G\text{-}\mathbf{Plex} \longrightarrow \mathbf{Mod}\text{-}\mathrm{End}_R(G)$ is a fully faithful functor.

Because G is self-small Lemma 9.5 states that given a right $\mathrm{End}_R(G)$-module M there is a $\mathcal{Q} \in G\text{-}\mathbf{Plex}$ such that $\mathbf{H}_G(\mathcal{Q})$ is a projective resolution for M. Thus

$$\mathrm{h}_G(\mathcal{Q}) = H_0(\mathbf{H}_G(\mathcal{Q})) \cong M.$$

Let $\phi : M \longrightarrow M'$ be a map of right $\mathrm{End}_R(G)$-modules. By Lemma 9.5 there are projective resolutions $\mathbf{H}_G(\mathcal{Q})$ and $\mathbf{H}_G(\mathcal{Q}')$ for M and M' respectively for some

$Q, Q' \in G$-**Plex**. Then ϕ lifts to a commutative diagram

$$
\begin{array}{ccccccc}
\cdots \xrightarrow{\;\delta_2^*\;} & H_G(Q_1) & \xrightarrow{\;\delta_1^*\;} & H_G(Q_0) & \xrightarrow{\;\partial_0\;} & M \\
 & \downarrow{\scriptstyle g_1 = f_1^*} & & \downarrow{\scriptstyle g_0 = f_0^*} & & \downarrow{\scriptstyle \phi = h_G([f])} \\
\cdots \xrightarrow{\;(\delta_2')^*\;} & H_G(Q_1') & \xrightarrow{\;(\delta_1')^*\;} & H_G(Q_0') & \xrightarrow{\;\partial_0'\;} & M'
\end{array}
\qquad (9.7)
$$

where ∂_0 and ∂_0' are the cokernel maps. Two applications of Theorem 1.13 show that for each integer $k \geq 0$ there are maps $f_k : Q_k \longrightarrow Q_k'$ such that $f_k^* = g_k$, and there is a commutative diagram (9.4) representing the chain map $f = (\cdots, f_1, f_0)$. By our definition of $h_G(\cdot)$, $h_G([f]) = \phi$, and so $h_G(\cdot) : G$-**Plex** $\longrightarrow$ **Mod**-End$_R(G)$ is a full functor.

To see that $h_G(\cdot)$ is faithful suppose that $\phi = h_G([f]) = 0$ for some chain map $f : Q \longrightarrow Q'$ as represented in (9.4). An application of $H_G(\cdot)$ yields the commutative diagram (9.7) where $\phi = 0$. Then

$$\text{image } g_0 \subset \ker \partial_0' = \text{image } (\delta_1')^*.$$

Because G is self-small, $H_G(Q_0)$ is a projective right R-module (Theorem 1.13), so there is a map $r_1 : H_G(Q_0) \longrightarrow H_G(Q_1')$ such that

$$(\delta_1')^* r_1 = g_0.$$

Another appeal to Theorem 1.13 produces a map $s_1 : Q_0 \longrightarrow Q_1'$ such that $s_1^* = r_1$, such that the square boundary in the diagram (9.6) commutes, (i.e. $f_0 \delta_1 = \delta_1' f_1$) and in which $\delta_1' s_1 = f_0$. Then by Lemma 9.10, $[f] = [0]$ and we conclude that $h_G(\cdot) : G$-**Plex** $\longrightarrow$ **Mod**-End$_R(G)$ is a faithful functor. Therefore $h_G(\cdot) : G$-**Plex** $\longrightarrow$ **Mod**-End$_R(G)$ is a category equivalence and the proof is complete.

Functors $F : \mathbf{C} \longrightarrow \mathbf{D}$ and $G : \mathbf{C} \longrightarrow \mathbf{D}$ are *inverse category equivalences* if the composites $F \circ G$ and $G \circ F$ are naturally equivalent to the respective identity functors.

Now that we know that $h_G(\cdot)$ is a category equivalence when G is self-small it is natural to ask for the inverse functor.

Theorem 9.12. [48, T. G. Faticoni] *Let G be a self-small right R-module. The functors* $H_G(\cdot)$ *and* $T_G(\cdot)$ *induce inverse category equivalences*

$$h_G(\cdot) : G\text{-}\mathbf{Plex} \longrightarrow \mathbf{Mod}\text{-End}_R(G)$$

$$t_G(\cdot) : \mathbf{Mod}\text{-End}_R(G) \longrightarrow G\text{-}\mathbf{Plex}.$$

Proof: Most of our effort in this proof is directed at defining $t_G(\cdot)$. For each $M \in$ **Mod**-End$_R(G)$ choose a fixed projective resolution $\mathcal{P}(M)$ of M. By Lemma 9.5, $T_G(\mathcal{P}(M))$ is a G-plex so define $t_G(M) = T_G(\mathcal{P}(M))$.

We are given a map $\phi : M \longrightarrow M'$ in **Mod**-$\text{End}_R(G)$. Since P_k is projective for $k \geq 0$, ϕ lifts to a chain map $f = (\cdots, f_1, f_0)$ as in diagram (9.8).

$$
\begin{array}{ccccccc}
\cdots \xrightarrow{\partial_2} & P_1 & \xrightarrow{\partial_1} & P_0 & \xrightarrow{\partial_0} & M \\
& \downarrow{f_1} & & \downarrow{f_0} & & \downarrow{\phi} \\
\cdots \xrightarrow{\partial'_2} & P'_1 & \xrightarrow{\partial'_1} & P'_0 & \xrightarrow{\partial'_0} & M'
\end{array}
\qquad (9.8)
$$

An application of $\mathbf{T}_G(\cdot)$ yields the commutative diagram (9.9) of right R-modules. We define

$$
t_G(f) = [(\cdots, \mathbf{T}_G(f_1), \mathbf{T}_G(f_0))] = [\mathbf{T}_G(f)]
$$

as the homotopy equivalence class of the chain map $\mathbf{T}_G(f)$.

$$
\begin{array}{ccccc}
\cdots \xrightarrow{\mathbf{T}_G(\partial_2)} & \mathbf{T}_G(P_1) & \xrightarrow{\mathbf{T}_G(\partial_1)} & \mathbf{T}_G(P_0) \\
& \downarrow{\mathbf{T}_G(f_1)} & & \downarrow{\mathbf{T}_G(f_0)} \\
\cdots \xrightarrow{\mathbf{T}_G(\partial'_2)} & \mathbf{T}_G(P'_1) & \xrightarrow{\mathbf{T}_G(\partial'_1)} & \mathbf{T}_G(P'_0)
\end{array}
\qquad (9.9)
$$

We then define $t_G(\cdot)$ as follows.

$$
\boxed{
\begin{array}{l}
t_G(M) = \mathbf{T}_G(\mathcal{P}(M)) \quad \text{for right } \text{End}_R(G)\text{-modules } M \\[4pt]
t_G(\phi) = [\mathbf{T}_G(f)] \text{ for maps } \phi : M \longrightarrow M' \text{ in } \textbf{Mod}\text{-}\text{End}_R(G).
\end{array}
}
$$

We have fixed $\mathcal{P}(M)$ for M, so the G-plex $t_G(M) \in G$-**Plex** is well-defined. Suppose that f and g are chain maps that are liftings of a single $\text{End}_R(G)$-module map $\phi : M \longrightarrow M'$. If we let $h_k = f_k - g_k$ then $h = (\cdots, h_1, h_0) : \mathcal{Q} \longrightarrow \mathcal{Q}'$ is a chain map lifting of $0 : M \longrightarrow M'$. Thus there is a diagram

$$
\begin{array}{ccccccc}
\cdots \xrightarrow{\partial_3} & P_2 & \xrightarrow{\partial_2} & P_1 & \xrightarrow{\partial_1} & P_0 \\
& \swarrow{s_3} \downarrow{h_2} & \swarrow{s_2} & \downarrow{h_1} & \swarrow{s_1} & \downarrow{h_0} \\
\cdots \xrightarrow{\partial'_3} & P'_2 & \xrightarrow{\partial'_2} & P'_1 & \xrightarrow{\partial'_1} & P'_0
\end{array}
\qquad (9.10)
$$

of right $\mathrm{End}_R(G)$-modules in which the square boundaries commute and $s = (\ldots, s_2, s_1)$ is a homotopy. An application of $\mathbf{T}_G(\cdot)$ to (9.10) yields the commutative diagram

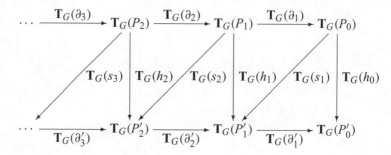

of right R-modules such that

1. The rows in this diagram are the G-plexes $\mathrm{t}_G(M)$ and $\mathrm{t}_G(M')$,
2. The square boundaries in this diagram commute, and
3. $\mathbf{T}_G(s) = (\ldots, \mathbf{T}_G(s_2), \mathbf{T}_G(s_1))$ is a homotopy for the chain map $\mathbf{T}_G(h)$.

Therefore, $\mathbf{T}_G(h) = \mathbf{T}_G(f - g) = \mathbf{T}_G(f) - \mathbf{T}_G(g)$ is null homotopic, or equivalently $[\mathbf{T}_G(f)] = [\mathbf{T}_G(g)]$. It follows that $\mathrm{t}_G(\phi)$ is independent of the choice of lifting $f : \mathcal{P}(M) \longrightarrow \mathcal{P}(M')$ and so $\mathrm{t}_G(\cdot) : \mathbf{Mod}\text{-}\mathrm{End}_R(G) \longrightarrow G\text{-}\mathbf{Plex}$ is a well-defined functor.

To complete the proof of Theorem 9.12 we will show that $\mathrm{h}_G \mathrm{t}_G(\cdot)$ is naturally equivalent to the identity functor on $\mathbf{Mod}\text{-}\mathrm{End}_R(G)$.

Consider $\mathrm{h}_G \mathrm{t}_G(\cdot)$. Given $M \in \mathbf{Mod}\text{-}\mathrm{End}_R(G)$ we have chosen a projective resolution $\mathcal{P}(M)$ for M. Then $\mathrm{t}_G(M) = \mathbf{T}_G(\mathcal{P}(M)) \in G\text{-}\mathbf{Plex}$. An application of $\mathbf{H}_G \mathbf{T}_G(\cdot)$ to $\mathcal{P}(M)$ yields the commutative square

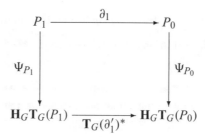

By Theorem 1.13, Ψ_{P_0} and Ψ_{P_1} are isomorphisms, so there are isomorphisms

$$M \cong \mathrm{coker}\, \partial_1$$
$$\cong \mathrm{coker}\, \mathbf{T}_G(\partial_1)^*$$
$$= H_0(\mathbf{H}_G \mathbf{T}_G(\mathcal{P}(M)))$$
$$= (H_0 \mathbf{H}_G)(\mathbf{T}_G(\mathcal{P}(M)))$$
$$\cong \mathrm{h}_G \mathrm{t}_G(M).$$

However, by Theorem 9.11, $h_G(\cdot) : G\text{-}\mathbf{Plex} \longrightarrow \mathbf{Mod}\text{-}\mathrm{End}_R(G)$ is a category equivalence, so $t_G(\cdot) : \mathbf{Mod}\text{-}\mathrm{End}_R(G) \longrightarrow G\text{-}\mathbf{Plex}$ is the inverse of $h_G(\cdot) : G\text{-}\mathbf{Plex} \longrightarrow \mathbf{Mod}\text{-}\mathrm{End}_R(G)$.

Without a self-small hypothesis $h_G(\cdot)$ and $t_G(\cdot)$ still restrict to inverse category equivalences on the category of *coherent right* $\mathrm{End}_R(G)$-*modules*. The right $\mathrm{End}_R(G)$-module M is *coherent* if M possesses a projective resolution $\mathcal{P}(M)$ in which P_k is finitely generated for each $k \geq 0$. Evidently coherent modules are finitely presented.

We say that $\mathcal{Q} \in G\text{-}\mathbf{Plex}$ is a *coherent G-plex* if $Q_k \in \mathbf{P}_o(G)$ for each $k \geq 0$. The right R-module H is *G-coherent* if there is a coherent G-plex for H. That is, there a coherent G-plex $\mathcal{Q}$ such that $H \cong \mathrm{coker}\, \delta_1$.

Let $\mathbf{C}$ be a subcategory of a category $\mathbf{C}$. Then $\mathbf{C}$ is a *full subcategory of* $\mathbf{D}$ if

$$\mathrm{Hom}_{\mathbf{C}}(X, Y) = \mathrm{Hom}_{\mathbf{D}}(X, Y)$$

for all objects $X, Y \in \mathbf{C}$. Then we let

> $G\text{-}\mathbf{CohPlx} = $ the full subcategory of $G\text{-}\mathbf{Plex}$ whose objects are the coherent G-plexes.
>
> $\mathbf{Coh}\text{-}\mathrm{End}_R(G) = $ the full subcategory of $\mathbf{Mod}\text{-}\mathrm{End}_R(G)$ of coherent right $\mathrm{End}_R(G)$-modules.

Theorem 9.13. *Let G be a right R-module (not necessarily self-small). The functors*

> $h_G(\cdot) : G\text{-}\mathbf{CohPlx} \longrightarrow \mathbf{Coh}\text{-}\mathrm{End}_R(G)$
>
> $t_G(\cdot) : \mathbf{Coh}\text{-}\mathrm{End}_R(G) \longrightarrow G\text{-}\mathbf{CohPlx}$

are inverse category equivalences.

Proof: Given $\mathcal{Q} \in G\text{-}\mathbf{CohPlx}$ define $h_G(\mathcal{Q})$ as we did in Theorem 9.11. Following the proof of Theorem 9.11 and using Theorem 1.14 instead of Theorem 1.13, one shows that $h_G(\mathcal{Q}) \in \mathbf{Coh}\text{-}\mathrm{End}_R(G)$ for each $\mathcal{Q} \in G\text{-}\mathbf{CohPlx}$. Employing the argument of Theorem 9.11 *verbatim* we show that $h_G(\cdot) : G\text{-}\mathbf{CohPlx} \longrightarrow \mathbf{Coh}\text{-}\mathrm{End}_R(G)$ is a category equivalence.

Furthermore, given $M \in \mathbf{Coh}\text{-}\mathrm{End}_R(G)$ we fix a projective resolution $\mathcal{P}(M)$ such that $P_k \in \mathbf{P}_o(\mathrm{End}_R(G))$ for $k \geq 0$. Then $\mathbf{T}_G(\mathcal{P}(M)) \in G\text{-}\mathbf{CohPlx}$ so that $t_G(\cdot) : \mathbf{Coh}\text{-}\mathrm{End}_R(G) \longrightarrow G\text{-}\mathbf{CohPlx}$ is well-defined. The rest of the proof proceeds along the lines of Theorem 9.12.

A ring E is *right coherent* iff each finitely presented right E-module is coherent. Right Noetherian rings are right coherent. Let

> $\mathbf{FP}\text{-}\mathrm{End}_R(G) = $ the full subcategory of $\mathbf{Mod}\text{-}\mathrm{End}_R(G)$ of finitely presented right $\mathrm{End}_R(G)$-modules.

Evidently $\mathbf{Coh}\text{-}\mathrm{End}_R(G) \subset \mathbf{FP}\text{-}\mathrm{End}_R(G)$.

A right R-module H is *finitely G-generated* if $H = Q/K$ for some $Q \in \mathbf{P}_o(G)$ and some R-submodule $K \subset Q$. A right R-module H is *finitely G-presented* if $H = Q/K$ for some $Q \in \mathbf{P}_o(G)$ and some finitely G-generated R-submodule K. Let

$$\boxed{\begin{array}{c} G\text{-}\mathbf{FP} = \text{the full subcategory of } \mathbf{Mod}\text{-}R \text{ of} \\ \text{finitely } G\text{-presented right } R\text{-modules.} \end{array}}$$

Evidently $H_0(\mathcal{Q}) \in G\text{-}\mathbf{FP}$ for each $\mathcal{Q} \in G\text{-}\mathbf{CohPlx}$.

Theorem 9.14. *Let G be a right R-module (not necessarily self-small). The following are equivalent.*

1. $\mathrm{End}_R(G)$ *is right coherent.*
2. $\mathrm{h}_G(\cdot) : G\text{-}\mathbf{CohPlx} \longrightarrow \mathbf{FP}\text{-}\mathrm{End}_R(G)$ *is a category equivalence.*
3. *Each finitely G-presented right R-module is G-coherent.*

Proof: $3 \Rightarrow 2$. Part 3 states that $\mathbf{FP}\text{-}\mathrm{End}_R(G) = \mathbf{Coh}\text{-}\mathrm{End}_R(G)$. Apply Theorem 9.13 to prove that $\mathrm{h}_G : G\text{-}\mathbf{CohPlx} \longrightarrow \mathbf{FP}\text{-}\mathrm{End}_R(G)$ is a category equivalence.

$2 \Rightarrow 1$. Suppose that $\mathrm{h}_G(\cdot) : G\text{-}\mathbf{CohPlx} \longrightarrow \mathbf{FP}\text{-}\mathrm{End}_R(G)$ is a category equivalence and let $M \in \mathbf{FP}\text{-}\mathrm{End}_R(G)$. By part 2 there is a $\mathcal{Q} \in G\text{-}\mathbf{CohPlx}$ such that $M \cong \mathrm{h}_G(\mathcal{Q})$ and, by Theorem 9.13, $M \cong \mathrm{h}_G(\mathcal{Q}) \in \mathbf{Coh}\text{-}\mathrm{End}_R(G)$. Then M is a coherent right $\mathrm{End}_R(G)$-module, so that $\mathrm{End}_R(G)$ is a right coherent ring.

$1 \Rightarrow 3$. Suppose that $\mathrm{End}_R(G)$ is right coherent and let $H \in G\text{-}\mathbf{FP}$. There is a short exact sequence

$$Q_1 \xrightarrow{\ \delta\ } Q_0 \longrightarrow H \longrightarrow 0$$

in which $Q_1, Q_0 \in \mathbf{P}_o(G)$. By Theorem 1.14, there is a short exact sequence

$$P_1 \xrightarrow{\ \partial\ } P_0 \longrightarrow M \longrightarrow 0$$

in which P_1, P_0 are finitely generated projective and such that $\mathbf{T}_G(P_1) \cong Q_1$, $\mathbf{T}_G(P_0) \cong Q_0$, and $\mathbf{T}_G(\partial) = \delta$. The right exactness of $\mathbf{T}_G(\cdot)$ and the uniqueness of the cokernel of δ shows that $\mathbf{T}_G(M) \cong H$. Furthermore, M is finitely presented and $\mathrm{End}_R(G)$ is right coherent so M is right coherent. Choose a projective resolution $\mathcal{P}(M)$ in which P_k is finitely generated for each $k \geq 0$. By Theorem 1.14, $\mathbf{T}_G(P_k) \in \mathbf{P}_o(G)$ for each $k \geq 0$ and, by Theorem 9.13, $\mathbf{T}_G(\mathcal{P}(M)) \in G\text{-}\mathbf{CohPlx}$, so that $\mathrm{coker}\, \mathbf{T}_G(\partial_1) \cong \mathbf{T}_G(M) \cong H$ is G-coherent. This completes the logical cycle.

$$\boxed{\begin{array}{c} \mathbf{FG}\text{-}\mathrm{End}_R(G) = \text{the full subcategory of } \mathbf{Mod}\text{-}\mathrm{End}_R(G) \text{ of} \\ \text{finitely generated right } \mathrm{End}_R(G)\text{-modules.} \end{array}}$$

Theorem 9.15. *Let G be a right R-module (not necessarily self-small). The following are equivalent.*

1. $\operatorname{End}_R(G)$ *is right Noetherian.*
2. $h_G(\cdot) : G\text{-}\mathbf{CohPlx} \longrightarrow \mathbf{FG}\text{-}\operatorname{End}_R(G)$ *is a category equivalence.*
3. *Each finitely G-generated right R-module is G-coherent.*

Proof: The proof, which is left to the reader, follows in a manner similar to that of Theorem 9.14.

D. Orsatti and C. Menini [88] have characterized the category equivalences between the category $\overline{\mathbf{P}}(G)$ of right R-modules that are *subgenerated* by G and certain full subcategories of $\mathbf{Mod}\text{-}\operatorname{End}_R(G)$ that are of the form $\mathbf{H}_G(\cdot)$. The next lemma is a special case of their work.

Theorem 9.16. *Let E be a ring, and assume that there is a category equivalence $F : \mathbf{C} \longrightarrow \mathbf{Mod}\text{-}E$ for some additive category $\mathbf{C}$. Then F is naturally equivalent to $\operatorname{Hom}_{\mathbf{C}}(G, \cdot)$ for some small projective generator $G \in \mathbf{C}$. If C is any small projective generator in $\mathbf{C}$ then $\operatorname{Hom}_{\mathbf{C}}(C, \cdot) : \mathbf{C} \longrightarrow \mathbf{Mod}\text{-}\operatorname{End}_C(C)$ is a category equivalence.*

Proof: Let G denote the object such that $F(G) = E$. Because E is a small projective generator in $\mathbf{Mod}\text{-}E$ and because F is a category equivalence, G is a small projective generator in $\mathbf{C}$. Thus

$$\operatorname{Hom}_{\mathbf{C}}(G, \oplus_{\mathcal{I}} G) \cong \operatorname{Hom}_E(E, \oplus_{\mathcal{I}} E) \cong \oplus_{\mathcal{I}} E \cong \oplus_{\mathcal{I}} F(G) \cong F(\oplus_{\mathcal{I}} G)$$

for each index set I. Hence

$$F(Q) \cong \operatorname{Hom}_{\mathbf{C}}(G, Q)$$

for each $Q \in \mathbf{P}(G)$.

Now let $C \in \mathbf{C}$. Since G is a small projective generator in $\mathbf{C}$ there is a G-plex

$$\mathcal{Q} = \quad \cdots \longrightarrow Q_1 \xrightarrow{\delta_1} Q_0 \longrightarrow C \longrightarrow 0$$

Apply $F(\cdot)$ and $\operatorname{Hom}_{\mathbf{C}}(G, \cdot)$ to $\mathcal{Q}$ to produce the commutative diagram

$$
\begin{array}{ccccccccc}
\cdots & \longrightarrow & F(Q_1) & \xrightarrow{\delta_1} & F(Q_0) & \xrightarrow{\delta} & F(C) & \longrightarrow & 0 \\
 & & \downarrow{\cong} & & \downarrow{\cong} & & \downarrow & & \\
\cdots & \longrightarrow & \operatorname{Hom}_{\mathbf{C}}(G, Q_1) & \xrightarrow{\delta_1} & \operatorname{Hom}_{\mathbf{C}}(G, Q_0) & \xrightarrow{\delta} & \operatorname{Hom}_{\mathbf{C}}(G, C) & \longrightarrow & 0
\end{array}
$$

with exact rows. Because the rows are exact, a diagram chase shows us that the map

$$F(C) \longrightarrow \mathrm{Hom}_{\mathbf{C}}(G, C)$$

is an isomorphism. The rest is left as an exercise for the reader.

Theorem 9.17. *Let G be a module and let $\mathcal{G}$ be the G-plex $0 \longrightarrow G$ such that $Q_0 = G$ and $Q_k = 0$ for each $k \geq 1$. Then G is self-small iff $\mathcal{G}$ is a small projective generator in G-Plex. In this case, $h_G(\cdot)$ and $\mathrm{Hom}_{G\text{-}\mathbf{Plex}}(\mathcal{G}, \cdot)$ are naturally equivalent category equivalences G-Plex $\longrightarrow$ Mod-$\mathrm{End}_R(G)$.*

Proof: Clearly, G is self-small if $\mathcal{G}$ is small in G-**Plex**. Conversely, if G is self-small then, by Theorem 9.11, $h_G : G$-**Plex** $\longrightarrow$ **Mod**-$\mathrm{End}_R(G)$ is a category equivalence. By Theorem 9.16, h_G is naturally equivalent to $\mathrm{Hom}_{G\text{-}\mathbf{Plex}}(\mathcal{Q}, \cdot)$ for some small projective generator $\mathcal{Q} \in G$-**Plex**. Examination of the proof of Theorem 9.16 reveals that $\mathcal{Q} \cong \mathcal{G}$ in G-**Plex**.

Example 9.18. The finitely generated right R-modules G are self-small. In fact they are small R-modules. Thus $0 \longrightarrow G$ is a small projective generator in G-**Plex** whenever G is a finitely generated right R-module. Moreover G-**Plex** is equivalent to **Mod**-$\mathrm{End}_R(G)$ for each finitely generated $G \in$ **Mod**-R.

9.2 Two commutative triangles

In this section we will construct commutative triangles

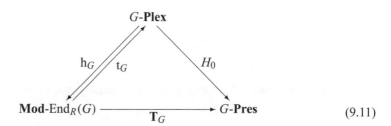

$$(9.11)$$

and

$$(9.12)$$

of categories and (inverse) functors.

We explore the consequences for G that occur when there is a category equivalence between full subcategories of **Mod**-$\mathrm{End}_R(G)$ and full subcategories of G-**Gen**. In the process we characterize several module theoretic properties for G and ring theoretic properties for $\mathrm{End}_R(G)$.

G-**Pres** $=$ the category of right R-modules of the form Q/K for some $Q \in \mathbf{P}_o(G)$ and some G-generated R-submodule $K \subset Q$. G-**Pres** is called the category of *G-presented modules.*

G-**Coh** $=$ the category of right R-modules of the form coker δ_1 for some $Q \in G$-**CohPlx**. G-**Coh** is called the category of *G-coherent modules.*

9.2.1 G-Solvable R-modules

We say that H is G-*solvable* if there is a G-balanced short exact sequence

$$0 \longrightarrow K \longrightarrow Q \overset{\delta}{\longrightarrow} H \longrightarrow 0 \tag{9.13}$$

of right R-modules such that $Q \in \mathbf{P}(G)$ and K is G-generated. Let

---

G-**Sol** $=$ the category of *G-solvable* right R-modules.

We will show presently that Θ_H is an isomorphism for each $H \in G$-**Sol** so the natural transformation Θ induces a natural equivalance between $\mathbf{T}_G\mathbf{H}_G(\cdot)$ and the identity functor on G-**Sol**. Thus G-**Sol** is a good starting place if we wish to investigate equivalences induced by $\mathbf{T}_G(\cdot)$ and $\mathbf{H}_G(\cdot)$.

G. Azumaya [18] used G-solvable right R-modules to aid in his discussion of K. Fuller's theorem [61] for finitely generated quasi-projective self-generators. (See Section 10.2.) M. Sato [98] also used G-solvable modules to discuss category equivalences surrounding self-small Σ-quasi-projective right R-modules. D. Orsatti and C. Menini [88] used G-solvable right R-modules to investigate their **-modules.* (See Section 10.4.) The results in this subsection will show that G-solvable modules will play an important role in our study of category equivalences. U. Albrecht [3] proves the following important result.

Lemma 9.19. *Let G be a right R-module, and let H be G-generated. If G is self-small then the following are equivalent.*

1. *H is G-solvable.*
2. *If (9.13) is a G-balanced short exact sequence such that $Q \in \mathbf{P}(G)$ then K is G-generated.*
3. *$\Theta_H : \mathbf{T}_G\mathbf{H}_G(H) \longrightarrow H$ is an isomorphism.*

Proof: $2 \Rightarrow 1$. By Lemma 9.4.1 there is a G-balanced short exact sequence (9.13), in which $Q \in \mathbf{P}(G)$. By part 2, K is G-generated, so H is G-solvable.

$1 \Rightarrow 3$. By definition there is a G-balanced short exact sequence (9.13) such that $Q \in \mathbf{P}(G)$, and such that K is G-generated. An application of $\mathbf{H}_G(\cdot)$ then yields the short exact sequence

$$0 \longrightarrow \mathbf{H}_G(K) \longrightarrow \mathbf{H}_G(Q) \overset{\delta^*}{\longrightarrow} \mathbf{H}_G(H) \longrightarrow 0$$

so a subsequent application of $\mathbf{T}_G(\cdot)$ produces the commutative diagram

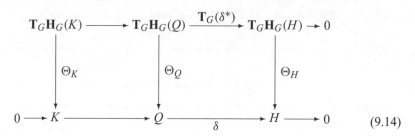

$$(9.14)$$

with exact rows. Since K is G-generated Θ_K is a surjection and since G is self-small, Theorem 1.13 states that Θ_Q is an isomorphism. A diagram chase through (9.14) shows us that Θ_H is an isomorphism. This proves part 3.

$3 \Rightarrow 2$. Let (9.13) be a G-balanced short exact sequence such that $Q \in \mathbf{P}(G)$. Since (9.13) is G-balanced $\mathbf{H}_G(\cdot)$ is right exact on (9.13), so an application of $\mathbf{T}_G\mathbf{H}_G(\cdot)$ leads us to the commutative diagram (9.14) with exact rows. By part 3, Θ_H is an isomorphism and because G is self-small Theorem 1.13 states that Θ_Q is an isomorphism. A diagram chase shows us that Θ_K is a surjection, thus proving that K is G-generated. This proves part 2 and completes the logical cycle.

The right R-module G is a Σ-*self-generator* if each R-submodule of each $Q \in \mathbf{P}(G)$ is G-generated. Free modules are Σ-self-generators, as are reduced torsion abelian groups. The right R-modules of the form $R \oplus H$ for cardinals c are Σ-self-generators. $\mathbf{Q}$ is not a Σ-self-generator as an abelian group. See page 155 for more properties of Σ-self-generators. For Σ-self-generators G the G-solvable R-modules are easily recognized.

Proposition 9.20. *Let G be a self-small Σ-self-generator. Each G-generated right R-module is G-solvable.*

Proof: Proof: Let H be a G-generated right R-module. By Lemma 9.4.1, H fits into a G-balanced short exact sequence (9.13) in which $Q \in \mathbf{P}(G)$. Since G is a Σ-self-generator K is G-generated so H is G-solvable.

The right R-module is *E-flat* if G is a flat left $\mathrm{End}_R(G)$-module.

Lemma 9.21. *Let G be an E-flat right R-module and let*

$$0 \longrightarrow K \longrightarrow H \overset{\delta}{\longrightarrow} L \longrightarrow 0 \qquad (9.15)$$

be a G-balanced short exact sequence of right R-modules.

1. *If K and L are G-solvable then H is G-solvable.*
2. *Suppose that H is G-solvable. The following are equivalent.*

   (a) *Θ_K is a surjection.*
   (b) *Θ_L is an injection.*
   (c) *K is G-solvable.*
   (d) *L is G-solvable.*

Proof: An application of $\mathbf{H}_G(\cdot)$ to the G-balanced short exact sequence (9.15) yields the exact sequence

$$0 \longrightarrow \mathbf{H}_G(K) \longrightarrow \mathbf{H}_G(H) \overset{\delta^*}{\longrightarrow} \mathbf{H}_G(L) \longrightarrow 0.$$

A subsequent application of the left exact functor $\mathbf{T}_G(\cdot)$, (G is E-flat), yields the exact top row of the commutative diagram

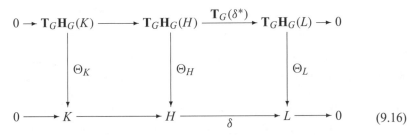

$$(9.16)$$

Now let us see how this diagram is effected by the hypotheses in parts 1 and 2.

1. Suppose that K and L are G-solvable. Then Θ_K and Θ_L are isomorphisms. A diagram chase (or an application of the Three Lemma [97]) shows us that Θ_H is then an isomorphism. Then H is G-solvable.

2. Suppose that H is G-solvable. Then Θ_H is an isomorphism. We will prove that d $\Leftrightarrow$ b $\Leftrightarrow$ a $\Leftrightarrow$ c. By commutativity of (9.16) and the Snake Lemma [94], Θ_L is a surjection and Θ_K is an injection. Thus L is G-solvable iff Θ_L is an isomorphism iff Θ_L is an injection. Moreover, Θ_K is an injection. Then Θ_K is a surjection iff Θ_K is an isomorphism iff K is G-solvable. Finally, a diagram chase shows us that Θ_L is an isomorphism iff Θ_K is an isomorphism. This completes the logical cycle.

Corollary 9.22. *Let G be a self-small, E-flat right R-module. Each G-generated R-submodule of $Q \in \mathbf{P}(G)$ is G-solvable.*

Proof: Since G is self-small, Theorem 1.13 implies that each $Q \in \mathbf{P}(G)$ is G-solvable. Then by Lemma 9.21, the G-generated $K \subset Q$ is G-solvable.

9.2.2 A factorization of the tensor functor

The results in this section are from [50]. The main result in this section is a factorization of the *tensor functor*

$$\mathbf{T}_G(\cdot) : \mathbf{Mod}\text{-}\mathrm{End}_R(G) \longrightarrow \mathbf{Mod}\text{-}R.$$

The following result shows that the bottom row of commutative triangle (9.11) is well-defined.

Lemma 9.23. *Let G be a right R-module. If M is a right*
$\text{End}_R(G)$*-module then* $\mathbf{T}_G(M)$ *is G-presented.*

Proof: Write down a projective resolution

$$P_1 \longrightarrow P_0 \longrightarrow M \longrightarrow 0$$

for M. An application of $\mathbf{T}_G(\cdot)$ produces the exact sequence

$$\mathbf{T}_G(P_1) \xrightarrow{\ \delta\ } \mathbf{T}_G(P_0) \longrightarrow \mathbf{T}_G(M) \longrightarrow 0$$

of right R-modules. Because $\mathbf{T}_G(\cdot)$ commutes with arbitrary direct sums and because $\mathbf{T}_G(\text{End}_R(G)) \cong G$, $\mathbf{T}_G(P_1), \mathbf{T}_G(P_0) \in \mathbf{P}(G)$. Thus $\mathbf{T}_G(M) \cong \mathbf{T}_G(P_o)/\text{image } \delta$ is G-presented, which proves the lemma.

The zero-th homology functor $H_0(\cdot)$ is a functor from G-**Plex** into **Mod-**R. Specifically, given $\mathcal{Q} \in G$-**Plex**, $H_0(\mathcal{Q}) = \mathcal{Q}_0/\text{image } \delta_1$. By part 1 of Lemma 9.9, a given chain map $f = (\cdots, f_1, f_0) : \mathcal{Q} \longrightarrow \mathcal{Q}'$ induces a unique R-module map $H_0(f) : H_0(\mathcal{Q}) \longrightarrow H_0(\mathcal{Q}')$. Moreover, if f and g are homotopic then $H_0(f) = H_0(g)$ by part 2 of Lemma 9.9. The identification $H_0([f]) = H_0(f)$ completes our definition of $H_0(\cdot) : G$-**Plex** $\longrightarrow$ G-**Pres**. Hence we have a functor

$$\boxed{H_0(\cdot) : G\text{-}\mathbf{Plex} \longrightarrow G\text{-}\mathbf{Pres}.}$$

Lemma 9.24. *Let G be a self-small right R-module. Then for each* $H \in G$-**Pres** *there is a* $\mathcal{Q} \in G$-**Plex** *such that* $H_0(\mathcal{Q}) = H$.

Proof: Let $H \in G$-**Pres**. There is a short exact sequence (9.13) in which $Q \in \mathbf{P}(G)$ and K is G-generated. By Lemma 9.4.2 there is a $\mathcal{Q} \in G$-**Plex** such that $Q_0 = Q$ and image $\delta_1 = K$. Then

$$H \cong Q/K = Q_0/\text{image } \delta_1 = H_0(\mathcal{Q})$$

so that $H_0(\mathcal{Q}) \cong H$. This proves the lemma.

We have constructed the commutative triangle (9.11) once we use Theorem 9.12 to embed the inverse category equivalences $h_G(\cdot)$ and $t_G(\cdot)$. However, we have yet to prove that (9.11) is commutative.

Theorem 9.25. *If G is a self-small right R-module then the tensor functor factors as*

$$\boxed{\mathbf{T}_G(\cdot) \cong H_0 \circ t_G(\cdot).}$$

Proof: Let M be a right $\text{End}_R(G)$-module and let $\mathcal{P}(M)$ be the projective complex we chose in defining $t_G(\cdot)$. Then $t_G(M) = \mathbf{T}_G(\mathcal{P}(M)) \in G\text{-}\mathbf{Plex}$. Since $\mathbf{T}_G(\cdot)$ is a right exact functor and since coker $\partial_1 \cong M$

$$H_0 t_G(M) = H_0(\mathbf{T}_G(\mathcal{P}(M))) = \text{coker } \mathbf{T}_G(\partial_1) \cong \mathbf{T}_G(M).$$

Next let $\phi : M \longrightarrow M'$ be a right $\text{End}_R(G)$-module map. Because the P_k are projective right $\text{End}_R(G)$-modules ϕ lifts to a chain map

$$\begin{array}{ccccccccc}
\cdots \longrightarrow & P_1 & \xrightarrow{\ \partial_1\ } & P_0 & \xrightarrow{\ \partial_0\ } & M & \longrightarrow & 0 \\
& \downarrow{\scriptstyle f_1} & & \downarrow{\scriptstyle f_0} & & \downarrow{\scriptstyle \phi} & & \\
\cdots \longrightarrow & P_1' & \xrightarrow[\ \partial_1'\ ]{} & P_0' & \xrightarrow[\ \partial_0'\ ]{} & M' & \longrightarrow & 0
\end{array}$$

which is unique up to homotopy equivalence. An application of $\mathbf{T}_G(\cdot)$ to f yields the commutative diagram

$$\begin{array}{ccccccccc}
\cdots \longrightarrow & \mathbf{T}_G(P_1) & \xrightarrow{\ \mathbf{T}_G(\partial_1)\ } & \mathbf{T}_G(P_0) & \xrightarrow{\ \mathbf{T}_G(\partial_0)\ } & \mathbf{T}_G(M) & \longrightarrow & 0 \\
& \downarrow{\scriptstyle \mathbf{T}_G(f_1)} & & \downarrow{\scriptstyle \mathbf{T}_G(f_0)} & & \downarrow{\scriptstyle \mathbf{T}_G(\phi)} & & \\
\cdots \longrightarrow & \mathbf{T}_G(P_1') & \xrightarrow[\ \mathbf{T}_G(\partial_1')\ ]{} & \mathbf{T}_G(P_0') & \xrightarrow[\ \mathbf{T}_G(\partial_0')\ ]{} & \mathbf{T}_G(M') & \longrightarrow & 0
\end{array}$$

whose rows are G-plexes by Lemma 9.5. By definition of $t_G(\cdot)$, $t_G(\phi) = [\mathbf{T}_G(f)]$, so that

$$H_0 t_G(\phi) = H_0[\mathbf{T}_G(f)] = \mathbf{T}_G(\phi).$$

Therefore $H_0 \circ t_G(\cdot) = \mathbf{T}_G(\cdot)$, which completes the proof.

Theorem 9.26. *If G is a self-small right R-module, there is a commutative triangle (9.11) in which opposing arrows indicate inverse category equivalences.*

Theorem 9.27. *If G is a right R-module, there is a commutative triangle (9.12) in which opposing arrows indicate inverse category equivalences.*

Corollary 9.28. *Let G be a self-small right R-module and let H be G-presented. There is a right $\text{End}_R(G)$-module M such that $\mathbf{T}_G(M) \cong H$.*

Proof: By Lemma 9.24 there is a $\mathcal{Q} \in G\text{-}\mathbf{Plex}$ such that $H_0(\mathcal{Q}) = H$ and by Theorem 9.12 there is an $M \in \mathbf{Mod}\text{-}\text{End}_R(G)$ such that $t_G(M) \cong \mathcal{Q}$. Then by Theorem 9.25,

$$\mathbf{T}_G(M) \cong H_0 t_G(M) \cong H_0(\mathcal{Q}) \cong H.$$

This completes the proof.

9.3 Left endomorphism modules

These results are from [42]. One of the interesting consequences of Theorems 9.11 and 9.12, and of the category equivalences $h_G : G\text{-}\mathbf{Plex} \longrightarrow \mathbf{Mod}\text{-}\mathrm{End}_R(G)$ and $t_G : \mathbf{Mod}\text{-}\mathrm{End}_R(G) \longrightarrow G\text{-}\mathbf{Plex}$, is that they readily dualize. By dualizing G-plexes to G-coplexes we are able to characterize the category $\mathrm{End}_R(G)\text{-}\mathbf{Mod}$ of *left* $\mathrm{End}_R(G)$-modules in terms of a category in which G is a *slender injective cogenerator*. The surprise is that the dualization is straightforward once you get the definitions right. For this reason we will leave most of the proofs in this section as exercises for the reader. We will not leave you dry, though, as we offer many hints on how to proceed with the proofs.

In this chapter we will let $\mathcal{P}$ denote a projective resolution of projective *left* $\mathrm{End}_R(G)$-modules. Let

> $\mathrm{End}_R(G)\text{-}\mathbf{Mod} =$ the category of left $\mathrm{End}_R(G)$-modules
>
> $\mathbf{coP}(\mathrm{End}_R(G)) =$ the category of projective left $\mathrm{End}_R(G)$-modules.

Recall that for cardinals c, G^c denotes the product of c copies of G, while $G^{(c)}$ is the direct sum of c copies of G.

> $\mathbf{coP}(G) =$ the category of right R-modules W such that
> $\qquad W \oplus W' \cong G^c$ for some cardinal c.
>
> $\mathbf{coP}_o(G) =$ the category of right R-modules W such that
> $\qquad W \oplus W' \cong G^n$ for some integer $n > 0$.

We require a set-theoretic hypothesis. A cardinal λ is *measurable* if there is an additive nonzero measure μ on the collection of cardinals such that $\mu(\lambda) \neq 0$. In order to give a complete dual to Theorem 9.11 we will assume

> (μ) Measurable cardinals do not exist.

The statement (μ) is true under Gödel's Constructibility Hypothesis $V = L$.

Assume (μ). Given a right R-module G, and the canonical map (9.17), then for *self-slender* right R-modules G (see below), (9.17) is an isomorphism for each cardinal c up to the first measurable cardinal. By assuming (μ) we are assuming that (9.17) is an isomorphism for each cardinal c if it is an isomorphism for some countable cardinal. By assuming (μ) we are allowing for the complete dualization of Theorem 1.13, and thus Theorem 9.11 can be dualized. Let us be more specific about the details.

Dualizing the self-small property we arrive at the underlying hypothesis for this section. The right R-module G is *slender* if for each index set $\mathcal{I}$ of nonmeasurable cardinality, and for each set $\{R_i \mid i \in \mathcal{I}\}$ of right R-modules such that $R \cong R_i$ for each

$i \in \mathcal{I}$, the canonical map

$$\bigoplus_{i \in \mathcal{I}} \operatorname{Hom}_R(R_i, G) \longrightarrow \operatorname{Hom}_R\left(\prod_{i \in \mathcal{I}} R_i, G\right)$$

is an isomorphism. We say that G is *self-slender* if the canonical map

$$\operatorname{Hom}_R(G, G)^{(c)} \longrightarrow \operatorname{Hom}_R(G^c, G) \tag{9.17}$$

is an isomorphism for each cardinal c.

Theorem 9.29. [59, Theorem 95.3] *Assume* (μ). *Rtffr groups are slender groups and self-slender groups.*

Thus there are plenty of self-slender right R-modules. However, there will be fewer examples of self-slender R-modules than there are of self-small R-modules. We know that finitely generated modules are (self-)small right R-modules. Thus by dualizing small modules we might expect that there are many slender modules. This is misled intuition. There are no nonzero (self-)slender injective right *R-modules*. Thus, being injective, **Q** and $\mathbf{Z}(p^\infty)$ are not self-slender groups.

Define contravariant functors

$$\boxed{\begin{aligned} (\cdot)^* : \textbf{Mod-}R &\longrightarrow \operatorname{End}_R(G)\textbf{-Mod} \\ (\cdot)^* : \operatorname{End}_R(G)\textbf{-Mod} &\longrightarrow \textbf{Mod-}R \end{aligned}}$$

by

$$(\cdot)^* = \operatorname{Hom}_R(\cdot, G) \quad \text{and} \quad (\cdot)^* = \operatorname{Hom}_{\operatorname{End}_R(G)}(\cdot, G).$$

It is traditional to denote these different functors with the same notation $(\cdot)^*$. Thus

$$X^{**} = (X^*)^*$$

is an unambiguous application of these functors. There are natural maps Φ_X given by

$$\boxed{\Phi_X : X \longrightarrow X^{**} : x \longmapsto \lambda_x,}$$

where $\lambda_x(f) = f(x)$ for each $f \in X^*$.

The following theorem, which is due to M. Huber and R. B. Warfield, Jr. [70, Theorem 1.2], will replace Theorem 1.13 in our deliberations as we change our focus from **Mod-**$\operatorname{End}_R(G)$ to $\operatorname{End}_R(G)$**-Mod**. Its proof is dual to that of Theorem 1.13. A *duality* is a category equivalence **D** that takes a map $\phi : X \longrightarrow Y$ to $\mathbf{D}(\phi) : \mathbf{D}(Y) \longrightarrow \mathbf{D}(X)$. Note the change in the direction of the arrow.

M. Huber and R. B. Warfield, Jr. [70, Theorem 1.2] extend a deep result of Łoś ([59, Theorem 94.4]) and thus dualize the Arnold–Lady theorem 1.13. We state their theorem in a manner that is useful to our purposes.

Theorem 9.30. [70, M. Huber and R. B. Warfield, Jr.] *Assume* (μ) *and let G be a self-slender group. Then* $(\cdot)^*$ *induces inverse dualities*

$$(\cdot)^* : \mathbf{coP}(G) \longrightarrow \mathbf{coP}(\mathrm{End}_R(G))$$

$$(\cdot)^* : \mathbf{coP}(\mathrm{End}_R(G)) \longrightarrow \mathbf{coP}(G).$$

Moreover, Ψ_W *is an isomorphism for each* $W \in \mathbf{coP}(G)$ *and* Ψ_P *is an isomorphism for each* $P \in \mathbf{P}(\mathrm{End}_R(G))$.

Proof: Since we are assuming that there are no measurable cardinals, [70, Theorem 1.2] applies to all cardinals. Moreover, by examining the proof of [70, Theorem 1.2] we see that

$$\Psi : 1_{\mathbf{coP}(G)} \longrightarrow (\cdot)^{**}$$

and

$$\Psi : 1_{\mathbf{P}(\mathrm{End}_R(G))} \longrightarrow (\cdot)^{**}$$

are the natural transformations associated with the duality. Thus Ψ_W and Ψ_P are isomorphisms for each $W \in \mathbf{coP}(G)$ and each $P \in \mathbf{P}(\mathrm{End}_R(G))$. This completes the proof.

A complex of right R-modules

$$\boxed{\mathcal{W} = W_0 \xrightarrow{\sigma_1} W_1 \xrightarrow{\sigma_2} W_2 \xrightarrow{\sigma_3} \cdots}$$

is called a *G-coplex* if

1. $W_k \in \mathbf{coP}(G)$ for each $k \geq 0$, and
2. For each map $\phi : W_k \longrightarrow G$ such that $\phi \sigma_{k-1} = 0$ there is a map $\psi : W_{k+1} \longrightarrow G$ such that $\phi = \psi \sigma_k$ as in the triangle below.

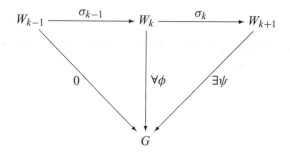

Let

$$\boxed{\mathcal{W}^* = \cdots \longrightarrow W_2^* \xrightarrow{\sigma_2^*} W_1^* \xrightarrow{\sigma_1^*} W_0^*}$$

denote the complex that is formed by applying $(\cdot)^*$ to the complex $\mathcal{W}$.

A *chain map* $f : \mathcal{W} \longrightarrow \mathcal{W}'$ is a commutative diagram

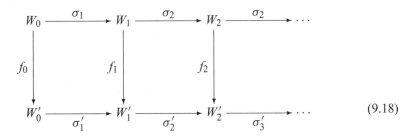

$$(9.18)$$

The homotopy of chain maps for these complexes is exactly the definition we used for homotopy of G-plexes on page 113.

9.3.1 Duality

In this section we will dualize Theorem 9.11 and in doing so we show that these dualities are effected by the foundations of mathematics.

Let

> G-**Coplx** = the category whose objects are G-coplexes $\mathcal{W}$ and whose morphisms $[f] : \mathcal{W} \longrightarrow \mathcal{W}'$ are homotopy classes of chain maps $f : \mathcal{W} \longrightarrow \mathcal{W}'$. G-**Coplx** is called *the category of G-coplexes.*

Recall that for a complex the zero-th homology functor of

$$\mathcal{C} = \quad \cdots \xrightarrow{\phi_3} C_2 \xrightarrow{\phi_2} C_1 \xrightarrow{\phi_1} C_0$$

is defined to be

$$H_0(\mathcal{C}) = \operatorname{coker} \phi_1.$$

Define a functor

> $h^G : G\text{-}\mathbf{coPlex} \longrightarrow \operatorname{End}_R(G)\text{-}\mathbf{Mod}$

by

$$h^G(\mathcal{W}) = H_0(\mathcal{W}^*)$$
$$h^G([f]) = H_0(f^*)$$

for objects $\mathcal{W} \in G\text{-}\mathbf{coPlex}$ and maps $[f] \in G\text{-}\mathbf{coPlex}$.

Lemma 9.31. *Assume (μ). Let $\mathcal{W}$ be a complex of right R-modules with $W_k \in \mathbf{coP}(G)$ for each integer k.*

1. W is a G-complex iff W^* is an exact sequence of left $\operatorname{End}_R(G)$-modules.
2. If $f, g : W \longrightarrow W'$ are homotopic chain maps from W to W' then $H_0(f^*) = H_0(g^*)$. Thus $\mathrm{h}^G([f]) = \mathrm{h}^G([g])$.
3. If G is a self-slender group and W is a G-complex then W^* is a projective resolution of the left $\operatorname{End}_R(G)$-module $H_0(W^*) = \mathrm{h}^G(W)$.

Proof: 1 follows immediately from the lifting property enjoyed by a G-complex.

2. If f and g are homotopic chain maps then the induced maps f^* and g^* are homotopic chain maps. Hence the homology maps $H_k(f^*)$ and $H_k(g^*)$ are equal for each integer k, so that $\mathrm{h}^G([f]) = H_0(f^*) = H_0(g^*) = \mathrm{h}^G([g])$.

3. Let W be a G-complex. By part 1, W^* is exact, and because G is self-slender, Theorem 9.30 implies that $W_k^* \in \mathbf{P}(E)$ for each term W_k of W. Thus W^* is a projective resolution of coker $\sigma_1^* = H_0(W^*) = \mathrm{h}^G(W)$. This completes the proof.

Theorem 9.32. [42, T.G. Faticoni] *Assume (μ). Let G be a self-slender right R-module. Then $(\cdot)^*$ lifts to a duality*

$$\mathrm{h}^G(\cdot) : G\text{-}\mathbf{Coplx} \longrightarrow \operatorname{End}_R(G)\text{-}\mathbf{Mod}.$$

Proof: By Lemma 9.31, $\mathrm{h}^G(\cdot) : G\text{-}\mathbf{coPlex} \longrightarrow \operatorname{End}_R(G)\text{-}\mathbf{Mod}$ is a well defined functor.

Let $M \in \operatorname{End}_R(G)\text{-}\mathbf{Mod}$. There is a projective resolution

$$\mathcal{P}(M) = \qquad \cdots \xrightarrow{\partial_2} P_1 \xrightarrow{\partial_1} P_0.$$

for M, and by Theorem 9.30, Ψ_{P_k} is an isomorphism for each $k \geq 0$. Then Ψ induces an isomorphism of complexes $\mathcal{P}(M)^{**} \cong \mathcal{P}(M)$. By Theorem 9.30, $P_k^* \in \mathbf{coP}(G)$ for each integer $k \geq 0$, so by Lemma 9.31.1, $\mathcal{P}(M)^*$ is a G-complex. Hence

$$\mathrm{h}^G(\mathcal{P}(M)^*) \cong H_0(\mathcal{P}(M)^{**}) \cong H_0(\mathcal{P}(M)) = M.$$

We show that $\mathrm{h}^G(\cdot)$ is a faithful functor. Suppose that $\mathrm{h}^G([f]) = 0$. We will show that f is nullhomotopic. We have the commutative diagram

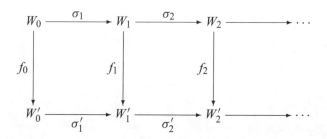

of $\mathrm{End}_R(G)$-modules, which is sent into the commutative diagram

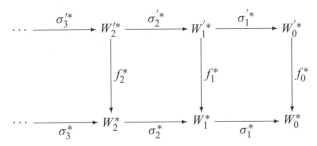

with exact rows. By Lemma 9.31.1, the rows of this diagram W^* and W'^* are projective resolutions of left $\mathrm{End}_R(G)$-modules, and, by hypothesis, $0 = \mathrm{h}^G([f]) = H_0(f^*)$. Then f^* is a nullhomotopic map. Let z be a homotopy for f^*. By Theorem 9.30, $z = s^*$ for some homotopy s for f. Thus f is null homotopic, so $[f] = 0$, whence $\mathrm{h}^G(\cdot)$ is a faithful functor.

The final item necessary is to show that $\mathrm{h}^G(\cdot)$ is a full functor. Let $\phi : M \longrightarrow M'$ be a map in $\mathrm{End}_R(G)$-**Mod**. There are projective resolutions $\mathcal{P}(M)$ and $\mathcal{P}(M')$ for M and M' respectively. Then ϕ lifts to a chain map

$$f = (\cdots, f_1, f_0) : \mathcal{P}(M) \longrightarrow \mathcal{P}(M')$$

in the usual way. Thus $\phi = H_0(f)$.

By our standing hypothesis (μ) and by Theorem 9.30, there is a chain map

$$g = (g_0, g_1, \cdots) : \mathcal{W}' \longrightarrow \mathcal{W}$$

of G-coplexes such that

$$\mathcal{W}^* = \mathcal{P}(M), \ \ \mathcal{W}'^* = \mathcal{P}(M') \ \text{ and } \ g^* = f.$$

Then $\mathrm{h}^G([g]) = H_0(g^*) = H_0(f) = \phi$. Therefore $\mathrm{h}^G(\cdot)$ is a full functor and the proof is complete.

Following tradition we let

$$\boxed{\mathrm{h}^G(\cdot) : \mathrm{End}_R(G)\text{-}\mathbf{Mod} \longrightarrow G\text{-}\mathbf{coPlex}}$$

denote the inverse of $\mathrm{h}^G(\cdot) : G\text{-}\mathbf{coPlex} \longrightarrow \mathrm{End}_R(G)\text{-}\mathbf{Mod}$.

Theorem 9.33. *Assume (μ) and let G be a countable reduced torsion-free group. Then $(\cdot)^*$ lifts to a duality*

$$\mathrm{h}^G(\cdot) : G\text{-}\mathbf{coPlex} \longrightarrow \mathrm{End}_R(G)\text{-}\mathbf{Mod}.$$

Proof: By Theorem 9.29, countable reduced torsion-free groups are self-slender. Now apply Theorem 9.32 to complete the proof.

An examination of the proof of Theorem 9.32 reveals that the inverse of $h^G(\cdot)$ is given as follows. For each left $\text{End}_R(G)$-module M fix a projective resolution $\mathcal{P}(M)$ for M. Then $\mathcal{P}(M)^*$ is a G-complex. Moreover, a map $\phi : M \longrightarrow M'$ of left $\text{End}_R(G)$-modules lifts to a chain map $f : \mathcal{P}(M) \longrightarrow \mathcal{P}(M')$ of projective resolutions, which induces a chain map $f^* : \mathcal{P}(M)^* \longrightarrow \mathcal{P}(M')^*$. Then the inverse of h^G is defined by the rules

$$
\boxed{
\begin{aligned}
h^G(M) &= \mathcal{P}(M)^* \\
h^G(\phi) &= [f^*].
\end{aligned}
}
$$

A straightforward dualization of Theorem 9.16 will prove the next result.

Theorem 9.34. *Assume* (μ). *Let G be a right R-module and let G denote the G-complex $G \longrightarrow 0$. Then G is self-slender iff G is a slender injective cogenerator for G-**coPlex**. In this case $h^G(\cdot)$ and $\text{Hom}_{G\text{-}\mathbf{coPlex}}(\cdot, G)$ are naturally equivalent dualities G-**coPlex** $\longrightarrow \text{End}_R(G)$-**Mod**.*

Remark 9.35. By combining Theorems 9.11 and 9.32 we have a complete survey of modules over endomorphism rings of self-small and self-slender right R-modules.

The G-complex W is a *coherent G-complex* if $W_k \in \mathbf{coP}_o(G)$ for each integer $k \geq 0$. The left E-module M is a *coherent left E-module* if M possesses a projective resolution $\mathcal{P}$ in which P_k is finitely generated for each integer $k \geq 0$. We collect these objects into the following categories.

G-**Cohcoplx** = the category of coherent G-coplexes.

$\text{End}_R(G)$-**Coh** = the category of coherent left $\text{End}_R(G)$-modules.

Observe that lack of a self-slender hypothesis in the following theorem.

Theorem 9.36. *Let G be a right R-module. There are inverse dualities*

$$h^G(\cdot) : G\text{-}\mathbf{Cohcoplx} \longrightarrow \text{End}_R(G)\text{-}\mathbf{Coh}$$

$$h^G(\cdot) : \text{End}_R(G)\text{-}\mathbf{Coh} \longrightarrow G\text{-}\mathbf{Cohcoplx}.$$

Proof. The canonical maps Ψ_P and Ψ_W are isomorphisms for P and W such that $P \oplus P' \cong \text{End}_R(G)^n$ and $W \oplus W' \cong G^n$ for integers $n > 0$, some left $\text{End}_R(G)$-module P', and some right R-module W'. Now follow the proof of Theorem 9.32, making the obvious changes.

The ring E is *left coherent* if each finitely presented left E-module is a coherent left E-module. The right R-module H is *finitely G-copresented* if there are integers $n, m > 0$ such that $H \subset G^n$ and $G^n/H \subset G^m$. Thus H is finitely G-copresented iff H fits into an exact sequence

$$0 \longrightarrow H \longrightarrow G^n \overset{\sigma}{\longrightarrow} G^m \tag{9.19}$$

for some integers $n, m > 0$. The right R-module H is *G-cocoherent* if there is a G-coplex $\mathcal{W}$ such that

1. $H = \ker \sigma_1 = H_0(\mathcal{W})$ and
2. $W_k \in \mathbf{coP}_o(G)$ for each $k \geq 0$.

Let

$$
\boxed{\begin{array}{c} \text{End}_R(G)\text{-}\mathbf{FP} = \text{the category of finitely presented left} \\ \text{End}_R(G)\text{-modules.} \end{array}}
$$

The proofs of the next results are dualizations of the proofs to Theorems 9.14 and 9.15.

Theorem 9.37. *Let G be a right R-module. The following are equivalent.*

1. $\text{End}_R(G)$ *is a left coherent ring.*
2. $\text{h}^G(\cdot) : G\text{-}\mathbf{Cohcoplx} \longrightarrow \text{End}_R(G)\text{-}\mathbf{FP}$ *is a duality.*
3. *Each finitely G-copresented right R-module is G-cocoherent.*

One translation of condition 3 above is that each exact sequence (9.19) embeds as the first two maps of a G-complex

$$
G^n \xrightarrow{\ \sigma\ } G^m \xrightarrow{\ \sigma_2\ } W_2 \xrightarrow{\ \sigma_3\ } \cdots
$$

in which $W_k \in \mathbf{coP}_o(G)$ for each $k \geq 2$.
Let

$$
\boxed{\begin{array}{c} \text{End}_R(G)\text{-}\mathbf{FG} = \text{the category of finitely generated left} \\ \text{End}_R(G)\text{-modules.} \end{array}}
$$

Observe that if E is left Noetherian then each finitely generated left E-module is coherent. The right R-module H is *finitely G-cogenerated* if there is an integer $n > 0$ such that $H \subset G^n$.

Theorem 9.38. *Let G be a right R-module. The following are equivalent.*

1. $\text{End}_R(G)$ *is a left Noetherian ring.*
2. $\text{h}^G(\cdot) : G\text{-}\mathbf{Cohcoplx} \longrightarrow \text{End}_R(G)\text{-}\mathbf{FG}$ *is a duality.*
3. *Each finitely G-cogenerated left R-module is G-cocoherent.*

One translation of condition 3 above is that if we can write down an inclusion $H \subset G^n$ then we can write down a G-complex

$$
G^n \xrightarrow{\ \sigma_1\ } W_1 \xrightarrow{\ \sigma_2\ } W_2 \xrightarrow{\ \sigma_3\ } \cdots
$$

in which $W_k \in \mathbf{coP}_o(G)$ for each $k \geq 0$.

9.4 Self-small self-slender modules

Theorems 9.11 and 9.32 imply that when G is a self-small and self-slender right R-module, we can characterize **Mod**-$\text{End}_R(G)$ and $\text{End}_R(G)$-**Mod** in terms of categories in which G is very pleasantly endowed.

Theorem 9.39. *Assume* (μ). *Let G be a self-small and self-slender right R-module.*

1. *There is a category G-**Plex** in which G is a small projective generator. The functors* $\mathbf{H}_G(\cdot)$ *and* $\mathbf{T}_G(\cdot)$ *induce inverse category equivalences*

$$h_G(\cdot) : G\text{-}\mathbf{Plex} \longrightarrow \mathbf{Mod}\text{-}\text{End}_R(G)$$

$$t_G(\cdot) : \mathbf{Mod}\text{-}\text{End}_R(G) \longrightarrow G\text{-}\mathbf{Plex}.$$

2. *There is a category G-**Coplx** in which G is a slender injective cogenerator. The functors* $\text{Hom}_R(\cdot, G)$ *and* $\text{Hom}_{\text{End}_R(G)}(\cdot, G)$ *lift to inverse dualities*

$$h^G(\cdot) : G\text{-}\mathbf{Coplx} \longrightarrow \text{End}_R(G)\text{-}\mathbf{Mod}$$

$$h^G(\cdot) : \text{End}_R(G)\text{-}\mathbf{Mod} \longrightarrow G\text{-}\mathbf{Coplx}.$$

For the remainder of this section we let G denote an abelian group, hereafter referred to simply as a *group*. The group G is said to be *reduced* if $\cap_{n>0} nG = 0$, or in other words, if $\text{Hom}_{\mathbf{Z}}(\mathbf{Q}, G) = 0$. We say that G is *torsion-free* if for each $n \in \mathbf{Z}$ and $x \neq 0 \in G$, $nx = 0$ implies that $n = 0$. Torsion-free groups are subgroups of $\mathbf{Q}$-vector spaces. The torsion-free group G has finite rank if each maximal linearly independent subset of G is finite. That is, G is contained in a finite-dimensional $\mathbf{Q}$-vector space.

Theorem 9.40. *Assume* (μ). *The* rtffr *groups are self-slender and self-small groups.*

Proof: By [59, Theorem 95.3] the torsion-free group G is slender if G does not contain a copy of $\mathbf{Q}$, $\widehat{\mathbf{Z}}_p$, or $\mathbf{Z}^{(\aleph_o)}$. Thus the reduced torsion-free groups G of finite rank are slender groups. Thus G is a self-slender group. We leave it as an exercise for the reader to show that a torsion-free group G of finite rank is a self-small group. This completes the proof.

Thus the reduced torsion-free groups of finite rank are ($\mathbf{Z}$-) modules on which Theorems 9.11 and 9.32 are readily applied. We should expect good results from our techniques on these groups.

Theorem 9.41. *Assume* (μ). *Let G be a reduced torsion-free finite-rank group.*

1. *There is a category G-**Plex** in which G is a small projective generator. The functors* $\mathbf{H}_G(\cdot)$ *and* $\mathbf{T}_G(\cdot)$ *induce inverse category equivalences*

$$h_G(\cdot) : G\text{-}\mathbf{Plex} \longrightarrow \mathbf{Mod}\text{-}\text{End}_R(G)$$

$$t_G(\cdot) : \mathbf{Mod}\text{-}\text{End}_R(G) \longrightarrow G\text{-}\mathbf{Plex}.$$

2. *There is a category G-**Coplx** in which G is a slender injective cogenerator. The functors* $\mathrm{Hom}_R(\cdot, G)$ *and* $\mathrm{Hom}_{\mathrm{End}_R(G)}(\cdot, G)$ *lift to inverse dualities*

$$\mathrm{h}^G(\cdot) : G\text{-}\mathbf{Coplx} \longrightarrow \mathrm{End}_R(G)\text{-}\mathbf{Mod}$$

$$\mathrm{h}^G(\cdot) : \mathrm{End}_R(G)\text{-}\mathbf{Mod} \longrightarrow G\text{-}\mathbf{Coplx}.$$

Proof: By Theorem 9.40 we can apply Theorem 9.39 to G.

Remark 9.42. At the time of this writing, the only known example of a nonzero self-small and self-slender right R-module is the R-module whose additive structure is a reduced torsion-free finite-rank group.

9.5 (μ) Implies slender injectives

Assume (μ) throughout this section. We show that with the assumption of (μ), each self-slender right R-module G is a nonzero slender injective object in G-**coPlex**. Thus there are many nonzero slender injective objects. Moreover, if we assume (μ) and if G is self-slender then there are no nonzero small projective objects in G-**coPlex**.

Example 9.43. Let $G \neq 0$ be an injective right R-module, and define a map ϕ : $\bigoplus_{i=1}^{\infty} G \longrightarrow G$ that is the identity in each component of the direct sum. Since G is injective, ϕ lifts to a map $\psi : \prod_{i=1}^{\infty} G \longrightarrow G$ that takes each component G identically to G. Such a $G \neq 0$ is not slender. Thus nonzero slender injective R-*modules* do not exist.

Assume (μ) and let G be a self-slender right R-module. Let $\mathcal{W} \in G$-**coPlex**. From Theorem 9.32 we conclude that

1. $\mathrm{h}^G(\mathcal{W})$ is small in $\mathrm{End}_R(G)$-**Mod** iff $\mathcal{W}$ is slender in G-**coPlex**.
2. $\mathrm{h}^G(\mathcal{W})$ is slender in $\mathrm{End}_R(G)$-**Mod** iff $\mathcal{W}$ is small in G-**coPlex**.
3. $\mathrm{h}^G(\mathcal{W})$ is projective in $\mathrm{End}_R(G)$-**Mod** iff $\mathcal{W}$ is injective in G-**coPlex**.
4. $\mathrm{h}^G(\mathcal{W})$ is injective in $\mathrm{End}_R(G)$-**Mod** iff $\mathcal{W}$ is projective in G-**coPlex**.

Under $\mathrm{h}^G(\cdot)$, $\mathrm{End}_R(G)$ corresponds to the G-coplex

$$G \longrightarrow 0 \longrightarrow 0 \longrightarrow \cdots,$$

which we continue to call G through an abuse of notation. Then $\mathrm{h}^G(G) = \mathrm{End}_R(G)$. Since $\mathrm{End}_R(G)$ is a small projective generator in $\mathrm{End}_R(G)$-**Mod**, G is a slender injective cogenerator in G-**coPlex**.

Theorem 9.44. *Assume* (μ). *Suppose that G is a self-slender right R-module. Then G is a slender injective cogenerator in G-**coPlex**.*

Corollary 9.45. *Assume* (μ). *Suppose that G is a countable reduced torsion-free group. Then G is a slender injective object in G-**coPlex**.*

Thus there are many slender injective objects. Notice that we are not proving that there exists some exceptional slender injective. Our work states that assuming (μ)

then every self-slender right R-module is a slender injective object in a category that we construct.

Compare the next result with the fact that finitely generated (projective) right R-modules are small.

Theorem 9.46. *Assume* (μ). *Suppose that G is a self-slender right R-module. There are no nonzero small projective objects in G-**coPlex**. Specifically, nonzero indecomposable projectives in G-**coPlex** are not small.*

Corollary 9.47. *Assume* (μ). *Suppose that G is a countable reduced torsion-free group. There are no nonzero small projective objects in G-**coPlex**.*

Corollary 9.48. *Assume* (μ). *Suppose that G is a self-slender right R-module. Further assume that $\mathrm{End}_R(G)$ is a semi-perfect ring. Then indecomposable injective objects in G-**coPlex** are slender.*

Proof: Let $\mathcal{W}$ be an indecomposable injective object in G-**coPlex**. The duality in Theorem 9.32 takes $\mathcal{W}$ to the indecomposable projective $\mathrm{h}^G(\mathcal{W})$. Since $\mathrm{End}_R(G)$ is semi-perfect $\mathrm{h}^G(\mathcal{W})$ is cyclic, hence small. Thus $\mathcal{W} \cong \mathrm{h}^G\mathrm{h}^G(\mathcal{W})$ is slender. This completes the proof.

Corollary 9.49. *Assume* (μ). *Suppose that G is a self-slender right R-module. Further assume that M is an indecomposable injective left $\mathrm{End}_R(G)$-module. Then the indecomposable projective object $\mathrm{h}^G(M)$ is not small.*

9.6 Exercises

Fix a ring R, let G, H, K, and L be right R-modules.

1. Let K be a finitely generated right $\mathrm{End}(G)$-module. Show that $\mathbf{T}_G(K)$ is finitely G-generated. Show that if K is finitley presented then $\mathbf{T}_G(K)$ is finitely G-presented.
2. Prove that $\Theta_{\mathbf{T}_G(M)} \circ \mathbf{T}_G(\Phi_M) = 1_{\mathbf{T}_G(M)}$ for each right $\mathrm{End}(G)$-module M.
3. Formulate and prove a dual identity to the previous problem for $\Phi_{\mathbf{H}_G(H)}$ and Θ_H for right R-modules H.
4. The following are equivalent.
   (a) G is projective relative to $0 \longrightarrow K \longrightarrow L \longrightarrow H \longrightarrow 0$.
   (b) Each $Q \in \mathbf{P}(G)$ is projective relative to $0 \longrightarrow K \longrightarrow L \longrightarrow H \longrightarrow 0$.
5. Suppose that $\mathcal{D}$ is a commutative diagram in a category and let F be an additive functor on this category. Show that $F(\mathcal{D})$ is a commutative diagram.
6. If F is an additive functor, if Q is a complex, and if $s = (\cdots, s_1, s_0)$ is a null homotopy for Q, then $F(s)$ is a null homotopy for $F(Q)$.
7. If F is an additive functor and if Q and Q' are homotopic complexes, then $F(Q)$ and $F(Q')$ are homotopic complexes.
8. Show that any two projective complexes $\mathcal{P}$ and $\mathcal{P}'$ of a single right R-module M have the same homotopy type.

9. Let M and M' be right $\text{End}_R(G)$-modules and let $\mathcal{P}$ and $\mathcal{P}'$ be projective complexes of M and M' respectively. Suppose that f and g are chain map liftings of $\phi : M \longrightarrow M'$. Show that $[f] = [g]$.

10. Suppose that $(\cdots, f_1, f_0) : \mathcal{Q} \longrightarrow \mathcal{Q}'$ is a chain map. If f_1 and f_0 are isomorphisms then $\mathcal{Q}$ and $\mathcal{Q}'$ have the same homotopy type.

11. Prove that H is G-generated iff $\mathbf{S}_G(H) = H$.

12. Prove that H is G-generated iff $\Theta_H : \mathbf{T}_G\mathbf{H}_G(H) \longrightarrow H$ is a surjection.

13. The R-module G is *quasi-projective* if G is projective relative to each surjection $G \longrightarrow H \longrightarrow 0$. G is Σ-*quasi-projective* if for each cardinal c the direct sum $G^{(c)}$ is projective relative to each surjection $G^{(c)} \longrightarrow H \longrightarrow 0$. A finitely generated quasi-projective right R-module is self-small and Σ-quasi-projective.

14. Let G be a Σ-quasi-projective module and let

$$0 \longrightarrow K \longrightarrow H \longrightarrow L \longrightarrow 0$$

be a short exact sequence of right R-modules. Show that H is G-generated if K and L are G-generated.

15. Let $n > 0$ be an integer and let G be a Σ-quasi-projective module and let

$$0 \longrightarrow K \longrightarrow G^{(n)} \longrightarrow L \longrightarrow 0$$

be a short exact sequence of right R-modules. Show that L is G-solvable iff K is G-generated.

16. Suppose that G is Σ-quasi-projective. Suppose that H and H' possess G-plexes $\mathcal{Q}$ and $\mathcal{Q}'$ respectively and let $\phi : H \longrightarrow H'$ be a map of right R-modules. Show that ϕ lifts to a chain map $f : \mathcal{Q} \longrightarrow \mathcal{Q}'$.

17. Give an example of a right R-module G and a G-plex $\mathcal{Q}$ such that $\mathcal{Q}$ is not exact.

18. Give an example of a right R-module G and a G-plex $\mathcal{Q}$ such that G is not projective relative to δ_1.

19. Show that finitely generated modules are self-small. Show that if G is an abelian subgroup of $\mathbf{Q}^{(n)}$ then G is a self-small abelian group.

20. Show that if p is a prime and if G is an abelian subgroup of $(\mathbf{Z}(p^\infty))^{(n)}$ then G is a self-small abelian group.

21. Dualize the notion of finitely generated module to a cofinitely generated module.

22. Suppose G is an R-module with finitely generated essential socle. That is, there are finitely many simple R-modules $S_1, \ldots, S_n \subset G$ such that $S_1 \oplus \cdots \oplus S_n$ intersects each R-submodule of G in a nonzero module. Show that G is cofinitely generated.

23. Let $p \in \mathbf{Z}$. Show that $\mathbf{Z}(p^\infty)$ is not self-small but that it is self-slender.

24. [L. Fuchs, L. Salce] Let R be a commutative integral domain with field of quotients Q, and whose ideals form a chain of length $\aleph > \aleph_o$. Show that Q/R is (self-)small. Prove or disprove: G is self-slender.

25. Show that R is right coherent iff the class of finitely presented right R-modules is closed under finitely generated submodules.

26. Let

$$0 \longrightarrow K \longrightarrow G^{(n)} \overset{\delta}{\longrightarrow} L \longrightarrow 0$$

be a short exact sequence of right R-modules. Given conditions on L and δ for K to be G-solvable. Given conditions on K and δ for K to be G-solvable.

27. Prove or disprove: G is a Σ-self-generator iff each G-generated R-module is G-solvable.

28. Each G-solvable R-module is G-presented. Given an example of a G-presented R-module that is not G-solvable.

29. Give a factorization of $\mathbf{H}_G(\cdot)$ that is similar to the factorization of $\mathbf{T}_G(\cdot)$ given in Theorem 9.25.

30. Verify that the triangles in (9.11) and (9.12) are commutative.

31. Prove the dual results that appear in Section 9.3.

32. Using the primes in $\mathbf{Z}$ construct a set of infinitely many nonisomorphic subgroups of $\mathbf{Q}$.

33. Let E be a *rtffr* ring. **Corner's theorem 1.11** states that $E \cong \mathrm{End}_{\mathbf{Z}}(G)$ for some reduced torsion-free group G of finite rank. In fact G fits into a short exact sequence

$$0 \longrightarrow E \longrightarrow G \longrightarrow QE \longrightarrow 0$$

of *left E-modules*. Let E and G be as above.

(a) Prove that G is a flat left E-module.

(b) Show that $K \otimes_E G \neq 0$ for each right E-module $K \neq 0$.

(c) Show that the injective dimension of $\mathbf{Q}G$, $\mathrm{id}_E(\mathbf{Q}G)$, is n iff $\mathrm{id}_E(\mathbf{Q}E) = n$.

34. Let E be a *rtffr* ring and let M be a left E-module whose additive structure is a *rtffr* group. A theorem due to T. G. Faticoni and P. H. Goeters [56] states that $E \cong \mathrm{End}_{\mathbf{Z}}(G)$ for some reduced torsion-free group G of finite rank that fits into a short exact sequence

$$0 \longrightarrow M \oplus E \longrightarrow G \longrightarrow QE \oplus QE \longrightarrow 0$$

of *left E-modules*. Let E, M, and G be as above.

(a) Prove that G is a flat left E-module iff M is a flat left E-module.

(b) Show that $K \otimes_E G \neq 0$ for each right E-module $K \neq 0$ iff $K \otimes_E M \neq 0$ for each right E-module $K \neq 0$.

(c) Show that the projective dimension of G, $\mathrm{pd}_E(G)$, is n iff $\mathrm{fd}_E(M) = n$.

(d) Show that the flat dimension of G, $\mathrm{fd}_E(G)$, is n iff $\mathrm{fd}_E(M) = n$.

35. Show that G satisfies an injective and cogenerator property in G-**Cohcoplx**.

36. Show that G satisfies some type of slender property in G-**Cohcoplx**.

9.7 Problems for future research

Let R be an associative ring with identity, let G be a right R-module, let $\mathrm{End}_R(G)$ be the endomorphism ring of G, let $\mathcal{Q}$ be a G-plex, let $\mathcal{P}$ be a projective resolution of right $\mathrm{End}_R(G)$-modules.

1. Dualize the property *finitely generated*. If your dualization is property P then under a duality a finitely generated (projective) module M should correspond to an injective with property P.

2. Given a G-presented R-module H determine all of the homotopy classes of G-plexes for H.

3. Given a G-closed R-submodule H of G^c determine all of the homotopy classes of G-coplexes for H.

4. Fix a full subcategory $\mathcal{X}$ of **Mod**-$\text{End}_R(G)$ or of G-**Plex** and develope a specialized category equivalence associated with $\mathcal{X}$.

5. Dualize the previous problem.

6. Find a self-contained method for determining whether a complex is a G-plex or a G-coplex.

7. Maps of the form $R^n \longrightarrow R^m$ for integers n and m can occur as the first maps in R-plexes or R-coplexes. Investigate this situation. Specifically, is there a hidden duality lurking behind this observation?

8. Investigate the nature of monomorphisms and epimorphisms in G-**Plex** and G-**coPlex**.

9. Characterize the G-coherent R-modules. Specifically, what are the R-coherent R-modules? Dualize.

10. Characterize Artinian rings in terms of R-plexes. In general, choose a ring-theoretic property and characterize rings R that have that property in terms of R-plexes.

11. We have factored the tensor functor. Investigate this factorization more carefully.

12. Study the G-solvable modules.

13. Study the direct sums $G^{(c)}$, the products G^c, and the maps between them. This seems to be what G-plexes and G-coplexes are steering us toward.

14. More examples of self-small, self-slender R-modules are needed.

15. When is G self-small or self-slender as a left $\text{End}_R(G)$-module?

16. What more can we say about reduced torsion-free finite rank abelian groups?

10

Tensor functor equivalences

Let G be a self-small right R-module. By Theorem 9.25 the tensor functor factors as $\mathbf{T}_G(\cdot) = H_0 \circ t_G(\cdot)$, and by Theorem 9.12, $\mathbf{T}_G(\cdot)$ lifts to a category equivalence $t_G(\cdot) : \mathbf{Mod}\text{-}\mathrm{End}_R(G) \longrightarrow G\text{-}\mathbf{Plex}$. It is then natural to seek conditions under which $H_0(\cdot)$ is a category equivalence.

We will also start an additional theme corresponding to the following chain of categories. By defining

$$
\begin{aligned}
G\text{-}\mathbf{Gen} &= \text{ the category of } G\text{-generated } R\text{-modules} \\
G\text{-}\overline{\mathbf{Gen}} &= \text{ the category of } R\text{-submodules of } H \in G\text{-}\mathbf{Gen}
\end{aligned}
$$

we find an interesting chain of inclusions for a (not necessarily self-small) right R-module.

$$
\mathbf{P}(G) \subset G\text{-}\mathbf{Sol} \subset G\text{-}\mathbf{Pres} \subset G\text{-}\mathbf{Gen} \subset G\text{-}\overline{\mathbf{Gen}} \subset \mathbf{Mod}\text{-}R.
$$

The main theme of this chapter is to determine properties of G that allow us to replace an inclusion $\subset$ with an equation $=$ in the above chain. For instance, we ask for conditions on G such that $G\text{-}\mathbf{Sol} = G\text{-}\mathbf{Pres}$.

10.1 Small projective generators

An additive functor $F : \mathbf{C} \longrightarrow \mathbf{D}$ induces abelian group homomorphisms

$$
\mathrm{Hom}_{\mathbf{C}}(X, Y) \longrightarrow \mathrm{Hom}_{\mathbf{D}}(F(X), F(Y)) : f \longmapsto F(f) \tag{10.1}
$$

for each $X, Y \in \mathbf{C}$. We say that F is *faithful* functor if (10.1) is an injection for each $X, Y \in \mathbf{C}$. We say that F is a *full* functor if (10.1) is a surjection for each $X, Y \in \mathbf{C}$.

We say that a map $\rho : H \longrightarrow H'$ is *G-balanced onto its image* if G is projective relative to the induced short exact sequence

$$
0 \longrightarrow \ker \rho \longrightarrow H \xrightarrow{\ \rho\ } \text{image } \rho \longrightarrow 0.
$$

Lemma 10.1. *Let G be a self-small right R-module. Suppose that $H_0(\cdot) : G$-**Plex** $\longrightarrow$ G-**Pres** is a full functor. Then*

*1. Each map $\rho : H \longrightarrow H'$ in G-**Pres** is G-balanced onto its image, and*
2. $H_0(\cdot)$ is a faithful functor.

Proof: 1. Suppose that $\rho : H \longrightarrow H'$ is a map in G-**Pres** and let $\phi : G \longrightarrow$ image δ be an R-module map. Let $\mathcal{G}$ denote the G-plex $0 \longrightarrow G$. There are G-plexes $\mathcal{Q}$ and $\mathcal{Q}'$ for H and H', respectively. Because $H_0(\cdot)$ is a full functor ρ and ϕ lift to chain maps

$$g = (\cdots, g_1, g_0) : \mathcal{Q} \longrightarrow \mathcal{Q}'$$
$$f = (\cdots, 0, f_0) : \mathcal{G} \longrightarrow \mathcal{Q}'$$

such that $H_0([g]) = \phi$ and $H_0([f]) = \phi$.

Since G is a self-small right R-module, Theorem 9.17 states that $\mathcal{G}$ is a small projective generator in G-**Plex**. Thus there is a chain map $h : \mathcal{G} \longrightarrow \mathcal{Q}$ such that $[g][h] = [gh] = [f]$. Let $\psi = H_0([h])$. An application of $H_0(\cdot)$ shows us that

$$\rho\psi = H_0([g])H_0([h]) = H_0([gh]) = H_0([f]) = \phi.$$

This proves part 1.

2. Suppose that $f : \mathcal{Q} \longrightarrow \mathcal{Q}'$ is a chain map such that $H_0([f]) = 0$ as in the commutative diagram below.

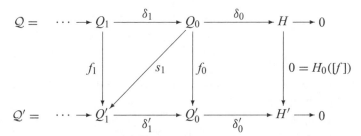

A chase through this diagram shows us that

$$\text{image } f_0 \subset \mathbf{S}_G(\ker \delta_0') = \text{image } \delta_1'.$$

By part 1, δ_1' is G-balanced onto its image, so δ_1' is Q-balanced onto its image for each $Q \in \mathbf{P}(G)$. Thus there is a map $s_1 : Q_0 \longrightarrow Q_1'$ such that $\delta_1' s_1 = f_0$. Lemma 9.10 then implies that f is null homotopic, or equivalently that $[f] = [0]$. Thus $H_0(\cdot)$ is faithful, which completes the proof.

Note the lack of a self-small hypothesis in the next result.

Theorem 10.2. *The following are equivalent for the right R-module G.*

*1. $H_0 : G$-**Plex** $\longrightarrow$ G-**Pres** is a full functor.*
*2. $H_0 : G$-**Plex** $\longrightarrow$ G-**Pres** is a category equivalence.*
*3. Each map $\rho : H \longrightarrow H'$ in G-**Pres** is G-balanced onto its image.*

Proof: $1 \Rightarrow 2$. By part 1, $H_0(\cdot)$ is a full functor and by Lemma 10.1.2, $H_0(\cdot)$ is a faithful functor. Furthermore, by Lemma 9.24, to each $H \in G\text{-}\mathbf{Pres}$ there is a $Q \in G\text{-}\mathbf{Plex}$ such that $H_0(Q) \cong H$. Thus $H_0(\cdot)$ is a category equivalence.

$2 \Rightarrow 3$. By part 2, $H_0(\cdot)$ is a full functor, so by Lemma 10.1 each map $\phi : H \longrightarrow H'$ in $G\text{-}\mathbf{Pres}$ is G-balanced onto its image.

$3 \Rightarrow 1$. Let $\rho : H \longrightarrow H'$ be a map in $G\text{-}\mathbf{Pres}$. By Lemma 9.24 there are Q, $Q' \in G\text{-}\mathbf{Plex}$ such that $H_0(Q) = H$ and $H_0(Q') = H'$. Since $Q_0 \in \mathbf{P}(G)$, part 3 shows us that δ'_0 is Q_0-balanced onto its image. Then $\rho\delta_0 : Q_0 \longrightarrow H'$ lifts to a map $f_0 : Q_0 \longrightarrow Q'_0$ such that the diagram

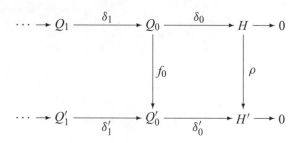

commutes.

Tracing the diagram and using the fact that Q is a complex we arrive at the equations

$$\delta'_0 f_0 \delta_1 = \rho\delta_0\delta_1 = 0.$$

Hence

$$f_0\delta_1 \subset \mathbf{S}_G(\ker \delta'_0) = \text{image } \delta'_1$$

by Lemma 9.6. Operating inductively, the projective property in part 3 lifts $f_0\delta_1 : Q_1 \longrightarrow Q'_0$ to a chain map

$$f = (\cdots, f_1, f_0) : Q \longrightarrow Q'$$

as in the commutative diagram below.

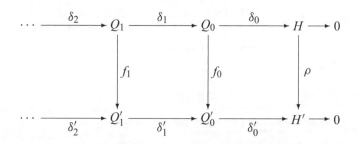

Then $H_0([f]) = \rho$, thus proving that $H_0(\cdot) : G\text{-}\mathbf{Plex} \longrightarrow G\text{-}\mathbf{Pres}$ is a full functor. This proves part 1 and completes the logical cycle.

The Morita theorems state that if G is an S-R-bimodule then $\mathbf{T}_G(\cdot) : \mathbf{Mod}\text{-}S \longrightarrow \mathbf{Mod}\text{-}R$ and $\mathbf{H}_G(\cdot) : \mathbf{Mod}\text{-}R \longrightarrow \mathbf{Mod}\text{-}S$ are inverse category equivalences iff G is a *progenerator* (= finitely generated projective generator) as a left S-module iff G is a progenerator as a right R-module. The next series of results combine to form a generalization of the Morita theorems.

Lemma 10.3. *Let G be a right R-module, and let $\mathbf{C}$ be a full subcategory of G-$\mathbf{Pres}$ that contains G.*

1. *If $\mathbf{H}_G(\cdot) : \mathbf{C} \longrightarrow \mathbf{Mod}\text{-}\mathrm{End}_R(G)$ is a category equivalence then the tensor functor $\mathbf{T}_G(\cdot) : \mathbf{Mod}\text{-}\mathrm{End}_R(G) \longrightarrow \mathbf{C}$ is the inverse of $\mathbf{H}_G(\cdot)$.*
2. *If $\mathbf{T}_G(\cdot) : \mathbf{Mod}\text{-}\mathrm{End}_R(G) \longrightarrow \mathbf{C}$ is a category equivalence then $\mathbf{H}_G(\cdot) : \mathbf{C} \longrightarrow \mathbf{Mod}\text{-}\mathrm{End}_R(G)$ is the inverse of $\mathbf{T}_G(\cdot)$.*

Proof: 1. Let $H \in \mathbf{C}$. Since $G \in \mathbf{C}$ and since $\mathbf{H}_G(G) = \mathrm{End}_R(G)$, G is a small projective generator in $\mathbf{C}$. Hence $\mathbf{T}_G\mathbf{H}_G(G^c) \cong G^{(c)}$ for each cardinal c. Since G is a small projective generator in $\mathbf{C}$, $\mathbf{T}_G\mathbf{H}_G(H) \cong H$. Hence $\mathbf{T}_G(\cdot)$ is the (left) inverse of $\mathbf{H}_G(\cdot)$.

2. Let $M \in \mathbf{Mod}\text{-}\mathrm{End}_R(G)$. The isomorphisms below follow from an application of the adjoint isomorphism. Since $M \in \mathbf{Mod}\text{-}\mathrm{End}_R(G)$,

$$\mathbf{H}_G\mathbf{T}_G(M) \cong \mathrm{Hom}_R(G, \mathbf{T}_G(M))$$
$$\cong \mathrm{Hom}_{\mathrm{End}_R(G)}(\mathbf{H}_G(G), M)$$
$$\cong \mathrm{Hom}_{\mathrm{End}_R(G)}(\mathrm{End}_R(G), M)$$
$$\cong M.$$

Then $\mathbf{H}_G(\cdot)$ is the (left) inverse of the category equivalence $\mathbf{T}_G(\cdot)$. This completes the proof.

Lemma 10.4. *Suppose that $\mathbf{H}_G(\cdot) : G\text{-}\mathbf{Pres} \longrightarrow \mathbf{Mod}\text{-}\mathrm{End}_R(G)$ is a category equivalence. Then G is a small object in G-$\mathbf{Pres}$.*

Proof: Let $\{H_j \,|\, j \in J\}$ be an indexed subset of G-$\mathbf{Pres}$. Because $\mathbf{H}_G(\cdot)$ is a category equivalence on G-$\mathbf{Pres}$, $\mathbf{H}_G(\cdot)$ takes direct sums to direct sums. Thus

$$\mathrm{Hom}_R(G, \oplus_j H_j) \cong \mathbf{H}_G(\oplus_j H_j)$$
$$\cong \oplus_j \mathbf{H}_G(H_j)$$
$$\cong \oplus_j \mathrm{Hom}_G(G, H_j)$$

which proves that G is a small object in G-$\mathbf{Pres}$.

Recall Lemma 9.23, which shows that given $M \in \mathbf{Mod}\text{-}\mathrm{End}_R(G)$ then $\mathbf{T}_G(M) \in G$-$\mathbf{Pres}$. Moreover, by Corollary 9.28, each $H \in G$-$\mathbf{Pres}$ can be written as $\mathbf{T}_G(M) \cong H$ for some $M \in \mathbf{Mod}\text{-}\mathrm{End}_R(G)$. Thus the functor $\mathbf{T}_G(\cdot) : \mathbf{Mod}\text{-}\mathrm{End}_R(G) \longrightarrow G$-$\mathbf{Pres}$ has restricted range.

Compare the following theorem, which characterizes when $\mathbf{T}_G(\cdot) : \mathbf{Mod}\text{-}\mathrm{End}_R(G)$ $\longrightarrow$ G-**Pres** is a category equivalence, with Theorem 10.2, which categorizes when $H_0(\cdot) : G$-**Plex** $\longrightarrow$ G-**Pres** is a category equivalence.

Theorem 10.5. *The following are equivalent for the right R-module G.*

1. *G is a small projective generator in G-**Pres**.*
2. *$\mathbf{T}_G(\cdot) : \mathbf{Mod}\text{-}\mathrm{End}_R(G) \longrightarrow G$-**Pres** is a category equivalence.*
3. *$\mathbf{T}_G(\cdot) : \mathbf{Mod}\text{-}\mathrm{End}_R(G) \longrightarrow G$-**Pres** and $\mathbf{H}_G(\cdot) : G$-**Pres** $\longrightarrow \mathbf{Mod}\text{-}\mathrm{End}_R(G)$ are inverse category equivalences.*
4. *G is self-small and $H_0(\cdot) : G$-**Plex** $\longrightarrow G$-**Pres** is a category equivalence.*
5. *G is a self-small right R-module and each map $\rho : H \longrightarrow H'$ in G-**Pres** is G-balanced onto its image.*

*In this case G-**Sol** $= G$-**Pres**.*

Proof: $1 \Rightarrow 3$. Because G is a small projective generator in G-**Pres**, Lemma 9.16 states that

$$\mathrm{Hom}_{G\text{-}\mathbf{Pres}}(G, \cdot) = \mathbf{H}_G(\cdot) : G\text{-}\mathbf{Pres} \longrightarrow \mathbf{Mod}\text{-}\mathrm{End}_R(G)$$

is a category equivalence. By Lemma 10.3, $\mathbf{T}_G(\cdot)$ is the inverse of $\mathbf{H}_G(\cdot)$. This is part 3.

$3 \Rightarrow 2$ is clear.

$2 \Rightarrow 1$. Since $\mathrm{End}_R(G)$ is a small projective generator in $\mathbf{Mod}\text{-}\mathrm{End}_R(G)$ and since $\mathbf{T}_G(\cdot)$ is a category equivalence, $G = \mathbf{T}_G(\mathrm{End}_R(G))$ is a small projective generator in G-**Pres**. This proves part 1.

$2 \Rightarrow 4$. Assume part 3. By Lemma 10.4, G is a small object in G-**Pres**, so that G is self-small. By Theorem 9.25 we can write $\mathbf{T}_G(\cdot) = H_0 \circ \mathrm{t}_G(\cdot)$, and, by Theorem 9.12, $\mathrm{t}_G(\cdot)$ is a category equivalence with inverse $\mathrm{h}_G(\cdot)$. By hypothesis, $\mathbf{T}_G(\cdot)$ is a category equivalence so

$$H_0(\cdot) = \mathbf{T}_G \circ \mathrm{h}_G(\cdot)$$

is a category equivalence. This is part 4.

$4 \Rightarrow 2$. Reverse the argument used to prove $2 \Rightarrow 4$.

$4 \Leftrightarrow 5$. follows from Theorem 10.2, which completes the logical cycle.

To prove that G-**Sol** $= G$-**Pres** we need only show that each G-presented R-module is G-solvable. Let H be G-presented. There is a short exact sequence

$$0 \longrightarrow K \longrightarrow Q \xrightarrow{\rho} H \longrightarrow 0$$

in which $Q \in \mathbf{P}(G)$ and K is G-generated. By part 1, G is projective relative to ρ so by definition H is G-solvable. This completes the proof.

Theorem 10.6. *Let G be a self-small right R-module. The following are equivalent.*

1. *G is a projective object in G-**Pres**.*
2. *$\mathbf{T}_G(\cdot) : \mathbf{Mod}\text{-}\mathrm{End}_R(G) \longrightarrow G$-**Pres** is a category equivalence.*

3. $\mathbf{T}_G(\cdot)$: $\mathbf{Mod}\text{-}\mathrm{End}_R(G) \longrightarrow G\text{-}\mathbf{Pres}$ *and* $\mathbf{H}_G(\cdot)$: $G\text{-}\mathbf{Pres} \longrightarrow \mathbf{Mod}\text{-} \mathrm{End}_R(G)$
   are inverse category equivalences.
4. H_0 : $G\text{-}\mathbf{Plex} \longrightarrow G\text{-}\mathbf{Pres}$ *is a category equivalence.*
5. *Each map* $\rho : H \longrightarrow H'$ *is G-balanced onto its image.*

In this case $G\text{-}\mathbf{Sol} = G\text{-}\mathbf{Pres}$.

Proof: The proof follows from the previous theorem.

Corollary 10.7. *If G is a small projective object in* $G\text{-}\mathbf{Pres}$ *then* $\mathbf{T}_G(K) \neq 0$ *for each right* $\mathrm{End}_R(G)$*-module* $K \neq 0$.

Example 10.8. Let c be an infinite cardinal and let $G = R^{(c)}$. Then G is a projective generator in $G\text{-}\mathbf{Pres} = \mathbf{Mod}\text{-}R$ but G is not small. (For instance, 1_G is not the sum of finitely many canonical idempotents $e_j : G \longrightarrow R$.) In fact, if I is the right ideal of $f \in \mathrm{End}_R(G)$ such that $f(G) \subset R^{(n)}$ for some integer $n > 0$, then $G = IG$. Hence $\mathbf{T}_G(\mathrm{End}_R(G)/I) = 0$, so that $\mathbf{T}_G(\cdot)$ is not a category equivalence.

10.2 Quasi-projective modules

The above theorems relate the tensor functor to the projective property. Much literature (see the references) has been devoted to showing that the right adjoint

$$\mathbf{H}_G(\cdot) : G\text{-}\mathbf{Gen} \longrightarrow \mathbf{Mod}\text{-}\mathrm{End}_R(G)$$

of $\mathbf{T}_G(\cdot)$ is a category equivalence iff G is small and satisfies some kind of projective property. We will pursue this line of investigation by extending a result due to G. Azumaya [18], K. Fuller [61], and M. Sato [98].

The right R-module G is *quasi-projective* if G is projective relative to each short exact sequence

$$0 \longrightarrow K \longrightarrow G \overset{\rho}{\longrightarrow} H \longrightarrow 0$$

of right R-modules. We say that G is Σ*-quasi-projective* if for each cardinal $c > 0$, G is projective relative to each short exact sequence

$$0 \longrightarrow K \longrightarrow G^{(c)} \overset{\rho}{\longrightarrow} H \longrightarrow 0 \tag{10.2}$$

of right R-modules. The reader will show as an exercise that if G is quasi-projective then for each *integer* $c > 0$, G is projective relative to (10.2). An additional exercise (which can be found in [7]) is the following. If G is quasi-projective and if $L \subset G^{(n)}$ is an R-submodule then G is projective relative to each short exact sequence

$$0 \longrightarrow K \longrightarrow L \overset{\rho}{\longrightarrow} H \longrightarrow 0 \tag{10.3}$$

of right R-modules. Furthermore, a finitely generated quasi-projective right R-module is self-small and Σ-quasi-projective.

In Theorem 10.5 we characterized the right R-modules G for which G-**Sol** = G-**Pres**. In the next result we characterize the right R-modules for which G-**Sol** = G-**Pres** = G-**Gen**.

To prove the following lemma, apply Lemma 10.3.

Lemma 10.9. *Let* G *be a right* R*-module such that* $\mathbf{H}_G(\cdot)$: G-**Gen** $\longrightarrow$ **Mod**-$\mathrm{End}_R(G)$ *is a category equivalence. Then* $\mathbf{T}_G(\cdot)$: **Mod**-$\mathrm{End}_R(G) \longrightarrow G$-**Gen** *is the inverse of* $\mathbf{H}_G(\cdot)$*. Specifically* G-**Pres** = G-**Gen**.

Theorem 10.10. *The following are equivalent for a right R-module G.*

1. *G is a small projective generator in G-**Gen**.*
2. *G is self-small, each map in G-**Pres** is G-balanced onto its image, and G-**Pres** = G-**Gen**.*
3. *G is self-small, G is Σ-quasi-projective, and G-**Pres** = G-**Gen**.*
4. *$\mathbf{T}_G(\cdot)$: **Mod**-$\mathrm{End}_R(G) \longrightarrow G$-**Gen** is a category equivalence.*
5. *$\mathbf{T}_G(\cdot)$: **Mod**-$\mathrm{End}_R(G) \longrightarrow G$-**Gen** and $\mathbf{H}_G(\cdot)$: G-**Gen** $\longrightarrow$ **Mod**-$\mathrm{End}_R(G)$ are inverse category equivalences.*

*In this case G-**Sol** = G-**Pres** = G-**Gen**.*

Proof: $3 \Rightarrow 2$ is clear.

$2 \Rightarrow 1$. By part 2 and Theorem 10.5, G is a small projective generator in G-**Pres** = G-**Gen**. This proves part 1.

$1 \Rightarrow 5$. Suppose that G is a small projective generator in G-**Gen**. By Theorem 9.16,

$$\mathbf{H}_G(\cdot) : G\text{-}\mathbf{Gen} \longrightarrow \mathbf{Mod}\text{-}\mathrm{End}_R(G)$$

is a category equivalence, and then by Lemma 10.3,

$$\mathbf{T}_G(\cdot) : \mathbf{Mod}\text{-}\mathrm{End}_R(G) \longrightarrow G\text{-}\mathbf{Gen}$$

and $\mathbf{H}_G(\cdot)$: G-**Gen** $\longrightarrow$ **Mod**-$\mathrm{End}_R(G)$ are inverse category equivalences. This is part 5.

$5 \Rightarrow 4$ is clear.

$4 \Rightarrow 3$. By Lemma 9.28, $\mathbf{T}_G(M) \in G$-**Pres** for each $M \in$ **Mod**-$\mathrm{End}_R(G)$ so $\mathbf{T}_G(\cdot)$: **Mod**-$\mathrm{End}_R(G) \longrightarrow G$-**Gen** = G-**Pres** is a category equivalence. By Theorem 10.5, G is self-small and each map $\rho : H \longrightarrow H'$ in G-**Pres** is G-balanced onto its image. This easily implies that G is a Σ-quasi-projective module, so we have proved part 3. This completes the logical cycle.

In case G satisfies any and hence all of the above conditions, then by Theorem 10.5 and part 3 above we have G-**Sol** = G-**Pres** = G-**Gen**. This completes the proof.

The right R-module K is G-*subgenerated* if there is a G-generated right R-module H such that $K \subset H$. The abelian group $\mathbf{Z}/n\mathbf{Z}$ is $\mathbf{Z}/n^2\mathbf{Z}$-subgenerated for each integer $n > 0$. Each torsion-free abelian group is $\mathbf{Q}$-subgenerated, and each torsion abelian p-group (p a fixed prime) is $\mathbf{Z}(p^\infty)$-subgenerated.

Let G be a right R-module. Then

1. G is a *self-generator* if G generates each R-submodule of G.
2. G is a σ-*self-generator* if for each integer $n \geq 1$, G generates each R-submodule of $G^{(n)}$.
3. G is a Σ-*self-generator* if for each cardinal c, G generates each R-submodule of $G^{(c)}$.

Evidently, Σ-self-generator $\Longrightarrow$ σ-self-generator $\Longrightarrow$ self-generator. Free modules are Σ-self-generating, as are right R-modules of the form $G = R \oplus H$ for some right R-module H.

Lemma 10.11. *Let G be a right R-module. Then G is a Σ-self-generator iff G is a σ-self-generator.*

Proof: Assume that G is a σ-self-generator, let $\mathcal{I}$ be an index set, and let $x \in G^{(\mathcal{I})}$. There is a finite subset $\mathcal{F} \subset \mathcal{I}$ such that $x \in G^{(\mathcal{F})} \subset G^{(\mathcal{I})}$. Since $\mathcal{F}$ is finite and since G is a σ-self-generator there is a set $\mathcal{J}$ and a surjection $\rho : G^{(\mathcal{J})} \longrightarrow xR$. Thus each cyclic R-submodule of $G^{(\mathcal{I})}$ is G-generated and so G is a Σ-self-generator. The converse is clear, so the proof is complete.

Lemma 10.12. *Let G be a right R-module. If G is a quasi-projective self-generator then G is a Σ-self-generator.*

Proof: Suppose that G is a quasi-projective self-generator. By Lemma 10.11 it suffices to show that G is a σ-self-generator.

Using induction on $n \geq 1$ suppose that G generates each R-submodule of $G^{(n)}$, let $x \in G^{(n+1)} = G^{(n)} \oplus G$, and consider the natural short exact sequence

$$0 \longrightarrow G^{(n)} \longrightarrow G^{(n+1)} \overset{\rho}{\longrightarrow} G \longrightarrow 0$$

of right R-modules. Since G is a self-generator there is an integer $k > 0$ and a map $\phi : G^{(k)} \longrightarrow G$ such that $\rho(xR) = \text{image } \phi$. Since G is quasi-projective there is a map $\psi : G^{(k)} \longrightarrow xR$ such that $\rho\psi = \phi$. Then $xR = \text{image } \psi + (xR \cap \ker \rho)$. Since $xR \cap \ker \rho \subset G^{(n)}$ the induction hypothesis implies that $xR \cap \ker \rho$ is G-generated. Thus xR is also G-generated and hence G is a σ-self-generator. This completes the proof.

Continuing our walk through the inclusions on page 148, it is now natural to ask for a characterization of the right R-modules G such that G-**Sol** is equal to

---

G-$\overline{\textbf{Gen}}$ = the category of G-subgenerated right R-modules.

Theorem 10.13. [61, K. Fuller] *Let G be a right R-module. The following are equivalent.*

1. G *is a small projective generator in G-$\overline{\textbf{Gen}}$.*
2. G *is a finitely generated quasi-projective self-generating right R-module.*

3. The right R-module G is self-small, Σ-quasi-projective, and a Σ-self-generator.
4. $\mathbf{T}_G(\cdot) : \mathbf{Mod}\text{-}\mathrm{End}_R(G) \longrightarrow G\text{-}\overline{\mathbf{Gen}}$ *is a category equivalence*
5. $\mathbf{T}_G(\cdot) : \mathbf{Mod}\text{-}\mathrm{End}_R(G) \longrightarrow G\text{-}\overline{\mathbf{Gen}}$ *and* $\mathbf{H}_G(\cdot) : G\text{-}\overline{\mathbf{Gen}} \longrightarrow \mathbf{Mod}\text{-}\mathrm{End}_R(G)$ *are inverse category equivalences.*

In this case $G\text{-}\mathbf{Sol} = G\text{-}\mathbf{Pres} = G\text{-}\mathbf{Gen} = G\text{-}\overline{\mathbf{Gen}}$.

Proof: $5 \Rightarrow 4$ is clear.

$4 \Rightarrow 1$. By part 4, $\mathbf{T}_G(\cdot) : \mathbf{Mod}\text{-}\mathrm{End}_R(G) \longrightarrow G\text{-}\overline{\mathbf{Gen}}$ is a category equivalence. Since $\mathrm{End}_R(G)$ is a small projective generator in $\mathbf{Mod}\text{-}\mathrm{End}_R(G)$, $\mathbf{T}_G(\mathrm{End}_R(G)) = G$ is a small projective generator in $G\text{-}\overline{\mathbf{Gen}}$.

$1 \Rightarrow 3$. Since G is a small projective generator in $G\text{-}\overline{\mathbf{Gen}}$, and since $G\text{-}\overline{\mathbf{Gen}}$ is closed under the formations of direct sums and R-submodules, it is readily shown that G is a self-small, Σ-quasi-projective, Σ-self-generator. This proves part 3.

$3 \Rightarrow 2$. Assume 3. Evidently G is a quasi-projective self-generator. We show that G is finitely generated.

Let $\oplus_{i \in \mathcal{I}} C_i$ be the direct sum of the cyclic R-submodules of G. Then $C_i \subset G \in G\text{-}\overline{\mathbf{Gen}}$. Since G generates $G\text{-}\overline{\mathbf{Gen}}$, there are $Q_i \in \mathbf{P}(G)$ and surjections $\rho_i : Q_i \longrightarrow C_i$ for each $i \in \mathcal{I}$. (Actually the ρ_i are epimorphisms in $G\text{-}\overline{\mathbf{Gen}}$, but the reader will show that an epimorphism in $G\text{-}\overline{\mathbf{Gen}}$ is a surjection.) Since G is projective in $G\text{-}\overline{\mathbf{Gen}}$, $1_G : G \longrightarrow G$ can be lifted over the canonical surjection

$$\rho = \bigoplus_{i \in \mathcal{I}} \rho_i : \bigoplus_i Q_i \longrightarrow \sum_i C_i = G$$

to a map $\phi : G \longrightarrow \oplus_i Q_i$. Since G is self-small there is a finite subset $\mathcal{F} \subset \mathcal{I}$ such that $\phi(G) \subset \oplus_{i \in \mathcal{F}} Q_i$. Then

$$G = 1_G(G) = \rho\phi(G) \subset \rho(\bigoplus_{i \in \mathcal{F}} Q_i) = \sum_{i \in \mathcal{F}} C_i \subset G$$

whence $G = \sum_{i \in \mathcal{F}} C_i$ is a finitely generated right R-module. This proves part 3.

$2 \Rightarrow 5$. By Lemma 10.3, it suffices to prove that

$$\mathbf{T}_G(\cdot) : \mathbf{Mod}\text{-}\mathrm{End}_R(G) \longrightarrow G\text{-}\overline{\mathbf{Gen}}$$

is a category equivalence.

Let $H \in G\text{-}\overline{\mathbf{Gen}}$. There is a G-generated module L such that $H \subset L$. By part 2 and Lemma 10.12, G is a Σ-self-generator, so H is G-generated. Write $H = Q/K$ for some $Q \in \mathbf{P}(G)$ and $K \subset Q$. Since G is a Σ-self-generator, K is G-generated, whence H is G-presented.

Now, by Corollary 9.28, there is an $M \in \mathbf{Mod}\text{-}\mathrm{End}_R(G)$ such that $H = \mathbf{T}_G(M)$. Hence we have proved that $G\text{-}\mathbf{Pres} = G\text{-}\overline{\mathbf{Gen}}$.

Also, by part 2, G is finitely generated and quasi-projective, so that each map $\rho : H \longrightarrow H'$ in $G\text{-}\mathbf{Pres} = G\text{-}\overline{\mathbf{Gen}}$ is G-balanced onto its image. By Theorem 10.6, $\mathbf{T}_G(\cdot) : \mathbf{Mod}\text{-}\mathrm{End}_R(G) \longrightarrow G\text{-}\mathbf{Pres} = G\text{-}\overline{\mathbf{Gen}}$ is a category equivalence, which is what we wanted to prove. This completes the logical cycle.

In case G satisfies any and hence all of the above conditions, we have shown that G-$\overline{\textbf{Gen}} = G$-**Pres**. An appeal to Theorem 10.10 shows us that

$$G\text{-}\textbf{Sol} = G\text{-}\textbf{Pres} = G\text{-}\textbf{Gen} = G\text{-}\overline{\textbf{Gen}}.$$

This completes the proof.

Corollary 10.14. *The following are equivalent for the right R-module G.*

1. *G is a small projective generator in G-$\overline{\textbf{Gen}}$.*
2. *G is a self-small Σ-quasi-projective Σ-self-generator.*
3. *G is a finitely generated quasi-projective self-generator.*
4. *Each map $\rho : H \longrightarrow H'$ in G-$\overline{\textbf{Gen}}$ is G-balanced onto its image.*

10.3 Flat endomorphism modules

A right R-module G is said to be *E-flat* if G is flat as a left $\text{End}_R(G)$-module and *faithful* if $IG \neq G$ for each right ideal $I \subset \text{End}_R(G)$. These conditions are not as restrictive as they might seem. That is, each generator (as a right R-module) is projective (= flat) over its endomorphism ring. Specifically $G = R \oplus H$ is projective over $\text{End}_R(G)$ where H is a right R-module. A finitely generated projective generator right R-module G is faithful and E-flat. By Corner's theorem 1.11 each ring E whose additive structure is a reduced countable torsion-free group is the endomorphism ring $\text{End}_R(G)$ for some group G. This G is faithful and E-flat by [56]. F. Richman and E. Walker [96] have show that a torsion p-group G is *not* E-flat iff $G = B \oplus D$ where B is bounded and $D \neq 0$ is divisible. Examples from [56] show that there are E-flat modules that are not faithful and faithful modules that are not E-flat. Given a prime $p \in \textbf{Z}$, the divisible group $\textbf{Z}(p^\infty)$ is neither faithful nor E-flat.

Given an E-flat right R-module G we determine the relationship between the G-generated submodules of $G^{(n)}$ for integers $n > 0$ and the right $\text{End}_R(G)$-submodules of finitely generated free right $\text{End}_R(G)$. If G is self-small and E-flat then we can delete the finitely generated hypothesis above.

10.3.1 A category equivalence for submodules of free modules

In this section R is an associative ring with identity and G is a self-small right R-module. Recall that $\textbf{P}(G)$ is the category of direct summands of $G^{(c)}$ for some cardinal $c > 0$, and that $\textbf{P}(\text{End}_R(G))$ is the category of projective right $\text{End}_R(G)$-modules. Let H be a right R-module. Given a subset $X \subset \textbf{H}_G(H)$ then $XG = \sum\{x(G) \mid x \in X\}$. The right $\text{End}_R(G)$-module M is called *G-torsion* if $\textbf{T}_G(M) = 0$ and M is called *G-torsion-free* if 0 is the largest G-torsion $\text{End}_R(G)$-submodule of M. Suppose that $N \subset M$. We say that N is *G-dense in M* if M/N is G-torsion. N is *G-closed in M* if M/N is G-torsion-free.

The next result is an exercise for the reader. It shows that the class of G-torsion $\text{End}_R(G)$-modules is a hereditary torsion class.

Lemma 10.15. *Let G be an E-flat R-module.*

1. *If* $0 \longrightarrow K \longrightarrow L \longrightarrow M \longrightarrow 0$ *is a short exact sequence of* $\mathrm{End}_R(G)$-modules *then L is G-torsion iff K and M are G-torsion.*
2. *If each* $M_j, j \in \mathcal{J}$ *is G-torsion then* $\oplus_{\mathcal{J}} M_j$ *is G-torsion.*
3. *If* $K \subset L$ *and if L is G-torsion-free then K is G-torsion-free.*

Proof: 1. Apply the exact functor $\mathbf{T}_G(\cdot)$ to the short exact sequence

$$0 \longrightarrow K \longrightarrow L \longrightarrow M \longrightarrow 0$$

to produce the short exact sequence

$$0 \longrightarrow \mathbf{T}_G(K) \longrightarrow \mathbf{T}_G(L) \longrightarrow \mathbf{T}_G(M) \longrightarrow 0.$$

Then $\mathbf{T}_G(L) = 0$ iff $\mathbf{T}_G(K) = \mathbf{T}_G(M) = 0$. This proves part 1.

The rest follows in the same manner.

Lemma 10.16. *Let G be an E-flat right R-module, let* $N \subset M \subset \mathbf{H}_G(G^{(n)})$ *be right* $\mathrm{End}_R(G)$-submodules for some integer $n > 0$.

1. $\mathbf{T}_G(M) \cong MA$ *canonically.*
2. *N is G-dense in M iff* $NG = MG$. *Specifically M is G-dense in* $\mathbf{H}_G(MG)$.

Proof: 1. This proof is an exercise.

2. Suppose that N is G-dense in M. Then $\mathbf{T}_G(M/N) = 0$. Apply the exact functor $\mathbf{T}_G(\cdot)$ to the short exact sequence

$$0 \longrightarrow N \longrightarrow M \longrightarrow M/N \longrightarrow 0 \qquad (10.4)$$

to show that $\mathbf{T}_G(N) \cong \mathbf{T}_G(M)$ canonically. Furthermore, by part 1,

$$NG \cong \mathbf{T}_G(N) \cong \mathbf{T}_G(M) \cong MG$$

canonically, so that $NG = MG$. The converse is an exercise for the reader.

The main theorem of this section requires the following notation. Given the right G-module G let

---

$\overline{\mathbf{P}}(G) =$ the set of G-generated R-modules H contained in some $Q \in \mathbf{P}(G)$

$\mathbf{Cl}(\mathrm{End}_R(G)) =$ the set of G-closed right $\mathrm{End}_R(G)$-sub-modules of free right $\mathrm{End}_R(G)$-modules.

Theorem 10.17. *Let G be a self-small E-flat right R-module. There are inverse category equivalences*

$$\mathbf{T}_G(\cdot) : \mathbf{Cl}(\mathrm{End}_R(G)) \longrightarrow \overline{\mathbf{P}}(G)$$

$$\mathbf{H}_G(\cdot) : \overline{\mathbf{P}}(G) \longrightarrow \mathbf{Cl}(\mathrm{End}_R(G)). \tag{10.5}$$

In particular, the maps Θ_H and Ψ_M are isomorphisms for each $H \in \overline{\mathbf{P}}(G)$ and $M \in \mathbf{Cl}(\mathrm{End}_R(G))$.

Proof: Consider the functors

$$\mathbf{T}_G(\cdot) : \mathbf{Mod}\text{-}\mathrm{End}_R(G) \longrightarrow \mathbf{Mod}\text{-}R$$

$$\mathbf{H}_G(\cdot) : \mathbf{Mod}\text{-}R \longrightarrow \mathbf{Mod}\text{-}\mathrm{End}_R(G).$$

In general, $\mathbf{H}_G(\cdot)$ is left exact and because G is E-flat, $\mathbf{T}_G(\cdot)$ is exact. Then $\mathbf{T}_G\mathbf{H}_G(\cdot)$ and $\mathbf{H}_G\mathbf{T}_G(\cdot)$ are left exact functors.

Given $H \in \overline{\mathbf{P}}(G)$, there is a $Q \in \mathbf{P}(G)$, an inclusion $H \subset Q$, and a commutative square

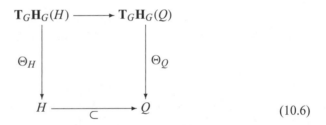

$$\tag{10.6}$$

of right R-modules. Since G is self-small, Theorem 1.13 states that Θ_Q is an isomorphism, and since $\mathbf{T}_G\mathbf{H}_G(\cdot)$ is left exact, the top row of (10.6) is an injection. Hence Θ_H is an injection. Because H is G-generated, Θ_H is a surjection, so that Θ_H is an isomorphism for each $H \in \overline{\mathbf{P}}(G)$.

Given $M \in \mathbf{Cl}(\mathrm{End}_R(G))$, there is a free $\mathrm{End}_R(G)$-module P such that M is a G-closed $\mathrm{End}_R(G)$-submodule of P. There is then a commutative square

$$
\begin{array}{ccc}
M & \overset{\subset}{\longrightarrow} & P \\
\Psi_M \downarrow & & \downarrow \Psi_P \\
\mathbf{H}_G\mathbf{T}_G(M) & \longrightarrow & \mathbf{H}_G\mathbf{T}_G(P)
\end{array}
\tag{10.7}
$$

of right $\mathrm{End}_R(G)$-modules. Since G is self-small, Theorem 1.13 states that Ψ_P is an isomorphism, so that Ψ_M is an injection. Now image $\Psi_M = \{m \otimes \cdot \mid m \in M\}$

so that

$$\Psi_M(M) \cdot G = \left\{ \sum_{\text{finite } i} m_i \otimes x_i \,\middle|\, m_i \in M, x_i \in G \right\} = \mathbf{T}_G(M).$$

Then Lemma 10.16.2, $\Psi_M(M)$ is G-dense in $\mathbf{H}_G\mathbf{T}_G(M)$.

Since G is E-flat, the four sides of the commutative square are injections. Thus we have

$$M = \Psi_M(M) \subset \mathbf{H}_G\mathbf{T}_G(M) \subset \mathbf{H}_G\mathbf{T}_G(P) = P.$$

Since M is chosen to be G-closed in P, M is also G-closed in $\mathbf{H}_G\mathbf{T}_G(M)$. Since $M = \Psi_M(M)$ is G-dense there, $M = \mathbf{H}_G\mathbf{T}_G(M)$. Hence Ψ_M is an isomorphism for each $M \in \mathbf{Cl}(\mathrm{End}_R(G))$.

We show that $\mathbf{H}_G(\cdot) : \overline{\mathbf{P}}(G) \longrightarrow \mathbf{Cl}(\mathrm{End}_R(G))$ is a well defined functor. Let $K \subset G^{(c)}$ for some cardinal c. Then $\mathbf{H}_G(K) \subset \mathbf{H}_G(G^{(c)})$. Let $M/\mathbf{H}_G(K)$ be the G-torsion $\mathrm{End}_R(G)$-submodule of $\mathbf{H}_G(G^{(c)})/\mathbf{H}_G(K)$. By Lemma 10.16.2 and because K is G-generated, $MG = \mathbf{H}_G(K)G = K$. Then $M \subset \mathbf{H}_G(K)$ so that $\mathbf{H}_G(G^{(c)})/\mathbf{H}_G(K)$ is G-torsion-free. In other words $\mathbf{H}_G(K)$ is G-closed in $\mathbf{H}_G(G^{(c)}) \cong \mathbf{H}_G(G)^{(c)}$. Inasmuch as $\mathbf{H}_G(G^{(n)})$ is a free right $\mathrm{End}_R(G)$-module, $\mathbf{H}_G(\cdot) : \overline{\mathbf{P}}(G) \longrightarrow \mathbf{Cl}(\mathrm{End}_R(G))$ is well defined.

Let $M \in \mathbf{Cl}(\mathrm{End}_R(G))$. Then by Lemma 10.16.1, $\mathbf{T}_G(M) = MG$. Hence $\mathbf{T}_G(\cdot) : \mathbf{Cl}(\mathrm{End}_R(G)) \longrightarrow \overline{\mathbf{P}}(G)$ is a well-defined functor, which completes the proof of the theorem.

The next result specializes Theorem 10.17 to the case where $\mathbf{T}_G(M) \neq 0$ for each right $\mathrm{End}_R(G)$-module $M \neq 0$. The right R-module G is *faithfully E-flat* if G is a flat left $\mathrm{End}_R(G)$-module such that $\mathbf{T}_G(M) \neq 0$ for each right $\mathrm{End}_R(G)$-module $M \neq 0$. Let

$$\boxed{\begin{array}{c} \overline{\mathbf{P}}(\mathrm{End}_R(G)) = \text{the category of right } \mathrm{End}_R(G)\text{-submodules} \\ \text{of free right } \mathrm{End}_R(G)\text{-modules.} \end{array}}$$

Theorem 10.18. *Let G be a self-small faithfully E-flat right R-module. There are inverse category equivalences*

$$\mathbf{T}_G(\cdot) : \overline{\mathbf{P}}(\mathrm{End}_R(G)) \longrightarrow \overline{\mathbf{P}}(G)$$
$$\mathbf{H}_G(\cdot) : \overline{\mathbf{P}}(G) \longrightarrow \overline{\mathbf{P}}(\mathrm{End}_R(G)).$$

Specifically, Θ_H and Ψ_M are isomorphisms for each $H \in \overline{\mathbf{P}}(G)$ and each right $\mathrm{End}_R(G)$-submodule of a free right $\mathrm{End}_R(G)$-module.

Proof: Observe that because G is faithfully E-flat, each $\mathrm{End}_R(G)$-submodule of a free right $\mathrm{End}_R(G)$-module is G-closed. Apply Theorem 10.17 to complete the proof.

Some notation is needed for the next result. Given a right R-module G let

$$\overline{\mathbf{P}}_o(G) = \text{the } G\text{-generated right } R\text{-submodules of } Q \text{ for some } Q \in \mathbf{P}_o(G)$$

$$\overline{\mathbf{P}}_o(\text{End}_R(G)) = \text{the category of right } \text{End}_R(G)\text{-submodules of} \\ \text{finitely generated free right } \text{End}_R(G)\text{-modules.}$$

$$\mathbf{Cl}_o(\text{End}_R(G)) = \text{the category of } G\text{-closed right } \text{End}_R(G)\text{-submodules} \\ \text{of finitely generated free right } \text{End}_R(G)\text{-modules.}$$

Without the self-small hypothesis there is a version of Theorem 10.17 for $\text{End}_R(G)$-submodules of finitely generated free right $\text{End}_R(G)$-modules. Since the proof for Theorem 10.20 is similar to the proof used for Theorem 10.17 (the only change is a use of Theorem 1.14 in place of Theorem 1.13), we will leave it as an exercise for the reader.

Theorem 10.19. *Let G be a (not necessarily self-small) E-flat right R-module. There are inverse category equivalences*

$$\mathbf{T}_G(\cdot) : \mathbf{Cl}_o(\text{End}_R(G)) \longrightarrow \overline{\mathbf{P}}_o(G)$$
$$\mathbf{H}_G(\cdot) : \overline{\mathbf{P}}_o(G) \longrightarrow \mathbf{Cl}_o(\text{End}_R(G)).$$

In particular, the maps Θ_H and Ψ_M are isomorphisms for each $H \in \overline{\mathbf{P}}_o(G)$ and $M \in \mathbf{Cl}_o(\text{End}_R(G))$.

For faithfully E-flat right R-modules there is the following theorem whose proof is left as an exercise. Follow the proof of Theorem 10.18.

Theorem 10.20. *Let G be a faithfully E-flat right R-module. There are inverse category equivalences*

$$\mathbf{T}_G(\cdot) : \overline{\mathbf{P}}_o(\text{End}_R(G)) \longrightarrow \overline{\mathbf{P}}_o(G)$$
$$\mathbf{H}_G(\cdot) : \overline{\mathbf{P}}_o(G) \longrightarrow \overline{\mathbf{P}}_o(\text{End}_R(G)).$$

In particular the maps Θ_H and Ψ_M are isomorphisms for each $H \in \overline{\mathbf{P}}_o(G)$ and $M \in \overline{\mathbf{P}}_o(\text{End}_R(G))$.

10.3.2 Right ideals in endomorphism rings

Given a (not necessarily self-small) E-flat right R-module G let

$$\mathcal{L}(G) = \text{the lattice of } G\text{-generated right } R\text{-submodules of } G$$

$$\mathcal{L}(\text{End}_R(G)) = \text{the lattice of right ideals of } \text{End}_R(G).$$

$$\mathcal{L}_c(\text{End}_R(G)) = \text{the lattice of } G\text{-closed right ideals in } \text{End}_R(G)$$

As usual, equivalence theorems spawn isomorphism theorems about the lattice of right ideals in $\text{End}_R(G)$.

Theorem 10.21. *Given a (not necessarily self-small) E-flat right R-module G, there is a functorial isomorphism of lattices*

$$\lambda_G : \mathcal{L}_c(\text{End}_R(G)) \longrightarrow \mathcal{L}(G)$$

defined by $\lambda_G(I) = IG$.

Proof: It is easy to see that λ_G is a homomorphism of posets, so all we have to do is prove that λ_G is a bijection.

Given $H \in \mathcal{L}(G)$ then $H = \mathbf{H}_G(H)G$. By Theorem 10.17, $\mathbf{H}_G(H) \in \text{Cl}(\text{End}_R(G))$, so that $\lambda_G(\mathbf{H}_G(H)) = H$, and hence λ_G is a surjection.

Suppose that we are given right ideals $I, J \subset \text{End}_R(G)$ such that $\lambda_G(I) = \lambda_G(J)$. Then

$$IG = JG = (I+J)G,$$

which by Lemma 10.16.2 implies that I and J are G-dense in $(I+J)$. Since I and J are G-closed in $\text{End}_R(G)$ we have $I = I + J = J$. Hence λ_G is an injection, whence $\lambda_G : \mathcal{L}_c(\text{End}_R(G)) \longrightarrow \mathcal{L}(G)$ is an isomorphism of lattices.

For a faithfully E-flat right R-module, the next result characterizes the lattice of G-generated R-submodules of G in terms of the right ideals in $\text{End}_R(G)$.

Theorem 10.22. *If G is a faithfully E-flat right R-module then there is a functorial isomorphism of lattices*

$$\lambda_G : \mathcal{L}(\text{End}_R(G)) \longrightarrow \mathcal{L}(G)$$

defined by $\lambda_G(I) = IG$.

Proof: Proceed as in the above proof, applying Theorem 10.20 where we applied Theorem 10.17.

10.3.3 A criterion for E-flatness

We will prove a result found in [115].

Theorem 10.23. *The Σ-self-generator G is E-flat.*

The proof of Theorem 10.23 is an application of an especially useful characterization of E-flat modules.

Lemma 10.24. *Let G be a self-small right R-module. The following are equivalent.*

1. G is E-flat.

2. Given an integer $n > 0$, K is G-generated in each exact sequence

$$0 \longrightarrow K \longrightarrow G^{(n)} \overset{\rho}{\longrightarrow} G. \tag{10.8}$$

3. Given cardinal c, K is G-generated in each exact sequence

$$0 \longrightarrow K \longrightarrow G^{(c)} \xrightarrow{\rho} G. \qquad (10.9)$$

Proof: $3 \Rightarrow 2$ is clear.

$2 \Rightarrow 1$. Assume 2, let $I \subset \mathrm{End}_R(G)$ be a finitely generated right ideal, and write $I = f_1 \mathrm{End}_R(G) + \cdots + f_n \mathrm{End}_R(G)$ for some $f_i \in \mathrm{End}_R(G)$. Let $\oplus_{i=1}^n G_i \cong G^{(n)}$ where $G_i = G$ for each i, and define a map $\rho : \oplus_{i=1}^n G_i \longrightarrow G$ such that $\rho(x_i) = f_i(x_i)$ for each $x_i \in G_i$ and $i = 1, \ldots, n$. Subsequently we have a short exact sequence (10.8). By part 2, K is G-generated.

An application of $\mathbf{H}_G(\cdot)$ to (10.8) produces the short exact sequence

$$0 \longrightarrow \mathbf{H}_G(K) \longrightarrow \mathbf{H}_G(G^{(n)}) \xrightarrow{\rho^*} I \longrightarrow 0.$$

A subsequent application of the exact functor $\mathbf{T}_G(\cdot)$ yields the commutative diagram

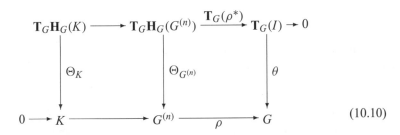

$$(10.10)$$

with exact rows. Since $\Theta_{G^{(n)}}$ is an isomorphism (Theorem 1.14), and since Θ_K is a surjection (K is G-generated), a diagram chase (or applying the Snake Lemma) shows that the canonical map $\theta : \mathbf{T}_G(I) \longrightarrow G$ is an injection. Thus G is E-flat.

$1 \Rightarrow 3$. Assume that G is a self-small E-flat right R-module and suppose that there is a short exact sequence (10.9) in which c is a cardinal. An application of the exact functor $\mathbf{T}_G\mathbf{H}_G(\cdot)$ produces the commutative diagram (10.10) with exact rows in which $n = c$, the center map $\Theta_{G^{(n)}}$ is an isomorphism, and where $I = \mathrm{image}\ \rho^*$. Since G is E-flat, θ is an injection, so a diagram chase (or the Snake Lemma) shows us that Θ_K is a surjection. Thus K is G-generated, which proves part 3. This completes the proof.

Remark 10.25. Won't someone please find a characterization of faithfully E-flat right R-modules along some other lines than what we have seen so far. Perhaps you could utilize the fact that the proof of Theorem 10.24 uses the Snake Lemma to connect the end maps of the commutative diagram (10.10).

10.4 Orsatti and Menini's *-modules

It is interesting to ask how E-flatness effects category equivalences $\mathbf{T}_G(\cdot)$ and $\mathbf{H}_G(\cdot)$. For this discussion we will consider the *-modules of D. Orsatti and C. Menini [88].

Given a right R-module G, let

$$\boxed{\begin{array}{c} G\text{-}\overline{\mathbf{Hom}} = \text{the category of right } \mathrm{End}_R(G)\text{-submodules of } \mathbf{H}_G(H) \\ \text{for } H \in \mathbf{Mod}\text{-}R. \end{array}}$$

Motivated by the work of D. Orsatti and C. Menini in [88], R. Colpi [24] coined the term *-module. We call G a *-module* if

$$\mathbf{T}_G(\cdot) : G\text{-}\overline{\mathbf{Hom}} \longrightarrow G\text{-}\mathbf{Gen}$$

and

$$\mathbf{H}_G(\cdot) : G\text{-}\mathbf{Gen} \longrightarrow G\text{-}\overline{\mathbf{Hom}}$$

are inverse category equivalences.

If G is a finitely generated quasi-projective self-generating right R-module then G is called a *quasi-progenerator*. By Theorem 10.13, G is a quasi-progenerator iff

$$\mathbf{T}_G(\cdot) : \mathbf{Mod}\text{-}\mathrm{End}_R(G) \longrightarrow G\text{-}\overline{\mathbf{Gen}}$$

is a category equivalence, so that a quasi-progenerator is a *-module. We will examine the converse.

Remark 10.26. It happens that *-modules are not too large. The *-module G is a small object in G-**Gen** since the category equivalence $\mathbf{H}_G(\cdot) : G\text{-}\mathbf{Gen} \longrightarrow G\text{-}\overline{\mathbf{Hom}}$ is a category equivalence that takes G to $\mathrm{End}_R(G)$. In fact, G is then a finitely generated object in G-**Gen**.

The E-flat *-modules are precisely the quasi-progenerators.

Theorem 10.27. *Suppose that the right R-module G is a *-module. The following are equivalent.*

1. *G is E-flat.*
2. *G is faithfully E-flat.*
3. *G is a Σ-self-generator.*
4. *G is quasi-projective.*
5. *G is a quasi-progenerator.*

Proof: $5 \Rightarrow 4$ is clear.

$4 \Rightarrow 3$. Let $K \subset Q \in \mathbf{P}_o(G)$ and let $H = Q/K$. Since G is quasi-projective, and since $Q \oplus Q' \cong G^{(n)}$ for some integer n, G is projective relative to the short exact sequence

$$0 \longrightarrow K \longrightarrow Q \overset{\rho}{\longrightarrow} H \longrightarrow 0.$$

An application of $\mathbf{H}_G(\cdot)$ yields the short exact sequence

$$0 \longrightarrow \mathbf{H}_G(K) \longrightarrow \mathbf{H}_G(Q) \overset{\rho^*}{\longrightarrow} \mathbf{H}_G(H) \longrightarrow 0$$

of right $\text{End}_R(G)$-modules. A subsequent application of the right exact functor $\mathbf{T}_G(\cdot)$ produces the exact first row of the commutative diagram with exact rows

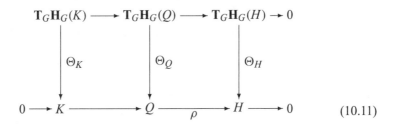

$$
\begin{array}{ccccccc}
\mathbf{T}_G\mathbf{H}_G(K) & \longrightarrow & \mathbf{T}_G\mathbf{H}_G(Q) & \longrightarrow & \mathbf{T}_G\mathbf{H}_G(H) & \to & 0 \\
\downarrow{\scriptstyle \Theta_K} & & \downarrow{\scriptstyle \Theta_Q} & & \downarrow{\scriptstyle \Theta_H} & & \\
0 \longrightarrow K & \longrightarrow & Q & \underset{\rho}{\longrightarrow} & H & \longrightarrow & 0
\end{array}
\qquad (10.11)
$$

Since G is a *-module, and since $Q, H \in G\text{-}\mathbf{Gen}$, there are natural homomorphisms Θ_Q and Θ_H that make the diagram commute. By Theorem 1.14, Θ_Q is an isomorphism, and by the *-module hypothesis Θ_H is an isomorphism. The naturality of Θ_K ensures that the diagram commutes. Then a diagram chase (or you can apply the Snake Lemma) shows us that Θ_K is a surjection. Hence K is G-generated, and so by Lemma 10.11, G is a Σ-self-generator. This proves part 3.

$3 \Rightarrow 1$ follows from Theorem 10.23.

$1 \Rightarrow 2$. Suppose that G is E-flat and for the purposes of this argument let $E = \text{End}_R(G)$. Let $I \subset E$ be a right ideal such that $IG = G$. Then $\mathbf{T}_G(E/I) = G/IG = 0$. An application of the exact functor $\mathbf{T}_G(\cdot)$ to the short exact sequence

$$
0 \longrightarrow I \overset{J}{\longrightarrow} E \longrightarrow E/I \longrightarrow 0
$$

yields the short exact sequence

$$
0 \longrightarrow \mathbf{T}_G(I) \overset{\mathbf{T}_G(J)}{\longrightarrow} \mathbf{T}_G(E) \longrightarrow 0
$$

in $\mathbf{Mod}\text{-}R$. That is, $\mathbf{T}_G(J)$ is an isomorphism. Since G is a *-module, $\mathbf{T}_G(J)$ is an isomorphism iff J is an isomorphism. Then $I = E = \text{End}_R(G)$, whence G is faithful. This proves part 2.

$2 \Rightarrow 5$. Let

$$
0 \longrightarrow K \longrightarrow Q \overset{\rho}{\longrightarrow} H \longrightarrow 0
\qquad (10.12)
$$

be a short exact sequence in $\mathbf{Mod}\text{-}R$ for some $Q \in \mathbf{P}(G)$. An application of $\mathbf{T}_G\mathbf{H}_G(\cdot)$ to (10.12) produces the diagram

$$
\begin{array}{ccccccc}
0 \to \mathbf{T}_G\mathbf{H}_G(K) & \longrightarrow & \mathbf{T}_G\mathbf{H}_G(Q) & \overset{\mathbf{T}_G(\rho^*)}{\longrightarrow} & \mathbf{T}_G\mathbf{H}_G(H) & \to & 0 \\
\downarrow{\scriptstyle \Theta_K} & & \downarrow{\scriptstyle \Theta_Q} & & \downarrow{\scriptstyle \Theta_H} & & \\
0 \longrightarrow K & \longrightarrow & Q & \underset{\rho}{\longrightarrow} & H & \longrightarrow & 0
\end{array}
\qquad (10.13)
$$

Since G is E-flat, $\mathbf{T}_G\mathbf{H}_G(\cdot)$ is a left exact functor, so the top row of (10.13) is left exact. Since G is a *-module, G is a small *-module, so that Θ_H and Θ_Q are isomorphisms. The top row of the diagram is then (right) exact. A diagram chase verifies that Θ_K is a surjection, and so G is a Σ-self-generator.

Now since $\mathbf{T}_G(\rho^*)$ is a surjection and since $\mathbf{T}_G(\cdot)$ is a right exact functor

$$\mathbf{T}_G(\mathrm{coker}\ \rho^*) = \mathrm{coker}\ \mathbf{T}_G(\rho^*) = 0.$$

Because G is faithfully E-flat, coker $\rho^* = 0$, or in other words, image $\rho^* = \mathbf{H}_G(H)$. Then G is projective relative to (10.12), and hence G is a Σ-quasi-projective right R-module. This proves that G is a small Σ-quasi-projective Σ-self-generator.

By Corollary 10.14, G is a finitely generated right R-module, so G satisfies part 5, which completes the logical cycle.

Remark 10.28. Jan Trlifaj [105] has proved that a *-module is a finitely generated right R-module.

10.5 Dualities from injective properties

In Section 10.1 we investigated the full subcategories **C** of **Mod**-R in which G is a small projective generator. In this section we dualize those results by considering categories in which G is a slender injective cogenerator. To make this happen in the largest possible setting we assume the statement

$$(\mu) \qquad \text{Measurable cardinals do not exist.}$$

We accomplish our stated goals by constructing commutative triangles

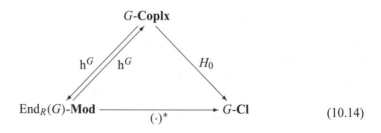

$$(10.14)$$

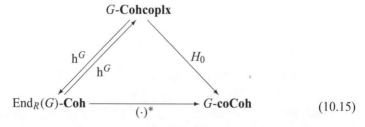

$$(10.15)$$

of categories and (inverse) functors. Let us define the elements of the diagrams.

The objects in $\mathbf{coP}(G)$ are direct summands of G^c for some cardinal c. The right R-module is *G-torsionless* if $H \subset G^c$ for some cardinal c. We say that H is *G-closed* if there is a $W \in \mathbf{coP}(G)$ such that $H \subset W$ and W/H is G-torsionless. Let

$$\boxed{G\text{-}\mathbf{Cl} = \text{ the category of } G\text{-closed right } R\text{-modules}}$$

and let

$$\boxed{\begin{array}{l} G\text{-}\mathbf{coCoh} = \text{ the right } R\text{-modules } H \text{ for which there exists} \\ \qquad \text{a } W \in G\text{-}\mathbf{Cohcoplx} \text{ such that } \ker \sigma_1 = H. \end{array}}$$

As usual, let

$$\boxed{H_0 : G\text{-}\mathbf{Coplx} \longrightarrow G\text{-}\mathbf{Cl}}$$

be the zero-th homology functor. That is,

$$\boxed{\begin{array}{c} H_0(\mathcal{W}) = \ker \sigma_1 \text{ for } G\text{-coplexes } \mathcal{W} \text{ and} \\ H_0([f]) = \text{ the restriction } f_0|\ker \sigma_1 \text{ for chain maps} \\ f = (f_0, f_1, \cdots) \text{ of } G\text{-coplexes.} \end{array}}$$

10.5.1 G-Cosolvable R-modules

We say that H is *G-cosolvable* if there is a short exact sequence

$$0 \longrightarrow H \longrightarrow W \overset{\rho}{\longrightarrow} C \longrightarrow 0 \tag{10.16}$$

of right R-modules such that G is injective relative to (10.16), $W \in \mathbf{coP}(G)$, and C is G-torsionless. As above, the G-cosolvable modules will play an important role in our study of category equivalences.

Let

$$\boxed{G\text{-}\mathbf{coSol} = \text{ the category of } G\text{-cosolvable right } R\text{-modules.}}$$

We require a lemma on the existence of G-balanced short exact sequences (9.13).

Lemma 10.29. *Assume* (μ). *Let G be a self-slender right R-module. Let H be a G-torsionless R-module. There is a short exact sequence (10.16) in which $W \in \mathbf{coP}(G)$.*

Proof. Dualize Lemma 9.4.

Lemma 10.30. *Assume* (μ). *Let G and H be right R-modules. The following are equivalent for H.*

1. *H is G-cosolvable.*
2. *If G is injective relative to the short exact sequence (10.16) in which* $W \in \mathbf{coP}(G)$ *then C is G-torsionless.*
3. $\Psi_H : H \longrightarrow H^{**}$ *is an isomorphism.*

Proof: Dualize the proof of Lemma 9.19.

We say that G is a Π-*self-cogenerator* if given $W \in \mathbf{coP}(G)$ and an R-submodule $K \subset W$ then W/K is G-torsionless. For a Π-self-cogenerator G the G-cosolvable right R-modules are easily recognized.

Proposition 10.31. *Assume* (μ). *Let G be a self-slender* Π-*self-cogenerator. Each G-torsionless right R-module is G-cosolvable.*

10.5.2 A factorization of $\mathrm{Hom}_R(\cdot, G)$

The main result in this subsection is a factorization of the functor

$$(\cdot)^* = \mathrm{Hom}_{\mathrm{End}_R(G)}(\cdot, G) \quad : \quad \mathrm{End}_R(G)\text{-}\mathbf{Mod} \longrightarrow \mathbf{Mod}\text{-}R$$

for right R-modules G.

Lemma 10.32. *Assume* (μ). *Let G be a right R-module. If M is a left* $\mathrm{End}_R(G)$-*module then* M^* *is G-torsionless. Thus the bottom rows of the triangles in (10.14) and (10.15) are well-defined.*

Lemma 10.33. *Assume* (μ). *Let G be a self-slender right R-module. Then* H_0 : G-**Coplx** $\longrightarrow$ G-**Cl** *is a functor such that for each* $H \in G$-**Cl** *there is a* $W \in G$-**Coplx** *such that* $H_0(W) = H$. *Thus the right-hand arrows of the triangles in (10.14) and (10.15) are well-defined.*

We have constructed the triangle (10.14) once we recall that the functors h^G are inverse dualities (Theorem 9.32).

Theorem 10.34. *Assume* (μ). *If G is a self-slender right R-module then*

$$\boxed{(\cdot)^* \cong H_0 \circ \mathrm{h}^G(\cdot).}$$

Proof: To define the duality

$$\mathrm{h}^G(\cdot) : \mathrm{End}_R(G)\text{-}\mathbf{Mod} \longrightarrow G\text{-}\mathbf{Coplx}$$

we choose for each $M \in \mathrm{End}_R(G)$-**Mod** a projective resolution $\mathcal{P}(M)$ and then we define $\mathrm{h}^G(M) = \mathcal{P}(M)^*$. By dualizing the proof of Theorem 9.25 we can prove that

$$h^G(\cdot) : \text{End}_R(G)\text{-}\mathbf{Mod} \longrightarrow G\text{-}\mathbf{Coplx}$$

is a duality and that the triangles (10.14) and (10.15) commute. This completes the proof.

Theorem 10.35. *Assume* (μ). *If* G *is a self-slender right* R-module, *there is a commutative triangle (10.14) in which opposing arrows indicate inverse category equivalences,* $H_0(\cdot)$ *is the zero-th homology functor, and* $(\cdot)^*$ *is the functor* $\text{Hom}_{\text{End}_R(G)}(\cdot, G)$.

Theorem 10.36. *Assume* (μ). *If* G *is any right* R-module, *there is a commutative triangle (10.15) in which opposing arrows indicate inverse category equivalences,* $H_0(\cdot)$ *is the zero-th homology functor, and* $(\cdot)^*$ *is the functor* $\text{Hom}_{\text{End}_R(G)}(\cdot, G)$.

10.5.3 Dualities for the dual functor

If G is a self-slender right R-module then $(\cdot)^* = H_0 \circ h^G(\cdot)$ by Theorem 10.34. Since in this case $h^G(\cdot)$ is a category equivalence, it is natural to ask about $H_0(\cdot)$. We will seek conditions under which $H_0(\cdot)$ is a full functor. Since these results are dual to the results in Section 10.1 we will leave the proofs in this section as exercises for the reader.

The next result shows us that there is some connection between injective properties for G, for the equivalence of $\text{End}_R(G)\text{-}\mathbf{Mod}$, and for $G\text{-}\mathbf{Cl}$.

Lemma 10.37. *Assume* (μ). *Let* G *be a self-slender right* R-module. *If* $H_0(\cdot)$: $G\text{-}\mathbf{Coplx} \longrightarrow G\text{-}\mathbf{Cl}$ *is a full functor, then*

1. G *is injective relative to each injection* $\phi : H \longrightarrow H'$ *in* $G\text{-}\mathbf{Cl}$ *and*
2. $H_0(\cdot)$ *is a faithful functor.*

The map $j : H \longrightarrow H'$ is said to be *G-cobalanced from its kernel* if given a map $\phi : H \longrightarrow G$ such that $\phi(\ker f) = 0$ then ϕ lifts to a map $\psi : H' \longrightarrow G$. That is, $f\psi = \phi$.

Note the lack of a self-slender hypothesis in the next result.

Theorem 10.38. *Assume* (μ). *The following are equivalent for the right* R-module G.

1. $H_0 : G\text{-}\mathbf{Coplx} \longrightarrow G\text{-}\mathbf{Cl}$ *is a full functor.*
2. $H_0 : G\text{-}\mathbf{Coplx} \longrightarrow G\text{-}\mathbf{Cl}$ *is a category equivalence.*
3. *Each map* $j : H \longrightarrow H'$ *in* $G\text{-}\mathbf{Cl}$ *is* G-cobalanced from its kernel.

Proof: Dualize the proof of Theorem 10.2.

The following series of results dualizes Theorems 10.5 and its consequences.

Theorem 10.39. *Assume* (μ). *The following are equivalent for the right R-module G.*

1. *G is a slender injective cogenerator in G-**Cl**.*
2. *G is self-slender and each map $\jmath : H \longrightarrow H'$ in G-**Cl** is G-cobalanced from its kernel.*
3. $(\cdot)^* :$ *End$_R(G)$-**Mod** $\longrightarrow$ G-**Cl** is a duality.*

Theorem 10.40. *Assume* (μ). *Let G be a self-slender right R-module. The following are equivalent.*

1. *G is an injective cogenerator in G-**Cl**.*
2. *Each map $\jmath : H \longrightarrow H'$ in G-**Cl** is G-cobalanced from its kernel.*
3. $H_0 :$ *G-**Coplx** $\longrightarrow$ G-**Cl** is a category equivalence.*
4. $(\cdot)^* :$ *End$_R(G)$-**Mod** $\longrightarrow$ G-**Cl** is a duality.*

Theorem 10.41. *Assume* (μ). *The following are equivalent for the right R-module G.*

1. *G is a slender injective cogenerator in G-**Cl**.*
2. $(\cdot)^* :$ *End$_R(G)$-**Mod** $\longrightarrow$ G-**Cl** and $(\cdot)^* :$ G-**Cl** $\longrightarrow$ End$_R(G)$-**Mod** are inverse dualities.*

Theorem 10.42. *Assume* (μ). *Let G be a self-slender right R-module. Then each left* End$_R(G)$-*module is torsionless. That is, each $M \in$ End$_R(G)$-**Mod** embeds in* Hom$_R(G, G^c) \cong$ End$_R(G)^c$ *for some cardinal c.*

Observe that in the above theorems we have related the functor $(\cdot)^*$ to the injective property. This partially dualizes Fuller's theorem 10.13 where it is shown that $\mathbf{H}_G(\cdot) :$ G-**Gen** $\longrightarrow$ **Mod**-End$_R(G)$ is a category equivalence iff G is a finitely generated ($=$ small) quasi-projective self-generator as a right R-module.

The right R-module G is Π-*quasi-injective* if for each cardinal c, G is injective relative to each short exact sequence

$$0 \longrightarrow H \overset{\sigma}{\longrightarrow} G^c \longrightarrow C \longrightarrow 0.$$

We say that G is a Π-*self-cogenerator* if for each cardinal c and R-submodule $H \subset G^c$, G^c/H is G-torsionless. G is a *cogenerator* if each quotient of G is G-torsionless. For example, if R is a local ring and if G is the injective hull of the unique simple R-module, then G is a Π-quasi-injective Π-self-cogenerator. Thus for primes $p \in \mathbf{Z}$, $\mathbf{Z}/p^2\mathbf{Z}$ is Π-quasi-injective and a Π-self-cogenerator but not a cogenerator. The abelian group $\mathbf{Z}(p^\infty)$ is an (Π-self-)injective (Π-self-)cogenerator. The additive group of rational numbers $\mathbf{Q}$ is an injective group. But $\mathbf{Q}$ is not a self-cogenerator since $\mathbf{Q}$ does not cogenerate its torsion factors.

Let

$$G\text{-}\overline{\mathbf{Cl}} = \text{the category of homomorphic images of} \atop G\text{-closed } R\text{-modules.}$$

Thus, given a self-slender right R-module G, we have defined a series of categories

$$\mathbf{coP}(G) \subset G\text{-}\mathbf{coSol} \subset G\text{-}\mathbf{Cl} \subset G\text{-}\overline{\mathbf{Cl}} \subset \mathbf{Mod}\text{-}R.$$

Dualizing the results in Chapter 9 will prove the following results. We are attempting to examine the conditions on G under which, for example, $G\text{-}\mathbf{Cl} = G\text{-}\overline{\mathbf{Cl}}$.

Theorem 10.43. *Assume* (μ). *Let* G *be a right* R-module. *The following are equivalent.*

1. *G is a slender injective cogenerator in G-**Cl**.*
2. *G is a self-slender Π-quasi-injective right R-module.*
3. *$(\cdot)^* : \mathrm{End}_R(G)\text{-}\mathbf{Mod}\longrightarrow G\text{-}\mathbf{Cl}$ is a category equivalence.*

*In this case, G-**coSol** $= G$-**Cl**.*

Theorem 10.44. *Assume* (μ). *Let* G *be a right* R-module. *The following are equivalent.*

1. *G is a slender injective cogenerator in G-$\overline{\mathbf{Cl}}$.*
2. *G is a self-slender Π-quasi-injective Π-self-cogenerator.*
3. *$(\cdot)^* : \mathrm{End}_R(G)\text{-}\mathbf{Mod} \longrightarrow G\text{-}\overline{\mathbf{Cl}}$ is a category equivalence*

*In this case, G-**coSol** $= G$-**Cl** $= G$-$\overline{\mathbf{Cl}}$.*

10.6 Exercises

1. Give examples of modules G for which one or more of the inclusions on page 148 is proper.
2. Give an example of a right R-module G for which the inclusions on page 148 are proper.
3. Show that $\mathbf{T}_G(\cdot) : \mathbf{Mod}\text{-}\mathrm{End}_R(G) \longrightarrow \mathbf{Mod}\text{-}R$ and $\mathbf{H}_G(\cdot) : \mathbf{Mod}\text{-}R \longrightarrow \mathbf{Mod}\text{-}\mathrm{End}_R(G)$ are inverse category equivalences iff G is a progenerator over either R or $\mathrm{End}_R(G)$.
4. If the right R-module G is a finitely generated generator, then G is a finitely generated projective left $\mathrm{End}_R(G)$-module.
5. If the right R-module G is finitely generated and projective, then the left $\mathrm{End}_R(G)$-module G is a generator.
6. Let G be a self-small right R-module. If G is a projective generator in G-**Pres**, then $\mathbf{T}_G(K) \neq 0$ for each right $\mathrm{End}_R(G)$-module $K \neq 0$.

7. The following are equivalent for the right R-module G.

   (a) G is a small projective generator in G-**Pres**
   (b) $\mathbf{T}_G(\cdot) : \mathbf{Mod}\text{-}\mathrm{End}_R(G) \longrightarrow G\text{-}\mathbf{Pres}$ and $\mathbf{H}_G(\cdot) : G\text{-}\mathbf{Pres} \longrightarrow \mathbf{Mod}\text{-}\mathrm{End}_R(G)$
   are inverse category equivalences.

8. Give an example of a G-generated right R-module H for which there *does not* exist a G-balanced short exact sequence

$$0 \longrightarrow K \longrightarrow G^{(c)} \xrightarrow{\ \rho\ } H \longrightarrow 0$$

   in which K is G-generated.
9. Fill in the details of the proof of Theorem 10.17 as indicated on page 160.
10. Show that Theorem 10.17 is false without the self-small hypothesis or the E-flat hypothesis.
11. Show that Theorem 10.18 is false without the faithful hypothesis.
12. Construct a self-generator that is not a σ-self-generator.
13. Give an example of a module G for which $0 \longrightarrow G$ fails to satisfy one or more of the following properties in G-**Plex**: (i) small, (ii) projective, (iii) generator.
14. Prove the Morita theorem. Let S and R be rings and let G be an S-R-bimodule. The following are equivalent.

   (a) $\mathbf{T}_G(\cdot) : \mathbf{Mod}\text{-}S \longrightarrow \mathbf{Mod}\text{-}R$ is a category equivalence.
   (b) G is a finitely generated projective generator.

15. Give an example of a module that is a self-generator but not a σ-self-generator.
16. Characterize the right R-modules G such that $\mathbf{H}_G(\cdot) : G\text{-}\overline{\mathbf{Gen}} \longrightarrow \mathbf{Mod}\text{-}\mathrm{End}_R(G)$ is a category equivalence.
17. Let G be an E-flat module and let $M \subset \mathbf{H}_G(H)$ be a right $\mathrm{End}_R(G)$-submodule. Then $\mathbf{T}_G(M) \cong MG$.
18. Show with a specific example that there is an E-flat module G and a proper ideal $I \subset \mathrm{End}_R(G)$ such that $\mathbf{T}_G(I) \cong IG = G = \mathbf{T}_G(\mathrm{End}_R(G))$. Compare this example with Theorems 10.17 and 10.21.
19. Let G be an E-flat module. State and prove a result like Theorem 10.21 for $\mathrm{End}_R(G)$-submodules of free right $\mathrm{End}_R(G)$-modules.
20. Construct an example of an E-flat module that is not a Σ-self-generator.
21. Prove that $H_0 : G\text{-}\mathbf{Coplx} \longrightarrow G\text{-}\mathbf{Cl}$ is a well-defined additive functor.
22. Assume (μ). Let G be a self-slender right R-module. If $H_0(\cdot) : G\text{-}\mathbf{Coplx} \longrightarrow G\text{-}\mathbf{Cl}$ is a full functor, then

   (a) G is injective relative to each injection $\phi : H \longrightarrow H'$ in G-**Cl** and
   (b) $H_0(\cdot)$ is a faithful functor.

23. Assume (μ). The following are equivalent for the right R-module G.

   (a) $H_0 : G\text{-}\mathbf{Coplx} \longrightarrow G\text{-}\mathbf{Cl}$ is a full functor.
   (b) $H_0 : G\text{-}\mathbf{Coplx} \longrightarrow G\text{-}\mathbf{Cl}$ is a category equivalence.
   (c) Each map $\jmath : H \longrightarrow H'$ in G-**Cl** is G-cobalanced from its kernel.

24. Assume (μ). The following are equivalent for the right R-module G.

    (a) G is a slender injective cogenerator in G-**Cl**.

    (b) G is self-slender and each map $\jmath : H \longrightarrow H'$ in G-**Cl** is G-cobalanced from its kernel.

    (c) $(\cdot)^* : \mathrm{End}_R(G)$-**Mod** $\longrightarrow G$-**Cl** is a duality.

25. Assume (μ). Let G be a self-slender right R-module. The following are equivalent.

    (a) G is an injective cogenerator in G-**Cl**.

    (b) Each map $\jmath : H \longrightarrow H'$ in G-**Cl** is G-cobalanced from its kernel.

    (c) $H_0 : G$-**Coplx** $\longrightarrow G$-**Cl** is a category equivalence.

    (d) $(\cdot)^* : \mathrm{End}_R(G)$-**Mod** $\longrightarrow G$-**Cl** is a duality.

26. Assume (μ). The following are equivalent for the right R-module G.

    (a) G is a slender injective cogenerator in G-**Cl**.

    (b) $(\cdot)^* : \mathrm{End}_R(G)$-**Mod** $\longrightarrow G$-**Cl** and $(\cdot)^* : G$-**Cl** $\longrightarrow \mathrm{End}_R(G)$-**Mod** are inverse dualities.

27. Assume (μ). Let G be a self-slender right R-module. Then each left $\mathrm{End}_R(G)$-module is torsionless. That is, each $M \in \mathrm{End}_R(G)$-**Mod** embeds in $\mathrm{Hom}_R(G, G^c) \cong \mathrm{End}_R(G)^c$ for some cardinal c.

28. Assume (μ). Let G be a right R-module. The following are equivalent.

    (a) G is a slender injective cogenerator in G-**Cl**.

    (b) G is a self-slender Π-quasi-injective right R-module.

    (c) $(\cdot)^* : \mathrm{End}_R(G)$-**Mod** $\longrightarrow G$-**Cl** is a category equivalence.

In this case, G-**coSol** $= G$-**Cl**.

29. Assume (μ). Let G be a right R-module. The following are equivalent.

    (a) G is a slender injective cogenerator in G-$\overline{\overline{\text{Cl}}}$.

    (b) G is a self-slender Π-quasi-injective Π-self-cogenerator.

    (c) $(\cdot)^* : \mathrm{End}_R(G)$-**Mod** $\longrightarrow G$-$\overline{\overline{\text{Cl}}}$ is a category equivalence

In this case, G-**coSol** $= G$-**Cl** $= G$-$\overline{\overline{\text{Cl}}}$.

10.7 Problems for future research

Let R be an associative ring with identity, let G be a right R-module, let $\mathrm{End}_R(G)$ be the endomorphism ring of G, let $\mathcal{Q}$ be a G-plex, let $\mathcal{P}$ be a projective resolution of right $\mathrm{End}_R(G)$-modules.

1. Determine when any two of the categories in the chain on page 148 are equal.
2. Extend the Orsatti–Mennini work on $*$-modules to include G-plexes.
3. Give an alternative description of R-modules G such that q_G is a category equivalence.
4. Characterize those G such that $\mathbf{P}(G) = G$-**Sol**.
5. Find a module-theoretic property that describes those G that are (small) projective generators in one of the categories $\mathbf{P}(G)$, G-**Sol**, G-**Pres**, etc.

6. More characterizations of E-flat, E-generator R-modules are needed.

7. Find a module-theoretic property that describes those G that are respectively, small, or projective, or generators in one of the categories $\mathbf{P}(G)$, G-**Sol**, G-**Pres**, etc.

8. Dualize Theorem 10.13.

9. Characterize the faithfully E-flat righht R-modules. At the time of this writing, none of the existing characterizations is satisfying.

10. Remove the E-flat hypothesis from Theorem 10.18.

11. Extend Theorem 10.22 by deleting the E-flat hypothesis.

12. Find the flat dimension of the $*$-module G as a left $\mathrm{End}_R(G)$-module.

13. Characterize the G-coherent R-modules and the G-cocoherent R-modules.

14. Given an internal characterization of G-closed modules.

15. Given an internal characterization of G-cosolvable modules.

16. Do some linear algebra with the factorization in Theorem 10.34.

17. Determine when any two of the terms in the framed sequence on page 171 are equal.

11

Characterizing endomorphisms

We fix for the duration of this chapter a ring R and a right R-module G.

The theorems in Chapter 2 raise a pair of interesting questions. (1) Which properties of G can be described in terms of $\text{End}_R(G)$. (2) Which properties of $\text{End}_R(G)$ can be described in terms of G?

In this chapter we will use the equivalences in Chapter 2 to provide partial answers to these questions. Specifically, we will consider the homological dimensions of G as a left $\text{End}_R(G)$-module, the right global dimension of $\text{End}_R(G)$, $\text{rgd}(\text{End}_R(G))$, the flat left dimension $\text{fd}(G)$, and of the injective dimension $\text{id}_E(G)$ of the left $\text{End}_R(G)$-module G. Satisfying results are given for semi-simple Artinian endomorphism rings ($\text{rgd}(\text{End}_R(G)) = 0$), right hereditary endomorphism rings ($\text{rgd}(\text{End}_R(G)) \leq 1$), for right Noetherian right hereditary endomorphism rings, and for rings $\text{End}_R(G)$ such that $\text{rgd}(\text{End}_R(G)) \leq 3$. The answers are most complete when G is a self-small or self-slender right R-module.

Furthermore, there are quite a few new terms used in this chapter. To help the reader organize these terms, we have included a small Glossary at the end of the chapter, and some exercises detailing implications among these new terms.

11.1 Flat endomorphism modules

Theorem 10.18 is an illuminating result. It explains among other things the right ideals of $\text{End}_R(G)$ in terms of the G-generated R-submodules of the self-small faithfully E-flat right R-module G. Thus, chain conditions on the right ideals of $\text{End}_R(G)$ correspond to chain conditions on the G-generated R-submodules of G. The proofs of the following results amount to verifying that the lattice isomorphism λ_G preserves the specified chain conditions. Ideally this is what an equivalence should do for $\text{End}_R(G)$ and G. A number of applications will demonstrate the point. These results can be found in [53].

Theorem 11.1. *The following are equivalent for the self-small faithfully E-flat right R-module G.*

1. *$\text{End}_R(G)$ is a right Noetherian ring.*
2. *G has the acc on the G-generated R-submodules of G.*
3. *Each G-generated R-submodule of G is finitely G-generated.*

Theorem 11.2. *Let G be a self-small faithfully E-flat right R-module. Then* $\text{End}_R(G)$ *is a right Artinian ring iff G has the dcc on the G-generated R-submodules of G.*

Theorem 11.3. *Let G be a self-small faithfully E-flat right R-module. The right ideals in* $\text{End}_R(G)$ *form a chain iff the finitely G-generated R-submodules of G form a chain.*

Theorem 11.4. *Let G be a self-small faithfully E-flat right R-module. Each right ideal in* $\text{End}_R(G)$ *is principal iff each G-generated R-submodule of G is a homomorphic image of G.*

I suppose some reader may begin to get the idea that these results are trivialities. Why not try to prove the next result without first peeking at the answer?

Recall that the ring E is *right hereditary* if each right ideal of E is a projective right E-module, which is equivalent to the condition that each E-submodule of a projective right E-module is projective.

Theorem 11.5. *Let G be a self-small faithfully E-flat right R-module. The following are equivalent.*

1. $\text{End}_R(G)$ *is right hereditary.*
2. *Each G-generated R-submodule of G is in* $\mathbf{P}(G)$.
3. *If H is a G-generated R-submodule of some* $Q \in \mathbf{P}(G)$ *then* $H \in \mathbf{P}(G)$.

Proof: $3 \Rightarrow 2$ is clear.

$2 \Rightarrow 1$. Assume part 2 and let $I \subset \text{End}_R(G)$ be a right ideal. Because G is E-flat $\mathbf{T}_G(I) \cong IG \subset G$ is a G-generated R-submodule. By part 2 there is a cardinal c such that $\mathbf{T}_G(I) \oplus H \cong G^{(c)}$, and because G is self-small and faithfully E-flat, Theorem 10.18 implies that

$$\text{End}_R(G)^{(c)} \cong \mathbf{H}_G(G^{(c)}) \cong \mathbf{H}_G\mathbf{T}_G(I) \oplus \mathbf{H}_G(H) \cong I \oplus \mathbf{H}_G(H).$$

Thus I is projective, which proves part 1.

$1 \Rightarrow 3$. Let $H \subset Q \in \mathbf{P}(G)$ be right R-modules and suppose that H is G-generated. Since $\mathbf{H}_G(\cdot)$ is left exact, $\mathbf{H}_G(H) \subset \mathbf{H}_G(Q)$, and since G is self-small , $\mathbf{H}_G(Q)$ is a projective right $\text{End}_R(G)$-module. Because $\text{End}_R(G)$ is right hereditary, $\mathbf{H}_G(H)$ is a projective right $\text{End}_R(G)$-module, so that $\mathbf{T}_G\mathbf{H}_G(H)$ is a direct summand of $G^{(c)}$ for some cardinal c.

Finally, because G is E-flat, Theorem 1.13 implies that

$$\mathbf{T}_G\mathbf{H}_G(H) \subset \mathbf{T}_G\mathbf{H}_G(Q) \cong Q.$$

Inasmuch as this chain of inclusions is the function $\Theta_H : \mathbf{T}_G\mathbf{H}_G(H) \longrightarrow H$, Θ_H is an injection. Finally, since H is G-generated Θ_H is a surjection. Hence $H \cong \mathbf{T}_G\mathbf{H}_G(H)$ is a direct summand of $G^{(c)}$, and therefore $H \in \mathbf{P}(G)$.

This completes the logical cycle.

Theorem 11.6. *Let G be a self-small faithfully E-flat right R-module.*

1. $\text{End}_R(G)$ *is right Noetherian right hereditary.*
2. *Each G-generated R-submodule of G is in* $\mathbf{P}_o(G)$.

We will find that some of these conditions carry the self-small or the E-flat condition with them.

11.2 Homological dimension

We will study several homological dimensions of G and $\mathrm{End}_R(G)$.

11.2.1 Definitions and examples

We say that *the projective dimension of M is at most k* if there is a long exact sequence

$$0 \longrightarrow P_k \xrightarrow{\partial_k} \cdots \xrightarrow{\partial_1} P_0 \xrightarrow{\partial_0} M \longrightarrow 0 \qquad (11.1)$$

whose entries P_j are projective left E-modules for each $j = 0, \ldots, k$. The sequence (11.1) is called a *projective resolution of M*. Then the *projective dimension of M* is

> $\mathrm{pd}_E(M) =$ the least integer k for which a projective resolution
>
> (11.1) exists, or ∞ if this minimum does not exist.

Equivalently, $\mathrm{pd}_E(M) = k$ if $\mathrm{Ext}_E^{k+1}(M,N) = 0$ for all left E-modules N. Notice that M is a projective left E-module iff $\mathrm{pd}_E(M) = 0$ iff $\mathrm{Ext}_E^1(M,N) = 0$ for each left E-module N.

The *left global dimension of E* is

> $\mathrm{lgd}(E) = \sup\{\mathrm{pd}_E(M) \,\big|\, M \text{ is a left } E\text{-module }\}.$

For example, E is semi-simple Artinian iff every left E-module M is projective iff $\mathrm{lgd}(E) = 0$. Similarly, E is right hereditary iff each right ideal of $\mathrm{End}_R(G)$ is projective iff $\mathrm{lgd}(E) \leq 1$.

Let E be a ring and let M be a left E-module. We say that *the flat dimension of M is at most k* if there is a long exact sequence

$$0 \longrightarrow F_k \xrightarrow{\partial_k} \cdots \xrightarrow{\partial_1} F_0 \xrightarrow{\partial_0} M \longrightarrow 0 \qquad (11.2)$$

whose entries F_j are flat left E-modules for each $j = 0, \ldots, k$. The sequence (11.2) is called a *a flat resolution of M*. The *flat dimension of M* is

> $\mathrm{fd}_E(M) =$ the least integer k for which there is a flat resolution (11.2),
>
> or ∞ if this minimum does not exist.

Equivalently, $\text{fd}_E(M) = k$ if $\text{Tor}_E^{k+1}(K, M) = 0$ for all right E-modules K. Notice that M is a flat left E-module iff $\text{fd}_E(M) = 0$ iff $\text{Tor}_E^1(K, M) = 0$ for each right E-module K.

The *weak left global dimension of E* is

$$\text{wlgd}(E) = \sup\{\text{fd}_E(M) \,|\, M \text{ is a left } E\text{-module }\}.$$

For example, $\text{wlgd}(E) = 0$ iff each left E-module is flat. Similarly, $\text{wlgd}(E) \le 1$ iff the E-submodules of flat left E-modules are flat.

We begin by showing that G can have almost any homological dimension over its endomorphism ring.

Let $p \in \mathbf{Z}$ be a prime and let $G = \mathbf{Z}/p^2\mathbf{Z}$. Then $\text{pd}_E(G) = 0$. If G is a Σ-self-generator then $\text{fd}_E(G) = 0$ (Theorem 10.23). If $G = \mathbf{Z}(p^\infty)$ for some prime $p \in \mathbf{Z}$ then $\text{End}(G) = \widehat{\mathbf{Z}}_p$. Since $\widehat{\mathbf{Z}}_p$ is a *pid*, and since flat modules over *pids* are torsion-free, $\text{fd}_E(\mathbf{Z}(p^\infty)) \ne 0$. Thus $\text{fd}_E(\mathbf{Z}(p^\infty)) = \text{pd}_E(\mathbf{Z}(p^\infty)) = 1$. However, $\mathbf{Z}(p^\infty)$ is not a self-small group.

Example 11.7. [60, L. Fuchs and L. Salce] There is a small right R-module G such that $\text{fd}_E(G) = 1$.

Proof: Let R be a valuation domain, let $\mathbf{k}$ denote the classical ring of quotients of R, and let $G = \mathbf{k}/R$. Furthermore, assume that G requires at least $\aleph_1$ generators as a module. Then Fuchs and Salce [60] show that G is a small module. (The reader should try to prove this on their own.) Furthermore, because $R \subset \text{End}_R(G)$, G is not a torsion-free $\text{End}_R(G)$-module, so $\text{fd}_E(G) \ne 0$. However, using the fact that any proper subchain of submodules of $G = \mathbf{k}/R$ is countable, we show that any R-module map $\delta : G^{(n)} \longrightarrow G$ is a surjection. Then image $\delta = G$ is G-solvable, and hence by Corollary 11.23, $\text{fd}_E(G) \le 1$ whence $\text{fd}_E(G) = 1$.

We call the left E-module M a *Corner module* if the additive structure $(M, +)$ of M is a reduced countable torsion-free abelian group. The ring E is a *Corner ring* if E is a Corner module over itself. Thus $\mathbf{Z}$ is a Corner module as is any reduced subgroup of $\mathbf{Q}^{(\aleph_o)}$. A proof of the theorem can be found in [46, 49].

Theorem 11.8. *Let E be a Corner ring.*

1. *Suppose that there is a left E-module N such that $\text{pd}_E(N) = n + 1$. Then for each integer $0 < k \le n$ there is a Corner group G_k such that $\text{End}(G_k) \cong E$ and $\text{pd}_E(G_k) = k$.*
2. *Suppose that there is a left E-module K such that $\text{fd}(K) = n + 1$. Then for each integer $0 \le k \le n$ there is a Corner group G_k such that $\text{End}(G_k) \cong E$ and $\text{fd}_E(G_k) = k$.*

For example, let E be a Corner ring that contains a left ideal I such that $\text{pd}_E(I) = k > 0$. Such a ring E and left ideal I are relatively easy to construct as rings of triangular martrices over $\mathbf{Z}$. Then there is a group G_k such that $E = \text{End}_\mathbf{Z}(G_k)$ and $\text{pd}_E(G_k) = k$.

With a little more effort we can construct a ring E of finite weak left global dimension $n + 1$ that contains a left ideal I such that $0 \leq \mathrm{fd}_E(I) = k \leq n$. Thus there is a group G_k such that $E = \mathrm{End}(G_k)$ and $\mathrm{fd}_E(G_k) = k$.

Corollary 11.9. *Let E be a Corner ring. Then*

1. $\mathrm{lgd}(E) = \sup\{\mathrm{pd}_E(G) \mid G$ *is a Corner group and* $E \cong \mathrm{End}(G)\}$.
2. $\mathrm{wlgd}(E) = \sup\{\mathrm{fd}_E(G) \mid G$ *is a Corner group and* $E \cong \mathrm{End}(G)\}$.

The above construction shows that the left $\mathrm{End}_R(G)$-module structure of G can be varied. There seems to be no obvious way of characterizing the homological dimension of G strictly in terms of the ideal structure of $\mathrm{End}_R(G)$.

11.2.2 The exact dimension of a G-plex

We will use G-plexes to characterize some homological dimensions of $\mathrm{End}_R(G)$ and of the left $\mathrm{End}_R(G)$-module G. For the remainder of this section let $\mathcal{Q}$ denote a G-plex and let

$$I_k = \text{image } \delta_k \quad \text{and} \quad K_k = \ker \delta_k.$$

Recall that for integers $k \geq 0$, $H_k(\cdot)$ denotes the k-th homology functor.

We say that $\mathcal{Q}$ *is exact at* k if $I_{k+1} = K_k$ or equivalently if $H_k(\mathcal{Q}) = 0$. A G-plex $\mathcal{Q}$ is always exact at $\mathcal{Q}_0$. Given $\mathcal{Q} \in G$-**Plex** then *the exact dimension of* $\mathcal{Q}$ is

$$
\boxed{
\begin{aligned}
\mathrm{ed}_G(\mathcal{Q}) = \min\{\text{integers } k > 0 \mid \mathcal{Q} \text{ is exact at } k + j \text{ for each} \\
\text{integer } j \geq 1\}, \text{ or } \infty \text{ if this minimum does not exist.}
\end{aligned}
}
$$

Lemma 11.10. *Let G be a right R-module and let $\mathcal{Q} \in G$-**Plex**. Then the following integers are equal*

1. $\mathrm{ed}_G(\mathcal{Q})$.
2. $\min\{\text{integer } k > 0 \mid H_{k+j}(\mathcal{Q}) = 0 \text{ for each integer } j \geq 1\}$, or ∞ *if this minimum does not exist.*
3. $\min\{\text{integer } k > 0 \mid K_{k+j} \text{ is } G\text{-generated for each integer } j \geq 1\}$, or ∞ *if this minimum does not exist. (See Lemma 9.6.)*

Proof: Let $\mathcal{Q}$ be a G-plex, and let $j \geq 0$ be an integer. Fix an integer $k \geq 0$.

$1 \Leftrightarrow 2$. $H_{k+j}(\mathcal{Q}) = 0$ iff $I_{k+j+1} = K_{k+j}$ iff $\mathcal{Q}$ is exact at $k + j$. Hence $H_{k+j}(\mathcal{Q}) = 0$ for each integer $j \geq 1$ iff $\mathrm{ed}(\mathcal{Q}) = k$.

$2 \Leftrightarrow 3$. Recall that $\mathbf{S}_G(K_k) = I_{k+1}$ for each integer $k > 0$. Then $H_{k+j}(\mathcal{Q}) = 0$ iff $I_{k+j} = K_{k+j}$ iff $\mathbf{S}_G(K_{k+j}) = K_{k+j}$ for each integer $j > 0$. This completes the proof.

It is traditional to take a supremum of the dimensions $\text{ed}_G(\mathcal{Q})$ to arrive at a dimension for G. The *exact dimension of G* is

$$\text{ed}(G) = \sup\{\text{ed}_G(\mathcal{Q}) \,\big|\, \mathcal{Q} \text{ is a } G\text{-plex }\}.$$

The next result is clear. It follows from Lemma 11.10.

Lemma 11.11. *Let G be a self-small right R-module. The following integers are equal.*

1. $\text{ed}(G)$.
2. $\min\{k > 0 \,\big|\, \text{each } \mathcal{Q} \in G\text{-\textbf{Plex} is exact at } k + j \text{ for each integer } j \geq 1\}$, *or ∞ if such a minimum does not exist.*
3. $\min\{\text{integer } k \leq 0 \,\big|\, K_{k+j} = K_{k+j}(\mathcal{Q}) \text{ is } G\text{-generated for each integer } j \geq 1 \text{ and each } \mathcal{Q} \in G\text{-\textbf{Plex}}\}$, *or ∞ if such a minimum does not exist.*

11.2.3 *The projective dimension of a G-plex*

Let

$$\mathcal{Q}(k) = 0 \longrightarrow Q_k'' \xrightarrow{\delta_k''} \cdots \xrightarrow{\delta_2''} Q_1'' \xrightarrow{\delta_1''} Q_0''$$

be a G-plex. An obvious but important observation is that $Q_{k+j}'' = 0$ for each integer $j \geq 1$.

Given $\mathcal{Q} \in G\text{-\textbf{Plex}}$, let

$$\text{pd}_G(\mathcal{Q}) = \min\{\text{integer } k \geq 0 \,\big|\, \mathcal{Q} \text{ has the homotopy type of}$$
$$\text{some } \mathcal{Q}(k) \in G\text{-\textbf{Plex}}\}, \text{ or } \infty \text{ if this minimum does not exist.}$$

We call $\text{pd}_G(\mathcal{Q})$ the *G-projective dimension of $\mathcal{Q}$*.

Lemma 11.12. *Let G be a self-small right R-module and let $\mathcal{Q} \in G\text{-\textbf{Plex}}$. Let $\mathcal{Q}_1 = \cdots \xrightarrow{\delta_3} Q_2 \xrightarrow{\delta_2} Q_1$ be formed by deleting Q_0 and δ_1 from $\mathcal{Q}$. Then*

$$\text{pd}(\mathcal{Q}) = \text{pd}(\mathcal{Q}_1) + 1.$$

Proof. The proof is the traditional one for dimensional shifting as in [94]. The details are left to the reader.

The results of this section will characterize $\text{pd}_G(\mathcal{Q})$ in terms of the homology groups $H_k(\mathcal{Q})$ associated with $\mathcal{Q}$.

Lemma 11.13. *Let G be a right R-module and let $Q \in G$-**Plex***. Then*

$$\mathrm{ed}_G(Q) \leq \mathrm{pd}_G(Q).$$

Proof: Write $k = \mathrm{pd}_G(Q)$. Then Q has the homotopy type of a $Q(k) \in G$-**Plex**. Specifically, $Q''_{k+j} = 0$ for each $j \geq 0$. Since the homology groups of complexes having the same homotopy type are isomorphic $H_{k+j}(Q) \cong H_{k+j}(Q(k)) = 0$ for each $j \geq 1$. That is, $\mathrm{ed}_G(Q) \leq k$. This completes the proof.

Let $Q \in G$-**Plex**. There is a short exact sequence

$$0 \longrightarrow K_k \longrightarrow Q_k \xrightarrow{\delta_k} I_k \longrightarrow 0 \tag{11.3}$$

for each integer $k \geq 0$.

Let M be a right E-module. It is traditional to write $\mathrm{pd}_E(M) \leq k$ iff there is a projective resolution $\mathcal{P}$ for M such that image ∂_k is projective iff there is a projective resolution $\mathcal{P}$ for M such that image ∂_{k+j} is a direct summand of P_{k+j} for each integer $j \geq 1$. For G-plexes, the situation is slightly more complicated.

Theorem 11.14. *Let G be a right R-module, let $Q \in G$-**Plex***, *and let $k > 0$ be an integer. The following are equivalent.*

1. $\mathrm{pd}_G(Q) \leq k$.
2. I_{k+j} *is a direct summand of Q_{k+j-1} for each integer $j \geq 1$.*
3. I_{k+1} *is a direct summand of Q_k.*

Proof: $1 \Rightarrow 2$. Fix $k > 0$ and assume that $\mathrm{pd}_G(Q) \leq k$. Then Q has the homotopy type of some $Q(k)$. Specifically, the terms of $Q(k)$ are such that $Q''_{k+j} = 0$ for each $j \geq 1$. Choose chain maps $(\cdots, f_0) : Q \longrightarrow Q(k)$ and $(\cdots, g_0) : Q(k) \longrightarrow Q$ and a homotopy $(\cdots, s_1)$ such that $s_{j+1} : Q_j \longrightarrow Q_{j+1}$ and

$$1_{Q_j} - g_j f_j = \delta_{j+1} s_{j+1} + s_j \delta_j$$

for each integer $j \geq 0$ where by convention we have $f_0 = s_0 = \delta_0 = 0$. The chain map $f : Q \longrightarrow Q(k)$ is represented by the commutative diagram

$$\tag{11.4}$$

and the homotopy is represented by the diagram

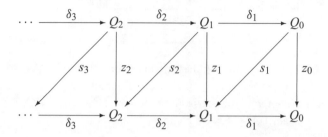

where

$$z_k = 1_{Q_k} - g_k f_k \text{ for each } k \geq 0.$$

Recall $k > 0$ from the beginning of this proof, and let $\iota_{k+1} : I_{k+1} \longrightarrow Q_k$ be the inclusion map. Because Q is a complex

$$\delta_k \iota_{k+1} = 0. \tag{11.5}$$

The commutativity of (11.4) shows us that

$$\delta_k'' f_k \iota_{k+1} = f_{k-1} \delta_k \iota_{k+1} = 0,$$

so that image $f_k \iota_{k+1} \subset \mathbf{S}_G(K_k'')$. By Lemma 9.6 and by definition of $Q(k)$, $\mathbf{S}_G(K_k'') = H_{k+1}'' = 0$, whence

$$f_k \iota_{k+1} = 0. \tag{11.6}$$

Then

$$\iota_{k+1} \overset{(11.6)}{=\!=\!=} (1_{Q_k} - g_k f_k) \iota_{k+1} = (\delta_{k+1} s_{k+1} + s_k \delta_k) \iota_{k+1} \overset{(11.5)}{=\!=\!=} (\delta_{k+1} s_{k+1}) \iota_{k+1}.$$

Inasmuch as image $(\delta_{k+1} s_{k+1}) \subset I_{k+1}$, we have proved that $\delta_{k+1} s_{k+1}$ splits ι_{k+1}. Hence I_{k+1} is a direct summand of Q_k. This proves part 2.

2 $\Rightarrow$ 3 is clear.

3 $\Rightarrow$ 1. By part 3 and Lemma 9.6, $I_{k+1} = \mathbf{S}_G(K_k)$ is a direct summand of Q_k, so that

$$I_{k+1} \oplus Q_k'' = Q_k \text{ for some } Q_k'' \in \mathbf{P}(G).$$

Let $\delta_k'' = \delta_k | Q_k''$. There are maps $f_k : Q_k \longrightarrow Q_k''$ and $g_k : Q_k'' \longrightarrow Q_k$ such that $f_k g_k = 1_{Q_k''}$. Let $f_{k+\ell} = g_{k+\ell} = 0$ for each integer $\ell \geq 1$, and let $f_{k-\ell} = g_{k-\ell} = 1_{Q_{k-\ell}}$ for each integer $k \geq \ell \geq 1$. Let $Q(k)$ denote the G-plex

$$Q(k) = 0 \longrightarrow Q_k'' \overset{\delta_k''}{\longrightarrow} Q_{k-1} \cdots \overset{\delta_1}{\longrightarrow} Q_0.$$

Then $f = (\ldots, f_{k+1}, f_k, f_{k-1}, \ldots)$ and $g = (\ldots, g_{k+1}, g_k, g_{k-1}, \ldots)$ are chain maps $f : \mathcal{Q} \longrightarrow \mathcal{Q}(k)$ and $g : \mathcal{Q}(k) \longrightarrow \mathcal{Q}$ such that

$$1_{Q_i''} - f_i g_i = 1_{Q_i} - g_i f_i = 0 \quad \text{for } i = 0, 1, \cdots, k - 1.$$

By Lemma 9.10.2, $1_{\mathcal{Q}''} - fg$ and $1_\mathcal{Q} - gf$ are null homotopic, so that $\mathcal{Q}$ is homotopic to $\mathcal{Q}(k)$. Therefore $\mathrm{pd}_G(\mathcal{Q}) \le k$. This completes the logical cycle.

Let H be a G-generated right R-module. Recall that a map $\rho : H \longrightarrow H'$ is *G-balanced onto its image* if G is projective relative to the canonical surjection $\rho : H \longrightarrow$ image ρ.

We say that H is *weakly balanced G-split* if given $Q \in \mathbf{P}(G)$ and a G-balanced surjection $\delta : Q \longrightarrow H$ then $Q = \mathbf{S}_G(\ker \delta) \oplus Q'$ for some right R-module Q'.

Theorem 11.15. *Let G be a self-small module, let $k \ge 2$ be an integer, and let $\mathcal{Q} \in G$-**Plex**. Then $\mathrm{pd}_G(\mathcal{Q}) \le k$ iff I_k is weakly balanced G-split.*

Proof: If I_k is weakly balanced G-split then $\mathbf{S}_G(K_k) = I_{k+1}$ is a direct summand of Q_k. By Theorem 11.14, $\mathrm{pd}_G(\mathcal{Q}) \le k$.

Conversely, suppose that $\mathrm{pd}_G(\mathcal{Q}) \le k$ and let $\delta_k'' : Q_k'' \longrightarrow I_k$ be any G-balanced surjection. Using Lemma 9.4.2, extend

$$Q_k'' \xrightarrow{\delta_k''} Q_{k-1} \xrightarrow{\delta_{k-1}} \cdots \xrightarrow{\delta_1} Q_0$$

to a G-plex $\mathcal{Q}''$. By Lemma 9.6 and Theorem 11.14, $I_{k+1} = \mathbf{S}_G(K_k)$ is a direct summand of Q_k'', so that I_k is weakly balanced G-split.

The remaining case where $\mathrm{pd}_G(\mathcal{Q}) \le 1$ is handled in the next theorem. We say that a G-generated right R-module H is *weakly G-split* if given *any* surjection $\delta : Q \longrightarrow H$ in which $Q \in \mathbf{P}(G)$ then $\mathbf{S}_G(\ker \delta) \oplus Q' = Q$ for some right R-module Q'.

Theorem 11.16. *Let G be a self-small module and let $\mathcal{Q} \in G$-**Plex**. Then $\mathrm{pd}_G(\mathcal{Q}) \le 1$ iff I_1 is weakly G-split.*

Proof: Suppose that $\mathcal{Q}$ is a G-plex and that I_1 is weakly G-split. By definition and Lemma 9.6, $I_2 = \mathbf{S}_G(K_1)$ is a direct summand of Q_1. An application of Theorem 11.14 shows that $\mathrm{pd}_G(\mathcal{Q}) \le 1$.

Conversely, choose $\mathcal{Q} \in G$-**Plex** such that $\mathrm{pd}_G(\mathcal{Q}) \le 1$. Notice that image $\delta_1 = I_1 \subset Q_0$. Let

$$Q_1' \xrightarrow{\delta_1'} Q_0$$

be any surjection such that $Q_1' \in \mathbf{P}(G)$ and image $\delta_1' = I_1$. Using Lemma 9.4.2 extend δ_1' to a G-plex $\mathcal{Q}'$. By Theorem 11.14, I_2 is a direct summand of Q_1', and by Lemma 9.6, $I_2 = \mathbf{S}_G(\ker \delta_1')$. Thus $\mathbf{S}_G(\ker \delta_1')$ is a direct summand of Q_1', and hence I_1 is weakly G-split. This completes the proof.

Example 11.17. This is an example where $\mathrm{ed}_G(\mathcal{Q}) = \mathrm{pd}_G(\mathcal{Q})$. Let $p \in \mathbf{Z}$ be a prime, and let $G = \mathbf{Z}(p^\infty)$. Let $\mathcal{Q}$ be the G-plex $\mathcal{Q} = 0 \longrightarrow G \xrightarrow{p} G$. Inasmuch as $\ker p = \mathbf{Z}/p\mathbf{Z} \neq 0$, $\mathrm{ed}_G(\mathcal{Q}) = 1$. Furthermore, $\mathrm{pd}_G(\mathcal{Q}) = 1$ because $I_2 = 0$ is a direct summand of Q_1.

Example 11.18. An elementary example of infinite projective dimension is given as follows. Let $R = \mathbf{Z}/p^2\mathbf{Z} = G$ and observe that

$$\mathcal{Q} = \cdots \longrightarrow \mathbf{Z}/p^2\mathbf{Z} \xrightarrow{p} \mathbf{Z}/p^2\mathbf{Z} \xrightarrow{p} \mathbf{Z}/p^2\mathbf{Z}$$

is a $\mathbf{Z}/p^2\mathbf{Z}$-plex. Then $\mathrm{ed}(\mathcal{Q}) = 0$ and $\mathrm{pd}(\mathcal{Q}) = \infty$ since at each $k > 0$, image p is not a direct summand of the indecomposable group $\mathbf{Z}/p^2\mathbf{Z}$.

11.3 The flat dimension

Let

$$\mathrm{Tor}_E^k(\cdot, \cdot) = \mathrm{Tor}_{\mathrm{End}_R(G)}^k(\cdot, \cdot).$$

We characterize the flat dimension of the left $\mathrm{End}_R(G)$-module G in terms of vanishing homology groups $H_k(\mathcal{Q})$ of G-plexes $\mathcal{Q}$. Subsequently, we prove that for each integer $k \geq 0$ the triangles

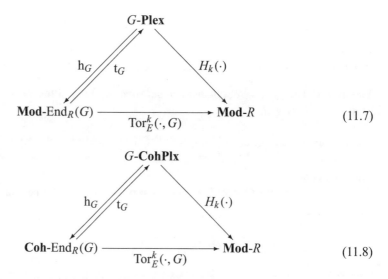

$$(11.7)$$

$$(11.8)$$

of categories and functors are commutative, where $H_k(\cdot)$ denotes the k-th homology functor on G-plexes.

Recall that given a G-plex $\mathcal{Q}$ then $K_k = \ker \delta_k$ and $I_k = \mathrm{image}\ \delta_k$. Recall the short exact sequences (11.3) associated with each $\mathcal{Q} \in G$-**Plex**.

Lemma 11.19. *Let G be a self-small right R-module, and let* $Q \in G$-**Plex**. *Then*

$$\mathrm{Tor}^k_E(\mathrm{h}_G(Q), G) \cong K_k/\mathbf{S}_G(K_k) = H_k(Q)$$

for each integer $k > 0$.

Proof: Let $Q \in G$-**Plex** and let $\mathrm{h}_G(Q) = M$. By Lemma 9.5, $\mathbf{H}_G(Q)$ is a projective resolution of M such that $Q \cong \mathbf{T}_G\mathbf{H}_G(Q)$ in G-**Plex**. Then Lemma 9.6.2 shows us that

$$\mathrm{Tor}^k_E(M, G) \cong H_k(\mathbf{T}_G\mathbf{H}_G(Q)) \cong H_k(Q) = K_k/\mathbf{S}_G(K_k)$$

for each $k > 0$. This proves the lemma.

The exact dimension of a G-plex is connected to G-solvable right R-modules. Recall Section 9.2.1 where we showed that H is *G-solvable* if there is a G-balanced exact sequence

$$0 \longrightarrow K \longrightarrow Q \longrightarrow H \longrightarrow 0$$

in which $Q \in \mathbf{P}(G)$ and $K \in G$-**Gen**.

Lemma 11.20. *Let G be a self-small right R-module, let $k > 0$ be an integer, and let* $Q \in G$-**Plex**. *Then* $\mathrm{ed}_G(Q) \leq k$ *iff* I_{k+j} *is G-solvable for each integer* $j \geq 1$.

Proof: Let $k > 0$, let $j \geq 1$ be given, and suppose that $\mathrm{ed}_G(Q) \leq k$. By Lemma 9.2, G is projective relative to the induced short exact sequence

$$0 \longrightarrow K_{k+j} \longrightarrow Q_{k+j} \longrightarrow I_{k+j} \longrightarrow 0$$

of right R-modules. By definition, $\mathrm{ed}_G(Q) \leq k$ implies that K_{k+j} is G-generated. Then also by definition, I_{k+j} is G-solvable. The converse is left as an exercise for the reader.

Theorem 10.23 states that G is E-flat if for each $Q \in \mathbf{P}(G)$ the kernel of each map $f : Q \longrightarrow G$ is G-generated. We will generalize Theorem 10.23 by characterizing $\mathrm{fd}_E(G)$ in terms of G-plexes Q such that $Q_0 = G$. Recall that $\mathrm{ed}(G)$ is the supremum of the integers $\mathrm{ed}_G(Q)$ over all G-plexes Q.

Theorem 11.21. *Let G be a self-small right R-module. The following integers are equal.*

1. $\mathrm{fd}_E(G)$
2. $\mathrm{ed}(G)$
3. $\sup\{\mathrm{ed}_G(Q) \,\big|\, Q \in G$-**Plex**, $Q_1 \in \mathbf{P}_o(G)$, *and* $Q_0 \cong G\}$
4. $\min\{$ *integers* $k \geq 0 \,\big|\,$ *each* $Q \in G$-**Plex** *is exact at* $k + 1\}$.

Proof: Suppose that G is self-small and let the natural number $\ell \geq 0$ be defined by

$$\ell = \sup\{\mathrm{ed}_G(Q) \,\big|\, Q \in G\text{-}\mathbf{Plex}, Q_1 \in \mathbf{P}_o(G), \text{ and } Q_0 \cong G\}.$$

Choose a G-plex $\mathcal{Q}$ and let us write $k = \mathrm{fd}_E(G)$. Then $\mathrm{h}_G(\mathcal{Q})$ is a right $\mathrm{End}_R(G)$-module, so that by Lemma 11.19,

$$0 = \mathrm{Tor}_E^{k+1}(\mathrm{h}_G(\mathcal{Q}), G) \cong H_{k+1}(\mathcal{Q}).$$

Since $\mathcal{Q}$ was arbitrarily selected, $\mathrm{ed}_G(\mathcal{Q}) \le k = \mathrm{fd}_E(G)$. Then

$$\mathrm{ed}(G) \le \mathrm{fd}_E(G).$$

Let M be a cyclic finitely presented right $\mathrm{End}_R(G)$-module. There is a projective resolution $\mathcal{P}$ for M such that P_1 is finitely generated and $P_0 \cong \mathrm{End}_R(G)$. By Lemma 9.5.1, $\mathbf{T}_G(\mathcal{P})$ is a G-plex such that $\mathbf{T}_G(P_1) \in \mathbf{P}_o(G)$, and $\mathbf{T}_G(P_0) \cong G$. Then $\mathrm{ed}_G(\mathbf{T}_G(\mathcal{P})) \le \ell$, so that

$$\mathrm{Tor}_E^{\ell+1}(M, G) = H_{\ell+1}(\mathbf{T}_G(\mathcal{P})) = 0$$

by Lemma 11.19. Since our choice of M was arbitrary,

$$\mathrm{fd}_E(G) \le \ell.$$

It follows immediately from the definitions and some elementary dimension shifting that

$$\ell \le \mathrm{ed}(G)$$
$$= \min\{\text{integers } k > 0 \,\big|\, \text{each } \mathcal{Q} \in G\text{-}\mathbf{Plex} \text{ is exact at } k + 1\}.$$

This completes the proof.

The next result extends Theorem 10.23 to flat dimensions.

Theorem 11.22. *Let G be a self-small right R-module and let $k > 0$ be an integer. The following are equivalent.*

1. $\mathrm{fd}_E(G) \le k$.
2. K_{k+1} is G-generated for each $\mathcal{Q} \in G$-**Plex**.
3. I_{k+1} is G-solvable for each $\mathcal{Q} \in G$-**Plex**.
4. $\mathbf{S}_G(K_k)$ is G-solvable for each $\mathcal{Q} \in G$-**Plex**.

Proof: $1 \Leftrightarrow 2$ holds for all integers $k > 0$ by Theorem 11.21.

$2 \Leftrightarrow 3$. Let $\mathcal{Q} \in G$-**Plex**. There is a short exact sequence

$$0 \longrightarrow K_{k+1} \longrightarrow Q_{k+1} \overset{\delta_{k+1}}{\longrightarrow} I_{k+1} \longrightarrow 0$$

of right R-modules. By hypothesis $k + 1 \ge 2$, so that G is projective relative to the above sequence. Then by definition and Lemma 9.19, I_{k+1} is G-solvable iff K_{k+1} is G-generated.

$3 \Leftrightarrow 4$. By Lemma 9.6, $I_{k+1} = \mathbf{S}_G(K_k)$ so this biconditional statement is clear. This completes the proof.

The next result extends Theorem 10.23 by characterizing the right R-modules G with flat dimension at most 1 over $\mathrm{End}_R(G)$. Its proof is left as an exercise for the reader.

Corollary 11.23. *Let G be a self-small right R-module. The following are equivalent.*

1. $\mathrm{fd}_E(G) \leq 1$.
2. $\mathrm{ed}(G) \leq 1$.
3. *Given any map $\delta : Q \longrightarrow G$ in which $Q \in \mathbf{P}(G)$ then $\mathbf{S}_G(\ker \delta)$ is G-solvable.*

Recall that G is a Σ-*quasi-projective* right R-module if for each cardinal c, G is projective relative to each surjection $G^{(c)} \longrightarrow H \longrightarrow 0$. The following result shows us that a self-small Σ-quasi-projective right R-module is almost E-flat.

Corollary 11.24. *Let G be a self-small Σ-quasi-projective right R-module. If M is a right $\mathrm{End}_R(G)$-module then*

$$\mathrm{Hom}_R(G, \mathrm{Tor}_E^k(M, G)) = 0$$

for each integer $k > 0$.

Proof: Given M and $k > 0$, let $\mathcal{Q} \in G$-**Plex** be such that $\mathrm{h}_G(\mathcal{Q}) = M$. Since G is self-small, Lemma 11.19 states that

$$\mathrm{Tor}_E^k(M, G) \cong K_k / \mathbf{S}_G(K_k).$$

Since G is Σ-quasi-projective each map $f : G \longrightarrow K_k / \mathbf{S}_G(K_k)$ lifts to one $g : G \longrightarrow K_k$, so that $g(G) \subset \mathbf{S}_G(K_k)$. Hence $f = 0$, whence $\mathrm{Hom}_R(G, K_k / \mathbf{S}_G(K_k)) = 0$.

Compare the following result with Theorem 10.13 where we show that a *-module is E-flat iff it is a *quasi-progenerator* (= finitely generated quasi-projective and a self-generator). The module K is *subgenerated* by G or is *G-subgenerated* if there is a cardinal c and a submodule $H \subset G^{(c)}$ such that $K \subset G^{(c)}/H$.

Corollary 11.25. *If G is a quasi-progenerator as a right R-module then G is E-flat.*

Proof: A finitely generated quasi-projective module G is self-small and Σ-quasi-projective. By Lemma 10.12, G is Σ-self-generating. Thus each G-subgenerated module is G-generated. Specifically since the Tor groups are G-subgenerated $\mathrm{Tor}_E^1(M, G)$ is G-generated for each right $\mathrm{End}_R(G)$-module M. But by Corollary 11.23,

$$\mathrm{Hom}_R(G, \mathrm{Tor}_E^1(M, G)) = 0,$$

so $\mathrm{Tor}_E^1(M, G) = 0$ for each right $\mathrm{End}_R(G)$-module M. Hence G is E-flat, which completes the proof.

11.4 Global dimensions

Let E be a ring and let $k \geq 0$ be an integer. Let

$$\mathcal{P}(k) = 0 \longrightarrow P_k'' \xrightarrow{\partial_k} \cdots \xrightarrow{\partial_2} P_1'' \xrightarrow{\partial_1} P_0''$$

denote a projective resolution of right E-modules such that $P_{k+j}'' = 0$ for each integer $j \geq 1$. Let $\mathcal{P}$ be a projective resolution of right E-module and let

$$\mathrm{pd}_E(\mathcal{P}) = \min\{\text{integers } k \geq 0 \,\big|\, \mathcal{P} \text{ has the homotopy type of a}$$
$$\mathcal{P}(k)\}, \text{ or } \infty \text{ if the minimum does not exist.}$$

The *projective dimension of* a right E-module M is $\mathrm{pd}_E(\mathcal{P}(M))$ for some (any) projective resolution $\mathcal{P}(M)$ of M. The *right global dimension of* E is

$$\mathrm{rgd}(E) = \sup\{\mathrm{pd}_E(M) \,\big|\, M \text{ is a right } E\text{-module}\}.$$

A classic result from homological algebra shows us that to calculate $\mathrm{rgd}(E)$ we need not consider *each* right E-module.

Lemma 11.26. *Let E be a ring. The following integers are equal.*

1. $\mathrm{rgd}(E)$
2. $\sup\{\mathrm{pd}_E(E/I) \,\big|\, I \text{ is a right ideal in } E\}$
3. $\sup\{\mathrm{pd}_E(I) + 1 \,\big|\, I \text{ is a right ideal in } E\}$

Proof: See [97, Theorem 9.12].

We characterize $\mathrm{rgd}(E)$ in terms of G-plexes. This will lead us to a characterization of the self-small modules G whose endomorphism rings have small right global dimension.

The *global dimension of G* is defined to be

$$\mathrm{gd}(G) = \sup\{\mathrm{pd}_G(\mathcal{Q}) \,\big|\, \mathcal{Q} \in G\text{-}\mathbf{Plex}\}.$$

Recall that

$$\mathcal{Q}(k) = 0 \longrightarrow Q_k'' \xrightarrow{\delta_k''} \cdots Q_1'' \xrightarrow{\delta_1''} Q_0''.$$

Theorem 11.27. *Suppose that G is a self-small right R-module. Then*

$$\mathrm{gd}(G) = \mathrm{rgd}(\mathrm{End}_R(G))$$
$$= \sup\{\mathrm{pd}_G(\mathcal{Q}) \,\big|\, \mathcal{Q} \in G\text{-}\mathbf{Plex} \text{ and } P_0 \cong G\}.$$

Proof: Let $E = \text{End}_R(G)$, let

$$\ell = \sup\{\text{pd}_G(\mathcal{Q}) \mid \mathcal{Q} \in G\text{-}\textbf{Plex} \text{ and } P_0 \cong G\}$$

and let $k = \text{rgd}(E)$. Evidently,

$$\ell \leq \text{gd}(G).$$

Given $\mathcal{Q} \in G\text{-}\textbf{Plex}$, then by Theorem 9.5.2, $\textbf{H}_G(\mathcal{Q})$ is a projective resolution, so that $\textbf{H}_G(\mathcal{Q})$ has the homotopy type of a projective complex $\mathcal{P}(k)$. An application of $\textbf{T}_G(\cdot)$ yields the G-plex $\textbf{T}_G\textbf{H}_G(\mathcal{Q}) \cong \mathcal{Q}$ which has the homotopy type of

$$\textbf{T}_G(\mathcal{P}(k)) = 0 \longrightarrow \textbf{T}_G(P_k'') \xrightarrow{\textbf{T}_G(\partial_k)} \cdots \xrightarrow{\textbf{T}_G(\partial_1)} \textbf{T}_G(P_0'').$$

Then

$$\text{pd}_G(\mathcal{Q}) = \text{pd}_E(\textbf{T}_G(\mathcal{P}(k))) \leq k = \text{rgd}(E),$$

so that

$$\text{gd}(G) \leq \text{rgd}(\text{End}_R(G)).$$

The same kind of argument will show that $\text{rgd}(E) \leq \ell$, which completes the proof.

The next result shows that the flat dimension of G is at most the global dimension of $\text{End}_R(G)$. The interesting thing about this inequality is that $\text{fd}_E(G)$ is the flat dimension of the *left* $\text{End}_R(G)$-module G while $\text{rgd}(\text{End}_R(G))$ is the *right* global dimension of $\text{End}_R(G)$. Here is an example where the left module structure of G influences the right module structure of the endomorphism ring.

Theorem 11.28. *Let G be a self-small right R-module. Then*

$$\boxed{0 \leq \text{fd}_E(G) \leq \text{rgd}(\text{End}_R(G)).}$$

Proof: Lemma 11.13 shows us that $\text{ed}(\mathcal{Q}) \leq \text{pd}(\mathcal{Q})$. Thus $\text{ed}(G) \leq \text{gd}(G)$. Theorems 11.21 and 11.27 then show us that

$$\text{fd}_E(G) = \text{ed}(G) \leq \text{gd}(G) = \text{rgd}(\text{End}_R(G)).$$

Corollary 11.29. *Let G be a self-small right R-module. Then*

$$\text{fd}_E(G) = \infty \implies \text{rgd}(\text{End}_R(G)) = \infty.$$

Recall the definitions of H is *weakly balanced G-split*, and H is *weakly G-split* from page 183.

Theorem 11.30. *Let G be a self-small module, and let $k \geq 2$ be an integer. Then* $\mathrm{rgd}(\mathrm{End}_R(G)) \leq k$ *iff given $Q \in G$-**Plex** then I_k is weakly balanced G-split.*

Proof: Suppose that $\mathrm{rgd}(\mathrm{End}_R(G)) \leq k$. By Theorem 11.27, $\mathrm{gd}(G) \leq k$, so given $Q \in G$-**Plex**, $\mathrm{pd}_G(Q) \leq k$. By Theorem 11.15, I_k is weakly balanced G-split.

Conversely, suppose that given $Q \in G$-**Plex**, then I_k admits a G-cover. By Theorem 11.15, $\mathrm{pd}_G(Q) \leq k$. Then by Theorem 11.27, $\mathrm{rgd}(\mathrm{End}_R(G)) \leq k$.

Corollary 11.31. *Let G be a self-small module. Then $\mathrm{rgd}(\mathrm{End}_R(G)) \leq 1$ iff given $Q \in G$-**Plex** then I_1 is weakly G-split.*

Corollary 11.32. *Let G be a self-small module, and let $k > 0$ be an integer. Then $\mathrm{rgd}(\mathrm{End}_R(G)) \leq k$ iff $\mathbf{S}_G(K_k)$ is a direct summand of Q_k for each G-plex Q.*

Proof: By reading Theorem 11.30 and Corollary 11.31 the reader will justify for themselves that $\mathrm{rgd}(\mathrm{End}_R(G)) \leq k$ iff $\mathrm{gd}(G) \leq k$ iff $\mathbf{S}_G(K_k)$ is a direct summand of Q_k for each G-plex Q.

Example 11.33. Let $X \subset Y \subset \mathbf{Q}$ be groups such that $\mathrm{End}_{\mathbf{Z}}(X) \cong \mathrm{End}_{\mathbf{Z}}(Y) \cong \mathbf{Z}$ and $\mathrm{Hom}(Y, X) = 0$. For example, let

$$X = \left\{ \frac{n}{p} \,\middle|\, n \in \mathbf{Z} \text{ and } p \in \mathbf{Z} \text{ is a prime} \right\}$$

and

$$Y = \left\{ \frac{n}{p^2} \,\middle|\, n \in \mathbf{Z} \text{ and } p \in \mathbf{Z} \text{ is a prime} \right\}.$$

The reader will show that if $G = X \oplus Y$ then

$$\mathrm{End}_R(G) = \begin{pmatrix} \mathbf{Z} & 0 \\ X & \mathbf{Z} \end{pmatrix}.$$

Also G is the directed union of the cyclic projective $\mathrm{End}_R(G)$-modules

$$\begin{pmatrix} \mathbf{Z} & 0 \\ X & 0 \end{pmatrix} \cong \mathrm{End}_R(G) \begin{pmatrix} 1 \\ 0 \end{pmatrix}$$

as left $\mathrm{End}_R(G)$-modules. Thus G is a flat left $\mathrm{End}_R(G)$-module. But G is not a projective left $\mathrm{End}_R(G)$-module since the projective left $\mathrm{End}_R(G)$-modules are, as abelian groups, uniquely a direct sum of copies of $\mathbf{Z}$ and X.

Recall that E is a *Corner ring* if the additive structure $(E, +)$ of E is a reduced countable torsion-free abelian group. The right E-module M is a *Corner module* if the additive structure $(M, +)$ of M is a reduced countable torsion-free abelian group.

Example 11.34. Let E be a Corner ring and suppose there is a Corner module M such that $\mathrm{pd}_E(M) = n \geq 0$. Theorem 11.8 states that for each $0 \leq k \leq n$ there is a group G_k such that $E \cong \text{End}_R(G_k)$ and $\text{pd}_E(G_k) = \max\{1, k\}$. Thus there is no obvious connection between $\text{pd}_E(G)$ and $\text{rgd}(\text{End}_R(G))$.

11.5 Small global dimensions

Let G be a self-small right R-module. Right hereditary endomorphism rings of self-small R-modules have been the subject of a number of papers in the literature, including [3, 12, 48, 49, 63, 77]. By Corollary 11.28, $\text{fd}_E(G) \leq \text{rgd}(\text{End}_R(G))$, so in studying right hereditary endomorphism rings we are studying those G such that

$$0 \leq \text{fd}_E(G) \leq \text{rgd}(\text{End}_R(G)) \leq 1.$$

By interpretting Theorems 11.14 and 11.27 as splitting conditions on R-modules, the results of the previous sections provide us with a unified approach to right hereditary endomorphism rings. By dualizing from G-plexes to G-coplexes we will characterize *left* hereditary endomorphism rings. With these techniques we will extend our investigation of right hereditary endomorphism rings to those G for which

$$0 \leq \text{fd}_E(G) \leq \text{rgd}(\text{End}_R(G)) \leq 3.$$

11.5.1 Baer's lemma

The G-generated module H is G-split if each short exact sequence

$$0 \longrightarrow K \longrightarrow Q \overset{\rho}{\longrightarrow} H \longrightarrow 0 \tag{11.9}$$

in which $Q \in \mathbf{P}(G)$, is split exact. Equivalently, H is G-split iff given any surjection $\delta : Q \longrightarrow H$ in which $Q \in \mathbf{P}(G)$, we have $Q = \ker \delta \oplus Q'$ for some right R-module Q'. We say that H is *fp G-split* if each short exact sequence (11.9) in which $Q \in \mathbf{P}_o(G)$, is split exact. That is, given a surjection $\delta : Q \longrightarrow H$ in which $Q \in \mathbf{P}_o(G)$, we have $Q = \ker \delta \oplus Q'$ for some right R-module $Q' \in \mathbf{P}(G)$. For example, a G-generated projective right R-module is G-split. In this section we investigate variations on the G-split property.

We say that G is *fp faithful* if $IG \neq G$ for each finitely generated proper right ideal I in $\text{End}_R(G)$, and that G is *faithful* if $IG \neq G$ for each proper right ideal I in $\text{End}_R(G)$. For each integer $n > 0$, $R^{(n)}$ is a faithful self-small right R-module. Each reduced torsion-free finite-rank abelian group G constructed according to Corner's theorem is faithful. (See [46, 56].) The free right R-module $G = R^{(\aleph_o)}$ is not faithful since the proper right ideal $\Delta \subset \text{End}_R(G)$ whose elements f are such that image f is finitely generated satisfies $\Delta G = G$.

R. Baer originally proved the following result for subgroups G of the rationals $\mathbf{Q}$. Now called Baer's lemma, this result is important in the modern literature on the study of properties of endomorphism rings.

Lemma 11.35. [Baer's lemma] *The following are equivalent for a right R-module G.*

1. *G is fp faithful.*
2. *G is fp G-split.*
3. *Each short exact sequence*

$$0 \longrightarrow K \longrightarrow G^{(n)} \overset{\rho}{\longrightarrow} G \longrightarrow 0 \tag{11.10}$$

in which $n > 0$ is an integer, is split exact.

Proof: $1 \Rightarrow 2$. Assume part 1 and let (11.10) be an exact sequence for some integer n. Write

$$G^{(n)} = G_1 \oplus \cdots \oplus G_n$$

for some groups G_i such that $G_i \cong G$ for each $i = 1, \cdots, n$. We will write

$$\rho = \sum_i \rho_i$$

for some maps $\rho_i : G_i \longrightarrow G$, and let

$$I = \rho \mathrm{Hom}_R(G, G^{(n)})$$

$$= \left\{ \sum_{\text{finite } i} \rho f_i(G) \,\middle|\, f_i : G \longrightarrow G_i \text{ for each } i = 1, \ldots, n \right\}.$$

Being a quotient of $\mathrm{Hom}_R(G, G^{(n)}) = \mathrm{Hom}_R(G, G)^{(n)}$, I is a finitely generated right ideal in $\mathrm{End}_R(G)$. Let $y \in G$. There is an $x \in G^{(n)}$ such that $\rho(x) = y$. Thus there are maps $f_i \in \mathrm{Hom}_R(G, G_i)$ and elements $x_i \in G$ for each $i = 1, \ldots, n$ such that

$$x = \sum_{i=1}^{n} f_i(x_i).$$

Hence

$$y = \rho(x) = \rho \left(\sum_{i=1}^{n} f_i(x_i) \right) = \sum_{i=1}^{n} \rho f_i(x_i) \in IG,$$

and thus $IG = G$. Inasmuch as G is fp faithful, $I = \mathrm{End}_R(G)$. Subsequently there is a map $f \in \mathrm{Hom}(G, G^{(n)})$ such that $\rho f = 1_G$. This proves part 3.

$3 \Rightarrow 2$. We leave it as an exercise for the reader to show that if each short exact sequence (11.10) in which n is an integer is split exact then each short exact sequence

$$0 \longrightarrow K \longrightarrow Q \longrightarrow G \longrightarrow 0$$

in which $Q \in \mathbf{P}_o(G)$ is split exact. Then part 3 implies that G is fp G-split.

$2 \Rightarrow 1$. Assume part 1 is false. There is a proper finitely generated right ideal $I \subset \text{End}(G)$ such that $IG = G$. Suppose that I is generated by $\{\rho_1, \ldots, \rho_m\}$ over $\text{End}_R(G)$, and let $G_i = G$ for each $i = 1, 2, \ldots, m$. There is a map

$$\rho : G_1 \oplus \cdots \oplus G_m \longrightarrow G$$

such that for each $x \in G_i$, $\rho(x) = \rho_i(x)$. Given any map

$$g : G \longrightarrow G_1 \oplus \cdots \oplus G_m$$

there are maps $g_i : G \longrightarrow G_i = G$ such that $g = g_1 \oplus \cdots \oplus g_m$. Then

$$\rho g = \rho_1 g_1 + \cdots + \rho_m g_m \in I \neq \text{End}(G).$$

Specifically $\rho g \neq 1_G$ for any g, so that ρ is not split. Thus part 2 is false, which completes the proof.

Lemma 11.36. *The following are equivalent for a self-small right R-module G.*

1. *G is faithful.*
2. *G is G-split.*
3. *Each short exact sequence*

$$0 \longrightarrow K \longrightarrow G^{(c)} \xrightarrow{\rho} G \longrightarrow 0 \qquad (11.11)$$

in which c is a cardinal, is split exact.

Proof: The proof is analogous to that of Lemma 11.35 and is left as an exercise for the reader.

Given G-split modules H and H', there are several questions we can ask. For instance, when are the direct sums $H \oplus H'$ and $H^{(c)}$, G-split, where c is a cardinal? The next result answers these questions.

Lemma 11.37. *Let G be a self-small right R-module.*

1. *If H and H' are G-split then $H \oplus H'$ is G-split.*
2. *If G is fp G-split then $G^{(n)}$ is finitely G-split for each integer $n > 0$.*
3. *If G is G-split then $G^{(c)}$ is G-split for each cardinal c.*

Proof: 1. Suppose that H and H' are G-split and consider the short exact sequence

$$0 \longrightarrow K \longrightarrow G^{(c)} \xrightarrow{\rho} H \oplus H' \longrightarrow 0, \qquad (11.12)$$

where c is a cardinal. We must show that (11.12) is split exact.

Let $\pi : H \oplus H' \longrightarrow H$ be the canonical projection with $\ker \pi = H'$. Since H is G-split there is an injection $g : H \longrightarrow G^{(c)}$ such that $1_H = (\pi \rho) g = \pi(\rho g)$. Then

$$H \oplus H' = \rho g(H) \oplus H' \quad \text{and} \quad G^{(c)} = g(H) \oplus X,$$

where $X = \ker \pi \rho$ and where $\rho g(H) \cong g(H) \cong H$. Let

$$p : \rho g(H) \longrightarrow g(H) \subset G^{(c)}$$

be the isomorphism with inverse $\rho | g(H)$. Then

$$\rho p(z) = z \text{ for each } z \in \rho g(H).$$

Since $G^{(c)} = g(H) \oplus X$ we have

$$\rho g(H) \oplus H' = H \oplus H' = \rho(G^{(c)}) = \rho g(H) \oplus \rho(X)$$

and so $H' \cong \rho(X)$.

There is a short exact sequence

$$0 \longrightarrow K \longrightarrow X \overset{\rho | X}{\longrightarrow} \rho(X) \longrightarrow 0,$$

where $\rho | X$ is the restriction of ρ to X. Since $\rho(X) \cong H'$ is G-split there is an injection
$q : \rho(X) \longrightarrow X \subset G^{(c)}$ such that

$$\rho q(w) = w \text{ for each } w \in \rho(X).$$

We have thus defined a map $p \oplus q : \rho g(H) \oplus \rho(X) \longrightarrow G^{(c)}$ such that

$$\rho(p \oplus q)(z \oplus w) = \rho p(z) + \rho q(w) = z \oplus w$$

for each $z \in \rho g(H)$ and $w \in \rho(X)$. Thus,

$$\rho(p \oplus q) = 1_{\rho g(H) \oplus \rho(X)} = 1_{H \oplus H'},$$

so that $H \oplus H'$ is G-split.

2. Since G is fp G-split, an induction argument using part 1 shows us that $G^{(n)}$ is G-split for each integer $n > 0$.

3. Proceed as in part 2 but use a transfinite induction argument to prove that $G^{(c)}$ is G-split for each cardinal $c > 0$. This completes the proof.

11.5.2 Semi-simple rings

The ring E is said to be *semi-simple Artinian* if $\operatorname{rgd}(E) = 0$.

Theorem 11.38. *The following are equivalent for a ring E.*

1. *E is semi-simple Artinian.*
2. *$E = S_1^{(n_1)} \oplus \cdots \oplus S_t^{(n_t)}$ for some simple left (right) E-modules $S_1, \ldots, S_t$ and integers $n_1, \ldots, n_t > 0$ such that $S_i \ncong S_j$ for each $i \neq j$.*
3. *There is a finite set $\{S_1, \ldots, S_t\}$ of simple left (right) E-modules such that*
   (a) *$S_i \ncong S_j$ for each $i \neq j$ and*
   (b) *Each left (right) E-module M is a direct sum of copies of the S_i.*

4. *Each (cyclic) left E-module is projective.*
5. *Each (cyclic) right E-module is projective.*
6. *Each (cyclic) left E-module is injective.*
7. *Each (cyclic) right E-module is injective.*

Proof: See [7].

In particular, if $E = \text{End}_R(G)$ is semi-simple Artinian then G is a projective left $\text{End}_R(G)$-module, so that $\text{fd}_E(G) = \text{pd}_E(G) = 0$.

Theorem 11.39. *Let G be a self-small right R-module. The following are equivalent.*

1. $\text{End}_R(G)$ *is a semi-simple ring.*
2. $\text{gd}(G) = 0$.
3. *Given a G-generated R-submodule $H \subset G$ then H is G-split and H is a direct summand of G.*

Proof: $1 \Leftrightarrow 2$. Say $\text{rgd}(\text{End}_R(G)) = 0$ and let Q be a G-plex. By Theorem 9.12, there is a projective resolution $\mathcal{P}$ over $\text{End}_R(G)$ such that $\mathbf{T}_G(\mathcal{P}) \cong Q$. Since $\text{rgd}(\text{End}_R(G)) = 0$, $\mathcal{P}$ is homotopic to a projective resolution $0 \longrightarrow P_0$, so that Q is homotopic to $0 \longrightarrow \mathbf{T}_G(P_0)$. It follows that $\text{gd}(G) = 0$. The converse follows in the same manner.

$2 \Rightarrow 3$. Say $\text{gd}(G) = 0$ and let $H \subset G$ be a G-generated R-submodule. There is by Lemma 9.2 a G-plex Q such that $Q_0 = G$ and image $\delta_1 = H$. Since $0 = \text{gd}(G) = \text{pd}_G(Q)$, the reader will make the obvious changes in the proof of Theorem 11.14 to prove that H is a G-split direct summand of G. This proves $2 \Rightarrow 3$.

$3 \Rightarrow 2$ follows in the converse manner, which completes the proof.

Example 11.40. There is a module G such that $\text{gd}(G) = 0$, but E is not a semi-simple ring. Let $R = \mathbf{Z}$, and let $G = \oplus_p \mathbf{Z}/p\mathbf{Z}$ where p ranges over the primes in $\mathbf{Z}$. Then $\text{End}(G) = \prod_p \mathbf{Z}/p\mathbf{Z}$ is not a semi-simple ring. Furthermore, given $Q \in \mathbf{P}(G)$ and a G-generated subgroup $H \subset Q$, then H is a direct summand of Q. Thus $\text{gd}(G) = 0$, and hence our use of the self-small hypothesis in Theorem 11.39 is necessary.

11.5.3 Right hereditary rings

Since we have already considered the self-small right R-modules G such that $\text{gd}(G) = 0$, we focus our attention on those G for which

$$0 \leq \text{fd}_E(G) \leq \text{rgd}(\text{End}_R(G)) \leq 1.$$

These are the self-small right R-modules G with *right hereditary* endomorphism rings.

Recall the definition of *weakly G-split* from page 183. The lemma is left as an exercise for the reader.

Lemma 11.41. *Let G be a self-small right R-module and let $H \subset G$ be a G-generated R-submodule. Then H is weakly G-split iff given a G-plex Q such that $Q_0 = G$ and image $\delta_1 = H$ then $\mathrm{pd}(Q) \leq 1$.*

Proof: Follows immediately from Theorem 11.16.

Theorem 11.42. *Let G be a self-small right R-module. The following are equivalent.*

1. $\mathrm{End}_R(G)$ is right hereditary.
2. Each G-generated R-submodule of G is weakly G-split.

Proof: $1 \Rightarrow 2$. Suppose that $\mathrm{rgd}(\mathrm{End}_R(G)) \leq 1$ and let $H \subset G$ be a G-generated R-submodule. There is a G-plex Q such that $Q_0 = G$ and image $\delta_1 = H$. By Corollary 11.31, H is weakly G-split.

$2 \Rightarrow 1$. Suppose that each G-generated R-submodule of G is weakly G-split and let Q be a G-plex such that $Q_0 = G$. By hypothesis, image δ_1 is weakly G-split so by Corollary 11.31, $\mathrm{rgd}(\mathrm{End}_R(G)) \leq 1$. This completes the proof.

We consider the case

$$0 = \mathrm{fd}_E(G) < \mathrm{rgd}(\mathrm{End}_R(G)) = 1.$$

We say that the G-generated module H is *G-split* if given $Q \in \mathbf{P}(G)$ and a surjection $\delta : Q \longrightarrow H$ then $Q = \ker \delta \oplus Q'$ for some right R-module Q'.

Theorem 11.43. *Let G be a self-small right R-module. The following are equivalent.*

1. G is E-flat and $\mathrm{End}_R(G)$ is right hereditary.
2. Each G-generated R-submodule of G is G-split.
3. Each G-generated R-submodule of each $Q \in \mathbf{P}(G)$ is G-split.

Proof: $1 \Rightarrow 2$. Let $H \subset G$ be a G-generated R-submodule and let $\delta : Q \longrightarrow H$ be a surjection in which $Q \in \mathbf{P}(G)$. Since $H \subset G$ there is a G-plex Q such that $Q_0 = G$, $Q_1 = Q$, and $\delta_1 = \delta$. (See Lemma 9.4.) Since $\mathrm{rgd}(\mathrm{End}_R(G)) \leq 1$, Theorem 11.42 states that H is weakly G-split. Thus $\mathbf{S}_G(\ker \delta) \oplus Q' = Q$ for some R-module Q'. Since $\mathrm{fd}_E(G) = 0$, Theorem 11.22.2 implies that $\ker \delta = \mathbf{S}_G(\ker \delta)$. Then $\ker \delta \oplus Q' = Q$ and hence H is G-split.

$2 \Rightarrow 3$. Assume part 2. Let $Q \in \mathbf{P}(G)$, and let $H \subset Q$ be a G-generated R-submodule. There is an index set $\mathcal{I}$ and a containment

$$H \subset Q \subset \bigoplus_{i \in \mathcal{I}} G_i \subset \prod_{i \in \mathcal{I}} G_i$$

of right R-modules with $G \cong G_i$ for each $i \in \mathcal{I}$. For each $i \in \mathcal{I}$ there is a projection

$$\rho_i : \prod_{i \in \mathcal{I}} G_i \longrightarrow G_i.$$

Inasmuch as $\rho_i(H)$ is isomorphic to a G-generated R-submodule of G, part 2 states that $\rho_i(H)$ is G-split. Since each homomorphic image of H in G is G-split, for any map $\phi : H \longrightarrow G$, $H \cong \phi(H) \oplus \ker \phi$.

Well order the set $\mathcal{I}$. We use transfinite induction on $\mathcal{I}$. The projection $\rho_1 : H \longrightarrow G_1$ splits since by hypothesis the G-generated submodule $\rho_1(H) \subset G$ is G-split. Then with

$$K_1 = H \cap \left(\bigoplus_{i>1} G_i \right)$$

there is an R-submodule $\rho_1(H) \cong I_1 \subset H$ such that

$$H = I_1 \oplus K_1.$$

Assume that there is some ordinal $\alpha \in \mathcal{I}$ such that

$$H = \left(\bigoplus_{i<\alpha^+} I_i \right) \oplus K_\alpha, \tag{11.13}$$

where

$$K_\alpha = \bigcap_{i<\alpha} \ker \rho_i.$$

By hypothesis, $\rho_\alpha(K_\alpha) \subset G_\alpha$ is G-split so that

$$\rho_\alpha(K_\alpha) \oplus K_{\alpha^+} \cong K_\alpha = I_\alpha \oplus K_{\alpha^+} \tag{11.14}$$

for some $I_\alpha \subset K_\alpha \subset H$. Hence,

$$H \stackrel{(11.13)}{=\!=\!=} \left(\bigoplus_{i<\alpha^+} I_i \right) \oplus K_\alpha$$

$$\stackrel{(11.14)}{=\!=\!=} \left(\bigoplus_{i<\alpha^+} I_i \right) \oplus I_\alpha \oplus K_{\alpha^+}$$

$$= \left(\bigoplus_{i<\alpha^{++}} I_i \right) \oplus K_{\alpha^+}.$$

By transfinite induction

$$H = \left(\bigoplus_{i\in\mathcal{I}} \rho_i(H) \right) \oplus K,$$

where $K = \bigcap_{i\in\mathcal{I}} \ker \rho_i = 0$. Thus,

$$H = \bigoplus_{i\in\mathcal{I}} \rho_i(H),$$

which proves part 3.

$3 \Rightarrow 1$. Assume that part 3 is true. By Corollary 11.31, $\mathrm{rgd}(\mathrm{End}_R(G)) \leq 1$. Consider the G-plex Q. By part 3, $\delta_1 : Q_1 \longrightarrow Q_0$ has G-split image, so $Q_1 = \ker \delta_1 \oplus Q$ for some right R-module Q. Since then $\ker \delta_1$ is G-generated, Lemma 10.24 implies that $\mathrm{fd}_E(G) = 0$. This proves part 1 and completes the logical cycle.

The next theorem illustrates the effect of the ascending chain condition on the hereditary property in endomorphism rings. We say that H is *weakly finite G-split* if given $Q \in \mathbf{P}(G)$ and a surjection $\delta : Q \longrightarrow H$ then $Q = \mathbf{S}_G(\ker \delta) \oplus Q_o$ for some $Q_o \in \mathbf{P}_o(G)$.

Theorem 11.44. *The following are equivalent for a self-small right R-module G.*

1. *$\mathrm{End}_R(G)$ is right Noetherian, right hereditary.*
2. *Each G-generated submodule of G is weakly finite G-split.*

Proof: Assume part 2. By Theorem 11.42, $\mathrm{End}_R(G)$ is right hereditary. To see that $\mathrm{End}_R(G)$ is right Noetherian, let $I \subset \mathrm{End}_R(G)$ be a right ideal and let

$$ j : I \longrightarrow \mathrm{End}_R(G) $$

be the inclusion map. Then I is a projective right $\mathrm{End}_R(G)$-module and image $\mathbf{T}_G(j) = IG$. Since IG is weakly finite G-split and since $\mathbf{T}_G(I) \in \mathbf{P}(G)$, there is a $Q_o \in \mathbf{P}_o(G)$ such that

$$ \mathbf{T}_G(I) = \mathbf{S}_G(\ker \mathbf{T}_G(j)) \oplus Q_o. $$

However, by Theorem 1.13 the category equivalence $\mathbf{T}_G(\cdot)$ takes the injection j of projective right $\mathrm{End}_R(G)$-modules to a monomorphism $\mathbf{T}_G(j)$ in $\mathbf{P}_o(G)$. Thus $\mathbf{S}_G(\ker \mathbf{T}_G(j)) = 0$ and hence $\mathbf{T}_G(I) = Q_o \in \mathbf{P}_o(G)$. Another appeal to Theorem 1.13 shows us that

$$ I \cong \mathbf{H}_G \mathbf{T}_G(I) \cong \mathbf{H}_G(Q_o) $$

is a finitely generated projective right $\mathrm{End}_R(G)$-module. Then $\mathrm{End}_R(G)$ is right Noetherian. This proves part 1.

Conversely, assume part 1 and let $H \subset G$ be a G-generated R-submodule. Choose a $Q \in \mathbf{P}(G)$ and a surjection

$$ \delta : Q \longrightarrow H $$

of right R-modules. By Theorem 11.42, H is weakly G-split, so

$$ Q = \mathbf{S}_G(\ker \delta) \oplus Q_o $$

for some $Q_o \in \mathbf{P}(G)$. Then

$$ \mathbf{H}_G(Q_o) \cong \mathbf{H}_G(H) \subset \mathbf{H}_G(G) $$

via the restriction of $\mathbf{H}_G(\delta)$ to $\mathbf{H}_G(Q_o)$. Since $\mathrm{End}_R(G) = \mathbf{H}_G(G)$ is right Noetherian right hereditary, $\mathbf{H}_G(Q_o)$ is a finitely generated projective right $\mathrm{End}_R(G)$-module. By Theorem 1.13 we see that

$$Q_o \cong \mathbf{T}_G\mathbf{H}_G(Q_o) \in \mathbf{P}_o(G).$$

This proves part 2 and completes the proof.

The right R-module G is said to be *Noetherian hereditary* if each G-generated R-submodule of G is in $\mathbf{P}_o(G)$. We say that G is *faithfully E-flat* if G is a flat left $\mathrm{End}_R(G)$-module and $IG \neq G$ for each proper right ideal $I \subset \mathrm{End}_R(G)$. The G-generated right R-module H is *finitely G-split* if given a surjection $\delta : Q \longrightarrow H$ in which $Q \in \mathbf{P}(G)$ then $Q = \ker \delta \oplus Q_o$ for some $Q_o \in \mathbf{P}_o(G)$.

Lemma 11.45. *Let G be a self-small right R-module.*

1. *If each G-generated R-submodule of G is G-split, then G is E-flat.*
2. *Let G be an E-flat right R-module. If a G-generated R-submodule $H \subset G$ is weakly (finite) G-split, then H is (finitely) G-split.*

Proof. 1. Given $Q \in \mathbf{P}(G)$ and a map $\delta : Q \longrightarrow G$, our hypothesis states that $\delta(Q)$ is G-split, so that $Q = \ker \delta \oplus Q_o$ for some Q_o. Specifically, $\ker \delta$ is G-generated, so by Lemma 10.24 G is E-flat.

2. Suppose that G is E-flat and that $H \subset G$ is weakly G-split. Let $\delta : Q \longrightarrow H$ be a surjection. Since H is weakly G-split, we can write

$$Q = \mathbf{S}_G(\ker \delta) \oplus Q_o$$

for some $Q_o \in \mathbf{P}(G)$, and since G is E-flat, $\ker \delta$ is G-generated (Lemma 10.24). Thus,

$$Q = \ker \delta \oplus Q_o$$

and hence H is G-split. The rest follows in the same manner. This completes the proof.

Theorem 11.46. [3, U. Albrecht] *The following are equivalent for a self-small right R-module G.*

1. *G is a faithfully E-flat Noetherian hereditary right R-module.*
2. *G is E-flat and $\mathrm{End}_R(G)$ is right Noetherian right hereditary.*
3. *Each G-generated submodule of G is finitely G-split.*

Proof. 3 $\Rightarrow$ 2. Assume part 3. By Theorem 11.44, $\mathrm{End}_R(G)$ is right Noetherian and right hereditary. By Lemma 11.45, G is E-flat. This proves part 2.

2 $\Rightarrow$ 1. Assume part 2. Let $H \subset G$ be a G-generated R-submodule. By Theorem 11.44, H is weakly finite G-split. Then given a $Q \in \mathbf{P}(G)$ and a surjection

$\delta : Q \longrightarrow H$ of right R-modules, $Q = \ker \delta \oplus Q_o$ for some $Q_o \in \mathbf{P}_o(G)$. Hence,

$$H = \delta(Q) = \delta(Q_o) \cong Q_o \in \mathbf{P}_o(G),$$

whence G is a Noetherian hereditary right R-module.

By part 2, G is E-flat. To see that G is faithful let $\delta : Q \longrightarrow G$ be a surjection in which $Q \in \mathbf{P}(G)$. By Theorem 11.44 and part 2, G is weakly finite G-split, so that

$$Q = \mathbf{S}_G(\ker \delta) \oplus Q_o$$

for some $Q_o \in \mathbf{P}_o(G)$. Since G is E-flat, Lemma 10.24 implies that $\ker \delta$ is G-generated. Hence

$$Q = \ker \delta \oplus Q_o$$

so that δ is split. Then by Lemma 11.36, G is faithful. This proves part 1.

$1 \Rightarrow 3$. Assume part 1. Then G is a self-small faithfully E-flat right R-module. By part 1 and Theorem 11.6, $\mathrm{End}_R(G)$ is right Noetherian and right hereditary, so by Theorem 11.44, each G-generated $H \subset G$ is weakly finite G-split. By Lemma 11.45.2, each such H is finitely G-split, which proves part 3. This completes the proof.

The above results take care of all of the possibilities for $0 \leq \mathrm{fd}_E(G) \leq \mathrm{rgd}$ $(\mathrm{End}_R(G)) \leq 1$. Let us investigate weaker conditions than Noetherian and hereditary. In our work we consider the effect that these weaker conditions have on the first two terms of G-plexes Q.

Recall that E is *right semi-hereditary* if each finitely generated right ideal of E is projective. Then E is right semi-hereditary iff $\mathrm{pd}_E(M) \leq 1$ for each (cyclic) finitely presented right $\mathrm{End}_R(G)$-module M.

Theorem 11.47. [12, D. Arnold and J. Hausen] *The following are equivalent for a self-small right R-module G.*

1. *$\mathrm{End}_R(G)$ is right semi-hereditary.*
2. *Given $Q \in \mathbf{P}_o(G)$ and a map $\delta : Q \longrightarrow G$ then $Q = \mathbf{S}_G(\ker \delta) \oplus Q'$ for some right R-module Q'.*

Proof: Assume part 2 and let M be a cyclic finitely presented right $\mathrm{End}_R(G)$-module. There is a projective resolution $\mathcal{P}(M)$ for M such that $P_0 \cong \mathrm{End}_R(G)$ and P_1 are finitely generated projective right $\mathrm{End}_R(G)$-modules. Since G is self-small, Lemma 9.2 states that $\mathbf{T}_G(\mathcal{P}(M)) \in G\text{-}\mathbf{Plex}$. Moreover, $\mathbf{T}_G(P_0) \cong G$, and $\mathbf{T}_G(P_1) \in \mathbf{P}_o(G)$. By part 2, $\mathbf{S}_G(\ker \mathbf{T}_G(\partial_1))$ is a direct summand of $\mathbf{T}_G(P_1)$, so that by Theorem 11.14, $\mathrm{pd}_G(\mathbf{T}_G(\mathcal{P}(M))) \leq 1$. Since $\mathbf{T}_G$ lifts to a category equivalence between $\mathbf{Mod}\text{-}\mathrm{End}_R(G)$ to $G\text{-}\mathbf{Plex}$ (Theorem 9.12), $\mathrm{pd}_E(M) \leq 1$, and hence $\mathrm{End}_R(G)$ is semi-hereditary.

Conversely, assume that $\mathrm{End}_R(G)$ is right semi-hereditary, let $Q \in \mathbf{P}_o(G)$, and let $\delta : Q \longrightarrow G$ be a map. By Theorem 1.13, there is a map $\partial : P \longrightarrow \mathrm{End}_R(G)$ such that $\mathbf{T}_G(\partial) = \delta$. If we let $I = \mathrm{image}\ \partial$, then by part 1, $\partial : P \longrightarrow I$ is a split surjection, so that $\mathbf{T}_G(\ker \partial)$ is a direct summand of $\mathbf{T}_G(P) = Q$. We leave it as an interesting exercise in diagram construction for the reader to prove that $\mathbf{T}_G(\ker \partial) = \mathbf{S}_G(\ker \delta)$. This proves part 2 and completes the proof.

We leave the proof of the following two results for the reader.

Corollary 11.48. *The following are equivalent for a self-small right R-module G.*

1. *G is E-flat and $\mathrm{End}_R(G)$ is right semi-hereditary.*
2. *Given $Q \in \mathbf{P}_o(G)$ and a map $\delta : Q \longrightarrow G$, then, $\ker \delta$ is a direct summand of Q.*

Corollary 11.49. *The following are equivalent for a self-small right R-module G.*

1. *Each cyclic right ideal in $\mathrm{End}_R(G)$ is projective.*
2. *Each endomorphic image of G is in $\mathbf{P}(G)$.*

11.5.4 Global dimension at most 3

Consider the dimensions

$$0 \leq \mathrm{fd}_E(G) \leq \mathrm{rgd}(\mathrm{End}_R(G)) \leq 3.$$

The proofs of the following theorems make excellent exercises in the uses of the above techniques. The reader is encouraged to prove these results for themselves.

Theorem 11.50. *The following are equivalent for a self-small right R-module G.*

1. *$\mathrm{rgd}(\mathrm{End}_R(G)) \leq 2$.*
2. *Given a map $\delta : Q \longrightarrow G$ in which $Q \in \mathbf{P}(G)$, then $\mathbf{S}_G(\ker \delta)$ is weakly balanced G-split.*

Proof: Assume part 1 and let $\delta : Q \longrightarrow G$ be a map in which $Q \in \mathbf{P}(G)$. By Lemma 9.4, there is a G-plex $\mathcal{Q}$ such that $\delta = \delta_1$ and, by Theorem 11.15, $I_2 = $ image δ_2 admits a G-cover. But by Lemma 9.6,

$$I_2 = \mathbf{S}_G(K_1) = \mathbf{S}_G(\ker \delta_1) = \mathbf{S}_G(\ker \delta),$$

so part 2 is proved.

Conversely, assume part 2. Let $\mathcal{Q}$ be an arbitrary G-plex in which $Q_0 = G$. By part 2 and Lemma 9.6, the module

$$I_2 = \mathbf{S}_G(K_1) = \mathbf{S}_G(\ker \delta)$$

is weakly balanced G-split, so by Theorem 11.15, $\mathrm{pd}_G(\mathcal{Q}) \leq 2$. Thus,

$$\mathrm{rgd}(\mathrm{End}_R(G)) = \mathrm{gd}(G) \leq 2$$

by Theorem 11.27, which completes the proof.

The surjection δ is *G-balanced* if G is projective relative to δ. We say that H is *balanced G-split* if given a G-balanced surjection $\delta : Q \longrightarrow H$ such that $Q \in \mathbf{P}(G)$, then $Q = \ker \delta \oplus Q'$ for some right R-module Q'.

Theorem 11.51. *The following are equivalent for a self-small right R-module G.*

1. $\mathrm{fd}_E(G) < 2$ and $\mathrm{rgd}(\mathrm{End}_R(G)) \leq 2$.
2. *Given a map $\delta : Q \longrightarrow G$ in which $Q \in \mathbf{P}(G)$, then $\mathbf{S}_G(\ker \delta)$ is balanced G-split.*

Proof: Assume part 1. Let $\delta : Q \longrightarrow G$ be a map in which $Q \in \mathbf{P}(G)$, and let $\delta_2 : Q_2 \longrightarrow Q$ be a G-balanced map such that image $\delta_2 = \mathbf{S}_G(\ker \delta)$. There is a G-plex Q whose maps are $\cdots, \delta_2, \delta$. Since $\mathrm{rgd}(\mathrm{End}_R(G)) \leq 2$, Theorem 11.50 states that $\mathbf{S}_G(\ker \delta)$ admits a G-cover. Thus $Q_2 = \mathbf{S}_G(\ker \delta_2) \oplus Q_2'$. Since $\mathrm{fd}_E(G) \leq 1$, $\ker \delta_2 = \mathbf{S}_G(\ker \delta_2)$, so $Q_2 = \ker \delta_2 \oplus Q_2'$. Therefore, $\mathbf{S}_G(\ker \delta)$ is balanced G-split, which proves part 2.

Conversely, assume part 2. Let Q be a G-plex in which $Q_0 = G$. As a consequence of part 2, $\mathbf{S}_G(\ker \delta_1)$ is weakly balanced G-split, so by Theorem 11.50, $\mathrm{rgd}(\mathrm{End}_R(G)) \leq 2$. Furthermore, since $\mathbf{S}_G(\ker \delta_1)$ is balanced G-split, $\ker \delta_2 \in \mathbf{P}(G)$ is G-generated. Hence $\mathrm{fd}_E(G) \leq 1$ by Theorem 11.21. This proves part 1 and completes the logical cycle.

Just to prove that it can be done, we present the following admittedly baroque characterization of right R-modules G whose endomorphism ring $\mathrm{End}_R(G)$ has right global dimension at most 3.

Corollary 11.52. *The following are equivalent for a self-small right R-module G.*

1. $\mathrm{rgd}(E) \leq 3$.
2. *Given a G-balanced map $\delta : Q_2 \longrightarrow Q_1$ in which $Q_1, Q_2 \in \mathbf{P}(G)$ and such that coker $\delta \subset Q_0$ for some $Q_0 \in \mathbf{P}(G)$, then $\mathbf{S}_G(\ker \delta)$ is weakly balanced G-split.*

Corollary 11.53. *Let G be a self-small right R-module. The following are equivalent.*

1. $\mathrm{fd}_E(G) < 3$ and $\mathrm{rgd}(E) \leq 3$.
2. *Given a G-balanced map $\delta : Q_2 \longrightarrow Q_1$ in which $Q_1, Q_2 \in \mathbf{P}(G)$ and such that coker $\delta \subset Q_0$ for some $Q_0 \in \mathbf{P}(G)$, then $\mathbf{S}_G(\ker \delta)$ is balanced G-split.*

11.6 Injective dimensions and modules

11.6.1 A review of G-coplexes

By dualizing our approach, our study of the flat dimension of the left $\mathrm{End}_R(G)$-module G becomes a study of the injective dimension of G. We characterize a number of injective properties for G as a left E-module. While the flat dimension of a ring R is not of much interest, our techniques yield a systematic way to study the injective dimension of R. The reader should review Section 9.3, which contains the notions and notations surrounding G-coplexes and the left $\mathrm{End}_R(G)$-modules.

Recall that by assuming that

(μ) Measurable cardinals do not exist

we can completely dualize Theorem 9.12. We let **coP**(G) denote the category of direct summands of Cartesian products G^c for cardinals c. The right R-module is *self-slender* if for each cardinal c the natural map

$$\mathrm{Hom}_R(G, G)^{(c)} \longrightarrow \mathrm{Hom}_R(G^c, G)$$

is an isomorphism of abelian groups. (This is where we use the hypothesis (μ). Without it, there are no self-slender modules.) There are functors

$$(\cdot)^* = \mathrm{Hom}_R(\cdot, G), \quad (\cdot)^* = \mathrm{Hom}_{\mathrm{End}_R(G)}, (\cdot, G)$$

and we let

$$\mathcal{W} \;=\; W_0 \xrightarrow{\;\sigma_1\;} W_1 \xrightarrow{\;\sigma_2\;} W_2 \xrightarrow{\;\sigma_3\;} \cdots$$

denote a G-coplex. For $X \in \{Q_k, W_k, \delta_k, \sigma_k\}$ we let $X^* = (X)^*$, where $(\cdot)^*$ is defined above. The *category of G-coplexes* is denoted by G-**Coplx**. Given a G-coplex $\mathcal{W}$,

$$\mathcal{W}^* = \cdots \xrightarrow{\;\sigma_3^*\;} W_2^* \xrightarrow{\;\sigma_2^*\;} W_1^* \xrightarrow{\;\sigma_1^*\;} W_0^*$$

is a projective resolution of *left* $\mathrm{End}_R(G)$-modules.

By Theorem 9.30, if G is self-slender then for each $W \in$ **coP**(G) the point evaluation map

$$\Psi_W \;:\; W \longrightarrow W^{**} \;:\; x \longmapsto \cdot(x)$$

is an isomorphim. In general, the map Ψ_W is an isomorphism for each direct summand W of G^n for each integer $n > 0$.

Assuming (μ), Theorem 9.32 states that if G is self-slender then $\mathrm{Hom}_R(\cdot, G)$ induces a category equivalence

$$\mathrm{h}^G(\cdot) : G\text{-}\mathbf{Cohcoplx} \longrightarrow \mathrm{End}_R(G)\text{-}\mathbf{Mod}$$
$$\mathrm{h}^G(\mathcal{W}) \;=\; \mathrm{coker}\, \sigma_1^*.$$

In other words, $\mathrm{h}^G(\mathcal{W})$ is the zero-th homology functor $H_0(\cdot)$ of the projective resolution $\mathcal{W}^*$.

A G-coplex $\mathcal{W}$ is said to be a *coherent* G-coplex if each W_k is a direct summand of G^n for some integer $n > 0$. We let G-**Cohcoplx** denote the full subcategory of G-**coPlex** whose objects are the coherent G-coplexes.

With each $\mathcal{W} \in G$-**Coplx** and each integer $k > 0$ there are short exact sequences

$$0 \longrightarrow K_k \longrightarrow W_{k-1} \xrightarrow{\;\sigma_k\;} J_k \longrightarrow 0. \tag{11.15}$$

Specifically

$$K_k = \ker \sigma_k \text{ and } J_k = \text{image } \sigma_k$$

for each integer $k > 0$.

Dualizing the G-socle $\mathbf{S}_G(\cdot)$ we arrive at the *G-reject*, $\mathbf{R}_G(\cdot)$. Given a right R-module X and an R-submodule $U \subset X$ then the *G-reject of U in X* is

$$\mathbf{R}_G(U,X) = \bigcap \{\ker f \,|\, f \in \text{Hom}_R(X,G) \text{ such that } f(U) = 0\}.$$

If $\mathbf{R}_G(0,X) = 0$ then X is said to be *G-torsionless*. Furthermore,

$$\mathbf{R}_G(0,X) = \bigcap \{\ker f \,|\, f \in \text{Hom}_R(X,G)\}$$

is the smallest R-submodule V of X such that X/V is G-torsionless. More generally $\mathbf{R}_G(U,X)$ is the smallest R-submodule $X \supset V \supset U$ such that X/V is G-torsionless. By dualizing Lemma 9.5 we can show that if X is G-torsionless then there is a $W \in \mathbf{coP}(G)$ and an injection $\iota : X \longrightarrow W$.

Lemma 11.54. *Let $\mathcal{W}$ be a G-coplex and let $k > 0$ be an integer. Then*

1. $K_{k+1} = \mathbf{R}_G(J_k, W_k)$.
2. $H_k(\mathcal{W}) = K_{k+1}/J_k = \mathbf{R}_G(J_k, W_k)/J_k$.

Proof: Dualize the proof of Lemma 9.6.

The *exact dimension of $\mathcal{W}$* is the least integer $\text{ed}_G(\mathcal{W})$ such that $\mathcal{W}$ is exact at $k+j$ for each integer $j \geq 1$. That is,

$$\text{ed}_G(\mathcal{W}) = \min\{\text{integer } k > 0 \,\big|\, H_{k+j}(\mathcal{W}) = 0 \,\forall \text{ integers } j \geq 1\}.$$

Theorem 11.55. *Assume (μ). Let G be a self-slender right R-module and let $\mathcal{W} \in G$-$\mathbf{Coplx}$. The following are equivalent.*

1. $\text{ed}_G(\mathcal{W}) \leq k$.
2. $J_{k+j} = \mathbf{R}_G(J_{k+j}, W_{k+j})$ *for each integer $j \geq 1$.*
3. W_{k+j}/J_{k+j} *is G-torsionless for each integer $j \geq 1$.*

Proof: The proof is a straightforward application of the definitions.

Lemma 11.56. *Assume (μ). Let G be a self-slender right R-module and let $\mathcal{W}$ be a G-coplex.*

1. $\mathcal{W}^*$ *is a projective resolution of left $\text{End}_R(G)$-modules.*
2. *The natural isomorphisms $\Psi_{W_k} : W_k \longrightarrow W_k^{**}$ lift to a natural isomorphism $\mathcal{W} \cong \mathcal{W}^{**}$ of complexes.*

Proof: Dualize the proof of Lemma 9.5.

Let

$$\text{Ext}_E^k(\cdot, G) = \text{Ext}_{\text{End}_R(G)}^k(\cdot, G).$$

There are commutative triangles

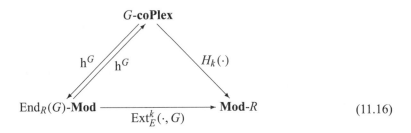

$$(11.16)$$

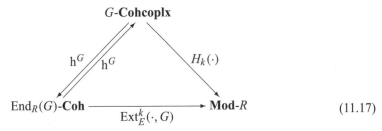

$$(11.17)$$

where $\text{End}_R(G)$-**Coh** denotes the category of coherent left $\text{End}_R(G)$-modules and where $H_k(\cdot)$ denotes the k-th homology functor on G-coplexes.

The commutativity of the triangles follows from Lemma 11.54 and the next lemma.

Lemma 11.57. *Assume* (μ). *Let G be a self-slender right R-module, let $W \in$ G-**Coplx**, and let $M = \text{h}^G(W)$. Then*

$$\text{Ext}_E^k(M, G) \cong \mathbf{R}_G(J_k, W_k)/J_k$$

for each integer $k > 0$.

Thus we have factored the extensions functors for each integer $k > 0$.

$$\text{Ext}_E^k(\cdot, G) \cong H_k \circ \text{h}^G(\cdot).$$

As a corollary to this fact, notice that W^* is a projective resolution of, say, the left $\text{End}_R(G)$-module M. Then, assuming (μ),

$$H_k(W) \cong H_k(W^{**}) \cong \text{Ext}_E^k(M, G)$$

for each integer $k > 0$.

The *injective dimension* of G is

$$\text{id}_E(G) = \text{the least integer } k \text{ such that } \text{Ext}_E^{k+j}(\cdot, G) = 0 \text{ for each}$$
$$\text{integer } j \geq 1, \text{ or } \infty \text{ if this minimum does not exist.}$$

We let

$$\text{coed}(G) = \sup\{\text{ed}_G(\mathcal{W}) \mid \mathcal{W} \in G\text{-}\mathbf{coPlex}\},$$

which is the least ordinal at which each G-complex $\mathcal{W}$ is exact. We call $\text{coed}(G)$ the *coexact dimension of G*.

Theorem 11.58. *Assume (μ) and let G be a self-slender right R-module. Then*

$$\text{id}_E(G) = \text{coed}(G).$$

Proof: Let $\mathcal{W} \in G\text{-}\mathbf{coPlex}$ and let $k = \text{ed}(\mathcal{W})$. By Lemma 11.57, $\text{h}^G(\mathcal{W}) = M$ is a left $\text{End}_R(G)$-module such that

$$\text{Ext}_E^{k+1}(M, G) = H_{k+1}(\mathcal{W}) = 0.$$

Hence $k \leq \text{id}_E(G)$, whence $\text{coed}(G) \leq \text{id}_E(G)$.

Conversely, let M be a left $\text{End}_R(G)$-module, and let $\text{Ext}_E^{k+1}(M, G) = 0$. Let $\mathcal{W} = \text{h}^G(M)$, so that by Lemma 11.57,

$$0 = \text{Ext}_E^{k+1}(M, G) = H_{k+1}(\mathcal{W}).$$

Hence $k \leq \text{coed}(G)$, whence $\text{id}_E(G) \leq \text{coed}(G)$. This completes the proof.

Corollary 11.59. *Assume (μ). Let G be a self-slender right R-module and let $k > 0$ be an integer. The following are equivalent.*

1. $\text{id}_E(G) \leq k$.
2. $\text{coed}(G) \leq k$.
3. *Given $\mathcal{W} \in G\text{-}\mathbf{Coplx}$ then $H_{k+1}(\mathcal{W}) = 0$.*
4. *Given $\mathcal{W} \in G\text{-}\mathbf{Coplx}$ then W_{k+1}/J_{k+1} is G-torsionless.*

11.6.2 Injective endomorphism rings

One of our goals is to characterize in a systematic fashion the properties of $\text{End}_R(G)$ in terms of G. The next series of results continues our march toward that goal.

The ring E is *left self-injective* if E is an injective left E-module. If we let $G = E = \text{End}_R(G)$ then we can characterize the injectivity of E as a left E-module. The left E-module M is *torsionless* if M is E-torsionless iff $M \subset \prod_{\mathcal{I}} E$ for some index set $\mathcal{I}$. We say that E is *quasi-Frobenius*, or a *QF-ring*, if E is *left* Artinian and left self-injective.

Let $F : \mathbf{C} \longrightarrow \mathbf{D}$ be a duality. That is, F is a contravariant category equivalence. It is worth the exercise to prove that $F(\cdot)$ applied to a small object in $\mathbf{C}$ is a slender left object in $\mathbf{D}$. Also, $F(\cdot)$ applied to a slender left object in $\mathbf{C}$ is a small object in $\mathbf{D}$.

Theorem 11.60. *Let G be a right R-module, identify G with the G-coplex $G \longrightarrow 0$, and assume that $\mathrm{End}_R(G)$ is left Noetherian. The following are equivalent.*

1. *$\mathrm{End}_R(G)$ is a left self-injective ring.*
2. *G is a projective object in G-**Cohcoplx**.*

Proof. $1 \Rightarrow 2$. Assume that $\mathrm{End}_R(G)$ is left self-injective. Then $\mathrm{End}_R(G)$ is an injective object in $\mathrm{End}_R(G)$-**Coh**. By Theorem 9.32 there is a duality

$$h^G(\cdot) : G\text{-}\mathbf{Cohcoplx} \longrightarrow \mathrm{End}_R(G)\text{-}\mathbf{Coh}$$

such that

$$h^G(G) = \mathrm{End}_R(G).$$

Then G is projective in G-**Cohcoplx**.

$2 \Rightarrow 1$. Let G be a projective object in G-**Cohcoplx**. Then $h^G(G) = \mathrm{End}_R(G)$ is an injective object in $\mathrm{End}_R(G)$-**Coh**. Since $\mathrm{End}_R(G)$ is left Noetherian, $\mathrm{End}_R(G)$ is injective relative to each injection $I \subset \mathrm{End}_R(G)$, where I is a left ideal. By Baer's Criterion for injectivity, $\mathrm{End}_R(G)$ is left self-injective. This completes the logical cycle.

If we let $R = G = \mathrm{End}_R(G)$ then we can characterize injective properties of rings in terms of projective properties in the *category of coherent R-coplexes R-**Cohcoplx***.

Corollary 11.61. *Let R be a left Noetherian ring and identify R with the R-coplex $R \longrightarrow 0$. The following are equivalent.*

1. *R is left self-injective.*
2. *R is a QF-ring.*
3. *R is projective in R-**Cohcoplx**.*
4. *R is a projective generator in R-**Cohcoplx**.*

Proof. $4 \Rightarrow 3 \Rightarrow 2 \Rightarrow 1$ is clear.

$1 \Rightarrow 4$. Assume part 1. Then R is left Noetherian and left self-injective, so that R is *QF*. *QF* rings are injective cogenerators, so that R is an injective cogenerator in $\mathrm{End}_R(G)$-**Coh** $= \mathrm{End}_R(G)$-**FG**. An application of the duality

$$h^R(\cdot) : R\text{-}\mathbf{Coh} \longrightarrow R\text{-}\mathbf{Cohcoplx}$$

shows us that $G = h^R(\mathrm{End}_R(G))$ is a projective generator in R-**Cohcoplx**. This completes the proof.

Using these ideas we can prove a cogenerator property for *QF*-rings. The *R*-module M is said to be *strongly finitely R-cogenerated* if there is an integer $n > 0$ and an embedding $f : M \longrightarrow R^{(n)}$ relative to which R is injective. In particular, the induced map $f^* : (R^{(n)})^* \longrightarrow M^*$ formed by applying $\operatorname{Hom}_R(\cdot, R)$ to f is a surjection.

Theorem 11.62. *Let R be a ring. If each finitely generated right and each finitely generated left R-module is strongly finitely R-cogenerated, then R is a QF-ring.*

Proof: We claim that R is left Noetherian. This is accomplished by proving that *each finitely generated left R-module is coherent*. Let M be a finitely generated left $\operatorname{End}_R(G)$-module. Since M is finitely generated and strongly finitely R-cogenerated, there are finitely generated projective left R-modules P_1, P_0, and a map

$$P_1 \xrightarrow{\partial_1} M \subset P_0$$

in which image $\partial_1 = M$. Then ∂_1 dualizes to a map

$$P_0^* \xrightarrow{\partial_1^*} P_1^*$$

between finitely generated projective right R-modules.

Let coker $\partial_1^* = C_1$. Inasmuch as C_1 is a finitely generated and a strongly finitely R-cogenerated right R-module our hypotheses imply that C_1 embeds in a finitely generated projective R-module, W_2, in such a way that R is injective relative to the inclusion

$$C_1 \subset W_2.$$

We then have the canonical map

$$\sigma_2 : P_1^* \longrightarrow W_2.$$

This is the first step in an induction that produces a complex

$$P_0^* \xrightarrow{\partial_1^*} P_1^* \xrightarrow{\sigma_2} W_2 \xrightarrow{\sigma_3} \cdots$$

in which each $W_k \cong R^{(n_k)}$ for some integer $n_k > 0$, and in which R is injective relative to the induced inclusions

$$\operatorname{coker} \sigma_k \subset W_k$$

for each integer $k \geq 2$. That is, we have constructed a coherent R-coplex.

An application of $(\cdot)^*$ yields the projective resolution

$$\cdots \xrightarrow{\sigma_3^*} W_2^* \xrightarrow{\sigma_2^*} P_1^{**} \xrightarrow{\partial_1^{**}} P_0^{**} \tag{11.18}$$

of left R-modules for coker ∂_1^{**}. Notice that W_k^* is a finitely generated projective left R-module for each $k \geq 2$. Since the maps $P_1^{**} \cong P_1$ and $P_0^{**} \cong P_0$ are natural isomorphisms, we have

$$\operatorname{coker} \partial_1^{**} \cong \operatorname{coker} \partial_1 = M.$$

Then (11.18) is a *coherent* projective resolution for M and consequently, as claimed, R is a left Noetherian ring by Theorem 9.38.

To prove that R is left self-injective, let M be a finitely generated left R-module and let $\mathcal{P}(M)$ be a projective resolution of M. Since R is left Noetherian, we can assume without loss of generality that in $\mathcal{P}(M)$ each term P_k is finitely generated projective. By Lemma 9.2, $\mathcal{P}(M)^* \in G\text{-}\mathbf{Cohcoplx}$. Let $J_1 = \text{image } \partial_1^* \subset P_1^*$. Then by Lemma 11.57,

$$\operatorname{Ext}_E^1(M, R) \cong \mathbf{R}_G(J_1, P_1^*)/J_1,$$

and since P_1^* is a finitely generated right R-module, our hypotheses imply that P_1^*/J_1 is strongly finitely R-cogenerated. Thus,

$$\mathbf{R}_G(J_1, P_1^*)/J_1 = \cap\{\ker f \,\big|\, f \in \operatorname{Hom}_R(P_1^*/J_1, R)\} = 0.$$

Since M was arbitrarily taken, $\operatorname{Ext}_R^1(\cdot, R) = 0$; hence R is left self-injective, and therefore R is a QF-ring. This completes the proof.

Let E be a ring. The left E-module G is called *FP-injective* if $\operatorname{Ext}_E^1(N, G) = 0$ for each finitely presented left E-module N. The right R-module H is *finitely G-presented* if $H = G^n/K$ for some integer n and some finitely G-generated R-submodule $K \subset G^n$. Identify G with the G-coplex $G \longrightarrow 0$ in $G\text{-}\mathbf{Coplx}$. The proof of Theorem 11.62 can be used as a template to characterize those G that are FP-injective left $\operatorname{End}_R(G)$-modules.

Theorem 11.63. *The following are equivalent for the right R-module G.*

1. G *is a left FP-injective* $\operatorname{End}_R(G)$*-module.*
2. *Each finitely G-presented right R-module is strongly finitely G-cogenerated.*

The proof is left as an exercise. The ring R is *left self-FP-injective* if R is an FP-injective left R-module.

Corollary 11.64. *The following are equivalent for a ring R.*

1. R *is a left self-FP-injective ring.*
2. *Each finitely presented right R-module is strongly finitely R-cogenerated.*

11.6.3 Left homological dimensions

By dualizing Theorem 11.27 and its corollaries we can prove a series of applications to homological dimensions of left $\operatorname{End}_R(G)$-modules and to the left global dimension of $\operatorname{End}_R(G)$. Specifically, when G is self-slender we will characterize the left global dimension of $\operatorname{End}_R(G)$ in terms of the projective dimension of $\mathcal{W} \in G\text{-}\mathbf{Coplx}$. As usual we leave it to the reader to dualize the appropriate proofs.

Let $\text{pd}_E(M)$ denote *the projective dimension* of the left E-module M. We need a notation to denote a G-coplex with at most k nonzero entries. Let

$$\mathcal{W}(k) = W_0'' \xrightarrow{\sigma_1''} W_1'' \xrightarrow{\sigma_2''} \cdots \xrightarrow{\sigma_k''} W_k'' \longrightarrow 0.$$

That is, $W_{k+j}'' = 0$ for each $j \geq 1$. For $\mathcal{W} \in G\text{-}\mathbf{Coplx}$ let

$$\text{pd}_G(\mathcal{W}) = \min\{\text{integers } k \geq 0 \,|\, \mathcal{W} \text{ has the homotopy type of a}$$
$$G\text{-coplex of the form } \mathcal{W}(k)\}.$$

A challenging exercise is to prove the next result by making the appropriate changes in Theorem 11.14. Recall that $J_k = \text{image } \sigma_k \subset W_k$ in a G-coplex $\mathcal{W}$.

Theorem 11.65. *Assume* (μ). *Let G be a self-slender right R-module, let $k > 0$ be an integer, and let $\mathcal{W} \in G\text{-}\mathbf{Coplx}$. Then following are equivalent.*

1. $\text{pd}_G(\mathcal{W}) \leq k$.
2. J_{k+j} *is a direct summand of W_{k+j} for each integer $j \geq 0$.*
3. J_k *is a direct summand of W_k.*

The *left global dimension* of the ring E is

$$\text{lgd}(E) = \sup\{\text{pd}_E(M) \,|\, M \text{ is a left } \text{End}_R(G)\text{-module}\}.$$

The *coglobal dimension of G* is

$$\text{cogd}(G) = \sup\{\text{pd}_G(\mathcal{W}) \,|\, \mathcal{W} \in G\text{-}\mathbf{Coplx}\}.$$

By dualizing Theorem 11.27 we can prove the following theorem.

Theorem 11.66. *Assume* (μ). *Let G be a self-slender right R-module. The following integers are equal.*

1. $\text{cogd}(G)$
2. $\text{lgd}(\text{End}_R(G))$
3. $\sup\{\text{pd}_G(\mathcal{W}) \,|\, \mathcal{W} \in G\text{-}\mathbf{Coplx} \text{ and } P_0 \cong G\}$.

Moreover, if we restrict our attention to coherent left $\text{End}_R(G)$-modules then we can generalize Theorem 11.66. *The coherent coglobal dimension of G is*

$$\text{cohcgd}(G) = \sup\{\text{pd}_G(\mathcal{W}) \,|\, \mathcal{W} \in G\text{-}\mathbf{Cohcoplx}\}.$$

If we assume that $\text{End}_R(G)$ is left Noetherian we can delete the need for the self-slender hypothesis in characterizing $\text{lgd}(\text{End}_R(G))$.

Theorem 11.67. *Let G be a right R-module and assume that $\text{End}_R(G)$ is left Noetherian. The following integers are equal.*

1. $\text{cohcgd}(G)$
2. $\text{lgd}(\text{End}_R(G))$
3. $\sup\{\text{pd}_G(\mathcal{W}) \,\big|\, \mathcal{W} \in G\text{-\textbf{Cohcoplx}} \text{ and } P_0 \cong G\}$.

Prove this theorem in the same manner that we proved Theorem 11.27.

Define $\text{cohgd}(G)$ as the dual of $\text{cohcgd}(G)$.

$$\text{cohgd}(G) = \sup\{\text{pd}_G(\mathcal{P}) \,\big|\, \mathcal{P} \in G\text{-\textbf{CohPlx}}\}.$$

A theorem of Auslander's [94, Theorem 9.23] states that $\text{lgd}(E) = \text{rgd}(E)$ when E is a (left and right) Noetherian ring. Then applications of Theorems 11.27 and 11.67 will prove the following theorem.

Theorem 11.68. *Let G be a right R-module and assume that E is Noetherian. The following integers are equal.*

1. $\text{cohgd}(G)$
2. $\text{rgd}(E)$
3. $\text{lgd}(E)$
4. $\text{cohcgd}(G)$.

Using our classification of $\text{lgd}(E)$ and by dualizing results from Section 11.4 we could characterize those G such that E is left hereditary, or left Noetherian left hereditary. We leave the statements of these results and their proofs as exercises. However, we list a pair of two-sided results that follow from Theorems 11.42 and 11.68.

Theorem 11.69. *Assume (μ). Let G be a self-small and self-slender right R-module and suppose that $\text{End}_R(G)$ is Noetherian. The following are equivalent.*

1. $\text{End}_R(G)$ *is left hereditary.*
2. $\text{End}_R(G)$ *is right hereditary.*
3. *Given a finitely G-generated R-submodule $L \subset G^n$, then $\mathbf{R}_G(L, G^n)$ is a direct summand of G^n.*
4. *For each map $\delta : G^{(n)} \longrightarrow G$, $\mathbf{S}_G(\ker \delta)$ is a direct summand of $G^{(n)}$.*
5. *For each pair of integers $n, m > 0$ and each map $\delta : G^{(n)} \longrightarrow G^{(m)}$, $\mathbf{S}_G(\ker \delta)$ is a direct summand of $G^{(n)}$ and $\mathbf{R}_G(G^{(n)}, \text{image } \delta)$ is a direct summand of $G^{(m)}$.*

Proof: Proof: $5 \Rightarrow 3$ and 4 is clear.

$4 \Rightarrow 2$. Let $\mathcal{Q} \in G\text{-\textbf{CohPlx}}$ be such that $Q_0 \cong G$. Then $\delta_1 : Q_1 \longrightarrow G$, where $Q_1 \in \mathbf{P}_o(G)$. By part 4, $\mathbf{S}_G(\ker \delta_1)$ is a direct summand of Q_1, so by Corollary 11.32, and since $\text{End}_R(G)$ is Noetherian, $\text{rgd}(\text{End}_R(G)) \leq 1$. This proves part 2.

$3 \Rightarrow 1$. Let $\mathcal{W} \in G\text{-\textbf{Cohcoplx}}$ be such that $W_0 \cong G$. Then $\sigma_1 : G \longrightarrow W_1$ where $W_1 \in \text{\textbf{coP}}_o(G)$. By part 3, $\mathbf{R}_G(W_1, J_1)$ is a direct summand of W_1, so by the dual to Corollary 11.32, and since $\text{End}_R(G)$ is Noetherian, $\text{rgd}(\text{End}_R(G)) \leq 1$. This proves part 1.

$1 \Leftrightarrow 2$ follows from Auslander's theorem [94, Theorem 9.23].

1 and $2 \Rightarrow 5$. Let $\delta : Q_1 \longrightarrow Q_0$ be a right R-module map. Then $\delta = \delta_1$ for some $Q \in G\text{-}\mathbf{Plex}$. By part 1, $\mathrm{rgd}(\mathrm{End}_R(G)) \leq 1$, so by Corollary 11.32, $\mathbf{S}_G(\ker \delta)$ is a direct summand of Q_1.

There is a G-coplex $\mathcal{W}$ such that $\sigma_1 = \delta$. Part 2 implies that $\mathrm{lgd}(\mathrm{End}_R(G)) \leq 1$, so the dual to Corollary 11.32 shows us that $\mathbf{R}_G(W_1, J_1)$ is a direct summand of W_1. This completes the proof.

By using Corollary 11.47 the reader can prove the following theorem.

Theorem 11.70. *The following are equivalent for a right R-module G.*

1. *$\mathrm{End}_R(G)$ is (left and right) semi-hereditary.*
2. *For each pair of integers $n, m > 0$ and each map $\delta : G^{(n)} \longrightarrow G^{(m)}$, $\mathbf{R}_G(G^{(m)},$ image $\delta)$ is a direct summand of $G^{(m)}$, and $\mathbf{S}_G(\ker \delta)$ is a direct summand of $G^{(n)}$.*

11.7 A glossary of terms

All terms are defined for *right R-modules* unless otherwise stated.

TERM	DEFINITION
B	
balanced G-split	given a G-balanced surjection $\delta : Q \to G$ with $Q \in \mathbf{P}(G)$ then $Q = \ker \delta \oplus Q'$ for some $Q' \in \mathbf{P}(G)$
C	
coexact dimension of G	the supremum of the exact dimensions over all G-coplexes
coglobal dimension of G	the least k such that each G-coplex is homotopic to some $\mathcal{W}(k)$
coherent G-coplex	a G-coplex whose terms are direct summands of $G^{(n)}$ for integers n
coherent coglobal dimension of G	the least k such that each G-coplex is homotopic to some coherent $\mathcal{W}(k)$
Corner module (ring)	A module (ring) whose additive structure is *rtffr*
E	
E-flat R-module G	G is a flat left $\mathrm{End}_R(G)$-module
(Q is) exact at k	image $\delta_{k+1} = \ker \delta_k$
exact dimension of Q	the least k for which Q is exact at $k + j$ for all $j > 0$
exact dimension of G	the sup of the exact dimensions over all G-plexes

TERM	DEFINITION
F	
faithful	$IG \neq G$ for each proper right ideal $I \subset \mathrm{End}_R(G)$
(M has) flat dimension at most k	M possesses a flat resolution whose terms $k+j$ for $j > 0$ are 0
(G is) faithfully E-flat	G is faithful and E-flat
(H is) finitely G-presented	$H = Q_o/K$ for some $Q_o \in \mathbf{P}_o(G)$ and some finitely G-generated $K \subset Q_o$
finitely G-split	given $Q \in \mathbf{P}(G)$ and any surjection $\delta : Q \to H$ then $Q = \ker \delta \oplus Q_o$ for some $Q_o \in \mathbf{P}_o(G)$
flat resolution	a long exact sequence whose terms are flat modules
fp faithful	$IG \neq G$ for each finitely generated proper right ideal $I \subset \mathrm{End}_R(G)$
(H is) fp G-split	each surjection $\delta : Q \to H$ with $Q \in \mathbf{P}_o(G)$ satisfies $Q = \ker \delta \oplus Q'$ for some $Q' \in \mathbf{P}_o(G)$
G	
($f : A \to B$ is) G-balanced onto its image	G is projective relative to $f : A \to \mathrm{image}\, f$
G-generated module H	H is a homomorphic image of $G^{(c)}$
(H is) G-split	each surjection $\delta : Q \to H$ with $Q \in \mathbf{P}(G)$ satisfies $Q = \ker \delta \oplus Q'$ for some $Q' \in \mathbf{P}(G)$
(H is) G-subgenerated	$H \subset K$ for some G-generated K
G-projective dimension of Q is at least k	Q is homotopic to some $Q(k)$
G-reject of U in W	the intersection of all kernels $\ker(W \to G)$ containing U
G-solvable	$\Theta_G : \mathbf{T}_G \mathbf{H}_G(G) \to G$ is an isomorphism
G-torsionless	a submodule of $\prod_c G$ for some cardinal c
global dimension of G	the supremum over the G-projective dimensions of all G-plexes
(right) global dimension of R	the supremum over the projective dimensions of all right R-modules
H	
(right) hereditary ring	each right ideal is projective
I	
(M has) injective dimension at most k	M possesses an injective resolution whose terms $k+j$ for $j > 0$ are 0
injective resolution	a long exact sequence whose terms are injective modules

TERM	DEFINITION
N	
(right) Noetherian ring	each right ideal is finitely generated
P	
(*M* has) projective dimension at most k	M possesses a projective resolution whose terms $k + j$ for $j > 0$ are 0
projective resolution	a long exact sequence whose terms are projective modules
Q	
$\mathcal{Q}(k)$	a G-plex of the form $0 \to Q_k \to Q_{k-1} \to \cdots \to Q_0$
quasi-progenerator	finitely generated self-generating quasi-projective module
S	
Σ-quasi-projective	G is projective relative to each surjection $\delta : G^{(c)} \to G$ for each cardinal c
$\mathbf{S}_G(H)$	$\sum \{ f(G) \big| f : G \to H \}$
self-slender	$\text{Hom}_R(G, G)^{(c)} \cong \text{Hom}_R(G^c, G) \forall$ cardinals c
(*G* is) self-small	$\text{Hom}_R(G, G)^{(c)} \cong \text{Hom}_R(G, G^{(c)}) \forall$ cardinals c
(right) semi-hereditary	each finitely generated right ideal is projective
(*M* is) strongly finitely R-cogenerated	there is an n and an embedding $M \to R^{(n)}$ relative to which R is injective
W	
weakly balanced G-split	given a G-balanced surjection $\delta : Q \to G$ with $Q \in \mathbf{P}(G)$ then $Q = \mathbf{S}_G(\ker \delta) \oplus Q'$ for some $Q' \in \mathbf{P}(G)$
weakly finite G-split	given $Q \in \mathbf{P}(G)$ and any surjection $\delta : Q \to H$ then $Q = \mathbf{S}_G(\ker \delta) \oplus Q_o$ for some $Q_o \in \mathbf{P}_o(G)$
weakly G-split	given $Q \in \mathbf{P}(G)$ and any surjection $\delta : Q \to H$ then $Q = \mathbf{S}_G(\ker \delta) \oplus Q'$ for some $Q' \in \mathbf{P}(G)$
weak global dimension of R	the supremum over the flat dimensions of all right R-modules
$\mathcal{W}(k)$	a G-coplex of the form $W_0 \to W_1 \to \cdots \to W_k \to 0$

11.8 Exercises

Let R be a ring, let G be a self-small right R-module, let $\mathcal{Q} \in G$-**Plex**.

1. Prove any of the results listed as exercises for the reader.
2. Prove the following implication for G and an R-module H.

$$H \text{ is finitely } G\text{-split} \Rightarrow H \text{ is fp } G\text{-split}$$

3. Prove the following implication for G and an R-module H.

$$H \text{ is finitely } G\text{-split} \Rightarrow H \text{ is weakly finite } G\text{-split}$$

4. Prove the following implication for G and an R-module H.

$$H \text{ is finitely } G\text{-split} \Rightarrow H \text{ is } G\text{-split} \Rightarrow H \text{ is weakly } G\text{-split}$$

5. Prove the following implication for G and an R-module H.

$$H \text{ is } G\text{-split} \Rightarrow H \text{ is balanced } G\text{-split} \Rightarrow H \text{ is weakly balanced } G\text{-split}$$

6. The *coherent flat dimension of G* is

$$\text{cofd}_E(G) = \min\{\text{integers } k > 0 \mid \text{Tor}_E^{k+j}(M, G) = 0 \text{ for}$$
$$\text{each integer } j \geq 1 \text{ and each coherent right}$$
$$\text{End}_R(G)\text{-module } M\}.$$

Prove that $\text{cofd}_E(G) \leq k$ iff I_{k+j} is G-solvable for each integer $j \geq 1$ and each $\mathcal{Q} \in G\text{-}\mathbf{CohPlx}$.

7. Suppose that G is a self-small E-flat right R-module. Find conditions on $\text{End}_R(G)$ that are equivalent to the following conditions on G.

   (a) G has the acc on G-generated R-submodules of G.
   (b) G has the dcc on G-generated R-submodules of G.
   (c) Each G-generated R-submodule of G is in $\mathbf{P}(G)$.
   (d) Each G-generated R-submodule of G is in $\mathbf{P}_o(G)$.
   (e) Each finitely G-generated R-submodule of G is in $\mathbf{P}_o(G)$.

8. Show that G is *fp faithful* iff $\mathbf{T}_G(M) \neq 0$ for each finitely presented right $\text{End}_R(G)$-module $M \neq 0$.

9. Let $\mathcal{Q} \in G\text{-}\mathbf{Plex}$. Show that $\text{ed}_G(\mathcal{Q}) \leq \text{pd}_G(\mathcal{Q})$.

10. Show that $\text{pd}_G(\mathcal{Q}) = 0$ iff I_1 is a direct summand of Q_0.

11. Let G be a self-small right R-module and let $E = \text{End}_R(G)$. Show that $\text{rgd}(E) = \text{gd}(G) = \sup\{\text{pd}_G(\mathcal{Q}) \mid \mathcal{Q} \in G\text{-}\mathbf{Plex} \text{ and } P_0 \cong G\}$.

12. Let G be a self-small module, and let $k > 0$ be an integer. Then $\text{rgd}(\text{End}_R(G)) \leq k$ iff given $\mathcal{Q} \in G\text{-}\mathbf{Plex}$ then $\mathbf{S}_G(\ker \delta_k)$ is a direct summand of Q_k.

13. Let G be a self-small module, and let $k > 0$ be an integer. Then $\text{rgd}(\text{End}_R(G)) \leq k$ iff given $\mathcal{Q} \in G\text{-}\mathbf{Plex}$ then image δ_k is weakly balanced G-split.

14. Let G be a self-small module, and let $k > 0$. Prove or disprove that $\text{fd}(G) < k$ and $\text{rgd}(E) \leq k$ iff image $\delta_k \in \mathbf{P}(G)$ for each $\mathcal{Q} \in G\text{-}\mathbf{Plex}$.

15. Let G be a self-small right R-module. Show that $\text{gd}(G) \leq k$ iff each $\mathcal{Q} \in G\text{-}\mathbf{Plex}$ is homotopic to a $\mathcal{Q}(k) \in G\text{-}\mathbf{Plex}$ of the form $0 \longrightarrow Q'_k \xrightarrow{\delta_k} \cdots \xrightarrow{\delta_1} Q'_0$.

16. Let G be a right R-module and let $S = \text{End}_{\text{End}_R(G)}(G)$. Then $\text{End}_R(G) = \text{End}_S(G)$.

17. Make a chart of the implications between the properties *G-split, balanced G-split, finitely G-split, weakly G-split, weakly finite G-split, admits a G-cover,*

admits a finite G-cover, and any other splitting property that might have been left out.

18. Let G be a self-small right R-module. The following are equivalent.

    (a) G is E-flat and $\text{End}_R(G)$ is right hereditary
    (b) Each G-generated submodule of H is G-split.

19. Let G be a self-small right R-module. The following are equivalent.

    (a) G is E-flat and $\text{End}_R(G)$ is right Noetherian right hereditary
    (b) Each G-generated R-submodule of G is finitely G-split.
    (c) Each G-generated R-submodule of G is in $\mathbf{P}(G)$.

20. Let G be a self-small right R-module. The following are equivalent.

    (a) Each cyclic right ideal in $\text{End}_R(G)$ is projective.
    (b) $\mathbf{S}_G(\ker \delta)$ is a direct summand of G for each map $\delta \in \text{End}(G)$.

21. Let G be a self-small right R-module and let $Q \in G\text{-}\mathbf{Plex}$. Then $\text{ed}_G(Q) \leq k$ iff I_k is G-solvable.

22. Let G be a self-small right R-module. The following are equivalent.

    (a) $\text{fd}(G) < 2$ and $\text{rgd}(E) \leq 2$
    (b) Given a map $\delta : Q \longrightarrow G$ in which $Q \in \mathbf{P}(G)$ then $\mathbf{S}_G(\ker \delta) \in \mathbf{P}(G)$.

23. Given a G-generated module $H \subset G$ prove the following implications. H is G-split and a direct summand of $G \Rightarrow H$ is finitely G-split $\Rightarrow H$ admits a finite G-cover $\Rightarrow H$ is weakly balanced G-split.

24. Show that Ψ_P and Ψ_W are isomorphisms if P is a finitely generated projective left $\text{End}_R(G)$-module and $W \oplus W' = G^n$ for some integer n.

25. Let R be a self-small self-slender right R-module. Let n, m be integers and let

$$0 \longrightarrow K \longrightarrow G^n \overset{\delta}{\longrightarrow} G^m \longrightarrow H \longrightarrow 0$$

be an exact sequence of right R-modules. Let $M = \text{coker } h_G(\delta)$ and let $N = \text{coker } h^G(\delta)$. Show the following.

    (a) $\text{Tor}^1_E(N, G) = K/\mathbf{S}_G(K)$.
    (b) $\text{Ext}^1_E(N, G) = \mathbf{R}_G(H, G^m)/H$.
    (c) Let G be a self-small right R-module. The following are equivalent.

        (i) $\text{fd}_E(G) \leq 1$
        (ii) $\text{ed}(G) \leq 1$
        (iii) Given any map $\delta : Q \longrightarrow G$ in which $Q \in \mathbf{P}(G)$ then $\mathbf{S}_G(\ker \delta)$ is G-solvable.

26. Let G be a self-small right R-module and let H be G-generated. Say $H \subset Q$ for some $Q \in \mathbf{P}(G)$. Prove or disprove that $\Theta_H : \mathbf{T}_G\mathbf{H}_G(H) \longrightarrow H$ is a G-balanced G-monomorphism.

27. Let C and G be right R-modules. Then C is G-injective iff for each integer $n > 0$, C is injective relative to each G-monomorphism $j : H \longrightarrow G^n$ in $G\text{-}\mathbf{Pres}$.

28. Let G and C be right R-modules and let G be self-small. Show that the following are equivalent.

    (a) C is injective relative to each G-monomorphism $\jmath : H \longrightarrow G$ in G-**Pres**.
    (b) For each integer $n > 0$, C is injective relative to each G-monomorphism $\jmath : H \longrightarrow G^n$ in G-**Pres**.

29. Suppose that $(\cdots, f_1, f_0) : \mathcal{Q} \longrightarrow \mathcal{Q}'$ is a chain map. If f_1 and f_0 are isomorphisms then $\mathcal{Q}$ and $\mathcal{Q}'$ have the same homotopy type.

30. Complexes of the same homotopy type have isomorphic homology groups.

31. Let G be a self-small E-flat right R-module. There is a fully faithful functor $\text{res}_G : G\text{-}\mathbf{Sol} \longrightarrow G\text{-}\mathbf{Plex}$ given by $\text{res}_G(H) = \mathcal{Q}$ where $\mathcal{Q}$ is any G-plex of H.

32. Let G be a self-small right R-module and let $\mathcal{Q} \in G$-**Plex**. Then $K_k/\mathbf{S}_G(K_k) \cong \ker \Theta_{I_k}$ for each integer $k > 0$.

33. Let G be a self-small right R-module and let $k > 0$ be an integer. The following are equivalent.

    (a) $\text{fd}(G) \leq k$.
    (b) I_{k+1} is G-solvable for each $\mathcal{Q} \in G$-**Plex**.
    (c) $\mathbf{S}_G(K_k)$ is G-solvable for each $\mathcal{Q} \in G$-**Plex**.

34. Let G be a self-small right R-module and let $\mathcal{Q} \in G$-**Plex**. Then $K_k/\mathbf{S}_G(K_k) \cong \ker \Theta_{I_k}$ for each integer $k \geq 2$.

35. Construct a group G and a $\mathcal{Q} \in G$-**Plex** such that $\text{ed}_G(\mathcal{Q}) < \text{pd}_G(\mathcal{Q}) < \infty$.

36. Say G is self-small. Then $\text{rgd}(\text{End}_R(G)) = \text{gd}(G) = \sup\{\text{pd}_G(\mathcal{Q}) \,\big|\, \mathcal{Q} \in G\text{-}\mathbf{Plex}$ and $P_0 \cong G\}$.

37. Let G be a self-small module, and let $k > 0$ be an integer. Then $\text{rgd}(\text{End}_R(G)) \leq k$ iff given $\mathcal{Q} \in G$-**Plex** then $\mathbf{S}_G(K_k)$ is a direct summand of Q_k.

38. Let G be a self-small module, and let $k > 0$ be an integer. Then $\text{rgd}(\text{End}_R(G)) \leq k$ iff given $\mathcal{Q} \in G$-**Plex** then I_k is weakly balanced G-split.

39. Let G be a self-small module, and let $k > 0$. Then $\text{fd}(G) < \text{rgd}(\text{End}_R(G)) \leq k$ iff $I_k \in \mathbf{P}(G)$ for each $\mathcal{Q} \in G$-**Plex**.

40. Let $\Omega(E) = \{$ groups $G \,\big|\, E \cong \text{End}(G)\}$. Show the following.

    (a) $\text{lgd}(E) = \sup\{\text{pd}_E(G) \,\big|\, G \in \Omega(E)\}$.
    (b) E is left hereditary iff each $G \in \Omega(E)$ is an E-left group.

41. Let G be a self-small right R-module. The following are equivalent.

    (a) G is E-flat and $\text{End}_R(G)$ is right hereditary
    (b) Each G-generated R-submodule of G is G-split.
    (c) Each G-generated R-submodule of G is in $\mathbf{P}(G)$.

42. Let G be a self-small right R-module, and let H be G-generated.

    (a) Show that H is weakly balanced G-split iff given a G-generated right R-module X and a surjection $\delta : X \longrightarrow H$ then $\mathbf{S}_G(\ker \delta)$ is a direct summand of X.
    (b) Let G be a self-small right R-module, and let H be a G-generated. Show that H is G-split iff given a G-generated right R-module X and a surjection $\delta : X \longrightarrow H$ then δ is split.

43. Let G be a self-small right R-module. The following are equivalent.

    (a) Each cyclic right ideal in $\mathrm{End}_R(G)$ is projective.
    (b) Each endomorphic image of G is in $\mathbf{P}(G)$.

44. Characterize those self-slender G for which $\mathrm{End}_R(G)$ is left hereditary.
45. Characterize those self-slender G for which $\mathrm{End}_R(G)$ is left Noetherian left hereditary.
46. Characterize those self-slender G for which $\mathrm{End}_R(G)$ is left semi-hereditary.

11.9 Problems for future research

Let R be an associative ring with identity, let G be a right R-module, let $\mathrm{End}_R(G)$ be the endomorphism ring of G, let $\mathcal{Q}$ be a G-plex, let $\mathcal{P}$ be a projective resolution of right $\mathrm{End}_R(G)$-modules.

1. Find more examples like those in Theorem 11.8.
2. Give a complete survey of the modules $K/\mathbf{S}_G(K)$ for right R-submodules K of modules $Q \in \mathbf{P}(G)$. See Lemma 9.6.
3. Discuss the dimension theory of G-plexes and G-coplexes.
4. Give a module theoretic description of the right R-modules H that admit a G-cover, or that possess a G-cover.
5. Study the set $\mathcal{D}(G)$ of right ideals I in $\mathrm{End}_R(G)$ such that $IG = G$. See [52].
6. Further characterize the (endlich) Baer splitting property.
7. Investigate the injective nature of G in G-**coPlex**.
8. Give an internal characterization of R-modules G whose homological dimension is at most n.
9. Give an internal characterization of R-modules G such that the global dimension of $\mathrm{End}_R(G)$ is at most n.
10. Extend Theorem 11.69 by deleting the Noetherian hypotheses.

12

Projective modules

An endomorphism module is a module over an endomorphism ring. In this chapter we continue to show that the machinery developed in Section 9.1.2 is an effective tool for studying right R-modules G as left modules over their endomorphism ring $\text{End}_R(G)$. While there are results in this chapter on right modules G over a general associative ring R with unit, some of our results assume that $R = \mathbf{Z}$ and that G is an *rtffr* group.

We investigate the existence and characterization of module-theoretic properties of an R-module G as a left $\text{End}_R(G)$-module. The properties we investigate are projective covers of G, projective modules G, finitely generated modules, finitely presented modules, coherent modules, and generator as a left $\text{End}_R(G)$-module. The most detailed characterizations occur when G is a *rtffr* group. In this case, the properties are characterized in terms of a direct sum decomposition.

12.1 Projectives

Throughout this chapter R denotes an associative ring with identity (which could possibly be $\mathbf{Z}$), G denotes a right R-module, $\text{End}_R(G)$ is the ring of R-module endomorphisms of G written on the left of G, and

$$S = \text{BiEnd}_R(G) = \text{End}_{\text{End}_R(G)}(G).$$

If $R = \mathbf{Z}$ then $\text{End}_R(G) = \text{End}(G)$ and $\text{BiEnd}_{\mathbf{Z}}(G) = S = $ the center of $\text{End}(G)$. We let $\mathbf{P}_o(G)$ denote the category of direct summands of direct sums G^n for integers $n > 0$. We let $\mathbf{coP}(G)$ denote the category of direct summands of direct products G^c for cardinals c. See Section 9.1.2 for the notation used here.

For $\text{End}_R(G)$, S, and $\text{End}_S(G)$ we have the following identities.

$$R \subset S. \tag{12.1}$$

$$\text{End}_S(G) = \text{End}_R(G). \tag{12.2}$$

The equation (12.2) is seen as follows. By (12.1), $\text{End}_S(G) \subset \text{End}_R(G)$. Note that G is an $\text{End}_R(G)$-S-bimodule. Thus $\text{End}_R(G) \subset \text{End}_S(G)$ and hence

$\text{End}_S(G) = \text{End}_R(G)$. Thus in $\mathcal{W}$ each map σ_k is an S-module homomorphism. Specifically, the natural inclusion $0 \to \ker \sigma_1 \to W_0$ is a sequence of S-*modules*.

If $N \subset M$ are left R-modules then N is *superfluous* in M if $N + L = M$ implies that $L = M$ for each right R-submodule $L \subset M$. If $N \subset M$ are left R-modules then N is *essential* in M if $N \cap L = 0$ implies that $L = 0$ for each right R-submodule $L \subset M$. The module U is uniform if $N \cap L \neq 0$ for each pair of nonzero submodules $L, N \subset U$. The module M has Goldie dimension n if there is an essential direct sum $U_1 \oplus \cdots \oplus U_n$ of uniform R-submodules U_i of M.

Lemma 12.1. *Let R be a ring with unit that possesses a semi-simple classical right ring of quotients Q. Let F be a free right R-module, and let $N \subset F$ be right R-modules. If N is superfluous in F then N has finite Goldie dimension.*

Proof: Suppose that N has infinite Goldie dimension. We will show that N is not superfluous in F.

We begin our argument by assuming that R is a prime ring and that Q is a simple Artinian ring.

Since the prime *rtffr* ring R, there is no loss of generality in assuming that F has countable basis $B = \{b_1, b_2, b_3, \dots\}$. Consider F as an R-submodule of $FQ = F \otimes_R Q$.

Since N has infinite rank, $N \not\subset b_1 Q$. Since Q is simple, we can choose a copy of Q in NQ, say $y_1 Q \cong Q$, such that

$$y_1 Q \cap b_1 Q = 0.$$

We may assume that $y_1 \in N$.

Using induction, there is a sequence $(y_1, y_2, y_3 \dots) \subset N$ such that

$$y_{n+1} Q \cap [b_1 Q + \cdots + b_n Q + y_1 Q + \cdots + y_n Q] = 0$$

for each integer $n > 0$.

Let

$$L = (b_1 + y_1)R + (b_2 + y_2)R + (b_3 + y_3)R + \cdots \subset F.$$

Then $L + N = F$ since the sum contains B.

We claim that $b_1 \notin L \neq F$. If so, then there are $r_1, \dots, r_n \in R$ such that

$$b_1 = (b_1 + y_1)r_1 + \cdots + (b_n + y_n)r_n$$

for some integer $n \neq 0$. Assume that n is least with respect to this sum. Then

$$b_1(1 - r_1) - b_2 r_2 - \cdots - b_n r_n - y_1 r_1 + \cdots + y_{n-1} r_{n-1} = y_n r_n$$

is in

$$y_n Q \cap [b_1 Q + \cdots + b_n Q + y_1 Q + \cdots + y_{n-1} Q] = 0.$$

Hence $b_1 = 0$, a contradiction to our choice of b_1 as a basis element.

Therefore $L \neq F$, and so N is not superfluous in F. That is, if N is superfluous in F then N has finite Goldie dimension as a right R-module.

The next case is to suppose that R is semi-prime, and to assume that N is superfluous right R-submodule in the free right R-module F. Write

$$Q = Q_1 \times \cdots \times Q_t$$

for some simple Artinian rings Q_i, and write

$$Q_i = Qe_i$$

for some central idempotent e_i and for each integer $i = 1, \ldots, t$. Then

$$N \subset Ne_1 \oplus \cdots \oplus Ne_t$$

is a direct sum of R-submodules of F, so that $Ne_1 \oplus \cdots \oplus Ne_t$ is superfluous in F. Furthermore, Ne_i is a superfluous Re_i-submodule of Fe_i. Since $Re_i \subset Qe_i$ is a prime ring, the first paragraph shows us that Ne_i has finite Goldie dimension as a right Re_i-module. Hence $N \subset Ne_1 \oplus \cdots \oplus Ne_t$ has finite Goldie rank as a right R-module.

We conclude that if R possesses a semi-simple classical right ring of quotients Q and if N is superfluous in F, then N has finite Goldie dimension. This completes the proof.

The right R-module G is E- *projective* iff G is a projective left $\text{End}_R(G)$-module. We say that G is E-*finitely generated* if G is finitely generated as a left $\text{End}_R(G)$-module.

Theorem 12.2. *Assume (μ), and let G be a self-slender right R-module. The following are equivalent.*

1. G *is a projective left* $\text{End}_R(G)$*-module.*
2. *There is a cardinal c such that S is a direct summand of G^c as right S-modules.*

In this case, G is generated by at most c elements as a left $\text{End}_R(G)$-module.

Proof: $1 \Rightarrow 2$. Suppose that G is E- projective. Then $\mathcal{P} = 0 \rightarrow G$ is a projective resolution of the left $\text{End}_R(G)$-module G. Form $\mathcal{P}^* = \text{Hom}_{\text{End}_R(G)}(\mathcal{P}, G)$ which as in the proof of Theorem 9.32 is a G-coplex

$$\mathcal{P}^* = \text{Hom}_{\text{End}_R(G)}(G, G) \rightarrow 0.$$

Since

$$\text{Hom}_{\text{End}_R(G)}(G, G) = S,$$

then $S \rightarrow 0$ is a G-coplex. Hence S is a direct summand of G^c for some cardinal c. This proves part 2.

$2 \Rightarrow 1$. Let S be a direct summand of $W = G^c$ for some cardinal c. Duals in this proof are

$$\text{Hom}_S(\cdot, G) = \text{Hom}_R(\cdot, G)$$

by (12.2). By Theorem 9.30, W^* is a free left $\text{End}_S(G)$-module on c generators and S^* is a direct summand of W^*. Since

$$S^* = \text{Hom}_S(S, G) \cong G$$

is a direct summand of W^* as left $\text{End}_S(G)$-modules G is a projective left $\text{End}_S(G)$-module generated by at most c elements. By (12.2), $\text{End}_S(G) = \text{End}_R(G)$, so G is a projective left $\text{End}_R(G)$-module. This completes the proof.

The situation for *rtffr* groups is more developed. The right R-module U is *uniform* if each nonzero submodule of U is essential in U. The right R-module M has *finite Goldie dimension* if M contains an essential R-submodule $U_1 \oplus \cdots \oplus U_t$ for some integer $t > 0$ and some (nonzero) uniform R-modules $U_1, \ldots, U_t$.

Lemma 12.3. *Let R be a ring that possesses a right Artinian classical right ring of quotients Q, and let P be a projective left R-module. If K is superfluous in P, and if P/K has finite Goldie dimension then P is a finitely generated left R-module.*

Proof: Assume for now that R is semi-prime, and let K be superfluous in P. Say $P \oplus P' = R^{(c)}$ for some cardinal c. By Lemma 12.1, K has finite Goldie dimension. Since P/K has finite Goldie dimension, P has finite Goldie dimension. Then

$$P \subset PQ \subset Q^{(n)} \subset Q^{(c)}$$

for some integer $n > 0$, so that

$$P \subset R^{(c)} \cap Q^{(n)} = R^{(n)}.$$

Hence

$$P = P \oplus (P' \cap R^{(n)}) = R^{(n)}$$

by the Modular Law, and therefore P is finitely generated.

Now for the general case. Since K maps onto a superfluous submodule $\widehat{K}$ of $P/\mathcal{N}(R)P$, and since $R/\mathcal{N}(R)$ is a semi-prime *rtffr* ring, the above paragraph shows us that $P/\mathcal{N}(R)P$ is a finitely generated right $R/\mathcal{N}(R)$-module. Finally, since Q is right Artinian,

$$\mathcal{N}(R) = R \cap \mathcal{J}(Q)$$

is nilpotent, and $\mathcal{N}(R)P$ is superfluous in P. Hence P is finitely generated, which completes the proof.

The right R-module G is *E-finitely generated* if G is a finitely generated left $\text{End}_R(G)$-module.

Corollary 12.4. *An rtffr E-projective group is E-finitely generated.*

Proof: Let G be E-projective. Since G has finite rank as a group, Lemma 12.3 states that G is E-finitely generated.

Theorem 12.5. *Let G be an rtffr group. The following are equivalent.*

1. G *is an rtffr E-projective group.*
2. G *is an rtffr E-projective group that is generated by $n \in \mathbf{Z}$ elements as a left* $\text{End}(G)$*-module.*
3. *There is an integer n such that S is a direct summand of G^n as right S-modules.*

Proof: $1 \Rightarrow 2$. Apply Corollary 12.4.

$\quad 2 \Rightarrow 3$ follows from Theorem 12.2.

$\quad 3 \Rightarrow 1$ is clear. This completes the logical cycle.

12.2 Finitely generated modules

Let A be a ring. We say that the injection $0 \to K \to W$ of left A-modules is *G-cobalanced* if an application of $\text{Hom}_A(\cdot, G)$ yields the surjection $W^* \to K^* \to 0$. We say that K is *G-cobalanced in W* if $0 \to K \to W$ is G-cobalanced. We say that a G-complex $\mathcal{W}$ is *doubly G-cobalanced* if the induced sequence $W_1^* \to W_0^* \to \ker \sigma_1^* \to 0$ is exact.

Lemma 12.6. *Let G be a right R-module. Let $\mathcal{P} = \cdots \to P_1 \to P_0 \to G \to 0$ be a projective resolution of the left $\text{End}_R(G)$-module G.*

1. *The induced map $0 \to G^* \to P_0^*$ of right S-modules is G-cobalanced.*
2. *Assume that the point evaluation maps Ψ_0 and Ψ_1 are isomorphisms. Then the induced G-coplex $\mathcal{P}^*$ is doubly G-cobalanced.*

Proof: We begin with a projective resolution

$$\mathcal{P} = P_1 \xrightarrow{\delta_1} P_0 \xrightarrow{\delta} G \to 0.$$

By Theorem 9.32, an application of $\text{Hom}_{\text{End}_R(G)}(\cdot, G)$ to $\mathcal{P}$ yields the G-coplex

$$0 \to G^* \xrightarrow{\delta^*} P_0^* \xrightarrow{\delta_1^*} P_1^*$$

for $G^* = S$. By (12.2), $\mathcal{P}^*$ is a complex of S-module homomorphisms. Apply $\text{Hom}_S(\cdot, G)$ to produce the complex

$$P_1^{**} \xrightarrow{\delta_1^{**}} P_0^{**} \xrightarrow{\delta^{**}} G^{**}$$

of left $\text{End}_S(G)$-modules. We have a commutative diagram

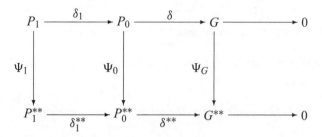

1. Ψ_G is an isomorphism by Theorem 9.30. Since δ is a surjection, a diagram chase shows us that δ^{**} is a surjection. Thus $0 \to G^* \to P_0^*$ is G-cobalanced.

2. By hypothesis, Ψ_0 and Ψ_1 are isomorphisms. Then the bottom row in the above diagram is exact, whence the complex $0 \to G^* \to P_0^* \to P_1^*$ is doubly G-cobalanced. This completes the proof.

Theorem 12.7. *Let G be a right R-module. The following are equivalent.*

1. *G is a finitely generated left $\text{End}_R(G)$-module, having at least n generators.*
2. *There is an integer n such that S is a G-cobalanced S-submodule of G^n.*

Proof: $1 \Rightarrow 2$. Suppose that G is E-finitely generated. There is a projective resolution of left $\text{End}_R(G)$-modules

$$\mathcal{P} = \cdots \to P_1 \to P_0 \to G \to 0$$

such that $P_0 = \text{End}_R(G)^n$ for some integer n. Apply $\text{Hom}_{\text{End}_R(G)}(\cdot, G)$ to form the G-coplex

$$\mathcal{P}^* = 0 \to G^* \to P_0^* \to P_1^* \to \cdots$$

where $G^* = S$ and where $P_0^* \cong G^n$. Then

$$0 \to S \to G^n \to P_1^*$$

is exact. By Lemma 12.6, the embedding $0 \to S \to G^n$ is G-cobalanced. This proves part 2.

$2 \Rightarrow 1$. Suppose that for some integer n, S embeds in G^n as a G-cobalanced S-submodule. Then an application of $\text{Hom}_S(\cdot, G)$ to $S \subset G^n$ yields the exact sequence

$$\text{Hom}_S(G^n, G) \to \text{Hom}_S(S, G) \to 0$$

which is just the sequence

$$\text{End}_R(G)^n \to G \to 0.$$

Then G is E-finitely generated, which completes the proof.

When $R = \mathbf{Z}$, then

$$\mathrm{BiEnd}_{\mathbf{Z}}(G) = S = \mathrm{center}(\mathrm{End}(G)).$$

A simple application of Theorem 12.7 proves the next result.

The *rtffr* group G is *E-finitely presented* if G is finitely presented as a left $\mathrm{End}_R(G)$-module.

Theorem 12.8. *Let G be an* rtffr *group. The following are equivalent.*

1. *G is E-finitely generated, having at least n generators.*
2. *There is an integer n such that S is a G-cobalanced S-submodule of G^n.*

Theorem 12.9. *Let G be a right R-module. The following are equivalent.*

1. *G is E-finitely presented.*
2. *There is a doubley G-cobalanced exact sequence $0 \to S \to W_0 \to W_1$ of right S-modules such that $W_0, W_1 \subset \mathbf{P}_o(G)$.*

Proof: $1 \Rightarrow 2$. Suppose that G is a finitely presented left $\mathrm{End}_R(G)$-module. There is a projective resolution of left $\mathrm{End}_R(G)$-modules

$$\mathcal{P} = P_1 \to P_0 \to G \to 0$$

in which $P_0 = \mathrm{End}_R(G)^n$ and $P_1 = \mathrm{End}_R(G)^m$ for some integers n, m. By Theorem 9.32, an application of

$$\mathrm{Hom}_{\mathrm{End}_R(G)}(\cdot, G)$$

to $\mathcal{P}$ forms the G-coplex

$$\mathcal{P}^* = 0 \to G^* \to P_0^* \to P_1^* \to \cdots$$

where $G^* = S$. Then

$$0 \to S \to P_0^* \to P_1^*$$

is exact. By our choice of P_0 and P_1, Theorem 9.30 shows us that $P_0^*, P_1^* \in \mathbf{P}_o(G)$, and that the point evaluation maps Ψ_0, Ψ_1 are isomorphisms. Then by Lemma 12.6, the sequence

$$\mathcal{P}^* \cong 0 \to S \to P_0^* \to P_1^*$$

is doubly G-cobalanced. This proves part 2.

$2 \Rightarrow 1$. Suppose that for some $W_0, W_1 \in \mathbf{P}_o(G)$, there is a doubly G-cobalanced exact sequence

$$0 \to S \to W_0 \to W_1$$

of right R-modules. Without loss of generality we will assume that $W_0 = G^n$ and $W_1 = G^m$. An application of $\mathrm{Hom}_S(\cdot, G)$ yields the exact sequence

$$\mathrm{Hom}_S(W_1, G) \to \mathrm{Hom}_S(W_0, G) \to \mathrm{Hom}_S(S, G) \to 0$$

which by (12.2) is isomorphic to the exact sequence

$$\mathrm{End}_R(G)^m \to \mathrm{End}_R(G)^n \to G \to 0.$$

Hence G is a finitely presented left $\mathrm{End}_R(G)$-module. This completes the proof.

Theorem 12.10. *Let G be an* rtffr *group. The following are equivalent.*

1. *G is E-finitely presented.*
2. *There is a doubly G-cobalanced exact sequence $0 \to S \to W_0 \to W_1$ of S-modules in which $W_0, W_1 \in \mathcal{P}_o(G)$.*

The left $\mathrm{End}_R(G)$-module G is *coherent* if there is a projective resolution $\cdots \to P_1 \to P_0 \to G \to 0$ each of whose terms P_k is a finitely generated projective left $\mathrm{End}_R(G)$-module.

Theorem 12.11. *Let G be a right R-module. The following are equivalent.*

1. *G is E-coherent.*
2. *There is a doubly G-cobalanced G-complex*

$$0 \to S \to W_0 \to W_1 \to \cdots$$

of right R-modules whose terms W_k are in $\mathcal{P}_o(G)$.

Proof: Proceed in the by now familiar manner.

Theorem 12.12. *Let G be an* rtffr *group. The following are equivalent.*

1. *G is E-coherent.*
2. *There is a doubly G-cobalanced G-complex $\mathcal{W}$ for S each of whose terms W_k is in $\mathcal{P}_o(G)$.*

Let A be a ring and assume that G is a left A-module. Then G has a *projective cover* if there is a projective A-module P and a surjection $P \to G \to 0$ such that if $P' \to G \to 0$ is a map from a projective left A-module P' onto G then the lifting $P' \to P$ is a (split) surjection. Also, G has a *finitely generated projective cover* if G has a projective cover $P \to G \to 0$ as a left A-module in which P is finitely generated.

We say that the *rtffr* group G has an *E-projective cover* if G has a projective cover as a left $\mathrm{End}_R(G)$-module. Also, G has a *finitely generated E-projective cover* if G has a finitely generated projective cover over $\mathrm{End}_R(G)$.

The right R-module K has a *finite G-hull* if there is a G-cobalanced injection $0 \to K \to W$ such that $W \in \mathbf{coP}_o(G)$ and such that if $W' \in \mathbf{coP}(G)$ and if $K \to W'$ is an injection then the lifting $W \to W'$ is a split injection.

Theorem 12.13. *Assume* (μ) *and let G be a self-slender right R-module. The following are equivalent.*

1. *G possesses a finitely generated projective cover as a left* $\text{End}_R(G)$*-module.*
2. *S possesses a finite G-hull as a left R-module.*

Proof: 1 $\Rightarrow$ 2. Suppose that G has a finitely generated projective cover $P_0 \xrightarrow{\delta} G \to 0$ as a left $\text{End}_R(G)$-module. Extend δ to a projective resolution $\mathcal{P}$ for G over $\text{End}_R(G)$. Applying $\text{Hom}_{\text{End}_R(G)}(\cdot, G)$ to $\mathcal{P}$ yields the G-coplex

$$\mathcal{P}^* = 0 \to G^* \xrightarrow{\delta^*} P_0^* \xrightarrow{\delta_1^*} P_1^* \longrightarrow \cdots,$$

where $G^* = S$. Since $\text{Hom}_{\text{End}_R(G)}(\cdot, G)$ is left exact,

$$0 \to S \to P_0^* \to P_1^*$$

is exact. By Lemma 12.6, the sequence $0 \to S \to P_0^*$ is G-cobalanced. Evidently $P_0^* \in \mathbf{coP}_o(G)$.

Finally, suppose there is a $W \in \mathcal{P}(G)$ such that there exists a G-cobalanced embedding $S \to W$. An application of $\text{Hom}_R(\cdot, G)$ then produces the surjection $W^* \to S^*$. By Theorem 9.30,

1. $P_0^{**} = P_0$,
2. $S^* = G^{**} = G$, and
3. $\delta^{**} = \delta : P_0 \to G$ is a finitely generated projective cover.

Then there is a split surjection $W^* \to P_0$ so that $P_0^* \to W^{**}$ is a split injection. Because G is self-slender, Ψ_W is an isomorphism, Theorem 9.30. Thus $P_0^* \to W$ is a split injection. That is, $0 \to S \to P_0^*$ is a finite G-hull. This proves part 2.

2 $\Rightarrow$ 1. Suppose that there is a finite G-hull $0 \to S \to W_0$. Dual to the above argument, show that the finite G-hull dualizes into a finitely generated projective cover of G. The rest of the proof follows in the by now familiar manner. This completes the proof.

Theorem 12.14. *Let G be an* rtffr *group. The following are equivalent.*

1. *G possesses an E-projective cover.*
2. *G possesses a finitely generated E-projective cover.*
3. *S possesses a finite G-hull.*

Proof: 3 $\Leftrightarrow$ 2 $\Rightarrow$ 1 by Theorem 12.13.

Assume part 1. Let G have an E-projective cover

$$0 \to K \to P \to G \to 0.$$

Then $P/K \cong G$ has finite rank and K is superfluous in P. By Lemma 12.3, P is finitely generated. This proves part 2 and completes the proof.

12.3 Exercises

Let G and H denote *rtffr* groups. Let $E(G) = \text{End}(G)/\mathcal{N}(\text{End}(G))$. Let E be an *rtffr* ring. Let $S = \text{center}(E)$. Given a semi-prime *rtffr* ring E let $\overline{E}$ be an integrally closed subring of $\mathbf{Q}E$ such that $E \subset \overline{E}$ and $\overline{E}/E$ is finite. Let τ be the *conductor* of E.

1. Prove (12.1) and (12.2).
2. Fill in the details of Theorem 12.2.
3. Prove Theorem 12.5.

12.4 Problems for future research

1. Characterize the E-finitely presented *rtffr* groups G by giving a quasi-direct sum decomposition of such groups. You may want first to read Reid's Theorem 13.8.
2. Give a direct sum decomposition of *rtffr* groups G that possess a projective cover over $\text{End}(G)$.
3. Let P be a module-theoretic property. Characterize those left R-modules G such that G has property P as a left $\text{End}_R(G)$-module.

13

Finitely generated modules

The right R-module G is *E-finitely generated* if G is a finitley generated left $\mathrm{End}_R(G)$-module. We will study the E-finitely generated R-modules.

13.1 Beaumont–Pierce

Let R be an *rtffr* ring with identity. There is a natural embedding of rings

$$\lambda : R \to \mathrm{End}_{\mathbf{Z}}(R, +)$$

such that for $f \in R$, $\lambda(f)(x) = fx$ for each $x \in R$. That is, $\lambda(f)$ is left multiplication by f on R. If λ is an isomorphism then R is called an *E-ring*. We will call the ring R an *rtffr Dedekind E-ring* if R is an *rtffr* group, a Dedekind domain, and an E-ring. The ring R is an *rtffr E-domain* if R is an integral domain and an *rtffr E*-ring.

We relate a quasi-direct sum decomposition of $R/\mathcal{N}(R)$ and center($R/\mathcal{N}(R)$). The symbols A, R, C_i, and R_i will be used throughout this chapter as they are in the next result.

Theorem 13.1. [9, Theorem 9.10 and Corollary 14.4] *Let R be a* rtffr *ring.*

1. *Let $C = \mathrm{center}(R/\mathcal{N}(R))$. There are integral domains $C_1, \ldots, C_t$ and an integer $m \neq 0$ such that*

$$m(C_1 \times \cdots \times C_t) \subset C \subset (C_1 \times \cdots \times C_t)$$

as rings.
2. *There are* rttfr *E-domains $R_i \subset C_i$ such that C_i is a finitely generated R_i-module. Thus $R_1 \times \cdots \times R_t$ is a subring of $C_1 \times \cdots \times C_t$.*
3. *Let*

$$A = (R_1 \times \cdots \times R_t) \cap C.$$

Then A is a subring of center$R/\mathcal{N}(R)$ *that has finite index in $R_1 \times \cdots \times R_t$. Furthermore, there are integers $n_1, \ldots, n_t > 0$ such that*

$$R/\mathcal{N}(R) \doteq R_1^{(n_1)} \oplus \cdots \oplus R_t^{(n_t)}$$

as A-modules.

Recall the Beaumont–Pierce theorem 1.8. The next several results form an extension of the Beaumont–Pierce theorem from rings to finitely generated *rtffr* modules.

Lemma 13.2. *Say $G \cong H$ are* rtffr *groups.*

1. *G is E-finitely generated iff H is E-finitely generated.*
2. *If G is finitely generated over some ring E then G is also E-finitely generated.*

Proof: This is an exercise.

If M is an R-module whose additive structure is a *rtffr* group, then M is called a *rtffr* R-module.

Lemma 13.3. *Let E be a semi-prime* rtffr *ring and let M and M' be torsion-free left E-modules with M' finitely generated. If $\pi : M \to M'$ is a surjection then $M \cong \ker \pi \oplus M'$.*

Proof: Observe that π lifts to a surjection $\pi : \mathbf{Q}M \to \mathbf{Q}M'$ of left modules over the semi-simple ring $\mathbf{Q}E$. Since every $\mathbf{Q}E$-module is projective $\mathbf{Q}M \cong \ker \pi \oplus \mathbf{Q}M'$. Thus there is a $\mathbf{Q}E$-module map $\sigma : \mathbf{Q}M' \to \mathbf{Q}M$ such that $\pi\sigma = 1_{\mathbf{Q}M'}$. Since M' is finitely generated there is an integer $n \neq 0$ such that $n\sigma(M') \subset M$. Inasmuch as

$$e = \sigma\pi = (\sigma\pi)^2 \in \mathrm{End}_{\mathbf{Q}T}(\mathbf{Q}M),$$

and since $ne \in \mathrm{End}_T(M)$, we have $n1_M = n(1 - e) \oplus ne$. Hence,

$$nM \subset n(1 - e)(M) \oplus (ne)(M)$$
$$\subset \ker(\pi|_M) \oplus (n\sigma)(M')$$
$$\subset M.$$

Since $n\sigma(M') \cong M'$, $M \doteq \ker \pi|_M \oplus M'$, and this completes the proof.

The mapping $f : M \to N$ of modules or groups is *quasi-split* if there is a mapping $g : N \to M$ and an integer $m \neq 0$ such that $gf = m1_M$.

Lemma 13.4. *Let R be an* rtffr *ring and let M be a finitely generated left $R/\mathcal{N}(R)$-module. If $f : N \to M$ is a mapping of left R-modules such that $\mathbf{Q}f(N) = \mathbf{Q}M$ then f is quasi-split as an R-module map.*

Proof: Let R be an *rtffr* ring, and let M be an *rtffr* left $R/\mathcal{N}(R)$-module. Suppose that N is a left R-module, and let $f : N \to M$ be an R-module map such that $\mathbf{Q}f(N) = \mathbf{Q}M$. By Theorem 1.8, there is a semi-prime subring $T \subset R$ such that $R \doteq T \oplus \mathcal{N}(R)$. Then $T \cong R/\mathcal{N}(R)$, so that M is a finitely generated *rtffr* left T-module.

Furthermore, f lifts to a surjection $\mathbf{Q}f : \mathbf{Q}N \to \mathbf{Q}M$ of left $\mathbf{Q}T$-modules. Since T is semi-prime, $\mathbf{Q}T$ is semi-simple, so that $\mathbf{Q}M$ is a projective $\mathbf{Q}T$-module. Thus there is a splitting $g : \mathbf{Q}M \to \mathbf{Q}N$ for f. Since M is finitely generated as a left

T-module and since $\mathbf{Q}N/N$ is a torsion abelian group, there is an integer $k \neq 0$ such that $kg(M) \subset N$. Then for each $x \in M$ we have

$$f(kg)(x) = kfg(x) = kx,$$

and hence $f : N \to M$ is quasi-split by kg. This completes the proof.

Corollary 13.5. *Let R be an rtffr ring, let M be a left R-module and let K be a pure R-submodule of M such that M/K is a finitely generated left $R/\mathcal{N}(R)$-module. Then*

1. $M \overset{\cdot}{\cong} \dfrac{M}{K} \oplus K$ *and*

2. There is an integer $n \neq 0$ and a left T-module M' such that $(M/K) \oplus M' \overset{\cdot}{\cong} T^{(n)}$.

Proof. 1. Apply Lemma 13.4.

2. Use the notation T as in the Beaumont–Pierce theorem 1.8. Since M/K is a finitely generated *rtffr* left T-module $\mathbf{Q}M/\mathbf{Q}K$ is a finitely generated projective left $\mathbf{Q}T$-module. There is a finitely generated projective $\mathbf{Q}T$-module P and an integer m such that

$$(\mathbf{Q}M/\mathbf{Q}K) \oplus P \cong \mathbf{Q}T^{(m)}.$$

Since P is finitely generated there is a finitely generated T-submodule $M' \subset P$ such that $\mathbf{Q}M' = P$. Then

$$(M/K) \oplus M' \subset \mathbf{Q}T^{(m)}$$

and since both M/K and M' are finitely generated T-modules there is an integer $n \neq 0$ such that $n(M/K \oplus M') \subset T^{(m)}$. Furthermore, since

$$\frac{T^{(m)}}{n(M/K \oplus M')}$$

is a finitely generated T-module that is also a torsion group there is an integer $n' \neq 0$ such that

$$n'T^{(m)} \subset n(M/K \oplus M') \subset T^{(m)}.$$

This proves part 2 and completes the proof.

Lemma 13.6. *Let R be a semi-prime rtffr and let M be a finitely generated rtffr left A-module. Choose A, R_i, and m for R as in Theorem 13.1. There are integers $n_1, \ldots, n_t \geq 0$ such that*

$$m(R_1^{(n_1)} \oplus \cdots \oplus R_t^{(n_t)}) \subset M \subset (R_1^{(n_1)} \oplus \cdots \oplus R_t^{(n_t)})$$

as A-modules.

Proof: Since R is semi-prime and since M is finitely generated, Corollary 13.5 states that there is an integer k and finitely generated left E-module M' such that $M \oplus M' \doteq R^k$. By (13.1), there are *rtffr* E-domains $R_1, \ldots, R_t$ and integers $m, t, m_1, \ldots, m_t > 0$ such that

$$m(R_1^{(m_1)} \oplus \cdots \oplus R_t^{(m_t)}) \subset R \subset (R_1^{(m_1)} \oplus \cdots \oplus R_t^{(m_t)})$$

as nonunital A-modules. The *rtffr* E-domains R_i are strongly indecomposable groups, so by Jónsson's theorem there are integers $n_1, \ldots, n_t \geq 0$ such that

$$M \cong R_1^{(n_1)} \oplus \cdots \oplus R_t^{(n_t)},$$

which completes the proof.

Theorem 13.7. *Let E be an* rtffr *ring, choose the semi-prime ring T for E as in the Beaumont–Pierce theorem 1.8, let $R_1, \ldots, R_t$ be the* rtffr *E-domains chosen in Theorem 13.1 for T, and let A be the subring of finite index in $R_1 \times \cdots \times R_t$ as in Theorem 13.1. If M is a finitely generated* rtffr *left E-module then $M \doteq M_o \oplus N$ as groups where*

1. *$M_o = R_1^{(n_1)} \oplus \cdots \oplus R_t^{(n_t)}$ for some integers $t, n_1, \ldots, n_t \geq 0$, and*
2. *$N = \mathcal{N}(E)M_o$.*

Proof: Recall that X_* denotes the purification of X in an unambiguous larger group. Let E be an *rtffr* ring and let M be a finitely generated *rtffr* left E-module. Inasmuch as $L_o = M/(\mathcal{N}(E)M)_*$ is a finitely generated left $E/\mathcal{N}(E)$-module, Lemma 13.3 implies that

$$M \cong L_o \oplus (\mathcal{N}(E)M)_*$$

as T-modules. Then by Lemma 13.6 there are integers $m, t, n_1, \ldots, n_t > 0$ such that

$$L_o \cong R_1^{(n_1)} \oplus \cdots \oplus R_t^{(n_t)}$$

as A-modules. We write

$$M_o = R_1^{(n_1)} \oplus \cdots \oplus R_t^{(n_t)}.$$

From Nakayama's theorem 1.1 we see that

$$\mathbf{Q}M = \mathbf{Q}EM_o + \mathcal{N}(\mathbf{Q}E)M = \mathbf{Q}EM_o,$$

and since M is a finitely generated *rtffr* left E-module, the torsion group M/EM_o is finite. Also, because $E \doteq T \oplus \mathcal{N}(E)$ and because L_o is a T-module, we have

$$M \doteq EM_o$$
$$\doteq TM_o \oplus \mathcal{N}(E)M_o$$
$$\doteq TL_o \oplus \mathcal{N}(E)M_o$$
$$= L_o \oplus \mathcal{N}(E)M_o$$
$$\doteq M_o \oplus \mathcal{N}(E)M_o$$

as *rtffr* groups. This proves the theorem.

We present J. D. Reid's classification of the E-cyclic and E-finitely generated *rtffr* groups.

Theorem 13.8. [92, J. D. Reid] *Let G be an* rtffr *group and write*

$$\text{End}(G) \doteq T \oplus \mathcal{N}(\text{End}(G))$$

as in the Beaumont–Pierce theorem 1.8. The following are equivalent.

1. *G is E-finitely generated.*
2. *There are* rtffr *E-domains $R_1, \ldots, R_t$ as in Theorem 13.1 for T, integers $t, n_1, \ldots$, $n_t > 0$, and a group N such that*
   (a) $G \doteq R_1^{(n_1)} \oplus \cdots \oplus R_t^{(n_t)} \oplus N$ as groups and
   (b) $N = \mathbf{S}_{R_1 \oplus \cdots \oplus R_t}(N)$ as groups.
3. *There are* rtffr *E-domains $R_1, \ldots, R_t$, integers $t, n_1, \ldots, n_t > 0$, and a group N such that*
   (a) $G \doteq R_1^{(n_1)} \oplus \cdots \oplus R_t^{(n_t)} \oplus N$ as groups and
   (b) $N = \mathbf{S}_{R_1 \oplus \cdots \oplus R_t}(N)$ as groups.

Proof: $1 \Rightarrow 2$. Assume part 1. Let $\text{End}(G)$ and suppose that G is an rtffr E-finitely generated group. By Theorem 13.7,

$$G \cong G_o \oplus \mathcal{N}(E)G_o$$

for some group G_o. Moreover, by Theorem 13.7 we can choose G_o, the *rtffr* E-domains $R_1, \ldots, R_t$, and the integers $n_1, \ldots, n_t > 0$ such that

$$G_o = R_1^{(n_1)} \oplus \cdots \oplus R_t^{(n_t)}.$$

Then G_o is an $A = R_1 \times \cdots \times R_t$-module. Let $N = \mathcal{N}(E)G_o$. Then

$$N = \mathbf{S}_{G_o}(N) = \mathbf{S}_{R_1 \oplus \cdots \oplus R_t}(N).$$

This proves part 2.

$2 \Rightarrow 3$ is clear.

$3 \Rightarrow 1$. Assume part 3. Suppose that $G \doteq G_o \oplus N$ where $G_o = R_1^{(n_1)} \oplus \cdots \oplus R_t^{(n_t)}$ and N satisfy the conditions in part 3. Since each R_i is an E-domain, each R_i is a cyclic R_i-module, whence G_o is naturally a finitely generated A-module. Let $x_1, \ldots, x_r \in G_o$ be generators for G_o over A, and hence over $\mathrm{End}(G_o)$. We have the following equations.

$$
\begin{aligned}
G_o \oplus N &= G_o \oplus \mathbf{S}_{G_o}(N) \\
&= G_o \oplus \mathrm{Hom}(G_o, N)G_o \\
&\subset \mathrm{End}(G_o)x_1 + \cdots + \mathrm{End}(G_o)x_r + \\
&\quad \mathrm{Hom}(G_o, N)(\mathrm{End}(G_o)x_1 + \cdots + \mathrm{End}(G_o)x_r) \\
&= \mathrm{End}(G_o)x_1 + \cdots + \mathrm{End}(G_o)x_r + \\
&\quad \mathrm{Hom}(G_o, N)x_1 + \cdots + \mathrm{Hom}(G_o, N)x_r \\
&\subset \mathrm{End}(G_o \oplus N)x_1 + \cdots + \mathrm{End}(G_o \oplus N)x_r \\
&\subset G_o \oplus N
\end{aligned}
$$

Thus $G_o \oplus N$, and hence any group quasi-isomorphic to $G_o \oplus N$ is E-finitely generated. This concludes the proof.

The group G is *E-cyclic* if $G = \mathrm{End}(G)x$ for some $x \in G$. Each ring is an E-cyclic group.

Corollary 13.9. *Let G be an* rtffr *group. Then G is E-finitely generated iff G is quasi-isomorphic to an E-cyclic group.*

Proof: Suppose that G is an rtffr E-finitely generated group. As in Theorem 13.8 there are E-domains $R_1, \ldots, R_t$, positive integers $t, n_1, \ldots, n_t$, and a group N such that $G \doteq G_o \oplus N$ where

$$
G_o = R_1^{(n_1)} \oplus \cdots \oplus R_t^{(n_t)}
$$

and

$$
N = \mathbf{S}_{G_o}(N).
$$

It suffices to show that $G_o \oplus N$ is an E-cyclic group. Let x_i be a generator of R_i as an R_i-module and let

$$
x = x_1 \oplus \cdots \oplus x_t \in G_o.
$$

One proves that $G_o = \mathrm{End}(G_o)x$. Since

$$
\mathrm{End}(G_o) \subset \mathrm{End}(G_o \oplus N),
$$

we have

$$
G_o = \mathrm{End}(G_o)x \subset \mathrm{End}(G_o \oplus N)x.
$$

Furthermore, by hypothesis,

$$N = \mathbf{S}_{G_o}(N)$$
$$= \mathrm{Hom}(G_o, N)G_o$$
$$= \mathrm{Hom}(G_o, N)\mathrm{End}(G_o)x$$
$$= \mathrm{Hom}(G_o, N)x,$$

so that

$$G_o \oplus N \subset \mathrm{End}(G_o \oplus N)x.$$

Inasmuch as

$$\mathrm{End}(G_o \oplus N)x \subset G_o \oplus N,$$

it follows that

$$G_o \oplus N = \mathrm{End}(G_o \oplus N)x.$$

The converse follows from Lemma 13.2, so the proof is complete.

13.2 Noetherian modules

The right R-module G is *E-Noetherian* if G is a Noetherian left $\mathrm{End}_R(G)$-module. Of course, G is E-finitely generated if G is E-Noetherian. In this section we consider the quasi-direct sum decomposition of the E-Noetherian rtffr groups *á la* A. Paras [91], showing that it is more specialized than the direct sum decomposition of an *rtffr* E-finitely generated group.

Lemma 13.10. *Let R be an* rtffr *ring. Then R is a right Noetherian ring iff for each integer $k > 0$, $\mathcal{N}(R)^k/\mathcal{N}(R)^{k+1}$ is a finitely generated right A-module.*

Proof: Suppose that for each integer $k > 0$, $\mathcal{N}(R)^k/\mathcal{N}(R)^{k+1}$ is a finitely generated left R-module. Then $\mathcal{N}(R)/\mathcal{N}(R)^2$ is finitely generated by the Noetherian ring $R/\mathcal{N}(R)$, so that $\mathcal{N}(R)/\mathcal{N}(R)^2$ and $R/\mathcal{N}(R)$ are Noetherian left R-modules. Hence $R/\mathcal{N}(R)^2$ is a Noetherian left R-module. Also, because R has finite rank, $\mathcal{N}(R)^k = 0$ for some integer $k > 0$. Continuing inductively on k we eventually show that $R/\mathcal{N}(R)^k = R$ is left Noetherian. The converse is clear so the proof is complete.

Lemma 13.11. *Say $G \overset{.}{\cong} H$ are* rtffr *groups.*

1. *G is E-Noetherian iff H is E-Noetherian.*
2. *If G is Noetherian over some ring R then G is also E-Noetherian.*

Proof: This is an exercise.

The following result can be viewed as a generalization of R. A. Beaumont and R. S. Pierce's classification [21] of the additive structure of semi-prime *rtffr* rings.

Theorem 13.12. *Let R be an* rtffr *ring, let $R \doteq T \oplus \mathcal{N}(R)$, and let $T \doteq R_1^{m_1} \oplus \cdots \oplus R_t^{m_t}$ as in Theorem 13.1. If M is an* rtffr *Noetherian left R-module then*

$$M \overset{\cdot}{\cong} R_1^{(n_1)} \oplus \cdots \oplus R_t^{(n_t)}$$

for some rtffr *E-domains $R_1, \ldots, R_t$, and integers $n_1, \ldots, n_t \geq 0$.*

Proof. The ring T is from the Beaumont–Pierce theorem 1.8; the *rtffr* E-domains $R_1, \cdots, R_t$ for T are from Theorem 13.7 applied to R.

Suppose that M is a Noetherian left rtffr R-module. Using Theorem 13.7, $M \overset{\cdot}{\cong} M_0 \oplus M_1$ as groups where

$$M_0 \overset{\cdot}{\cong} U_1^{(m_1)} \oplus \cdots \oplus U_t^{(m_r)} \text{ and } M_1 = \mathcal{N}(A)M_0$$

for some *rtffr* E-domains $U_1, \ldots, U_r$, and some integers $r, n_1, \ldots, n_r \geq 0$. Moreover, M_1 is an R-submodule of M. Since M is Noetherian as a left R-module, M_1 is Noetherian. Induction on the rank of M shows that

$$M_1 \overset{\cdot}{\cong} V_1^{(m_1)} \oplus \cdots \oplus V_t^{(m_s)}$$

for some *rtffr* E-domains $V_1, \ldots, V_s$, and some integers $s, m_1, \ldots, m_s \geq 0$. By choosing a unique representative R_i of the quasi-isomorphism class of $\{U_i, V_j \,|\, i, j\}$ we arrive at a set $\{R_1, \ldots, R_t\}$ of *rtffr* E-domains such that $M \overset{\cdot}{\cong} R_1^{(n_1)} \oplus \cdots \oplus R_t^{(n_t)}$. This proves the theorem.

The properties Noetherian and finite generation are linked in the following result.

Theorem 13.13. *Let R be an* rtffr *ring, and let M be an* rtffr *left A-module. Then M is a Noetherian left R-module iff M is a finitely generated module over some* rtffr *semi-prime commutative (not necessarily central) ring $A \subset R$.*

Proof. Write $R \doteq T \oplus \mathcal{N}(R)$ as in the Beaumont–Pierce theorem 1.8.

Suppose that M is a finitely generated module over some *rtffr* semi-prime commutative (not necessarily central) ring $A \subset R$. Then A is Noetherian, and thus M is Noetherian as a left A-module. Consequently, M is a Noetherian left R-module.

Conversely, if M is a Noetherian left R-module, then by Theorem 13.12, there are E-domains $R_1, \ldots, R_t$ chosen for T as in Theorem 13.1 such that

$$M \doteq R_1^{(n_1)} \oplus \cdots \oplus R_t^{(n_t)}.$$

Then M is finitely generated by $R \cap A$, where $A = R_1 \times \cdots \times R_t$ is a subring of the semi-prime ring center(T). This completes the proof.

The E-Noetherian *rtffr* groups are characterized up to quasi-isomorphism by A. Paras in [91], where she proves the following theorem.

Theorem 13.14. [91, A. Paras]. *Let G be an* rtffr *group. The following are equivalent.*

1. *G is E-Noetherian.*
2. *There are* rtffr *E-domains $R_1, \ldots, R_t$ and integers $t, n_1, \ldots, n_t > 0$ such that*

$$G \doteq R_1^{(n_1)} \oplus \cdots \oplus R_t^{(n_t)}. \qquad (13.1)$$

In this case $\text{End}(G)$ *is a left Noetherian ring.*

Proof: $1 \Rightarrow 2$. Assume part 1. By Theorem 13.12, G can be written as in (13.1).

$2 \Rightarrow 1$. Assume part 2. By Lemma 13.11 we can proceed by assuming that (13.1) is an equation. Then G is a finitely generated left $A = R_1 \times \cdots \times R_t$-module so that A is a subring of $\text{End}(G)$. Since $R_1, \ldots, R_t$ are E-domains, A is a semi-prime Noetherian ring, so that G is a Noetherian left A-module. Then by Lemma 13.11, G is a Noetherian left $\text{End}(G)$-module. This proves part 1.

In this case, because G has finite rank, $\text{End}(G) \subset G^n$ for some integer n. Then $\text{End}(G)$ is a Noetherian left $\text{End}(G)$-submodule of the Noetherian module G^n. This completes the proof.

Corollary 13.15. *If G is an* rtffr *E-Noetherian group then G is E-finitely presented and quasi-isomorphic to an E-cyclic E-projective* rtffr *group.*

13.3 Generators

The group G is an *E-generator* if G generates the category $\text{End}(G)$-**Mod**, and we say that G is an *E-progenerator* if G is a finitely generated projective generator in the category $\text{End}(G)$-**Mod**.

The following proposition is well known. We include it for completeness.

Proposition 13.16. *Let R be an* rtffr *ring.*

1. *If U is a generator in R-**Mod** then U is finitely generated projective left $\text{End}_R(U)$-module.*
2. *If U is a finitely generated projective left R-module then U is a generator in $\text{End}_R(U)$-**Mod**.*

Proof: 1. Since U is a generator in R-**Mod**, $U^n \cong R \oplus K$ for some left R-module K. Then

$$\text{End}_R(U)^n \cong \text{Hom}_R(U^n, U)$$

$$\cong \text{Hom}_R(R \oplus K, U)$$

$$\cong \text{Hom}_R(R, U) \oplus \text{Hom}_R(K, U)$$

$$\cong U \oplus \text{Hom}_R(K, U)$$

as left $\text{End}_R(U)$-modules. Thus U is a projective left $\text{End}_R(U)$-module.

2. Because U is a projective left R-module, write $U \oplus K = R^n$ for some integer n and R-module K. Then

$$
\begin{aligned}
U^n &\cong \mathrm{Hom}_R(R^n, U) \\
&\cong \mathrm{Hom}_R(U \oplus K, U) \\
&= \mathrm{Hom}_R(U, U) \oplus \mathrm{Hom}_R(K, U) \\
&= \mathrm{End}_R(U) \oplus \mathrm{Hom}_R(K, U),
\end{aligned}
$$

so that U is a generator in $\mathrm{End}_R(U)$-**Mod**. This completes the proof.

We are thus motivated to study the *rtffr* E-generator groups. For an *rtffr* group G, and with $R = \mathbf{Z}$, we have

$$
S = \mathrm{BiEnd}_R(G) = \mathrm{center}(\mathrm{End}(G)).
$$

Lemma 13.17. *Let G be an* rtffr *group. Then G is an E-generator iff $G = U \oplus V$ where U is an invertible ideal of S and V is some finitely generated projective S-module.*

Proof: Let $G = U \oplus V$ for some invertible ideal $U \subset S$ and some finitely generated projective S-module V. Then G is finitely generated projective over S. By Proposition 13.16, G is a cogenerator in $\mathrm{End}_S(G)$-**Mod**, and by (12.2), $\mathrm{End}(G) = \mathrm{End}_S(G)$. Thus G is an E-generator.

Conversely, suppose that G is an *rtffr* E-generator. By Proposition 13.16, G is a finitely generated S-module. Since

$$
S = S_1 \times \cdots \times S_t
$$

is a finite product of indecomposable commutative *rtffr* rings, we have

$$
G = GS_1 \oplus \cdots \oplus GS_t
$$

as finitely generated projective S-modules. Then for each $i = 1, \ldots, t$, Corollary 2.12 states that

$$
GS_i = U_i' \oplus U_{1i} \oplus \cdots \oplus U_{s_i i}
$$

for some indecomposable invertible ideals U_i' and U_{ji} of S_i. Since $\mathrm{ann}_S(G) = 0$, $U_i' \neq 0$ for each i. Hence $U = U_1' \oplus \cdots \oplus U_t'$ is an invertible ideal in S such that $G = U \oplus V$ where $V = \oplus_{i,j} U_{ji}$ of indecomposable ideals of S. This completes the proof.

Corollary 13.18. *An* rtffr *E-generator group is E-finitely generated.*

The group G is *E-self-generating* if G generates each $\mathrm{End}(G)$-submodule of $G^{(n)}$ for each $n > 0$. The following result shows us that the notions of an E-generating and E-self-generating *rtffr* are interchangeable.

Lemma 13.19. *Let G be an* rtffr *group. Then G is an E-self-generating group iff G is an E-generator group.*

Proof: Let $x_1, \ldots, x_n \in G$ be a maximal linearly independent subset of G. Then $f(x_1, \ldots, x_n) = 0$ iff $f(G) = 0$ iff $f = 0$. Thus

$$\text{End}(G) \cong \text{End}(G)(x_1 \oplus \cdots \oplus x_n) \subset G^{(n)}$$

as left $\text{End}(G)$-modules. Then G generates $\text{End}(G)$, so that G is a generator in $\text{End}(G)$-**Mod**. The converse is clear, so the proof is complete.

Several of the properties that we are studying coincide. Recall that $S = \text{center}(\text{End}(G))$ for abelian groups G.

Theorem 13.20. *Let G be an* rtffr *group. The following are equivalent.*

1. *G is an E-generator.*
2. *G is an E-progenerator.*
3. *G is a progenerator as an S-module.*
4. *G is a finitely generated projective S-module.*
5. *G is a projective right S-module.*
6. $G \cong U_1 \oplus \cdots \oplus U_t$ *for some indecomposable projective ideals* $U_1, \ldots, U_t \subset S$.

Proof: We will prove $1 \Rightarrow 2 \Rightarrow 3 \Rightarrow 4 \Rightarrow 5 \Rightarrow 6 \Rightarrow 1$.

$1 \Rightarrow 2$. Assume using part 1 that G is an E-generator group. By Lemma 13.17, there is an invertible ideal U of S such that $G = U \oplus V$. Then U is a generator of S-**Mod** and hence G is a generator in S-**Mod**. Thus by (12.2), G is a progenerator in $\text{End}_S(G) = \text{End}(G)$-**Mod**. Hence G is a progenerator in $\text{End}(G)$-**Mod**. This proves part 2.

$2 \Rightarrow 3$ follows from Proposition 13.16.

$3 \Rightarrow 4 \Rightarrow 5$ is clear.

$5 \Rightarrow 6$ follows from Corollary 2.12. (The integral domain hypothesis is not needed to prove Corollary 2.12.)

$6 \Rightarrow 1$. Since G is a finitely generated projective S-module, G is a generator in $\text{End}_S(G)$-**Mod**. By (12.2), $\text{End}(G) = \text{End}_S(G)$, so G is a generator in $\text{End}(G)$-**Mod**. Hence G is an E-generator. This proves 1 and completes the logical cycle.

Corollary 13.21. *Let G be an* rtffr *group. Then G is an E-generator group iff*

$$\text{Hom}_{\text{End}(G)}(G, \cdot) : \text{End}(G)\text{-}\mathbf{Mod} \to S\text{-}\mathbf{Mod}$$

is a Morita equivalence.

Here is another list of over lapping properties for rtffr groups.

Theorem 13.22. *Let G be an* rtffr *group, and assume that* $\text{End}(G)$ *is semi-prime. The following are equivalent for G.*

1. *G is an E-Noetherian group.*
2. *G is an E-finitely generated group.*

3. *G is quasi-isomorphic to an E-projective group.*
4. *G is quasi-isomorphic to an E-generator group.*
5. *G is quasi-isomorphic to an E-progenerator group.*

Proof: Let $\text{End}(G)$ be semi-prime. We will prove $5 \Rightarrow \{4, 3\} \Rightarrow 2 \Rightarrow 1 \Rightarrow 5$.

$5 \Rightarrow 4$ and 3 is clear.

4 or $3 \Rightarrow 2$ follows from Corollaries 12.4 and 13.18.

$2 \Rightarrow 1$. Suppose that G is E-finitely generated. Since $\text{End}(G)$ is semi-prime, $\mathbf{Q}\text{End}(G)$ is a semi-simple ring, and thus $\mathbf{Q}G$ is a projective left $\mathbf{Q}\text{End}(G)$-module. There is then an integer $n > 0$ and a split embedding $f : \mathbf{Q}G \to \mathbf{Q}\text{End}(G)^{(n)}$ of left $\mathbf{Q}\text{End}(G)$-modules. Let $g : \mathbf{Q}\text{End}(G)^n \to \mathbf{Q}G$ be such that $gf = 1_{\mathbf{Q}G}$. Since G is finitely generated, there is an integer $k > 0$ such that $kg(G) \subset \text{End}(G)^n$ and $kf(\text{End}(G)^n) \subset G$. Hence $k^2(gf) = k^2 1_G$ so that G is a quasi-summand of $\text{End}(G)^n$. Since $\text{End}(G)$ is semi-prime, $\text{End}(G)$ and hence $\text{End}(G)^n$, are Noetherian modules, [9]. Thus G is E-Noetherian.

$1 \Rightarrow 5$. Suppose that G is E-Noetherian. Since $\text{End}(G)$ is semi-prime there is a maximal order $\overline{E}$ in $\mathbf{Q}\text{End}(G)$ such that $\text{End}(G) \subset \overline{E}$ and $\overline{E}/\text{End}(G)$ is finite. Then $\overline{G} = \overline{E}G$ is a Noetherian left $\overline{E}$-module with $\text{ann}_{\overline{E}}(\overline{G}) = 0$. Because $\overline{E}$ is integrally closed (a classical maximal order), $\overline{G}$ is a progenerator in $\overline{E}$-**Mod** [94]. Furthermore, since $\overline{E}$ is a maximal order, since $\text{End}(\overline{G}) \supset \overline{E}$, and since $\text{End}(\overline{G})/\text{End}(G)$ is finite, $\text{End}(\overline{G})/\overline{E}$ is finite, whence $\text{End}(G) = \overline{E}$. Hence G is a progenerator in $\text{End}(\overline{G})$-**Mod**. This proves part 5 and completes the logical cycle.

For strongly indecomposable *rtffr* groups we can prove a strong relationship between these *E-properties*.

Theorem 13.23. *Let G be a strongly indecomposable* rtffr *group. The following are equivalent.*

1. *G is quasi-isomorphic to an* rtffr *E-domain.*
2. *G is quasi-isomorphic to an E-cyclic group.*
3. *G is an E-finitely generated group.*
4. *G is an E-finitely presented group.*
5. *G is quasi-isomorphic to an E-projective group.*
6. *G is quasi-isomorphic to an E-generator group.*
7. *G is quasi-isomorphic to an E-progenerator group.*

Proof: We will prove $1 \Rightarrow 2 \Rightarrow 3 \Rightarrow 4 \Rightarrow \{5, 6, 7\} \Rightarrow 1$.

$1 \Rightarrow 2 \Rightarrow 3$ is clear.

$3 \Rightarrow 4$. By Theorem 13.8, $G \doteq R$ for some *rtffr E*-domain R, so that $\text{End}(G) \doteq R$ is a Noetherian integral domain. The E-finitely generated G is thus E-finitely presented.

$4 \Rightarrow 5, 6$, and 7. By Theorem 13.8, $G \doteq R$ for some *rtffr E*-domain R. This proves part 5, 6, and 7.

$5, 6$, or $7 \Rightarrow 1$. Suppose that G is an E- projective *rtffr* group or an E-generator *rtffr* group. By Corollaries 12.4 and 13.18, G is E-finitely generated, and since G is a strongly indecomposable group, Theorem 13.8 implies that $G \doteq R$ for some *rtffr* E-domain R. This proves part 1 and completes the logical cycle.

13.4 Exercises

Let G and H denote *rtffr* groups. Let $E(G) = \text{End}(G)/\mathcal{N}(\text{End}(G))$. Let E be an *rtffr* ring (=a ring for which $(E, +)$ is a *rtffr* ring. Let $S = \text{center}(E)$. Given a semi-prime *rtffr* ring E let $\overline{E}$ be an integrally closed subring of $\mathbf{Q}E$ such that $E \subset \overline{E}$ and $\overline{E}/E$ is finite. Let τ be the conductor of E.

1. Prove Theorem 13.1.
2. Write a theorem for E-finitely presented R-modules like Corollary 13.5.
3. Let $G \subset H$. If G/H is *bounded* (annihilated by some nonzero integer) then G/H is finite.
4. Let M and N be E-modules. If M/N is a torsion group and if M is finitely generated then M/N is bounded.
5. Fill in the details to Theorem 13.8.
6. Fill in the details to Corollary 13.9.
7. Fill in the details to Theorem 13.20.
8. Fill in the details to Theorem 13.22.

13.5 Problems for future research

1. Characterize the E-finitely presented *rtffr* groups G using quasi-direct summands in a manner similar to the characterization of finitely generated torsion-free modules given by J. Reid in Theorem 13.8.
2. Characterize the Noetherian modules over an *rtffr* ring E.

14

Rtffr E-projective groups

Consider the problem of classifiying the E-projective rtffr groups. The investigation is complicated by the fact that if N is any *rtffr* group then $\mathbf{Z} \oplus N$ is E-cyclic and E-projective. In this chapter we consider these *rtffr* abelian groups.

14.1 Introduction

An important example of an E-projective group is as follows.

Let S be a commutative ring, and let N be an S-module. We say that N is an S-*linear* module if

$$\operatorname{Hom}_S(S, N) = \operatorname{Hom}(S, N).$$

Our S-linear modules are referred to as E-*modules* by A. Mader and C. Vinsonhaler [86]. Any group is a $\mathbf{Z}$-linear module, a $\mathbf{Q}$-vector space is a $\mathbf{Q}$-linear module, and one can prove that if I is an ideal of finite index in an *rtffr* E-ring S then I is an S-linear module.

Lemma 14.1. *Let S be an* rtffr *commutative ring and let $I \doteq S$ be an ideal. If I is a projective S-module then I is a progenerator in S-**Mod** and $I \oplus I = S \oplus K$ for some S-module K.*

Proof: By hypothesis, S/I is a finite commutative ring so that

$$S/I = T_1 \times \cdots \times T_r$$

for some local rings $T_1, \ldots, T_r$. Since S is commutative, since $S \doteq I$, and since I is a projective S-module, I is a progenerator in S-**Mod**. So I/I^2 is a progenerator in S/I-**Mod**. Thus

$$I/I^2 = L_1 \oplus \cdots \oplus L_r$$

where L_i is a projective T_i-module. Since the projective S-module I generates S, L_i generates T_i, and since T_i is a local ring, there is a surjection $L_i \to T_i$. Therefore there is a surjection

$$I \to I/I^2 \to S/I$$

that lifts to a map $f : I \to S$ such that $f(I) + I = S$. It follows that

$$I \oplus I \cong S \oplus \ker(f \oplus 1_I).$$

This proves the lemma.

Lemma 14.2. *Let S be a (commutative) rtffr E-ring, let $I \doteq S$ be a projective ideal, and let N be an S-linear module. Then*

$$G = I \oplus N$$

is an E-projective group that is generated by two elements as a left $\mathrm{End}(G)$*-module.*

Proof: Since S is an *rtffr* E-ring, I is S-linear, and since N is S-linear

$$I = \mathrm{Hom}_S(S, I) = \mathrm{Hom}(S, I)$$

and

$$N = \mathrm{Hom}_S(S, N) = \mathrm{Hom}(S, N).$$

The reader can show that then G is S-linear. Hence $G \cong \mathrm{Hom}_S(S, G) \cong \mathrm{Hom}(S, G)$ as a left $\mathrm{End}(G)$-module.

The next series of equations shows us that G is a direct summand of $\mathrm{End}(G)^{(2)}$. Since $I \doteq S$ is a projective ideal, Lemma 14.1 states that $I \oplus I \cong S \oplus K$ for some S-module K. Then because G is S-linear,

$$\begin{aligned}
\mathrm{End}(G) \oplus \mathrm{End}(G) &\cong \mathrm{Hom}(I \oplus I, G) \oplus \mathrm{Hom}(N \oplus N, G) \\
&\cong \mathrm{Hom}(S \oplus K, G) \oplus \mathrm{Hom}(N \oplus N, G) \\
&\cong \mathrm{Hom}(S, G) \oplus \mathrm{Hom}(K \oplus N \oplus N, G) \\
&= \mathrm{Hom}_S(S, G) \oplus \mathrm{Hom}(K \oplus N \oplus N, G) \\
&\cong G \oplus \mathrm{Hom}(K \oplus N \oplus N, G)
\end{aligned}$$

as left $\mathrm{End}(G)$-modules. Thus G is a projective left $\mathrm{End}(G)$-module that is generated by at most two elements as a left $\mathrm{End}(G)$-module.

Example 14.3. Let S be an *rtffr* E-ring and let N be an *rtffr* S-linear module. By the above example $G = S \oplus N$ is an E-projective *rtffr* group. Furthermore, $G = \mathrm{End}(G)(1_S \oplus 0)$ is an E-cyclic *rtffr* group.

Our work here will show that the above examples are the *only* kind of E-projective *rtffr* group. Recall that $S = \text{center}(\text{End}(G))$.

Theorem 14.4. [89, G. P. Niedzwecki, J. D. Reid] *Let G be an* rtffr *group. Then G is an E-cyclic E-projective group iff S is an* rtffr *E-ring and* $G \cong S \oplus N$ *for some S-linear module N.*

Proof: Suppose that S is an E-ring and that $G = S \oplus N$ for some S-linear module N. By Example 14.3, G is an *rtffr* an E-cyclic and an E-projective group.

Conversely, assume that G is an *rtffr* E-cyclic E-projective group. There is a split surjection

$$\pi : \text{End}(G) \to G$$

of left E-modules so there is an $e^2 = e \in \text{End}(G)$ such that $G \cong \text{End}(G)e$. Thus,

$$S = \text{End}_{\text{End}(G)}(G) = \text{End}_{\text{End}(G)}(\text{End}(G)e) \cong e\text{End}(G)e.$$

If we let $N = (1 - e)G$ then

$$G = eG \oplus (1 - e)G \cong e(\text{End}(G)e) \oplus N \cong S \oplus N.$$

In this case, because $G \cong S \oplus N$, each group endomorphism $f : S \to S$ is in $\text{End}(G)$, so that f commutes with each element of S. Thus,

$$S \cong \text{Hom}_S(S, S) = \text{Hom}(S, S)$$

is an *rtffr* E-ring. Similarly,

$$\text{Hom}_S(S, N) = \text{Hom}(S, N),$$

so that N is an S-linear module. This completes the proof of the theorem.

The module U is called *quasi-projective* if U is projective relative to each short exact sequence

$$0 \longrightarrow K \longrightarrow U \longrightarrow V \longrightarrow 0$$

of left E-modules. The group G is *E-quasi-projective* if G is a quasi-projective left $\text{End}(G)$-module.

Theorem 14.5. [111, C. Vinsonhaler, W. Wickless] *Let R be an* rtffr *ring, and let M be an* rtffr *left E-module such that* $\text{ann}_R(M) = 0$. *Then M is a quasi-projective left R-module iff M is a finitely generated projective left R-module.*

Proof: Let M be a quasi-projective left R-module. Because M is an *rtffr* group there is an integer $m \neq 0$ such that $\cap_{k>0} m^k M = 0$. Since M/mM is a finite group there is a finitely generated R-submodule $F \subset M$ such that

$$(F + mM)/mM = M/mM.$$

The usual properties of relative projectivity (see [7, Proposition 16.12]) show us that the quasi-projective left R-module M is projective relative to the short exact sequence

$$0 \to C \longrightarrow F \xrightarrow{\pi} M/mM \to 0$$

of left R-modules. Thus the natural projection $M \to M/mM$ lifts to a map $f : M \to F$ such that

$$\ker f \subset \ker \pi f = mM. \tag{14.1}$$

Since M is torsion-free $\ker f$ is pure in M and so by (14.1) and our choice of m, $\ker f = m(\ker f) = 0$. Thus f is an injection, and hence

$$M \cong f(M) \subset F \subset M.$$

Since M is an *rtffr* group, $M/f(M)$ is finite, thus M/F is finite and therefore M is generated as a left A-module by the finitely many representatives of M/F and the finitely many generators of F.

To see that M is projective, observe that because M has finite rank and because $\mathrm{ann}_R(M) = 0$, there is an integer $n > 0$ and an embedding $R \to M^n$ of left R-modules. The usual properties of relative projectivity show us that M is projective relative to each surjection $R \to N \to 0$. But because M is finitely generated there is an integer and a surjection $R^n \to M \to 0$. Since M is projective relative to this surjection, M is a direct summand of R^n. That is, M is a projective left R-module.

The converse is clear so the proof is complete.

14.2 The UConn '81 Theorem

Corollary 12.4 leaves open the question of just how many generators an *rtffr* E-projective group requires. That issue is addressed in the famous paper [16] written at the University of Connecticut during its Special Year in Algebra 1981. The main theorem of [16] is our next result. Let us agree to call the module M a *local direct summand of N* if given an integer $n \neq 0$ there is an integer m that is relatively prime to n, and maps $f_n : N \to M$ and $g_n : M \to N$ such that $f_n g_n = m 1_M$.

Theorem 14.6. [16, UConn '81 Theorem] *Let G be an* rtffr *group. The following are equivalent.*

1. *G is an E-projective group and generated by two elements.*
2. *G is an E-projective group.*
3. *G is an E-quasi-projective group.*
4. *G is a local direct summand of* End(G).

5. *G is E-finitely generated and G_p is a cyclic projective left* $\mathrm{End}(G)_p$-*module for each prime* $p \in \mathbf{Z}$.
6. *There is an* tffr *E-ring* Σ, *an invertible ideal I in Σ, and a Σ-linear module N such that $G = I \oplus N$.*

In this case, $\Sigma = \mathrm{center}(\mathrm{End}(G))$ *and* Σ *is an* rtffr *E-ring.*

Proof: We will prove the logical circuit $1 \Rightarrow 2 \Rightarrow 3 \Rightarrow 6 \Rightarrow 5 \Rightarrow 4 \Rightarrow 1$.

$1 \Rightarrow 2 \Rightarrow 3$ is clear.

$3 \Rightarrow 6$. By Theorem 14.5, the quasi-projective right $\mathrm{End}(G)$-module is a finitely generated projective left $\mathrm{End}(G)$-module, and by Theorem 12.5, G is a generator over S. We will proceed by proving a pair of lemmas.

Lemma 14.7. *Let G be an* rtffr *E-finitely generated group. Then $S = \mathrm{center}(\mathrm{End}(G))$ is a Noetherian semi-prime* rtffr *ring.*

Proof: Since G is E-finitely generated, Theorem 13.8 states that there are *rtffr* E-domains $R_1, \ldots, R_t$ and integers $t, n_1, \ldots, n_t > 0$ such that

$$G \doteq R_1^{(n_1)} \oplus \cdots \oplus R_t^{(n_t)} \oplus N, \tag{14.2}$$

where N is generated by

$$A = R_1 \times \cdots \times R_t.$$

Since $S = \mathrm{center}(\mathrm{End}(G))$, and since A is a quasi-summand of G, there is a mapping $S \to A$. Suppose there is an $f \in S$ such that $fA = 0$. Because A generates a subgroup of finite index in G,

$$fG \doteq f\mathrm{Hom}(A, G)A = \mathrm{Hom}(A, G)fA = 0.$$

Thus $f = 0$, and hence $S \subset A$.

Since A is a semi-prime ring, S is a semi-prime ring, whence a Noetherian ring. This proves the lemma.

Lemma 14.8. *Under the hypotheses of Theorem 14.6.3, there is an S-module mapping $f : G \to S$ such that $f(G) \doteq S$ and $f(G)$ is a progenerator in* **Mod**-*S.*

Proof: By Theorem 14.5, the quasi-projective left $\mathrm{End}(G)$-module G is a finitely generated projective left $\mathrm{End}(G)$-module. Then by Proposition 13.16, G is a generator in **Mod**-S.

By Lemma 14.7, S is semi-prime, so there is a maximal order $S \subset \overline{S} \subset \mathbf{Q}S$ such that $\overline{S}/S$ is finite. There is an integer $m \neq 0$ large enough so that $m\overline{S} \subset S$ and S_m is reduced.

Let $M_1, \ldots, M_s$ be a complete list of the maximal ideals of S that contain mS. Since G is a generator in **Mod**-S there are S-module maps $f_i : G \to S$ such that $f_i(G) \not\subset M_i$. By the Chinese Remainder Theorem there is a map $f : G \to S$ such that

$$f(G) \not\subset M_i \quad \text{for any } i = 1, \ldots, s.$$

Then $f(G) + mS$ is an ideal in S that is not included in any of the maximal ideals that contain mS, hence $f(G) + mS = S$. Then by Lemma 2.10, $f(G)$ is a progenerator ideal in S. It follows that $f(G)$ is a progenerator in **Mod**-S, which proves the lemma.

We continue with the proof of Theorem 14.6. By Lemma 14.8 there is an S-module mapping $G \to S$ whose image $I \doteq S$ is a progenerator in S-**Mod**. Hence $G \cong I \oplus N$ for some S-module N. Since S is then a quasi-summand of G, $\text{Hom}(S, G) = \text{Hom}_S(S, G)$. Hence S is an *rtffr* E-ring, $I \doteq S$ is an invertible ideal in S, and I and N are S-linear modules. This proves part 6.

$6 \Rightarrow 5$. Let $p \in \mathbf{Z}$ be prime. Choose $n \in \mathbf{Z}$ large enough that $p|n$ and S_n is a reduced group. Since $I \doteq S$ is an invertible ideal in S, I_n is an invertible ideal in S_n. Then S_n/nS_n is generated by I_n/nI_n. Write the finite commutative ring

$$S_n/nS_n = L_1 \times \cdots \times L_t$$

as a product of local finite rings. Also write

$$I_n/mI_n = K_1 \oplus \cdots \oplus K_t,$$

where K_i is a projective generator in L_i-**Mod**. Since L_i is local there is a surjection $h_i : K_i \to L_i$, so that

$$h = \oplus_i h_i : I_n/mI_n \to S_n/mS_n$$

is a surjection. Since I_n is projective, h lifts to a map $f : I_n \to S_n$ such that $f(I_n) + nS_n = S_n$. By our choice of n and S, $n \in \mathcal{J}(S_n)$, and hence $f(I_n) = S_n$. Since I_n has finite rank, $I_n \cong S_n$. Hence $I_p \cong S_p$. Because S and I are finitely presented S-modules we can apply Lemma 6.18 to show that

$$\text{Hom}_S(I, G)_p \cong \text{Hom}_{S_p}(I_p, G_p) \cong \text{Hom}_{S_p}(S_p, G_p) \cong \text{Hom}_S(S, G)_p.$$

Thus $G_p = \text{Hom}_S(S, G)_p$ is a direct summand of

$$\text{Hom}_S(I \oplus N, G)_p \cong \text{End}_S(G)_p$$

as a left $\text{End}_S(G)_p$-module. By (12.2), $\text{End}_S(G) = \text{End}(G)$, so we have proved part 5.

$5 \Rightarrow 4$. Let $p \in \mathbf{Z}$ be a prime. By part 5 there are maps $f_p : G_p \to \text{End}(G)_p$ and $g_p : \text{End}(G)_p \to G_p$ such that $g_p f_p = 1_{G_p}$. Because G is finitely generated as a left $\text{End}(G)$-module, there is an integer m that is relatively prime to p such that $mf_p(G) \subset \text{End}(G)$ and $mg_p(\text{End}(G)) \subset G$. Thus,

$$(mg_p)(mf_p) = m^2 1_{G_p} = m^2 1_G.$$

This proves part 4.

$4 \Rightarrow 1$. The trick that makes this argument work is due to E. L. Lady. Suppose that G is a local summand of $\text{End}(G)$. We claim that G is a projective left $\text{End}(G)$-module generated by two elements. Let $n \neq 0 \in \mathbf{Z}$. There is an integer m and $\text{End}(G)$-module maps $f_n : \text{End}(G) \to G$ and $g_n : G \to \text{End}(G)$ such that $\gcd(m, n) = 1$ and $f_n g_n = m1_G$. By part 4, there is an integer k and maps $f_m : \text{End}(G) \to G$ and

$g_m : G \to \text{End}(G)$ such that $f_m g_m = k 1_G$. Since $\gcd(k, m) = 1$ there are integers a, b such that $ak + bm = 1$. Then the maps

$$f : \text{End}(G) \oplus \text{End}(G) \longrightarrow G : (x, y) \mapsto af_m(x) + bf_n(y)$$
$$g : G \longrightarrow \text{End}(G) \oplus \text{End}(G) : z \mapsto g_m(z) \oplus g_n(z)$$

satisfy

$$fg(z) = af_m g_m(z) + bf_n g_n(z) = (ak + bm)z = z.$$

Thus,

$$\text{End}(G) \oplus \text{End}(G) \cong G \oplus \ker f,$$

and hence G is a projective left $\text{End}(G)$-module that is generated by two elements. This proves part 1 and completes the logical cycle.

14.3 Exercises

Let G and H denote *rtffr* groups. Let $E(G) = \text{End}(G)/\mathcal{N}(\text{End}(G))$. Let E be an *rtffr* ring. Let $S = \text{center}(E)$. Given a semi-prime *rtffr* ring E let $\overline{E}$ be an integrally closed subring of $\mathbf{Q}E$ such that $E \subset \overline{E}$ and $\overline{E}/E$ is finite. Let τ be the conductor of E.

1. Let M be a finitely presented R-module, let $S = \text{center}(R)$, and let $\mathcal{C}$ be a multiplicatively closed subset of S containing 1. Show that given an R-module N then

$$\text{Hom}_R(M, N)[\mathcal{C}^{-1}] \cong \text{Hom}_{R[\mathcal{C}^{-1}]}(M[\mathcal{C}^{-1}], N[\mathcal{C}^{-1}])$$

naturally.

2. Let M be a left E-module. The *trace ideal of M* in E is the ideal $\sum \{f(M) \mid f \in \text{Hom}_E(M, E)\}$
   (a) Show that the trace of M is $M^* \otimes_E M$ where $M^* = \text{Hom}_E(M, E)$.
   (b) Determine the trace ideal of G if G is a projective left $\text{End}(G)$-module.
   (c) Prove that the trace Δ of a finitely generated projective left E-module M is an idempotent ideal such that $\Delta M = M$.

3. Let I be a fractional right ideal of E such that $EI = E$.
   (a) $\{q \in \mathbf{Q}E \mid qI \subset E\} = E$.
   (b) Prove that such an I is not invertible.
   (c) Give an example of such an ideal I in the commutative case (if possible) and in the noncommutative case (most possible).

14.4 Problems for future research

1. Characterize those *rtffr* groups G such that $\mathbf{Q}G$ is an injective left $\mathbf{Q}\text{End}(G)$-module.

2. Characterize the trace ideal of G in $\text{End}(G)$, especially when G is a projective left $\text{End}(G)$-module.

15

Injective endomorphism modules

The right R-module M is *injective* if for each injection $0 \to K \xrightarrow{f} L$, each mapping $g : K \to M$ lifts to one $h : L \to M$. That is, there is an h such that $hf = g$. It is a theorem of R. Baer's that M is injective iff to each injection $0 \to I \xrightarrow{f} R$, each mapping $g : I \to M$ lifts to one $h : I \to M$. We say that M is injective relative to each injection $0 \to K \xrightarrow{f} L$ and to each injection $0 \to I \xrightarrow{f} R$. The injective property is dual to the projective property.

If G is self-small then the Arnold–Lady theorem 1.13 explains the existence of projective right $\text{End}_R(G)$-modules in terms of direct summands of $G^{(c)}$ for cardinals c. It is natural to ask whether we can similarly write an injective right $\text{End}_R(G)$-module M in terms of $\text{Hom}_R(G, H)$ for some right R-module H. That is the aim of this chapter.

15.1 *G*-Monomorphisms

Recall notation from Chapter 9. Specifically, the complex

$$\mathcal{Q} = \cdots \xrightarrow{\delta_3} \mathcal{Q}_2 \xrightarrow{\delta_2} \mathcal{Q}_1 \xrightarrow{\delta_1} \mathcal{Q}_0,$$

where $\mathcal{Q}_k \in \mathbf{P}(G)$ denotes a G-plex, and

$$(\cdots, f_2, f_1, f_0) : \mathcal{Q} \longrightarrow \mathcal{Q}'$$

denotes a chain map between G-plexes. Also

$$K_k = \ker \delta_k \quad \text{and} \quad I_k = \text{image } \delta_k.$$

Recall that the zero-th homology functor $H_0(\cdot) : G\text{-}\mathbf{Plex} \longrightarrow G\text{-}\mathbf{Pres}$ is defined by $H_0(\mathcal{Q} \to 0)$, where $\mathcal{Q} \to 0$ is the complex formed by adding the zero map $\mathcal{Q}_0 \to 0$ onto $\mathcal{Q}$.

Furthermore, G-**Pres** denotes the category of right R-modules of the form Q/K where $Q \in \mathbf{P}(G)$ and K is a G-generated R-submodule of Q.

The additive functor $F : \mathbf{C} \longrightarrow \mathbf{D}$ is a *full (or faithful) functor* if the induced homomorphism

$$\mathrm{Hom}_{\mathbf{C}}(X, Y) \longrightarrow \mathrm{Hom}_{\mathbf{D}}(F(X), F(Y))$$

is a surjection (or an injection) of abelian groups for all objects X, Y in $\mathbf{C}$.

The lemma shows that while $H_0(\cdot) : G$-**Plex** $\longrightarrow$ G-**Pres** is not in general a full functor, many interesting maps in G-**Pres** do lift to chain maps in G-**Plex**.

Lemma 15.1. *Let G be a self-small right R-module, and let $Q \in \mathbf{P}(G)$.*

1. *If $\hat{j} : H \longrightarrow Q$ is a map in G-**Pres** then $\hat{j} = H_0([j])$ for some chain map j of G-plexes.*
2. *If $\hat{j} : H \longrightarrow Q$ is a map in G-**Pres** then $\hat{j} = \mathbf{T}_G(\phi)$ for some $\mathrm{End}_R(G)$-module homomorphism $\phi : K \longrightarrow P$ for some right $\mathrm{End}_R(G)$-module K such that $\mathbf{T}_G(K) \cong H$ and projective right $\mathrm{End}_R(G)$-module $\mathbf{T}_G(P) = Q$.*

Proof: 1. Since $H \in G$-**Pres** there is by Lemma 9.5 a $Q \in G$-**Plex** such that $H_0(Q) = \mathrm{coker}\,\delta_1 \cong H$ and such that $\delta_1 : Q_1 \longrightarrow I_1$ is G-balanced. The map $\hat{j} : H \longrightarrow Q$ lifts to the chain map $j = (\cdots, 0, \hat{j}\delta)$ as in the diagram.

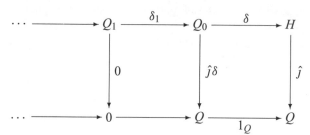

Hence j is a chain map such that

$$H_0([j]) = \hat{j}.$$

2. We are given the maps $\hat{j}$ and j as in part 1. By Theorem 1.13 there is a projective right $\mathrm{End}_R(G)$-module P such that $\mathbf{T}_G(P) = Q$. Because $H \in G$-**Pres**, Corollary 9.28 states that there is a $K \in \mathbf{Mod}$-$\mathrm{End}_R(G)$ such that $\mathbf{T}_G(K) = H$. By Theorem 9.12 there is a map $\phi : K \longrightarrow P$ in $\mathbf{Mod}$-$\mathrm{End}_R(G)$ such that $\mathrm{t}_G(\phi) = [j]$, so that

$$\mathbf{T}_G(\phi) = H_0 \mathrm{t}_G(\phi) = H_0([j]) = \hat{j}$$

by Theorem 9.25. Consequently, $\mathbf{T}_G(\phi) = \hat{j}$, $\mathbf{T}_G(K) = H$, and $\mathbf{T}_G(P) = Q$. This completes the proof.

Let G be a right R-module. The map $\hat{j} : H \longrightarrow H'$ in G-**Pres** is called a *G-monomorphism in G-**Pres*** if there is a monomorphism $[j] : Q \longrightarrow Q'$ in G-**Plex** such that $H_0([j]) = \hat{j}$.

Lemma 15.2. *Let G be a self-small right R-module. Then $\hat{\jmath}$ is a G-monomorphism in G-**Pres** iff $\hat{\jmath} = \mathbf{T}_G(\iota)$ for some injection ι in* **Mod**-$\text{End}_R(G)$.

Proof: If $\hat{\jmath}$ is a G-monomorphism in G-**Pres** then there is a monomorphism $[\jmath]$ in G-**Plex** such that $H_0([\jmath]) = \hat{\jmath}$. By Theorem 9.12 , there is an injection ι in **Mod**-$\text{End}_R(G)$ such that $t_G(\iota) = [\jmath]$. Then by the factorization of $\mathbf{T}_G(\cdot)$ in Theorem 9.25 we have

$$\mathbf{T}_G(\iota) = H_0 t_G(\iota) = H_0([\jmath]) = \hat{\jmath}.$$

The converse is proved in a similar manner.

Corollary 15.3. *Let G be a self-small right R-module. Each G-monomorphism is an injection iff G is E-flat.*

Thus a G-monomorphism in G-**Pres** is exactly the biproduct of the action of $\mathbf{T}_G$ on an injection. The next result continues along that line.

Lemma 15.4. *Let G be a self-small right R-module and let $Q \in \mathbf{P}(G)$. If $\hat{\jmath} : H \longrightarrow Q$ is a monomorphism in G-**Pres** then $\hat{\jmath}$ is a G-monomorphism .*

Proof: Assume that $\hat{\jmath}$ is a monomorphism in G-**Pres** with $Q \in \mathbf{P}(G)$. There is by Lemma 15.1 a chain map $\jmath : Q \longrightarrow 1_Q$ such that $h_0([\jmath]) = \hat{\jmath}$ and such that $\delta_1 : Q_1 \longrightarrow I_1$ in Q is G-balanced . We have a chain map of G-plexes

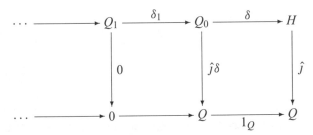

in which the top row is Q and the bottom row is 1_Q. The reader will show by stacking commutative diagrams and by using the G-balanced property that if $(\jmath\delta)z$ is null homotopic for some chain map $z : Q' \longrightarrow Q$ in G-**Plex** then z is null homotopic. Thus $[\jmath]$ is a monomorphism in G-**Plex** such that $h_0([\jmath]) = \hat{\jmath}$ and hence $\hat{\jmath}$ is a G-monomorphism.

15.2 Injective properties

To streamline our notation, let

$$\text{Hom}_{\text{End}_R(G)}(\cdot, \cdot) = \text{Hom}_E(\cdot, \cdot).$$

We will use a variety of injective properties to characterize those right R-modules G such that $\mathbf{H}_G(C)$ is an injective right $\text{End}_R(G)$-module.

The lemma is an immediate consequence of the adjoint isomorphism.

Lemma 15.5. *Let* G, C *be right* R*-modules. Let* $\phi : M \longrightarrow M'$ *be a map of right* $\mathrm{End}_R(G)$*-modules.*

1. There is a commutative diagram

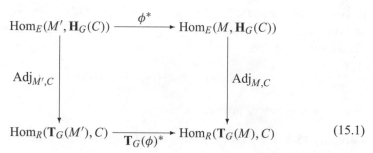

where $\mathrm{Adj}_{M',C}$ *and* $\mathrm{Adj}_{M,C}$ *are adjoint isomorphisms, and where* ϕ^* *and* $\mathbf{T}_G(\phi)^*$
are the canonical maps.

2. The bottom row of (15.1) is an injection iff the top row of (15.1) is an injection.

3. The bottom row of (15.1) is an surjection iff the top row of (15.1) is an surjection.

Ideally one wants to classify an injective module of the form $\mathbf{H}_G(C)$ in terms of an injective property for C. We feel that the following result meets this ideal. Notice that the second condition below resembles the Baer Criterion for Injectivity in modules.

Theorem 15.6. *Let* G, C *be right* R*-modules and assume that* G *is self-small. The following are equivalent.*

1. $\mathbf{H}_G(C)$ *is an injective right* $\mathrm{End}_R(G)$*-module.*

2. C *is injective relative to each* G*-monomorphism* $\hat{\jmath} : H \longrightarrow G$ *in* G*-***Pres***.*

3. C *is injective relative to each* G*-monomorphism* $\hat{\jmath} : H \longrightarrow H'$ *in* G*-***Pres***.*

Proof: $1 \Rightarrow 3$. Suppose that $\mathbf{H}_G(C)$ is an injective right $\mathrm{End}_R(G)$-module, and let $\hat{\jmath} : H \longrightarrow H'$ be a G-monomorphism in G-**Pres**. By Lemma 15.2 there is an injection $\iota : M \longrightarrow M'$ of right $\mathrm{End}_R(G)$-modules such that $\mathbf{T}_G(\iota) = \hat{\jmath}$. Since $\mathbf{H}_G(C)$ is injective,

$$\mathrm{Hom}_E(\iota, \mathbf{H}_G(C)) = \iota^*$$

is a surjection, so by Lemma 15.5 the bottom row of (15.1) is a surjection. That is,

$$\hat{\jmath}^* = \mathbf{T}_G(\iota)^* = \mathrm{Hom}_R(\mathbf{T}_G(\iota), C)$$

is a surjection, which proves that C is injective relative to each G-monomorphism in G-**Pres**.

$3 \Rightarrow 2$ is clear.

$2 \Rightarrow 1$. Suppose that C is injective relative to each G-monomorphism $\hat{\jmath} : H \longrightarrow G$ in G-**Pres**, and let $\iota : I \longrightarrow \mathrm{End}_R(G)$ be an injection of right $\mathrm{End}_R(G)$-modules. By Lemma 15.2, $\mathbf{T}_G(\iota) : \mathbf{T}_G(M) \longrightarrow G$ is a G-monomorphism in G-**Pres**, so that

$$\mathrm{Hom}_R(\mathbf{T}_G(\iota), C) = \mathbf{T}_G(\iota)^*$$

is a surjection. By Lemma 15.5, the top row ι^* of (15.1) is also a surjection, whence Baer's Criterion for Injectivity shows us that $\mathbf{H}_G(C)$ is an injective right $\text{End}_R(G)$-module. This completes the proof.

Theorem 15.7. *The following are equivalent for the self-small right R-module G.*

1. *$\text{End}_R(G)$ is a right self-injective ring.*
2. *G is injective relative to each G-monomorphism $\hat{j} : H \longrightarrow G$ in G-**Pres**.*
3. *G is injective relative to each G-monomorphism $\hat{j} : H \longrightarrow H'$ in G-**Pres**.*

Let us take the above theorems and see how various hypotheses effect the outcome.

Corollary 15.8. *Let G and C be right R-modules and let G be self-small and E-flat. The following are equivalent.*

1. *$\mathbf{H}_G(C)$ is an injective right $\text{End}_R(G)$-module.*
2. *C is injective relative to each injection $\hat{j} : H \longrightarrow G$ in which $H \in$ G-**Pres**.*

Proof. $2 \Rightarrow 1$. Let $\hat{j} : H \longrightarrow G$ be a G-monomorphism in G-**Pres**. By Lemma 15.2, there is an injection $\iota : I \longrightarrow \text{End}_R(G)$ such that $\mathbf{T}_G(\iota) = \hat{j}$. Since G is E-flat, $\mathbf{T}_G(\iota) = \hat{j}$ is an injection. Hence by part 2, C is injective relative to $\hat{j}$, so that by Theorem 15.6, $\mathbf{H}_G(C)$ is an injective right $\text{End}_R(G)$-module.

$1 \Rightarrow 2$. Suppose that $\mathbf{H}_G(C)$ is an injective right $\text{End}_R(G)$-module, and let $\hat{j} : H \longrightarrow G$ be an injection in which $H \in$ G-**Pres**. By Lemma 15.4, $\hat{j}$ is a G-monomorphism, so that by Theorem 15.6, C is injective relative to $\hat{j}$. This completes the proof.

Corollary 15.9. *The following are equivalent for the self-small E-flat right R-module G.*

1. *$\text{End}_R(G)$ is a right self-injective ring.*
2. *G is injective relative to each injection $\hat{j} : H \longrightarrow G$ in which $H \in$ G-**Pres**.*

Recall that G is Σ-quasi-projective if for each cardinal c, G is projective relative to each surjection $\delta : G^{(c)} \longrightarrow H$.

Theorem 15.10. *Let G and C be right R-modules and suppose that G is self-small Σ-quasi-projective. The following are equivalent.*

1. *$\mathbf{H}_G(C)$ is an injective right $\text{End}_R(G)$-module.*
2. *C is injective each injection $\hat{j} : H \longrightarrow G$ in which $H \in$ G-**Pres**.*
3. *C is injective relative to each injection $\hat{j} : H \longrightarrow H'$ in which $H \in$ G-**Pres**.*

Proof. $1 \Rightarrow 3$. Assume part 1. Let $\hat{j} : H \longrightarrow H'$ be a monomorphism in G-**Pres**. Because G is Σ-quasi-projective, Theorem 10.2 states that $H_0 : G$-**Plex** $\longrightarrow$ G-**Pres** is a category equivalence, so that there is a monomorphism $[j] \in$ G-**Plex** such that $H_0([j]) = \hat{j}$. Hence $\hat{j}$ is a G-monomorphism. By Theorem 15.6, C is injective relative to $\hat{j}$, which proves part 3.

$3 \Rightarrow 2$ is clear.

$2 \Rightarrow 1$. If $\hat{j} : H \longrightarrow G$ is an injection in which $H \in G$-**Pres** then $\hat{j}$ is a G-monomorphism in G-**Pres**. By part 2, C is injective relative to $\hat{j}$, so that by Theorem 15.6, $\mathbf{H}_G(C)$ is injective as a right $\text{End}_R(G)$-module. Given our reductions, the proof is complete.

Corollary 15.11. *Let G be a self-small Σ-quasi-projective right R-module. The following are equivalent.*

1. $\text{End}_R(G)$ *is a right self-injective ring.*
2. G *is injective relative to each injection $\hat{j} : H \longrightarrow G$ in G-**Pres**.*

The following results show that, as traditional injective modules, the above injective properties are associated with a splitting result. Recall the *natural transformations* Θ and Ψ and their identities

$$1_{\mathbf{H}_G(C)} = \mathbf{H}_G(\Theta_C) \circ \Psi_{\mathbf{H}_G(C)} \qquad (15.2)$$

$$1_{\mathbf{T}_G(M)} = \Theta_{\mathbf{T}_G(M)} \circ \mathbf{T}_G(\Psi_M)$$

from Chapter 1. Thus, e.g., the map $\Psi_{HG(C)}$ is a split injection for each $C \in \mathbf{Mod}\text{-}R$.

Theorem 15.12. *Let G be a self-small right R-module and let $C \in G$-**Pres**. The following are equivalent.*

1. $\mathbf{H}_G(C)$ *is an injective right $\text{End}_R(G)$-module.*
2. *Each G-monomorphism $\hat{j} : C \longrightarrow H$ in G-**Pres** is a split injection.*

Proof: Assume part 1. If $\mathbf{H}_G(C)$ is an injective right $\text{End}_R(G)$-module then by Theorem 15.6.3, C is injective relative to the G-monomorphism $\hat{j} : C \longrightarrow H$ in G-**Pres**. Thus there is an $f : H \longrightarrow C$ such that $f\hat{j} = 1_C$, and hence $\hat{j}$ is a split injection.

Conversely, assume part 2. Suppose that each G-monomorphism $\hat{j} : C \longrightarrow H$ in G-**Pres** is a split injection. Let $\iota : \mathbf{H}_G(C) \longrightarrow M$ be an injection. By Lemma 15.2, $\mathbf{T}_G(\iota) : \mathbf{T}_G\mathbf{H}_G(C) \longrightarrow \mathbf{T}_G(M)$ is a G-monomorphism in G-**Pres**, so by our hypothesis $\mathbf{T}_G(\iota)$ is a split injection. Hence $\mathbf{T}_G(\iota)^*$ is a split surjection.

There is a commutative square

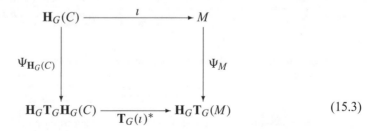

$$(15.3)$$

of right $\text{End}_R(G)$-modules. By (15.2), $\Psi_{\mathbf{H}_G(C)}$ and $\mathbf{T}_G(\iota)^*$ are split injections, so that ι is a split injection. Hence $\mathbf{H}_G(C)$ is injective, which completes the proof.

Recall the category **Coh**-E of *coherent* right E-modules. The right E-module N is said to be **Coh**-*injective* if N is injective relative to each injection $\iota : M \longrightarrow M'$

in **Coh**-*E*. We will characterize the right *R*-modules *C* for which $\mathbf{H}_G(C)$ is a **Coh**-injective right $\text{End}_R(G)$-module.

Recall that *G*-**CohPlx** is the category of $\mathcal{Q} \in G$-**Plex** such that $\mathcal{Q}_k \in \mathbf{P}_o(G)$ for each $k \geq 0$, and that *G*-**Coh** is the category of *G*-*coherent modules*. That is, $H \in G$-**Coh** iff there is a $\mathcal{Q} \in G$-**CohPlx** such that $H \cong H_0(\mathcal{Q})$. A *G*-**Coh**-*monomorphism* is a map $\hat{j} : H \longrightarrow H'$ in *G*-**Coh** for which there is a monomorphism $[J] : \mathcal{Q} \longrightarrow \mathcal{Q}'$ in *G*-**CohPlx** such that $H_0([J]) = \hat{j}$.

Theorem 15.13. *Let G, C be right R-modules. The following are equivalent.*

1. $\mathbf{H}_G(C)$ *is a* **Coh**-*injective right* $\text{End}_R(G)$-*module.*
2. *C is injective relative to each G-***Coh**-*monomorphism* $\hat{j} : H \longrightarrow H'$.

Proof: $1 \Rightarrow 2$. Suppose that $\mathbf{H}_G(C)$ is a **Coh**-injective right $\text{End}_R(G)$-module, let $\hat{j} : H \longrightarrow H'$ be a *G*-**Coh**-monomorphism. There is a monomorphism $[J] : \mathcal{Q} \longrightarrow \mathcal{Q}'$ in *G*-**CohPlx** such that

$$H_0([J]) = \hat{j}.$$

By Theorem 9.13, **Coh**-$\text{End}_R(G)$ and *G*-**CohPlx** are equivalent via $t_G(\cdot)$, so there is a monomorphism (= injection) $\iota : M \longrightarrow M'$ in **Coh**-$\text{End}_R(G)$ such that $t_G(\iota) = [J]$. By Theorem 9.25,

$$\mathbf{T}_G(\iota) = H_0(t_G(\iota)) = H_0([J]) = \hat{j}.$$

Examine (15.1) with the identification $\phi = \iota$. Since $\mathbf{H}_G(C)$ is **Coh**-injective the top row of (15.1), ι^*, is a surjection, so that the bottom row of (15.1), $\mathbf{T}_G(\iota)^* = \hat{j}^*$, is a surjection. Thus *C* is injective relative to $\hat{j}$, and hence *C* is *G*-**Coh**-injective.

$2 \Rightarrow 1$. Suppose that *C* is injective relative to each *G*-**Coh**-monomorphism, and let ι be an injection in **Coh**-$\text{End}_R(G)$. By Theorem 9.13, $t_G(\cdot) :$ **Coh**-$\text{End}_R(G) \longrightarrow$ *G*-**CohPlx** is a category equivalence, so $t_G(\iota) = [J]$ is a monomorphism in *G*-**CohPlx**. By Theorem 9.25, $H_0 t_G(\cdot) = \mathbf{T}_G(\cdot)$, so that

$$\mathbf{T}_G(\iota) = H_0(t_G(\iota)) = H_0([J]) = \hat{j}$$

is a *G*-**Coh**-monomorphism. By part 2, $\mathbf{T}_G(\iota)^*$ is a surjection, so by (15.1), ι^* is a surjection. Therefore $\mathbf{H}_G(C)$ is **Coh**-injective, which proves part 1, and completes the proof.

Corollary 15.14. *The following are equivalent for the right R-module G.*

1. $\text{End}_R(G)$ *is a* **Coh**-*injective right* $\text{End}_R(G)$-*module.*
2. *G is injective relative to each G-***Coh**-*monomorphism* $\hat{j} : H \longrightarrow H'$.

The following shows that like the usual notions of injectivity, **Coh**-injective properties enjoy a splitting property.

Theorem 15.15. *Let G be a right R-module. The following are equivalent.*

1. *$\text{End}_R(G)$ is a **Coh**-injective right $\text{End}_R(G)$-module.*
2. *Each injection $\iota : \text{End}_R(G) \longrightarrow M$ in which M is finitely presented, is a split injection.*
3. *Each G-**Coh**-monomorphism $\hat{\jmath} : G \longrightarrow H$ is a split injection.*

Proof: $1 \Rightarrow 2$ follows in the classic manner.

$2 \Rightarrow 3$ is left as an exercise.

$3 \Rightarrow 1$. Assume that each G-**Coh**-monomorphism $\hat{\jmath} : G \longrightarrow H$ is a split injection. Let $\iota : \mathbf{H}_G(G) \longrightarrow M$ be an injection in which M is a coherent right $\text{End}_R(G)$-module. Inasmuch as $\mathbf{T}_G(M)$ is a G-coherent right R-module,

$$\mathbf{T}_G(\iota) : G = \mathbf{T}_G\mathbf{H}_G(G) \longrightarrow \mathbf{T}_G(M)$$

is a G-**Coh**-monomorphism in G-**Pres**, so that by part 2, $\mathbf{T}_G(\iota)$ is a split injection. Then $\mathbf{T}_G(\iota)^*$ is a split surjection. There is a commutative square

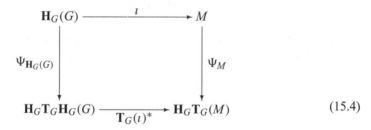

$$(15.4)$$

Since $\Psi_{\mathbf{H}_G(G)}$ and $\mathbf{T}_G(\iota)^*$ are split injections, (page 254), the reader can easily verify that ι is a split injection. That is, $\mathbf{H}_G(G) = \text{End}_R(G)$ is **Coh**-injective, which completes the proof.

Recall that $\text{End}_R(G)$ is *right coherent* iff each finitely presented right $\text{End}_R(G)$-module is coherent. The ring E is *right self-FP-injective* if E is injective relative to each injection $M \longrightarrow N$ of finitely presented right E-modules. It is then immediate that the right coherent ring E is right **Coh**-injective iff E is a right self-*FP*-injective ring. Recall that G-**FP** is the category of right R-modules of the form Q/K for some $Q \in \mathbf{P}_o(G)$ and finitely G-generated R-submodule $K \subset Q$.

Theorem 15.16. *Let G, C be right R-modules and suppose that $\text{End}_R(G)$ is right coherent. The following are equivalent.*

1. *$\mathbf{H}_G(C)$ is an FP-injective right $\text{End}_R(G)$-module.*
2. *C is injective relative to each G-**Coh**-monomorphism $\hat{\jmath} : H \longrightarrow H'$.*

Proof: Apply Theorem 15.15.

Corollary 15.17. *Let G be a right R-module such that $\text{End}_R(G)$ is right coherent. The following are equivalent.*

1. *$\text{End}_R(G)$ is right self FP-injective.*
2. *G is injective relative to each G-**Coh**-monomorphism $\jmath : H \longrightarrow H'$.*

The next results characterize the right R-modules G with right Noetherian right self-injective (= QF) endomorphism ring. We have characterized those self-small right R-modules G with left or right Noetherian endomorphism ring $\text{End}_R(G)$ in Theorems 9.15 and 9.38.

Theorem 15.18. *Let G, C be right R-modules and suppose that $\text{End}_R(G)$ is right Noetherian. The following are equivalent.*

1. $\mathbf{H}_G(C)$ *is an injective right $\text{End}_R(G)$-module.*
2. C *is injective relative to each G-**Coh**-monomorphism $\hat{\jmath} : H \longrightarrow H'$.*
3. C *is injective relative to each G-**Coh**-monomorphism $\hat{\jmath} : H \longrightarrow G$.*

Proof: $1 \Rightarrow 2$. Assume that $\mathbf{H}_G(C)$ is an injective right $\text{End}_R(G)$-module. Then $\mathbf{H}_G(C)$ is *FP*-injective, so by Theorem 15.16, C is injective relative to each G-**Coh**-monomorphism $\hat{\jmath} : H \longrightarrow H'$ in G-**Coh**. This is part 2.

$2 \Rightarrow 3$ is clear.

$3 \Rightarrow 1$. By Theorem 15.16, $\mathbf{H}_G(C)$ is *FP*-injective, and since $\text{End}_R(G)$ is right Noetherian, $\mathbf{H}_G(C)$ is then injective. This completes the logical cycle.

Theorem 15.19. *Let G be a right R-module such that $\text{End}_R(G)$ is right Noetherian. The following are equivalent.*

1. $\text{End}_R(G)$ *is a QF ring.*
2. G *is injective relative to each G-**Coh**-monomorphism $\jmath : H \longrightarrow H'$.*
3. *Each G-**Coh**-monomorphism $\jmath : G \longrightarrow H$ is a split injection.*

If G is E-flat then each G-**Coh**-monomorphism is an injection, so that we have the following result.

Theorem 15.20. *Let G be a self-small E-flat right R-module. The following are equivalent for the right R-module C.*

1. $\mathbf{H}_G(C)$ *is a **Coh**-injective right $\text{End}_R(G)$-module.*
2. C *is injective relative to each injection $\hat{\jmath} : H \longrightarrow H'$ in which $H \in G$-**Coh** and for which $\hat{\jmath}$ lifts to a monomorphism in G-**CohPlx**.*

Proof: Because G is self-small and E-flat, Corollary 15.3 states that $\hat{\jmath}$ is a G-monomorphism in G-**Plex** iff it is an injection in **Mod**-R. Hence part 2 is equivalent to $2'$: C is injective relative to each G-**Coh**-monomorphism $\hat{\jmath} : H \longrightarrow H'$ in G-**Coh**. Now apply Theorem 15.13 to complete the proof.

Corollary 15.21. *The following are equivalent for the self-small E-flat right R-module G.*

1. $\text{End}_R(G)$ *is a **Coh**-injective right $\text{End}_R(G)$-module.*
2. G *is injective relative to each injection $\hat{\jmath} : H \longrightarrow H'$ in which $H, H' \in G$-**Coh** and for which $\hat{\jmath}$ lifts to a monomorphism in G-**CohPlx**.*

The reader will show that if G is a Σ-quasi-projective right R-module then each G-monomorphism $H \longrightarrow H'$ in G-**Pres** is a monomorphism in G-**Pres**, and conversely. Thus we have the following results.

Theorem 15.22. *Let G be a Σ-quasi-projective right R-module. The following are equivalent for the right R-module C.*

1. $\mathbf{H}_G(C)$ *is a* **Coh**-*injective right* $\mathrm{End}_R(G)$-*module.*
2. *C is injective relative to each monomorphism $\hat{\jmath} : H \longrightarrow H'$ in G-**Coh**.*

Proof: Apply Theorem 15.13.

The ring E is *right* **Coh**-*self-injective* if E is a **Coh**-injective right E-module.

Corollary 15.23. *The following are equivalent for the Σ-quasi-projective right R-module G.*

1. $\mathrm{End}_R(G)$ *is a right* **Coh**-*self-injective ring.*
2. *G is injective relative to each monomorphism $\hat{\jmath} : H \longrightarrow H'$ in G-**Coh**.*
3. *G is injective in G-**Coh**.*

For a self-small Σ-quasi-projective there are category equivalences

$$\mathbf{Mod}\text{-}\mathrm{End}_R(G) \cong G\text{-}\mathbf{Plex} \cong G\text{-}\mathbf{Pres}$$

by Theorems 9.11 and 10.2. Moreover, $\mathrm{End}_R(G)$ corresponds to G under these equivalences. Thus we have the following theorem.

Theorem 15.24. *The following are equivalent for the self-small Σ-quasi-projective right R-module G.*

1. $\mathrm{End}_R(G)$ *is a right self-injective ring.*
2. *G is injective relative to each monomorphism $\hat{\jmath} : H \longrightarrow H'$ in G-**Pres**.*
3. *Each monomorphism $G \longrightarrow H$ in G-**Pres** is a split monomorphism .*
4. *G is injective in G-**Pres**.*
5. *G is a small projective, injective, generator in G-**Pres**.*

15.3 *G*-Cogenerators

Let E be a ring. We will say that the right E-module M is a *cogenerator* if $\mathrm{Hom}_E(N, M) \neq 0$ for each right E-module N. It is easily seen (see the exercises) that M is a cogenerator iff each right E-module embeds in a product of copies of M. For example, $\mathbf{Z}/p^2\mathbf{Z}$ is a cogenerator over $\mathbf{Z}/p^2\mathbf{Z}$, and $\mathbf{Z}(p^\infty)$ is a cogenerator over $\mathbf{Z}$. The ring E is a *right self-cogenerator* if E is a cogenerator as a right E-module. QF rings are right self-cogenerators. Our immediate goal is to characterize those right R-modules C for which $\mathbf{H}_G(C)$ is a cogenerator over $\mathrm{End}_R(G)$.

The right R-module C is a *G-cogenerator* if $\mathrm{Hom}_R(H, C) \neq 0$ for each $H \in G$-**Pres**.

Theorem 15.25. *Let G be a self-small right R-module. The following are equivalent for a right R-module C.*

1. $\mathbf{H}_G(C)$ *is a cogenerator over* $\mathrm{End}_R(G)$.
2. *(a)* $\mathbf{T}_G(M) \neq 0$ *for each right* $\mathrm{End}_R(G)$-*module* $M \neq 0$.
   (b) C is a G-cogenerator right R-module.

Proof: $1 \Rightarrow 2$. Suppose that $\mathbf{H}_G(C)$ is a cogenerator over $\mathrm{End}_R(G)$ and let $H \neq 0 \in G\text{-}\mathbf{Pres}$. By Corollary 9.28 there is a right $\mathrm{End}_R(G)$-module $M \neq 0$ such that $\mathbf{T}_G(M) \cong H$ and by the adjoint isomorphism,

$$\mathrm{Hom}_R(H, C) \cong \mathrm{Hom}_R(\mathbf{T}_G(M), C) \cong \mathrm{Hom}_E(M, \mathbf{H}_G(C)) \neq 0.$$

Then C is a G-cogenerator. If $N \neq 0$ is a right $\mathrm{End}_R(G)$-module then $\mathbf{T}_G(N)$ is a G-presented module, so that

$$\mathrm{Hom}_R(\mathbf{T}_G(N), C) \cong \mathrm{Hom}_E(N, \mathbf{H}_G(C)) \neq 0$$

since $\mathbf{H}_G(C)$ is a cogenerator. Thus $\mathbf{T}_G(N) \neq 0$ for each $N \in \mathbf{Mod}\text{-}\mathrm{End}_R(G)$.

$2 \Rightarrow 1$. Assume part 2 and let $M \neq 0$ be a right $\mathrm{End}_R(G)$-module. By condition 2(a) and Corollary 9.28, $\mathbf{T}_G(M) = H \neq 0 \in G\text{-}\mathbf{Pres}$, so that by condition 2(b),

$$0 \neq \mathrm{Hom}_R(H, C) = \mathrm{Hom}_R(\mathbf{T}_G(M), C) \cong \mathrm{Hom}_E(M, \mathbf{H}_G(C)).$$

Thus $\mathbf{H}_G(C)$ is a cogenerator over $\mathrm{End}_R(G)$ and the proof is complete.

It is clear that $C \in G\text{-}\mathbf{Pres}$ is a G-cogenerator iff C is a cogenerator in $G\text{-}\mathbf{Pres}$. Then using Theorem 15.26, we characterize the right R-modules whose endomorphism rings are cogenerator rings.

Corollary 15.26. *Let G be a self-small right R-module. The following are equivalent for $C \in G\text{-}\mathbf{Pres}$.*

1. *$\mathbf{H}_G(C)$ is a cogenerator over $\mathrm{End}_R(G)$.*
2. *(a) $\mathbf{T}_G(M) \neq 0$ for each right $\mathrm{End}_R(G)$-module $M \neq 0$.*
   (b) C is a cogenerator in $G\text{-}\mathbf{Pres}$.

Theorem 15.27. *The following are equivalent for a self-small right R-module G.*

1. *$\mathrm{End}_R(G)$ is a cogenerator as a right $\mathrm{End}_R(G)$-module.*
2. *(a) $\mathbf{T}_G(M) \neq 0$ for each nonzero right $\mathrm{End}_R(G)$-module M.*
   (b) G is a cogenerator in $G\text{-}\mathbf{Pres}$.

Seen from another point of view, we characterize those G for which each $\mathrm{End}_R(G)$-module is of the form $\mathbf{H}_G(C)$.

Corollary 15.28. *If G is self-small and if the right $\mathrm{End}_R(G)$-module $\mathbf{H}_G(C)$ is a cogenerator for some right R-module C then G is a fully faithful right R-module. That is, $\mathbf{T}_G(M) \neq 0$ for nonzero $M \in \mathbf{Mod}\text{-}\mathrm{End}_R(G)$.*

Corollary 15.29. *Let G be a self-small right R-module, and let M be a cogenerator in $\mathbf{Mod}\text{-}\mathrm{End}_R(G)$. If G is not faithful (i.e., if $IG = G$ for some proper right ideal $I \subset \mathrm{End}_R(G)$) then $M \neq \mathbf{H}_G(C)$ for any right R-module C.*

Corollary 15.30. *Let G be a self-small right R-module such that $\mathrm{End}_R(G)$ is a right self-cogenerator ring. Then $\mathbf{T}_G(M) \neq 0$ for each nonzero right $\mathrm{End}_R(G)$-module M.*

Corollary 15.31. *If E is a right self-cogenerator ring then given a self-small right R-module G such that $E \cong \mathrm{End}_R(G)$ we have $\mathbf{T}_G(M) \neq 0$ for each right E-module M.*

Corollary 15.32. *Let G be a self-small right R-module and suppose that $\mathrm{End}_R(G)$ is a right self-cogenerator ring. If c is a cardinal then each short exact sequence*

$$0 \longrightarrow K \longrightarrow G^{(c)} \longrightarrow G \longrightarrow 0$$

is a split exact.

Proof: Apply Corollary 15.30 and Baer's lemma 11.35.

By combining theorems we can characterize the injective cogenerators over $\mathrm{End}_R(G)$.

Theorem 15.33. *Let G be a self-small right R-module. The following are equivalent for an R-module $C \in G$-$\mathbf{Pres}$.*

1. *$\mathbf{H}_G(C)$ is an injective cogenerator in $\mathbf{Mod}$-$\mathrm{End}_R(G)$.*
2. *(a) $\mathbf{T}_G(M) \neq 0$ for each right $\mathrm{End}_R(G)$-module $M \neq 0$.*
   (b) C is injective relative to each G-monomorphism $H \longrightarrow G$ in G-$\mathbf{Pres}$, and
   (c) C is a G-cogenerator.

Proof: $1 \Leftrightarrow 2$ follows from Theorems 15.6 and 15.26. This completes the proof.

Assume that G is E-flat. Then a map $\hat{\jmath} : H \longrightarrow G$ is a G-monomorphism iff it is an injection. (See Corollary 15.3.) This proves the following result.

Theorem 15.34. *Let G be a self-small E-flat right R-module. The following are equivalent for an R-module $C \in G$-$\mathbf{Pres}$.*

1. *$\mathbf{H}_G(C)$ is an injective cogenerator in $\mathbf{Mod}$-$\mathrm{End}_R(G)$.*
2. *(a) $\mathbf{T}_G(M) \neq 0$ for each right $\mathrm{End}_R(G)$-module $M \neq 0$.*
   (b) C is injective relative to each injection $H \longrightarrow G$ in G-$\mathbf{Pres}$, and
   (c) C is a G-cogenerator.
3. *(a) $\mathbf{T}_G(M) \neq 0$ for each right $\mathrm{End}_R(G)$-module $M \neq 0$.*
   (b) C is an injective cogenerator object in G-$\mathbf{Pres}$.

Recall that E is a right *PF* ring if E is a right self-injective right cogenerator ring. By [113, 48.15(d)], E is a *QF* ring if E is a left self-injective and right cogenerator ring.

Theorem 15.35. *Let G be a self-small right R-module. The following are equivalent.*

1. *$\mathrm{End}_R(G)$ is a right PF-ring.*
2. *(a) $\mathbf{T}_G(M) \neq 0$ for each nonzero right $\mathrm{End}_R(G)$-module M.*
   (b) G is injective relative to each G-monomorphism $H \longrightarrow G$ in G-$\mathbf{Pres}$, and
   (c) G is a G-cogenerator.

3. (a) $\mathbf{T}_G(M) \neq 0$ for each nonzero right $\text{End}_R(G)$-module M.

   (b) G is an injective cogenerator object in G-**Pres**.

Proof: Apply Theorem 15.33.

Theorem 15.36. *Let G be a self-small right R-module. The following are equivalent.*

1. $\text{End}_R(G)$ *is a* QF-*ring.*
2. (a) G *is injective relative to each G-monomorphism $H \longrightarrow G$ in G-**Pres**, and*

   (b) *Each finitely G-generated right R-module is G-coherent.*

Proof: $1 \Rightarrow 2$. By part 1, $\text{End}_R(G)$ is right self-injective, so by Theorem 15.6, G is injective relative to each G-monomorphism $H \longrightarrow G$ in G-**Pres**. Also QF-rings are right Noetherian, so by Theorem 9.15, each finitely G-generated module is G-coherent. This proves part 2.

$2 \Rightarrow 1$. Since G is self-small , part 2(b) and Theorem 9.15 show that $\text{End}_R(G)$ is right Noetherian. Part 2(a) and Theorem 15.6 show us that $\text{End}_R(G)$ is a right self-injective ring. This proves part 1 and completes the proof.

15.4 Projective modules revisited

Since we know the conditions under which $\mathbf{H}_G(C)$ is a cogenerator or an injective right $\text{End}_R(G)$-module it is reasonable to expect that we can characterize those C for which $\mathbf{H}_G(C)$ is a generator or a projective right $\text{End}_R(G)$-module. Recall that a surjection ϕ is *G-balanced* if G is projective relative to ϕ. We say that H *admits a G-cover* if for each G-balanced surjection $\delta : Q \longrightarrow H$ in which $Q \in \mathbf{P}(G)$ then $Q = \mathbf{S}_G(\ker \delta) \oplus Q'$ for some $Q' \in \mathbf{P}(G)$. Of course, if the projective R-module H admits a G-cover then it is G-generated. Observe that given a prime $p \in \mathbf{Z}$ then $G = H = \mathbf{Z}(p^\infty)$ admits a $\mathbf{Z}(p^\infty)$-cover since $\mathbf{S}_{\mathbf{Z}(p^\infty)}(Q)$ is divisible for each group Q. Divisible subgroups of Q are direct summands of Q.

Theorem 15.37. *Let G be a self-small right R-module and let $C \in G$-**Pres**. The following are equivalent.*

1. $\mathbf{H}_G(C)$ *is a projective right $\text{End}_R(G)$-module.*
2. C *admits a G-cover.*

Proof: $2 \Rightarrow 1$. Let $\delta : Q \longrightarrow C$ be a G-balanced surjection in which $Q \in \mathbf{P}(G)$, and let $\delta^* = \mathbf{H}_G(\delta)$. Then $\delta^* : \mathbf{H}_G(Q) \longrightarrow \mathbf{H}_G(C)$ is a surjection with kernel $\ker \delta^* = \mathbf{H}_G(\ker \delta)$. By part 2 we can assume without loss of generality that $\mathbf{S}_G(\ker \delta) = 0$, so that $\ker \delta^* = 0$ and δ^* is an injection. Then $\delta^* : \mathbf{H}_G(Q) \longrightarrow \mathbf{H}_G(C)$ is an isomorphism. Theorem 1.13 thus implies that $\mathbf{H}_G(Q) \cong \mathbf{H}_G(C)$ is a projective right $\text{End}_R(G)$-module.

$1 \Rightarrow 2$. Assume part 1 and let

$$0 \longrightarrow K \longrightarrow Q \overset{\delta}{\longrightarrow} C \longrightarrow 0$$

be a G-balanced short exact sequence of right R-modules in which $Q \in \mathbf{P}(G)$. The induced sequence

$$0 \longrightarrow \mathbf{H}_G(K) \longrightarrow \mathbf{H}_G(Q) \xrightarrow{\delta^*} \mathbf{H}_G(C) \longrightarrow 0 \qquad (15.5)$$

is then exact. By part 1, $\mathbf{H}_G(C)$ is a projective right $\text{End}_R(G)$-module so that (15.5) is split exact. Thus the top row of the commutative diagram

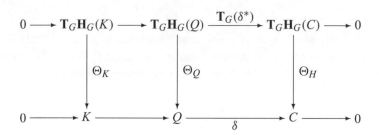

is split exact. By Theorem 1.13, Θ_Q is an isomorphism, by part 1, $Q' = \mathbf{T}_G\mathbf{H}_G(C) \in \mathbf{P}(G)$, and image $\Theta_K = \mathbf{S}_G(K)$. Thus $Q \cong \mathbf{T}_G\mathbf{H}_G(Q) = \text{image } \Theta_K \oplus Q'$. This proves part 2 and completes the logical cycle.

15.5 Examples

Example 15.38. Let $R = G = \mathbf{Z}/n\mathbf{Z} = \text{End}_R(G)$. Given $H \subset G$, then each map $H \longrightarrow G$ is multiplication by some integer, so G is injective relative to each inclusion $H \subset G$. By Theorem 15.6, $\text{End}_R(G)$ is a self-injective ring. More interesting is a result from abelian groups that follows from our work. If $nH = 0$, then each injection $\mathbf{Z}/n\mathbf{Z} \longrightarrow H$ splits. That is, if $nH = 0$, if $x \in H$, and if the order of x is n, then $x\mathbf{Z}$ is a direct summand of H.

Example 15.39. It is well known that $\mathbf{Z}(p^\infty)$ is an injective abelian group. The reader can show that $\mathbf{Z}(p^\infty)$ is neither self-small nor self-slender. However, $\mathbf{Z}(p^\infty) \xrightarrow{p} \mathbf{Z}(p^\infty)$ is a $\mathbf{Z}(p^\infty)$-monomorphism whose kernel is not a direct summand of $\mathbf{Z}(p^\infty)$. Thus $\text{End}(\mathbf{Z}(p^\infty)) \cong \widehat{\mathbf{Z}}_p$ is not self-injective.

Example 15.40. Let $G = \mathbf{Q}$ and $\text{End}_R(G) = \mathbf{Q}$. Then G is self-small and E-flat. Each G-monomorphism $\jmath : G \longrightarrow H$ is a split injection since G is divisible. Thus $\text{End}(G)$ is self-injective.

Example 15.41. Let G be a finitely generated semi-simple right R-module. Then G is self-small, and for each cardinal c, each inclusion $G \subset G^{(c)}$ is a direct summand of $G^{(c)}$. Hence $\text{End}_R(G)$ is self-injective and G is a projective and injective left $\text{End}_R(G)$-module. We observe that G-**Pres** $= \mathbf{P}(G)$. The reader can use Shur's lemma to prove that $\text{End}_R(G)$ is semi-simple Artinian.

Example 15.42. Let $p \in \mathbf{Z}$ be a prime and construct the commutative ring

$$\mathcal{O} = \left\{ \begin{pmatrix} x & 0 \\ y & x \end{pmatrix} \middle| x \in \widehat{\mathbf{Z}}_p, y \in \mathbf{Z}(p^{\infty}) \right\}.$$

B. Osofsky has proved (see [32]) that $\mathcal{O}$ is a *PF* ring whose ideals form a chain, but that $\mathcal{O}$ is not Noetherian. Hence $\mathcal{O}$ is a *PF* ring that is not a *QF* ring.

15.6 Exercises

Let R and E be rings, let G and H be a right R-module with endomorphism ring $\text{End}_R(G)$. The rest of the notation used in the chapter is in effect for these exercises.

1. The dual of E-flat would be E-injective. Characterize those self-slender G for which $\text{End}_R(G)$ is left hereditary and G is E-injective.
2. Fill in the details to any of the statements in this section that is not accompanied by a proof.
3. Prove that if $0 \neq h : M \longrightarrow \mathbf{H}_G(C)$ is a nonzero map of right $\text{End}_R(G)$-modules then $\mathbf{T}_G(h) \neq 0$.
4. The ring E is *right PF* if R is right self-injective and a cogenerator for the category of right E-modules. Characterize the right R-modules G that have right *PF* endomorphism rings.
5. Characterize the rings E that are left self-injective and right-cogenerator rings.
6. Construct a torsion-free example of a right self-injective ring R that is not right Noetherian.
7. Construct a torsion-free example of a right self-*FP*-injective ring R that is not right Noetherian and not right self-injective.
8. Fix a right R-module K and consider the right R-module $K \oplus R$. Let $S = \text{End}_R(K)$.
   (a) Show that $R = \text{End}_S(K \oplus R)$.
   (b) Determine the homological dimensions of $K \oplus R$, and so of K, as a right R-module.
   (c) Study the right or left R-modules in terms of the $K \oplus R$-plexes or $K \oplus R$-coplexes.
   (d) Study the functor $\text{Ext}^n(\cdot, K)$ by studying the homology of $K \oplus R$-coplexes.
9. Show that the following two conditions are equivalent.
   (a) $\mathbf{T}_G(M) \neq 0$ for each nonzero finitely generated $M \in \mathbf{Mod}\text{-}\text{End}_R(G)$.
   (b) $IG \neq G$ for each proper right ideal $I \subset \text{End}_R(G)$.
10. Recall that the right R-module G is *faithful* if $IG \neq G$ for each proper right ideal $I \subset \text{End}_R(G)$, iff $\mathbf{T}_G(M) \neq 0$ for each cyclic right $\text{End}_R(G)$-module M. Characterize those right R-modules C such that $\mathbf{H}_G(C)$ is a finite (cyclic) cogenerator.

15.7 Problems for future research

Let R be an associative ring with identity, let G be a right R-module, let $\text{End}_R(G)$ be the endomorphism ring of G, let Q be a G-plex, let $\mathcal{P}$ be a projective resolution of right $\text{End}_R(G)$-modules.

1. Study G-monomorphisms. Specifically, give more characterizations of G-monomorphisms.
2. Characterize those R-modules C such that $\text{Hom}_R(G, C)$ is an injective as a right $\text{End}_R(G)$-module.
3. Characterize those R-modules C such that $\text{Hom}_R(C, G)$ is an injective left $\text{End}_R(G)$-module.
4. Characterize those R-modules C such that $\text{Hom}_R(G, C)$ is a projective right $\text{End}_R(G)$-module.
5. Characterize those R-modules C such that $\text{Hom}_R(C, G)$ is a projective left $\text{End}_R(G)$-module.
6. Characterize those left $\text{End}_R(G)$-modules of the form $\text{Hom}_R(H, G)$.
7. Characterize the properties of $\text{End}_R(G)$ when G is an injective right R-module, a projective right R-module, or a cogenerator as a right R-module.
8. Study the biendomorphism ring

$$\text{End}_{\text{End}_R(G)}(G)$$

of G in terms of G-plexes over $\text{End}_R(G)$.
9. Let $i(R)$ denote the injective hull of the left R-module R, let $S = \text{End}_R(i(R))$, and consider $i(R)$ as an $\text{End}_S(i(R))$-S-bimodule. The *maximal right ring of quotients of R* is $\text{End}_S(i(R))$.

$$Q_{\max}(R) = \text{End}_{\text{End}_R(i(R))}(i(R)).$$

Study the $i(R)$-plexes and $i(R)$-coplexes. Then study the ring structure of $Q_{\max}(R)$.

16

A diagram of categories

16.1 The diagram

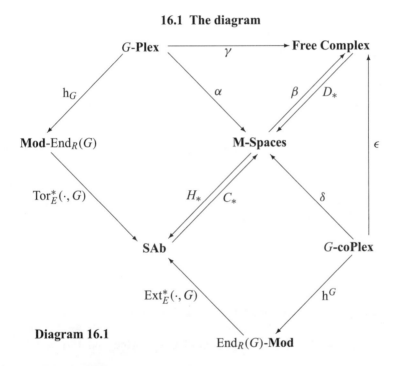

Diagram 16.1

The work in this chapter appears in [45].

Throughout this section G denotes a right R-module, not necessarily self-small or self-slender. We will construct a commutative diagram, Diagram (16.1), that contains the category of abelian groups, the categories of left modules and right modules over a ring, categories of complexes, and categories of point set topological spaces.

Let **Ab** denote the category of abelian groups, and let **SAb** denote the category of sequences $(\dots, A_2, A_1)$ of abelian groups A_k. A map f in **SAb** is a sequence

$$f = (\dots, f_2, f_1) : (\dots, A_2, A_1) \longrightarrow (\dots, B_2, B_1)$$

of abelian group maps $f_k : A_k \longrightarrow B_k$ for each integer $k \geq 1$.

Let **Complex** denote the category whose objects are complexes Q of right R-modules and whose maps are the homotopy equivalence classes $[f]$ of chain maps

$f : Q \longrightarrow Q'$ between complexes Q and Q'. As is usual, for an integer $k \geq 0$,

$$H_k(\cdot) : \textbf{Complex} \longrightarrow \textbf{Ab}$$

denotes the *k-th homology functor*. The *zero-th homology group* $H_0(Q)$ for a complex Q is

$$H_0(Q) = Q_0/\text{image } \delta_1 = \text{coker } \delta_1.$$

The *homology functor*

$$H_*(\cdot) : \textbf{Complex} \longrightarrow \textbf{SAb}$$

is defined by

$$H_*(\cdot) = (\cdots, H_2(\cdot), H_1(\cdot)).$$

This homology functor will be restricted to full subcategories of **Complex** without a change in notation. For example, we have the homology functors $H_*(\cdot) : G\text{-}\textbf{Plex} \longrightarrow$ **SAb** and

$$H_*(\cdot) : \textbf{Free Complex} \longrightarrow \textbf{SAb}$$

where **Free Complex** is the full subcategory of **Complex** whose objects are complexes whose terms are free abelian groups. For technical reasons this homology functor does not appear in Diagram (16.1).

For each G-plex Q,

$$\text{Hom}_R(G, Q) = \cdots \xrightarrow{\delta_2^*} \text{Hom}_R(G, Q_1) \xrightarrow{\delta_1^*} \text{Hom}_R(G, Q_0)$$

is a complex of right $\text{End}_R(G)$-modules. Define the functor

$$h_G(\cdot) = H_0 \circ \text{Hom}_R(G, \cdot) : \textbf{Complex} \longrightarrow \textbf{Mod-}\text{End}_R(G).$$

Dually, for each G-coplex W,

$$\text{Hom}_R(W, G) = \cdots \xrightarrow{\sigma_2^*} \text{Hom}_R(W_1, G) \xrightarrow{\sigma_1^*} \text{Hom}_R(W_0, G)$$

is a complex of left $\text{End}_R(G)$-modules. Define the functor

$$h^G(\cdot) = H_0 \circ \text{Hom}_R(\cdot, G) : \textbf{Complex} \longrightarrow \text{End}_R(G)\textbf{-Mod}.$$

The Torsion and Extension functors are naturally homology functors and they play a fundamental role in our discussions. Let

$$\text{Tor}_E^*(\cdot, G) : \textbf{Mod-}\text{End}_R(G) \longrightarrow \textbf{SAb}$$

denote the functor defined by

$$\mathrm{Tor}_E^*(\cdot, G) = (\ldots, \mathrm{Tor}_E^2(\cdot, G), \mathrm{Tor}_E^1(\cdot, G)),$$

where

$$\mathrm{Tor}_E^k(\cdot, G) = \mathrm{Tor}_{\mathrm{End}_R(G)}^k(\cdot, G)$$

for each integer $k > 0$. Dually define the functor

$$\mathrm{Ext}_E^*(\cdot, G) : \mathrm{End}_R(G)\text{-}\mathbf{Mod} \longrightarrow \mathbf{SAb}$$

as

$$\mathrm{Ext}_E^*(\cdot, G) = (\cdots, \mathrm{Ext}_E^2(\cdot, G), \mathrm{Ext}_E^1(\cdot, G)),$$

where

$$\mathrm{Ext}_E^k(\cdot, G) = \mathrm{Ext}_{\mathrm{End}_R(G)}^k(\cdot, G)$$

for each integer $k > 0$.

Now let us gravitate to the category **Spaces** of point set topological spaces. More precisely, the category of point set topological spaces is denoted by **Spaces**, and in a departure from the norm, the maps in **Spaces** are the homotopy equivalence classes $[f]$ of continuous maps $f : X \longrightarrow Y$ between topological spaces. We will write $X \sim Y$ when two topological spaces X and Y are *homotopic*. To each topological space X we fix a simplicial approximation $\sigma(X)$ of X and then form the associated complex $\beta(X)$ of free abelian groups from $\sigma(X)$. Inasmuch as $X \sim \sigma(X)$, the free complex $\beta(X)$ is unique up to homotopy. Thus we have defined a map

$$\beta(\cdot) : \mathbf{Spaces} \longrightarrow \mathbf{Free\ Complex}.$$

For each integer $k \geq 0$ we define the *k-th homology group of X* to be the k-th homology group of the free complex $\beta(X)$, or symbolically,

$$H_k(X) = H_k(\beta(X)).$$

We can then define the homology functor

$$H_*(\cdot) : \mathbf{Spaces} \longrightarrow \mathbf{SAb}$$

on the category of point set topological spaces by

$$H_*(\cdot) = (\cdots, H_2(\cdot), H_1(\cdot)).$$

Given an indexed set $\{X_i \mid i \in \mathcal{I}\}$ of topological spaces let

$$\bigvee_{i \in \mathcal{I}} X_i = \text{the one point union of the } X_i.$$

The following is a classic construction due to Moore. Given integers $k > 0, n \geq 0$ let S^k denote the k-sphere and fix a continuous function

$$f_n : S^k \longrightarrow S^k \text{ of degree } n.$$

Such functions exist in abundance. See [69, Example 2.31]. Let D^{k+1} be the $k+1$-disk and note that S^k is the boundary of D^{k+1}. Define

$$C_k(f_n) = \frac{S^k \cup D^{k+1}}{\{x \sim f_n(x) \mid x \in S^k\}}.$$

The constructed space is simply connected. However, it is possible by choosing different boundary functions f_n and f_n' that the end results $C_1(f_n)$ and $C_1(f_n')$ are not homotopic spaces. They are certainly not simply connected for $n \neq 1$. See [69, page 338]. Thus our applications will deal almost exclusively with the cases $k \geq 2$.

If A is a finitely generated abelian group then $A = A_1 \oplus \cdots \oplus A_t$ for some indecomposable cyclic groups A_j. Then we define

$$C_k(A) = C_k(A_1) \vee \cdots \vee C_k(A_t).$$

This construction is extended to any abelian group A by taking the direct limit of the finitely generated subgroups of A so that

$$C_k(A) = C_k(\varinjlim A_o) = \varinjlim C_k(A_o),$$

where A_o ranges over the finitely generated subgroups of A. See [69, page 286]. The space $C_k(A)$ is called a *Moore k-space*. For integers $k \geq 2$, $C_k(A)$ is simply connected. For a fixed group A and integer $k \geq 2$, $C_k(A)$ is unique up to homotopy. See [69, page 338]. We summarize uniqueness in the following two lemmas.

Lemma 16.1. [69, page 127] *Let A be an abelian group and fix an integer $k \geq 2$. The Moore space $C_k(A)$ is a simply connected space that satisfies the following group isomorphisms for each integer $p > 0$.*

$$H_p(C_k(A)) = \begin{cases} A & \text{if } p = k \\ 0 & \text{if } p \neq k \end{cases} \tag{16.1}$$

Lemma 16.2. [69, page 338] *Let $k \geq 2$ be an integer, and let X be a Moore k-space. Then*

$$C_k(H_k(X)) \sim X.$$

Let **M-Spaces** denote the full subcategory of **Spaces** whose objects are homotopic to the one point unions of Moore k-spaces $C_k(A)$ ranging over abelian groups A and integers $k > 0$. That is $X \in$ **M-Spaces** iff there is a sequence of abelian groups $S = (\ldots, A_2, A_1)$, and a fixed set of generators and relations for A_1, such that

$$X \sim \bigvee_{k>0} C_k(A_k).$$

The *function*

$$C_*(\cdot) : \mathbf{SAb} \longrightarrow \mathbf{M\text{-}Spaces}$$

sends a sequence of abelian groups $S = (\ldots, A_2, A_1)$ to

$$C_*(S) = \bigvee_{k \geq 2} C_k(A_k).$$

(Notice that the first subscript is $k = 2$, not 1.) The maps α, γ, δ, and ϵ are then the unique maps that make the diagram commute. A classic result in topology states that C_* is not a functor, so the maps α, γ, δ, ϵ are *not functors*.

There is a homology functor $H_*(\cdot) :$ **Free Complex** $\longrightarrow$ **SAb**, so there is a function

$$D_*(\cdot) = C_* \circ H_*(\cdot) : \mathbf{Free\ Complex} \longrightarrow \mathbf{M\text{-}Spaces}$$

on the objects of the categories.

This completes the construction of Diagram (16.1). It is interesting to note that we have included the category of right modules over a (not necessarily commutative) ring $\mathrm{End}_R(G)$ and its category of left modules, together with the usual derived functors $\mathrm{Tor}_E^*(\cdot, G)$ and $\mathrm{Ext}_E^*(\cdot, G)$ in a nontrivial way.

16.2 Smallness and slenderness

Initially, assume that G is a self-small right R-module. With this assumption we will be able to supplant some functors and functions in Diagram (16.1) with category equivalences and bijections, as in Diagram (16.2) below.

We assume that G is self-small. Then by Theorem 9.11,

$$h_G(\cdot) : G\text{-}\mathbf{Plex} \longrightarrow \mathbf{Mod\text{-}End}_R(G)$$

is a covariant category equivalence. By Theorem 9.12, the inverse of $h_G(\cdot)$ is the functor

$$t_G(\cdot) : \mathbf{Mod\text{-}End}_R(G) \longrightarrow G\text{-}\mathbf{Plex}.$$

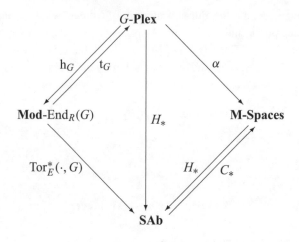

Diagram 16.2

Because G is self-small, Lemma 11.19 shows us that for each G-plex $\mathcal{Q}$

$$\mathrm{Tor}_E^k(\mathrm{h}_G(\mathcal{Q}), G) = H_k(\mathcal{Q})$$

for each integer $k > 0$. Hence

$$\mathrm{Tor}_E^*(\mathrm{h}_G(\cdot), G) = H_*(\cdot),$$

so that the triangles in Diagram (16.2) commute. The inverse relationship between $\mathrm{h}_G(\cdot)$ and $\mathrm{t}_G(\cdot)$ proves the following corollary.

Corollary 16.3. *Let G be a self-small right R-module. The k-th Torsion functor factors as*

$$\mathrm{Tor}_E^k(\cdot, G) = H_k \circ \mathrm{t}_G(\cdot)$$

for each integer $k > 0$.
We have thus described Diagram (16.2).

Now assume the statement (μ) that there are no measurable cardinals, and assume that G is a self-slender right R-module.

Since (μ) is assumed and since G is self-slender, Theorem 9.32 states that

$$\mathrm{h}^G(\cdot) : G\text{-}\mathbf{coPlex} \longrightarrow \mathrm{End}_R(G)\text{-}\mathbf{Mod}$$

is a contravariant category equivalence (= a duality), and its inverse $\mathrm{h}^G(\cdot)$ is induced by $\mathrm{Hom}_{\mathrm{End}_R(G)}(\cdot, G)$.

Furthermore, the commutative diagram (11.17) on page 270 shows that the homology functor

$$H^*(\cdot) : G\text{-}\mathbf{coPlex} \longrightarrow \mathbf{SAb}$$

is a factor of the functor $\mathrm{Ext}_E^*(\cdot, G)$

$$\mathrm{Ext}_E^k(\mathrm{h}^G(\cdot), G) = H_*(\cdot).$$

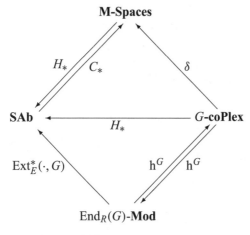

Diagram 16.3

Hence in an argument dual to that used in Diagram (16.2), the triangles in Diagram (16.3) commute.

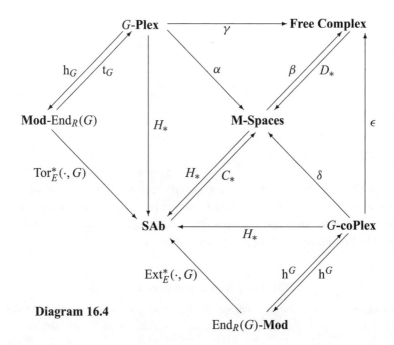

Diagram 16.4

Assume (μ). At the time of this writing the *rtffr* groups, (reduced torsion-free finite rank abelian groups), are essentially the only known examples of a module that is both self-small and self-slender. See Remark 17.16.

By combining Diagrams (16.1), (16.2), and (16.3) we form *the commutative diagram Diagram (16.4)*, in which opposing arrows indicate inverse functions on the objects of the categories.

16.3 Coherent objects

It is interesting that a large portion of Diagram (16.4) is preserved after we delete the hypotheses (μ), self-small, and self-slender. Let $\text{End}_R(G)$-**Coh** and **Coh**-$\text{End}_R(G)$ denote the categories of coherent left $\text{End}_R(G)$-modules and coherent right $\text{End}_R(G)$-modules, respectively.

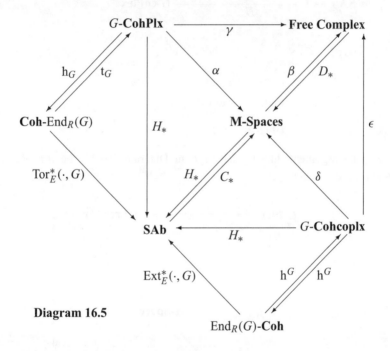

Diagram 16.5

The functors h_G and t_G in Diagram (16.5) are inverse equivalences by Theorem 9.13, and the functors h^G are inverse contravariant equivalences by Theorem 9.36. We leave it to the reader to mimic the proof of the commutativity of Diagram (16.4) to prove that *Diagram (16.5) commutes*. We note once again that Diagram (16.5) is constructed with very few hypotheses on G.

16.4 The construction function

In this section we investigate the nature of the construction map and the homology map. Homotopy type is an equivalence relation $\sim$ on the set **M-Spaces** so that

> **M-Spaces**/$\sim$ is the set of homotopy equivalence classes of M-spaces.

Two sequences of abelian groups $(\dots, A_2, A_1)$ and $(\dots, B_2, B_1)$ are isomorphic iff $A_k \cong B_k$ for each integer $k > 0$. Then

> **SAb**/$\cong$ is the set of isomorphism classes of sequences of abelian groups.

Theorem 16.4. *1. If $S \in$ **SAb** then there is an M-space X such that $H_*(X) \cong S$.*
2. If X and Y are simply connected M-spaces, and if $H_k(X) \cong H_k(Y)$ for each $k \geq 2$ then $X \sim Y$.

Proof: 1. Given $S = (\dots, A_2, A_1) \in$ **SAb** there are Moore k-spaces X_k, $k = 1, 2, \dots$ whose homology groups satisfy

$$H_k(X_k) \cong A_k$$

and such that $H_p(X_k) = 0$ for integers $p \neq k > 0$. Let $X = \bigvee_{k>0} X_k$ and then observe that for an integer $p > 0$,

$$H_p(X) \cong H_p \left(\bigvee_{k>0} X_k \right) \cong \bigoplus_{k>0} H_p(X_k) \cong H_p(X_p)$$

because $H_p(\cdot)$ changes one point unions into direct sums, [69, Corollary 2.25]. Hence

$$\begin{aligned}
H_*(X) &\cong (\dots, H_2(X), H_1(X)) \\
&\cong (\dots, H_2(X_2), H_1(X_1)) \\
&\cong (\dots, A_2, A_1) \\
&= S.
\end{aligned}$$

2. Let X and Y be simply connected M-spaces such that $H_k(X) \cong H_k(Y)$ for each integer $k \geq 2$. Since X and Y are M-spaces there are, for each integer $k > 0$, Moore k-spaces X_k and Y_k such that

$$X \sim \bigvee_{k>0} X_k \quad \text{and} \quad Y \sim \bigvee_{k>0} Y_k.$$

Since X and Y are simply connected, X_1 and Y_1 contract to a point. Thus $X_1 \sim Y_1$. Furthermore, for any integer $p \geq 2$,

$$H_p(X_p) \cong H_p(X) \cong H_p(Y) \cong H_p(Y_p),$$

so that $X_p \sim Y_p$ by Lemma 16.2. Hence $X \sim Y$.

16.5 The Greek maps

In this section we will determine the rules for the Greek maps α, β, γ, δ, and ϵ.

Theorem 16.5. *Suppose that G is a self-small right R-module.*

1. *Let $Q, Q' \in G$-**Plex** be such that $H_1(Q) = H_1(Q') = 0$. Then $\alpha(Q) \sim \alpha(Q')$ iff $H_*(Q) \cong H_*(Q')$.*
2. *Let $Q \in G$-**Plex** be such that $H_1(Q) = 0$ and let $X \in$ **M-Space** be simply connected. Then $\alpha(Q) \sim X$ iff $H_*(Q) \cong H_*(X)$.*

Proof: 1. Let $Q, Q' \in G$-**Plex** be such that $\alpha(Q) \sim \alpha(Q')$. Theorem 16.4.1 implies that $H_*(\alpha(Q)) \cong H_*(\alpha(Q'))$, so that $H_*(Q) \cong H_*(Q')$ by the commutativity of Diagram (16.2).

Conversely, suppose that $H_*(Q) \cong H_*(Q)$. Then $H_*(\alpha(Q)) \cong H_*(\alpha(Q'))$, so that $C_*(H_*(\alpha(Q))) \sim C_*(H_*(\alpha(Q')))$. Since $H_1(Q) = H_1(Q') = 0$, Lemma 16.2 implies that $\alpha(Q) \sim \alpha(Q')$.

2. If $\alpha(Q) \sim X$ then $H_*(\alpha(Q)) \cong H_*(X)$, so that $H_*(Q) \cong H_*(X)$ by the commutativity of Diagram (16.2).

Conversely, suppose that $H_*(Q) \cong H_*(X)$. By the commutativity of Diagram (16.2), because X is simply connected, and by Theorem 16.4,

$$\alpha(Q) \sim C_*(H_*(Q)) \sim C_*(H_*(X)) \sim X.$$

This completes the proof.

Reading the above result a different way we see that $\alpha^{-1}(X)$ is the set of G-plexes Q whose homology groups are the homology groups of X. A similar set of results is true for δ.

Theorem 16.6. *Assume (μ) and suppose that G is a self-slender right R-module.*

1. *Let $W, W' \in G$-**coPlex** be such that $H_1(W) = H_1(W') = 0$. Then $\delta(W) \sim \delta(W')$ iff $H^*(W) \cong H^*(W')$.*
2. *Let $W \in G$-**coPlex** be such that $H_1(W) = 0$ and let $X \in$ **M-Space** be simply connected. Then $\delta(W) \sim X$ iff $H^*(W) \cong H^*(X)$.*

Let **Complex**$/H_*$ denote the set of equivalence classes $[Q] = \{Q' \mid H_*(Q) \cong H_*(Q')\}$ for complexes Q. Similar quotients are defined for categories of complexes or topological spaces.

The homology functor in the next result is

$$H_*(\cdot) : \textbf{Free Complex} \longrightarrow \textbf{SAb}$$

and does not appear in Diagram (16.1). It is used, however, to define $D_* = C_* \circ H_*$.

Theorem 16.7. *Suppose that G is a right R-module.*

1. *Let $X \in$ **M-Spaces** be simply connected. There is an $\mathcal{F} \in$ **Free Complex** such that $D_*(\mathcal{F}) \sim X$.*
2. *Let $\mathcal{F}, \mathcal{F}' \in$ **Free Complex** and suppose that $D_*(\mathcal{F}) \sim D_*(\mathcal{F}')$. Then $H_*(\mathcal{F}) \cong H_*(\mathcal{F}')$.*

Proof: 1. Let $X \in$ **M-Spaces** be simply connected. The free complex $\beta(X)$ has homology groups $H_*(\beta(X)) = H_*(X)$ so that

$$D_*(\beta(X)) = C_*(H_*(\beta(X))) = C_*(H_*(X)) \sim X$$

by Lemma 16.2.

2. Let $\mathcal{F}, \mathcal{F}' \in$ **Free Complex** and suppose that $D_*(\mathcal{F}) \sim D_*(\mathcal{F}')$. Since $D_* = C_* \circ H_*$ we have $C_*(H_*(\mathcal{F})) \sim C_*(H_*(\mathcal{F}'))$ so that

$$H_*(\mathcal{F}) \cong H_*(C_*(H_*(\mathcal{F}))) \cong H_*(C_*(H_*(\mathcal{F}'))) \cong H_*(\mathcal{F}')$$

by Lemma 16.1. Hence $\mathcal{F}$ and $\mathcal{F}'$ have the same homology type.

Corollary 16.8. *Suppose that G is a right R-module. Let $\mathcal{F}, \mathcal{F}' \in$ **Free Complex** be such that $H_1(\mathcal{F}) = H_1(\mathcal{F}') = 0$. Then $D_*(\mathcal{F}) \sim D_*(\mathcal{F}')$ iff $H_*(\mathcal{F}) \cong H_*(\mathcal{F}')$.*

Proof: Suppose that $H_*(\mathcal{F}) \cong H_*(\mathcal{F}')$. Apply C_* and Lemma 16.2 to prove that $\mathcal{F} \sim \mathcal{F}'$. The converse is the above theorem.

Theorem 16.9. *Suppose that G is a right R-module.*

1. *Let $\mathcal{F} \in$ **Free Complex** be such that $H_1(\mathcal{F}) = 0$. There is an $X \in$ **M-Space** such that $\beta(X) = \mathcal{F}$.*
2. *Let $X, X' \in$ **M-Spaces** be simply connected. If $\beta(X) \sim \beta(X')$ then $X \sim X'$.*

Proof: Proceed as in Theorem 16.7.

16.6 Applications

16.6.1 Complete sets of invariants

Let **X** be a set and let $\sim$ be an equivalence relation on **X**. A *complete set of invariants for* **X** *up to* $\sim$ is a set **Y** for which there exists a bijection $\psi : \mathbf{X}/\!\sim \longrightarrow \mathbf{Y}$ from the equivalence classes in $\mathbf{X}/\sim$ onto the set **Y**. If **Y** is a set of groups then we say that **Y** is a *complete set of algebraic invariants for* **X** up to $\sim$. If **Y** is a set of topological spaces then we say that **Y** is a *complete set of topological invariants for* **X** up to $\sim$. For example, as a consequence of Jónsson's theorem, if G is an rtffr group then the strongly indecomposable quasi-summands of G and their multiplicities in G form a complete set of algebraic invariants for G up to *quasi-isomorphism*. See Theorem 18.53 and [9]. An important unanswered question in abelian group theory is to find a set of accessible *numeric* invariants for strongly indecomposable $A \in$ **Ab** up to quasi-isomorphism. The following theorem shows us that the class of homotopy equivalence classes of M-spaces is a complete set of topological invariants for **Ab**.

Lemma 16.10. *Let A and A' be abelian groups and let $k \geq 2$ be an integer.*

1. *$C_k(A) \sim C_k(A')$ iff $A \cong A'$.*
2. *If $C_k(A) \sim X \vee Y$ for some M-spaces X and Y then $A \cong H_k(X) \oplus H_k(Y)$.*
3. *If $A \cong B \oplus C$ for some groups B and C then $C_k(A) \sim C_k(B) \vee C_k(C)$.*

Proof: 1. By Lemma 16.1 and because $C_k(A) \sim C_k(A')$ we have

$$A \cong H_k(C_k(A)) \cong H_k(C_k(A')) \cong A'.$$

The converse is Lemma 16.2.

2. If $C_k(A) \sim X \vee Y$ then by Lemma 16.1,

$$A \cong H_k(C_k(A)) \cong H_k(X \vee Y) \cong H_k(X) \oplus H_k(Y)$$

since H_k takes one point unions to direct sums.

3. Consider $C_k(B) \vee C_k(C)$. By definition $C_k(B) \vee C_k(C)$ is a Moore k-space, so an application of H_k and Lemma 16.1 yields

$$H_k(C_k(B) \vee C_k(C)) \cong H_k(C_k(B)) \oplus H_k(C_k(C))$$
$$\cong B \oplus C \cong A$$
$$\cong H_k(C_k(A)).$$

Because $k \geq 2$ and by Theorem 16.4.2, $C_k(B) \vee C_k(C) \sim C_k(A)$. This completes the proof.

We interpret the next result as stating that $C_2(A)$ is a complete set of topological invariants for an abelian group A. Consequently, the topological invariants for **SAb** allow one to sculpt abelian groups into clay figures.

Theorem 16.11. *Let A and A' be abelian groups. Then*

$$A \cong A' \text{ iff } C_2(A) \sim C_2(A').$$

Proof: Apply Lemma 16.10.1.

16.6.2 Unique topological decompositions

Lemma 16.12 will lead us to an Azumaya–Krull–Schmidt theorem for topological spaces. We say that an M-space X is *M-indecomposable* if given M-spaces U and V such that $X = U \vee V$ then either U or V contracts to a point. The lemma follows immediately from the Theorem 16.4 and Lemma 16.10.

Lemma 16.12. *Let X, X' be simply connected M-spaces. Then $X \sim X'$ iff $H_*(X) \cong H_*(X')$.*

Lemma 16.13. *Suppose that X is a simply connected Moore k-space for some integer $k > 0$. Then X is M-indecomposable iff $H_k(X)$ is an indecomposable abelian group.*

The M-indecomposable Moore k-spaces yield a unique decomposition for M-spaces. We say that a decomposition

$$X \sim \bigvee_{i \in \mathcal{I}} X_i$$

of an M-space X into M-indecomposable M-spaces X_i is *unique for X up to homotopy* if given another decomposition

$$X \sim \bigvee_{j \in \mathcal{J}} Y_j$$

of X into a one point union of M-indecomposable M-spaces Y_j then there is a bijection $\sigma : \mathcal{I} \longrightarrow \mathcal{J}$ such that $X_i \sim Y_{\sigma(i)}$ for each $i \in \mathcal{I}$.

Theorem 16.14. *Let X be a simply connected M-space such that $H_k(X)$ is finite for each integer $k > 0$. There is a decomposition*

$$X \sim \bigvee_{k=1}^{\infty} (\vee_{i=1}^{t_k} X_{ik})$$

in which each t_k is an integer and in which each X_{ik} is an M-indecomposable Moore k-space. Any such decomposition for X is unique up to homotopy.

Proof. First we show that such an M-indecomposable decomposition exists for X. The Moore construction $C_k(\cdot)$ gives, for each integer $k \geq 2$, a Moore k-space X_k such that

$$H_k(X) \cong H_k(X_k).$$

By hypothesis $H_k(X_k)$ is a finitely generated abelian group so there are finitely many indecomposable cyclic abelian groups $A_{1k}, \ldots, A_{t_k k}$ such that

$$H_k(X_k) \cong A_{1k} \oplus \cdots \oplus A_{t_k k}. \tag{16.2}$$

By Lemmas 16.1 and 16.2 there are M-indecomposable simply connected Moore k-spaces $X_{1k}, \ldots, X_{t_k k}$ such that

$$H_k(X_{ik}) = A_{ik}.$$

Then

$$H_k(X_k) = \oplus_{i=1}^{t_k} A_{it_k} = H_k(\vee_{i=1}^{t_k} X_{it_k}). \tag{16.3}$$

Since X_k, X_{i,t_k} are simply connected spaces,

$$X_k \sim \vee_{i=1}^{t_k} X_{it_k}$$

by Lemma 16.12. Let

$$Y = \vee_{k>0} X_k = \bigvee_{k>0} (\vee_{i=1}^{t_i} X_{ik}).$$

This decomposition for Y is one into simply connected M-indecomposable Moore k-spaces with X_1 contracting to a point. (Recall that $H_1(X) = 0$.) Thus Y is simply connected. We will show that $X \sim Y$.

By (16.2) and because X is simply connected we see that

$$H_*(X) = (\dots, \oplus_{i=1}^{t_3} A_{i3}, \oplus_{i=1}^{t_2} A_{i2}, 0). \tag{16.4}$$

Then (16.3) and (16.4) show us that

$$H_*(X) \cong (\oplus_{i=1}^{t_k} A_{ik} \,\big|\, k > 0)$$
$$\cong (H_k(X_k) \,\big|\, k > 0)$$
$$\cong H_*(\vee_{k>0} X_k)$$
$$\cong H_*(Y).$$

Since X and Y are simply connected $X \sim Y$ by Lemma 16.10.

The uniqueness of this decomposition Y for X will follow from the following lemma.

Lemma 16.15. *Suppose that $k, t > 0$ are integers, and let $X \sim X_1 \vee \cdots \vee X_t$ be a topological space such that each X_i is an M-indecomposable simply connected Moore k-space for each $i = 1, \dots, t$. If $H_k(X)$ is a finitely generated abelian group then this decomposition of X is unique up to homotopy.*

Proof: Suppose we can write

$$X \sim X_1 \vee \cdots \vee X_t$$
$$\sim Y_1 \vee \cdots \vee Y_s$$

for some M-indecomposable M-spaces Y_j. We then have group isomorphisms

$$H_k(X) \cong H_k(X_1) \oplus \cdots \oplus H_k(X_t)$$
$$\cong H_k(Y_1) \oplus \cdots \oplus H_k(Y_s).$$

Since the X_i and the Y_j are M-indecomposable, Lemmas 16.1 and 16.2 show that $H_k(X_i)$ and $H_k(Y_j)$ are indecomposable abelian groups for each i, j. Azumaya's theorem holds for finitely generated indecomposable abelian groups so that $s = t$ and after a permutation of the subscripts $H_k(X_i) \cong H_k(Y_i)$ for each $i = 1, \dots, t$. Theorem 16.4.2 implies that $X_i \sim Y_i$ for each integer i. That is, the decomposition $X \sim X_1 \vee \cdots \vee X_t$ is unique for X, and this completes the proof.

Continue the proof of Theorem 16.14: By the above lemma it is sufficient to show that the decomposition $X = \bigvee_{k>0} X_k$ into Moore k-spaces X_k is unique. Suppose that

$$X \sim \bigvee_{k>0} X_k \sim \bigvee_{k>0} X_k'$$

for some simply connected Moore k-spaces X'_k. Then $H_p(X_k) = 0$ for each integer $k \neq p > 0$ so that

$$H_k(X) \cong \bigoplus_{p>0} H_k(X_p)$$

$$\cong H_k(X_k).$$

Similarly,

$$H_k(X) \cong H_k(X'_k)$$

and so

$$H_k(X_k) \cong H_k(X'_k).$$

By Theorem 16.4.2, $X_k \sim X'_k$, and thus the terms in the one-point union $\bigvee_{k=1}^{\infty} X_k$ are unique for X. Given our reductions, this completes the proof.

16.6.3 Homological dimensions

As usual, fd_E denotes the flat dimension of a left $\mathrm{End}_R(G)$-module. Also, id_E denotes the injective dimension of a left $\mathrm{End}_R(G)$-module. For a fixed integer $k > 0$, *projective Euclidean k-space* is a quotient space of a union of Euclidean k-Spaces, $\mathbf{R}^k$.

Theorem 16.16. *Let $k > 0$ be an integer and let G be a self-small right R-module. The following are equivalent.*

1. $\mathrm{fd}_E(G) \leq k$
2. *Each M-space $X \in$ image α is a subspace of projective Euclidean $k + 1$-space.*

Proof: Assume part 1. So that the notation agrees with that of **SAb**, let

$$A_k(\cdot) = \mathrm{Tor}^k_E(\cdot, G) \text{ for each integer } k > 0$$

and let $A_*(\cdot) = \mathrm{Tor}^*_E(\cdot, G)$. Observe that $k + j \geq 2$ for each integer $j > 0$. Follow this argument by tracing through Diagram (16.2).

If $\mathrm{fd}_E(G) \leq k$ then $A_{k+j}(\cdot) = 0$ for each integer $j > 0$. In our construction of $C_{k+j}(A_{k+j}(\cdot))$ we have

$$C_{k+j}(A_{k+j}(\cdot)) = C_{k+j}(0) = \text{ a point}$$

while for $p = 1, \ldots, k$, $C_p(A_p(\cdot))$ is a quotient space of a one point union of $p + 1$-disks. The $p + 1$-disk is a subspace of $\mathbf{R}^{p+1}$, so that $C_p(A_p(\cdot))$ is a subspace of projective Euclidean $k + 1$-spaces. Hence, for each right E-module M,

$$C_* \circ A_*(M) = \bigvee_{p>0} C_p(A_p(M)) = \bigvee_{p=1}^{k} C_p(A_p(M))$$

is a subspace of projective Euclidean $k + 1$ space. Finally since $h_G(\cdot)$ is a category equivalence, each G-plex Q satisfies

$$\alpha(Q) = C_* \circ A_* \circ h_G(Q) = C_* \circ A_*(h_G(Q)),$$

which is a subspace of projective Euclidean $k + 1$ space. This proves part 2.

The converse is proved by reversing the argument.

Corollary 16.17. *Let G be self-small. Then $\mathrm{fd}_E(G)$ is finite iff there is an integer k such that each $X \in$ image α embeds in a projective Euclidean k-space.*

Corollary 16.18. *Let G be self-small. Then G is E-flat iff each $X \in$ image α embeds in a projective Euclidean 1-space.*

Proof: If some $X \in$ image α embeds in some projective Euclidean k-space U for some integer $k \geq 2$ and for no smaller integer k then $\mathrm{fd}(G) \geq 1$ and G is not flat.

Conversely, if G is not flat then $\mathrm{fd}(G) \geq 1$ so that $\mathrm{Tor}_E^1(M, G) \neq 0$. Let $X = C_1(A_1(M))$. Then $X \in$ image α embeds in a projective Euclidean 2-space, and for no smaller dimension $k < 2$. This completes the proof.

Theorem 16.19. *Let G be self-small and let $k > 0$ be an integer. For each G-plex Q, $C_{k+1}(H_{k+1}(Q))$ is compact iff $\mathrm{fd}_E(G) \leq k$.*

Proof: $\mathrm{fd}_E(G) \not\leq k$ iff there is a right $\mathrm{End}_R(G)$-module M such that $\mathrm{Tor}_E^{k+1}(M, G) \neq 0$ iff $\mathrm{Tor}_E^{k+1}(M^{(\aleph_o)}, G)$ is infinite iff the Moore $k + 1$-space X corresponding to $\mathrm{Tor}_E^{k+1}(M, G)^{(\aleph_o)}$ is not compact.

Dualizing flat dimension we arrive at injective dimension.

Theorem 16.20. *Assume (μ) and let G be a self-slender right R-module. The following are equivalent for G.*

1. *$\mathrm{id}_E(G) \leq k$.*
2. *Each $X \in$ image δ is a subspace of projective Euclidean $k + 1$-space.*

Corollary 16.21. *Assume (μ) and let G be a self-slender right R-module. Then $\mathrm{id}_E(G)$ is finite iff there is an integer k such that each $X \in$ image δ embeds in projective Euclidean k-space.*

Corollary 16.22. *Assume (μ) and let G be a self-slender right R-module. Then $\mathrm{id}_E(G) \leq 1$ iff for each G-coplex W, $\delta(W)$ is a subspace of projective Euclidean 2-space.*

Proposition 16.23. *Let G be a self-small right R-module. If α is a surjection then $\mathrm{fd}_E(G) = \infty$.*

Proof: We use the fact derived from Theorem 11.21 that $\mathrm{fd}_E(G)$ is the supremum of the integers k such that $H_{k+1}(Q) = 0$ for each G-plex Q.

Let $M \in$ **Mod**-$\text{End}_R(G)$ and assume that α is a surjection. Let

$$X_1, X_2, X_3, \ldots$$

be a sequence in which X_k is a nontrivial Moore k-space, and let

$$X = X_1 \vee X_2 \vee X_3 \vee \cdots.$$

Then

$$H_k(X) = H_k(X_k) \neq 0.$$

Since α is a surjection, there is a $\mathcal{Q} \in G$-**Plex** such that

$$\alpha(\mathcal{Q}) = X.$$

By the commutativity of Diagram (16.2),

$$H_*(\mathcal{Q}) \cong H_*(\alpha(\mathcal{Q})) \cong H_*(X) \neq 0.$$

Therefore $\text{fd}_E(G) = \infty$.

Proposition 16.24. *Assume* (μ) *and let* G *be a self-slender right* R-module. If δ *is a surjection then* $\text{id}_E(G) = \infty$.

Proposition 16.25. *Assume* (μ) *and let* G *be a self-slender right* R-module. There is at least one noncompact $X \in$ image δ.

Proof: G, being nonzero and self-slender, is not injective. Thus there is an $M \neq 0$ such that $\text{Ext}^1_E(M, G) \neq 0$ and so

$$\text{Ext}^1_E(M^{(\aleph_o)}, G) \cong \text{Ext}^1_E(M, G)^{(\aleph_o)}$$

corresponds under C_* to a space that is the union of countably many copies of a nontrivial M-space X. Such a space is not compact.

Hence if the rtffr group G is a flat left $\text{End}_R(G)$-module then image $\alpha = \{0\} \neq$ image δ.

Remark 16.26. Diagram (16.1) has an empty space that can be filled. Notice that the upper right corner of Diagram (16.1) contains **Free Complex** and associated functions. But the lower left corner of Diagram (16.1) is empty. Let **Bim** denote the category of $\text{End}_R(G)$-$\text{End}_R(G)$-bimodules M such that

$$\text{Tor}^k_E(M, G) \cong \text{Ext}^k_E(M, G)$$

for all integers $k > 0$. **Bim** contains the category of free $\text{End}_R(G)$-$\text{End}_R(G)$-bimodules so that **Bim** is nonzero. The functors **Bim** $\longrightarrow$ **Mod**-$\text{End}_R(G)$ and **Bim** $\longrightarrow$ $\text{End}_R(G)$-**Mod** are the natural inclusion functors. The category **Bim** and these functors can be added to Diagram (16.1) to fill in the empty space. It is worth noting that this is a naturally occurring category defined by an unnatural isomorphism.

16.7 Exercises

1. Construct Diagram (16.k) for $k = 1, 2, 3, 4, 5$ assuming that G is
   (a) E-flat, or
   (b) Σ-quasi-projective, or
   (c) a finitely generated abelian group.

16.8 Problems for future research

Some of the more intriguing unanswered questions concerning these functors follow.

1. Examine Diagram (16.1) in the category **Torsion** of torsion abelian groups.
2. Examine Diagram (16.1) in the category **Torsion-free** of torsion-free abelian groups. Restrict attention to those torsion-free groups of finite rank.
3. Examine Diagram (16.1) in the category of quasi-homomorphisms **QAb** of rtffr groups and with homsets **QHom**$(\cdot, \cdot)$.
4. Investigate Diagram (16.1) assuming that $G = \mathbf{Z}$, or $\mathbf{Q}$, or a torsion group, or any other type of abelian group.
5. Complete G and E in the p-adic topology. Then examine Diagram (16.1).
6. Must image α be the set of one point unions of projective disks? Is there some other natural construction of topological spaces that we can substitute for C_*?
7. α is injective iff $G = 0$. Then determine the nice functional properties of α.
8. Investigate maps α', β', γ', δ', and ϵ' that exchange the domain and codomain of the maps α, β, γ, δ, and ϵ, respectively. E.g. $\alpha : G\text{-}\mathbf{Plex} \longrightarrow \mathbf{M\text{-}Spaces}$ is a function while $\alpha' : \mathbf{M\text{-}Spaces} \longrightarrow G\text{-}\mathbf{Plex}$ exchanges the domain and codomain of α.
9. If $\mathcal{F}$ is a free complex then $\mathcal{F} \otimes_{\mathbf{Z}} G$ is a complex whose terms are direct sums of copies of G. Because G is a flat abelian group

$$H_k(\mathcal{F} \otimes_{\mathbf{Z}} G) \cong H_k(\mathcal{F}) \otimes_{\mathbf{Z}} G$$

which is rarely isomorphic to $H_k(\mathcal{F})$. Furthermore, $\mathcal{F} \otimes_{\mathbf{Z}} G$ is not necessarily a G-plex. Is there a natural function $\gamma' : \mathbf{Free\ Complex} \longrightarrow G\text{-}\mathbf{Plex}$ that preserves homology groups.
10. $X \in \mathbf{M\text{-}Spaces}$ is point set topological. Is there a way to realize similar diagrams using the p-adic numbers $\mathbf{Q}_p$ and the p-adic integers $\widehat{\mathbf{Z}}_p$ for some prime p? There is ample evidence that this can happen since in $\mathbf{Q}_p$, the unit interval is $\widehat{\mathbf{Z}}_p$. Disks and spheres then appear to be a little more accessible than do the point set topological spaces we have used.
11. Investigate the functors $\mathbf{Mod}\text{-}\mathrm{End}_R(G) \longrightarrow \mathbf{M\text{-}Spaces}$ and $\mathrm{End}_R(G)\text{-}\mathbf{Mod} \longrightarrow \mathbf{M\text{-}Spaces}$ implied by Diagram (16.1).
12. Given $M \in \mathbf{M\text{-}Spaces}$ choose a G-plex Q that most accurately reflects the algebraic properties of M.

13. Further investigate γ and ϵ.
14. Under what conditions is a subcollection of **M-Spaces** the image of some function α for some group G?
15. Which sequences of abelian groups S are of the form $\operatorname{Tor}_E^*(\cdot, G)$?
16. Suppose G is self-small and self-slender as a right R-module. Then G is a left $\operatorname{End}_R(G)$-module. To what M-Space does G correspond in Diagram (16.2)?
17. Find a general relationship between Diagram (16.1) and homological dimensions.
18. Discuss similar types of diagrams for Lie algebras or for more general algebras.
19. Suppose we endow the sequences in **SAb** with the obvious grading. Discuss the changes that occur in Diagram (16.1).
20. Find a connection between the rtffr group G, Diagram (16.1), and the fundamental groups of topological spaces.
21. There should be a category and associated functors in the empty lower left part of Diagram (16.1). Find them.

17

Diagrams of abelian groups

In this chapter, we characterize quadratic number fields possessing unique factorization in terms of the power cancellation property of torsion-free rank two abelian groups, in terms of Σ-unique decomposition, in terms of a pair of point set topological properties of Eilenberg–Maclane spaces, and in terms of the sequence of rational primes. We give a complete set of topological invariants of abelian groups, we characterize those abelian groups that have the power cancellation property in the category of abelian groups, and we characterize those abelian groups that have Σ-unique decomposition. Our methods can be used to characterize any direct sum decomposition property of an abelian group.

Throughout this chapter A and G denote abelian groups, A *is variable, and G is fixed.* Recall G-**Plex** and G-**coPlex** from Chapter 9. Also ϕ is a group homomorphism, $X, Y, U,$ and V denote topological spaces, and f denotes a continuous map or a sequence of continuous maps. These are abstract topologies on topological spaces, *not linear topologies on groups.* For a space X, $\pi_1(X)$ is the fundamental group, $\prod_{i \in \mathcal{I}} X_\in$ is Cartesian product of spaces $\{X_i \mid i \in \mathcal{I}\}$, $X \sim Y$ denotes homotopic spaces X and Y, $f \sim g$ denotes homotopic continuous maps f and g, $[f]$ denotes the homotopy equivalence class of f, and given a continuous map f, $f^* = \pi_1(f)$.

Given G we fix a connected CW complex space X such that

$$G \cong \pi_1(X)$$

and such that $\pi_k(X) = 0$ for each $k > 1$, when written additively. We call X a $K(G, 1)$-*space.* For each abelian group G there is a $K(G, 1)$-space X. Any two $K(G, 1)$-spaces are homotopic and the implied mapping $K(\cdot, 1)$ is a functor. Then $\pi_1(X) \cong G$ and $X \sim K(G, 1)$. Since this is not a topology book, we will avoid the discussion of $K(G, 1)$ spaces in [69]. (But see [69, page 79].)

The questions we are interested in are *What properties of G (respectively, of X) can be studied in terms of X (respectively, of G)?* We will answer these questions by producing category equivalences between categories determined by G and categories determined by X.

Let **KAb** be the category of $K(A, 1)$-spaces where A ranges over abelian groups A. Let $\sim$ denote *is homotopic to.* Let $Y \in$ **KAb**. We say that Y is *K-indecomposable* if given $U, V \in \mathbf{KAb}$ such that $Y \sim U \times V$ then U or V contracts to a point. We say that Y has a *unique finite Cartesian K-decomposition* if

1. $Y \sim \prod_{i \in \mathcal{I}} U_i$ for some finite set $\{U_i \mid i \in \mathcal{I}\} \subset \mathbf{KAb}$ of K-indecomposable spaces, and
2. If $Y \sim \prod_{j \in \mathcal{J}} V_j$ for some finite set $\{V_j \mid j \in \mathcal{J}\} \subset \mathbf{KAb}$ of K-indecomposable spaces, then there is a bijection $\Sigma : \mathcal{I} \longrightarrow \mathcal{J}$ such that $U_i \sim V_{\Sigma(i)}$ for each $i \in \mathcal{I}$.

The topological space Y has the *power cancellation property in* **KAb** if for each integer $n > 0$ and topological space $Z \in \mathbf{KAb}$, $Y^n \sim Z^n \Rightarrow Y \sim Z$. We say that Y has a Σ-*unique decomposition in* **KAb** if for each integer $n > 0$, Y^n has a unique finite Cartesian K-decomposition. These direct sum properties appear surprisingly in the study of the unique factorization problem in algebraic number fields.

17.1 The ring End$_C$(X)

For topological spaces U and V let

$$\mathrm{Hom}_{\mathbf{C}}(U, V) = \{[f] \mid f : U \longrightarrow V \text{ is a continuous map}\}$$

for topological spaces U and V. Let $\mathrm{End}_{\mathbf{C}}(X) = \mathrm{Hom}_{\mathbf{C}}(X, X)$.

Our choice of X implies a lifting property.

Theorem 17.1. *[69, Proposition 1B.9] Let V be a $K(A, 1)$ for some abelian group A and let U be a connected CW-complex. A group homomorphism $\phi : \pi_1(U) \longrightarrow \pi_1(V)$ is induced by a continuous map $f : U \longrightarrow V$ that is unique up to homotopy equivalence.*

Lemma 17.2. *Let G be an abelian group, let $X = K(G, 1)$, and let $U = K(A, 1)$ space for some abelian group A.*

1. $\alpha : \mathrm{Hom}_{\mathbf{C}}(X, U) \longrightarrow \mathrm{Hom}(G, A) : [f] \mapsto f^*$ *is a functorial bijection.*
2. $\beta : \mathrm{Hom}_{\mathbf{C}}(U, X) \longrightarrow \mathrm{Hom}(A, G) : [f] \mapsto f^*$ *is a functorial bijection.*
3. $\rho : \mathrm{End}_{\mathbf{C}}(X) \longrightarrow \mathrm{End}_R(G) : [f] \mapsto f^*$ *is a functorial bijection.*

Proof: 1. α, β, and ρ are well defined functions, and certainly $(\cdot)^*$ is functorial, so that α, β, and ρ are functorial. Let $\phi \in \mathrm{Hom}(G, A)$. Since X is connected, Theorem 17.1 implies that $\phi = f^*$ for some continuous map $f : X \longrightarrow U$. Thus $\alpha : \mathrm{Hom}_{\mathbf{C}}(X, U) \longrightarrow \mathrm{Hom}(G, A)$ is a surjection.

Next assume that $f^* = g^* = \phi$ for some $f, g \in \mathrm{Hom}_{\mathbf{C}}(X, U)$. Since $U = K(A, 1)$, Theorem 17.1 shows us that ϕ is induced by a map in $\mathrm{Hom}_{\mathbf{C}}(X, U)$ that is unique up to homotopy equivalence. Thus $[f] = [g]$, whence α is a bijection. The maps β and ρ are handled in a similar manner. This completes the proof.

It follows from Lemma 17.2 that

$$[g] = \alpha^{-1}(g^*) \tag{17.1}$$

for each continuous map $g : X \longrightarrow U$. Similarly, $[f] = \rho^{-1}(f^*)$ for each continuous map $f : X \longrightarrow X$. Thus the bijection ρ makes $\text{End}_{\mathbf{C}}(X)$ an associative ring with identity, and α makes $\text{Hom}_{\mathbf{C}}(X, U)$ a right $\text{End}_{\mathbf{C}}(X)$-module.

Lemma 17.3. *Let G be an abelian group and let $X = K(G, 1)$.*

1. *The set $\text{End}_{\mathbf{C}}(X)$ possesses a natural ring structure and ρ is a ring isomorphism.*
2. *Let U be a $K(A, 1)$ space for some abelian group A. Then $\text{Hom}_{\mathbf{C}}(X, U)$ is a right $\text{End}_{\mathbf{C}}(X)$-module and α is a right $\text{End}_{\mathbf{C}}(X)$-module isomorphism.*
3. *Let U be a connected CW-complex. Then $\text{Hom}_{\mathbf{C}}(U, X)$ is a left $\text{End}_{\mathbf{C}}(X)$-module and β is an isomorphism of left $\text{End}_{\mathbf{C}}(X)$-modules.*

Proof: The proof of this lemma follows from Theorem 17.1 and Lemma 17.2.

Corollary 17.4. *Let G be an abelian group and let $X = K(G, 1)$. The ring isomorphism ρ induces category isomorphisms*

$$\rho_R : \mathbf{Mod}\text{-}\text{End}_{\mathbf{C}}(X) \longrightarrow \mathbf{Mod}\text{-}\text{End}_R(G)$$

and

$$\rho_L : \text{End}_{\mathbf{C}}(X)\text{-}\mathbf{Mod} \longrightarrow \text{End}_R(G)\text{-}\mathbf{Mod}$$

sending $\text{End}_{\mathbf{C}}(X)$ to $\text{End}_R(G)$.

Corollary 17.5. *Let G be an abelian group and let $X = K(G, 1)$. Let $\mathbf{KAb}$ be the category whose objects are $K(A, 1)$ spaces for abelian groups A, and whose maps are the homotopy equivalence classes of continuous maps. There is an additive left exact functor*

$$\text{Hom}_{\mathbf{C}}(X, \cdot) : \mathbf{KAb} \longrightarrow \mathbf{Mod}\text{-}\text{End}_{\mathbf{C}}(X).$$

17.2 Topological complexes

We showed above that there is a ring of endomorphisms associated with a $K(G, 1)$ space X. This suggests that we can study X via these endomorphisms. Specifically, we will diagram the connection between a category in which X is a small projective generator and the category of right $\text{End}_{\mathbf{C}}(X)$-modules. To do this we will parallel our study of G-**Plex** and the category of right $\text{End}(G)$-modules.

The ring $\text{End}_{\mathbf{C}}(X)$ suggests that we can study X in the same way we studied G using G-*plexes*. An X-*plex* is a complex

$$\mathcal{C} = \cdots \xrightarrow{\lambda_3} X_2 \xrightarrow{\lambda_2} X_1 \xrightarrow{\lambda_1} X_0 \tag{17.2}$$

of continuous maps λ_k, $k > 0$ such that

1. For each integer $k \geq 0$, X_k is a $K(Q_k, 1)$ space for some $Q_k \in \mathbf{P}(G)$,
2. For each integer $k > 0$, $\lambda_k \lambda_{k+1}$ is a null homotopic map.

3. For each integer $k > 0$ and for each continuous map $f : X \longrightarrow X_k$ such that $\lambda_k f$ is a null homotopic map there is a map $g : X \longrightarrow X_{k+1}$ such that $f \sim \lambda_{k+1} g$ as in the diagram below.

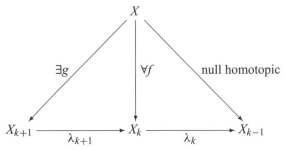

A *ladder map* $f : \mathcal{C} \longrightarrow \mathcal{C}'$ is a sequence $f = (\ldots, f_2, f_1, f_0)$ of continuous maps such that

$$f_{k-1} \lambda_k \sim \lambda'_k f_k \text{ for each integer } k > 0.$$

The way to remember this is by the familiar commutative diagram below.

$$\cdots \xrightarrow{\lambda_{k+1}} X_k \xrightarrow{\lambda_k} X_{k-1} \xrightarrow{\lambda_{k-1}} \cdots$$

$$\Big\downarrow f_k \qquad \Big\downarrow f_{k-1}$$

$$\cdots \xrightarrow{\lambda'_{k+1}} X'_k \xrightarrow{\lambda'_k} X'_{k-1} \xrightarrow{\lambda'_{k-1}} \cdots$$

We define *the homotopy equivalence of ladder maps* between X-plexes. Let f and $g : \mathcal{C} \longrightarrow \mathcal{C}'$ be ladder maps. Because $a^* = b^*$ whenever $a \sim b$, we see that $f^* = (\ldots, f_2^*, f_1^*, f_0^*) : \pi_1(\mathcal{C}) \longrightarrow \pi_1(\mathcal{C}')$ is a chain map between complexes of abelian groups. The classic homotopy equivalence class, $[f^*]$ is then defined. (See [97].) We say that f *is homotopic to* g, and we write $f \sim g$, if $[f^*] = [g^*]$. We could define the homotopy equivalence of ladder maps by an appropriate generalization of homotopy of chain maps between complexes, but that would take us too far afield. You are invited to try, though. The reader will show that *is homotopic to* is an equivalence relation on the set of ladder maps between X-plexes $\mathcal{C}$ and $\mathcal{C}'$.

> Let X-**Plex** denote the category whose objects are X-plexes, and whose maps $[f] : \mathcal{C} \longrightarrow \mathcal{C}'$ are the homotopy equivalence classes of ladder maps $f : \mathcal{C} \longrightarrow \mathcal{C}'$.

In the next few paragraphs we dualize the category X-**Plex**.
An X-*coplex* is a complex

$$\mathcal{U} = U_0 \xrightarrow{\mu_1} U_1 \xrightarrow{\mu_2} U_2 \xrightarrow{\mu_3} \cdots$$

of maps μ_k and integers $k > 0$, such that:

1. For each integer $k \geq 0$, U_k is a $K(W_k, 1)$ space for some $W_k \in \mathbf{coP}(G)$,
2. For each $k > 0$, $\mu_{k+1}\mu_k$ is a null homotopic map, and
3. For a given integer $k > 0$, and given a map $f : U_k \longrightarrow X$ such that $f\mu_k$ is null homotopic then there is a map $g : U_{k+1} \longrightarrow X$ such that $g\mu_{k+1} \sim f$ as in the accompanying commutative diagram.

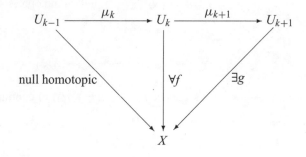

> Let X-**coPlex** denote the category whose objects are the
> X-coplexes and whose maps are the homotopy equivalence
> classes $[f]$ of ladder maps $f : \mathcal{U} \longrightarrow \mathcal{U}'$.

17.3 Categories of complexes

Lemma 17.6. *Let G be an abelian group and let $X = K(G, 1)$. Then $\pi_1(\cdot)$ lifts to a functor*

$$\widehat{\pi}_1(\cdot) : X\text{-}\mathbf{Plex} \longrightarrow G\text{-}\mathbf{Plex}$$

defined by

$$\widehat{\pi}_1(\mathcal{C}) = \cdots \xrightarrow{\lambda_2^*} \pi_1(X_1) \xrightarrow{\lambda_1^*} \pi_1(X_0) \tag{17.3}$$

for X-plexes $\mathcal{C}$ and

$$\widehat{\pi}_1([f]) = [(\ldots, f_2^*, f_1^*, f_0^*)]$$

for ladder maps $f = (\ldots, f_2, f_1, f_0) : \mathcal{C} \longrightarrow \mathcal{C}'$ in X-plex.

Proof: Define $\widehat{\pi}_1$ as in the lemma. We must show that $\widehat{\pi}_1(\mathcal{C})$ is a G-plex and then prove that $(\ldots, f_2^*, f_1^*, f_0^*)$ is a chain map between G-plexes.

Since $\mathcal{C}$ is an X-plex, for each integer $k > 0$ there is a $Q_k \in \mathbf{P}(G)$ such that $X_k = K(Q_k, 1)$ space. Then

$$\pi_1(X_k) \cong Q_k,$$

so that the terms of $\widehat{\pi}_1(\mathcal{C})$ are objects in $\mathbf{P}(G)$.

Suppose that $\phi : G \longrightarrow \pi_1(X_k)$ is a map such that $\lambda_k^* \phi = 0$. By Theorem 17.1 there is a continuous map $f : X \longrightarrow X_k$ unique up to homotopy equivalence such that $f^* = \phi$. Then

$$(\lambda_k f)^* = \lambda_k^* f^* = \lambda_k^* \phi = 0.$$

By the uniqueness statement in Theorem 17.1, $\lambda_k f$ is a null homotopic map, so by the lifting condition that defines an X-plex (page 287) there is a map $g : X \longrightarrow X_{k+1}$ such that $\lambda_{k+1} g \sim f$. Hence,

$$\lambda_{k+1}^* g^* = (\lambda_{k+1} g)^* = f^* = \phi$$

and therefore $\widehat{\pi}_1(\mathcal{C})$ is a G-plex.

Moreover, the ladder map $f = (\dots, f_2, f_1, f_0) : \mathcal{C} \longrightarrow \mathcal{C}'$ is mapped to the chain map $f^* = (\dots, f_2^*, f_1^*, f_0^*) : \pi_1(\mathcal{C}) \longrightarrow \pi_1(\mathcal{C}')$ between G-plexes. Thus $[f^*]$ is a mapping in G-**Plex**. This completes the proof of the lemma.

Theorem 17.7. *Let G be an abelian group and let $X = K(G,1)$. Then*

$$\widehat{\pi}_1(\cdot) : X\text{-}\mathbf{Plex} \longrightarrow G\text{-}\mathbf{Plex}$$

is a category equivalence.

Proof: Let $\mathcal{Q}$ be a G-plex. For each integer $k \geq 0$ choose a $K(\mathcal{Q}_k, 1)$ space X_k, whence $\pi_1(X_k) = \mathcal{Q}_k$. Since X_k is connected, Theorem 17.1 can be employed to show that for integers $k > 0$ there are continuous maps $\lambda_k : X_k \longrightarrow X_{k-1}$, $k > 0$, such that $\lambda_k^* = \delta_k$. Since

$$0 = \delta_k \delta_{k+1} = (\lambda_k \lambda_{k+1})^*$$

and since the lifting is unique up to homotopy equivalence, $\lambda_k \lambda_{k+1}$ is null homotopic. We have constructed a *sequence of continuous maps*

$$\mathcal{C} = \cdots \xrightarrow{\lambda_3} C_2 \xrightarrow{\lambda_2} C_1 \xrightarrow{\lambda_1} C_0$$

such that $\pi_1(\mathcal{C}) = \mathcal{Q}$.

Let $k > 0$ be an integer and let $f : X \longrightarrow X_k$ be a continuous map such that $\lambda_k f$ is null homotopic. Then $\lambda_k^* f^* = (\lambda_k f)^* = 0$ and $f^* : G \longrightarrow \pi_1(X_k) = \mathcal{Q}_k$. By the lifting property of a G-plex $\mathcal{Q}$ given on page 114, there is a map $\psi : G \longrightarrow \mathcal{Q}_{k+1}$ such that $f^* = \delta_{k+1} \psi = \lambda_{k+1}^* \psi$. Theorem 17.1 states that $\psi = g^*$ for some continuous map $g : X \longrightarrow X_{k+1}$. Then $f^* = \lambda_{k+1}^* g^* = (\lambda_{k+1} g)^*$. Since these liftings are unique up to homotopy equivalence, $f \sim \lambda_{k+1} g$. Thus $\mathcal{C}$ is an X-plex such that $\widehat{\pi}_1(\mathcal{Q}) = \mathcal{C}$.

If $\widehat{\pi}_1(f) = \widehat{\pi}_1(g)$ for some ladder maps f, g in X-**Plex** then $[f^*] = [g^*]$, which by definition implies that $f \sim g$. Therefore $\widehat{\pi}_1(\cdot) : X\text{-}\mathbf{Plex} \longrightarrow G\text{-}\mathbf{Plex}$ is a faithful functor.

Let $f = (\dots, f_2, f_1, f_0) : \mathcal{Q} \longrightarrow \mathcal{Q}'$ be a chain map between G-plexes. By the above construction, there are complexes $\mathcal{C}$ and $\mathcal{C}'$ such that $\pi_1(\mathcal{C}) = \mathcal{Q}$ and $\pi_1(\mathcal{C}') = \mathcal{Q}'$.

By Theorem 17.1, for each integer $k \geq 0$ there are maps g_k, unique up to homotopy, such that $g_k^* = f_k$. By the fact that $f : \pi_1(\mathcal{C}) \longrightarrow \pi_1(\mathcal{C}')$ is a chain map, we have

$$g_k^* \lambda_{k+1}^* = \lambda_{k+1}'^* g_{k+1}^*.$$

Again the uniqueness of the lifting implies that $g_k \lambda_{k+1} \sim \lambda_{k+1}' g_{k+1}$. We have constructed the ladder map $g = (\ldots, g_2, g_1, g_0) : \mathcal{C} \longrightarrow \mathcal{C}'$ such that $g^* = f$. Thus $\widehat{\pi}_1(\cdot) : X\text{-}\mathbf{Plex} \longrightarrow G\text{-}\mathbf{Plex}$ is a full functor, and therefore $\widehat{\pi}_1(\cdot) : X\text{-}\mathbf{Plex} \longrightarrow G\text{-}\mathbf{Plex}$ is a category equivalence. This completes the proof.

Recall that (μ) is the statement that *measurable cardinals do not exist*. The following result is dual to Theorem 17.7.

Theorem 17.8. *Assume* (μ). *Let G be an abelian group and let $X = K(G, 1)$. There is a category equivalence*

$$\widehat{\pi}_1^*(\cdot) : X\text{-}\mathbf{coPlex} \longrightarrow G\text{-}\mathbf{coPlex}$$

defined by

$$\widehat{\pi}_1^*(\mathcal{W}) = \pi_1(W_0) \xrightarrow{\lambda_1^*} \pi_1(W_1) \xrightarrow{\lambda_2^*} \pi_1(W_2) \xrightarrow{\lambda_3^*} \cdots$$

for each X-coplex

$$\mathcal{W} = W_0 \xrightarrow{\lambda_1} W_1 \xrightarrow{\lambda_2} W_2 \xrightarrow{\lambda_3} \cdots$$

and

$$\widehat{\pi}_1^*(f) = (\ldots, f_2^*, f_1^*, f_0^*)$$

for ladder maps

$$f : \mathcal{W} \longrightarrow \mathcal{W}'.$$

Corollary 17.9. *The inverse of $\widehat{\pi}_1 : X\text{-}\mathbf{Plex} \longrightarrow G\text{-}\mathbf{Plex}$ is the functor*

$$\widehat{K}(\cdot) : G\text{-}\mathbf{Plex} \longrightarrow X\text{-}\mathbf{Plex}$$

given by $\widehat{K}(A) = K(A, 1)$.

Proof: The construction of $K(A, 1)$ is functorial in A by [69] so $\widehat{K}(\cdot)$ is a functor. By its construction $\pi_1(K(A, 1)) \cong A$ so, by Theorem 17.7, $\widehat{K}(\cdot)$ is the inverse of $\widehat{\pi}_1$. This proves the corollary.

Dual to Theorem 17.7 is Theorem 17.8.

Theorem 17.10. *Let G be an abelian group and let* $X = K(G, 1)$. *Then*

$$\widehat{\pi}_1^*(\cdot) : X\text{-}\mathbf{coPlex} \longrightarrow G\text{-}\mathbf{coPlex}$$

is a category equivalence with inverse

$$\widehat{K}^*(\cdot) : G\text{-}\mathbf{coPlex} \longrightarrow X\text{-}\mathbf{coPlex}$$

given by $\widehat{K}^*(\cdot) = K(\cdot, 1)$.

Proof: Dualize the proof of Theorem 17.7.

17.4 Commutative triangles

We use the category equivalences as a basis for the contruction of the commutative triangles Diagram (17.1) and Diagram (17.3). These diagrams provide a functorial connection between the topological structure of X and the algebraic structure of G.

Let H_0 denote the zeroth homology functor, and define a functor

$$\boxed{h_X(\cdot) : X\text{-}\mathbf{Plex} \longrightarrow \mathbf{Mod}\text{-}\mathrm{End}_{\mathbf{C}}(X)}$$

by

$$h_X(\mathcal{C}) = H_0 \circ \mathrm{Hom}_{\mathbf{C}}(X, \mathcal{C})$$

for X-plexes $\mathcal{C}$ and

$$h_X([f]) = H_0 \circ \mathrm{Hom}_{\mathbf{C}}(X, f)$$

for ladder maps $f : \mathcal{C} \longrightarrow \mathcal{C}'$.

Theorem 17.11. *Let G be an abelian group and let* $X = K(G, 1)$. *There is a commutative triangle*

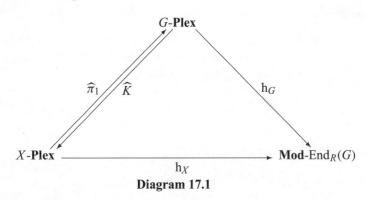

Diagram 17.1

in which opposing arrows denote inverse category equivalences.

Proof: Let $\mathcal{C} \in X$-**Plex**. The terms X_k in $\mathcal{C}$ satisfy $\widehat{\pi}_1(X_k) \cong Q_k$ for some $Q_k \in \mathbf{P}(G)$. Also the maps $\lambda_k : X_k \longrightarrow X_{k-1}$ in $\mathcal{C}$ satisfy $\widehat{\lambda}_k = \delta_k$ for some $\delta_k : Q_k \longrightarrow Q_{k-1}$. Hence $\widehat{\pi}_1(\mathcal{C}) = Q$. See page 114. Then by Corollary 17.3,

$$h_G\widehat{\pi}_1(\mathcal{C}) = H_0 \circ \mathrm{Hom}(G, \widehat{\pi}_1(\mathcal{C})) \cong \mathrm{coker}\, \mathrm{Hom}(X, \widehat{\pi}_1(\mathcal{C})) = h_X(\mathcal{C}).$$

Similarly, $h_X([f]) = h_G\widehat{\pi}_1([f])$, so the triangle commutes.

Theorem 17.12. *Let G be a self-small abelian group and let $X = K(G, 1)$. There is a commutative triangle*

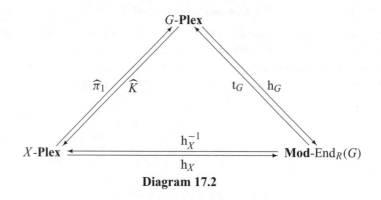

Diagram 17.2

in which opposing arrows represent inverse functors.

Proof: The functor $\widehat{\pi}_1(\cdot)$ has an inverse functor $\widehat{K}(\cdot)$ by Theorem 17.7. The functors $h_G(\cdot)$ and $t_G(\cdot)$ are inverse functors by Theorem 9.12. Then by Theorem 17.11, $h_X(\cdot)$ is a category equivalence. Let $h_X^{-1}(\cdot)$ denote its inverse functor. This completes the proof.

By dualizing the above arguments one can prove the following results. Define a functor

$$\boxed{h^X(\cdot) : X\text{-}\mathbf{coPlex} \longrightarrow \mathrm{End}_R(G)\text{-}\mathbf{Mod}}$$

by

$$h^X(\mathcal{W}) = H_0 \circ \mathrm{Hom}_{\mathcal{C}}(\mathcal{W}, X)$$
$$h^X([f]) = H_0 \circ \mathrm{Hom}_{\mathcal{C}}(f, X).$$

Theorem 17.13. *Let G be an abelian group and let* $X = K(G, 1)$. *There is a commutative diagram*

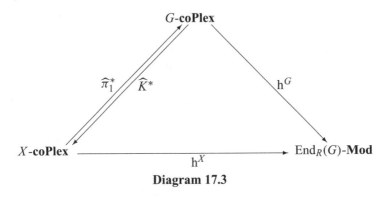

Diagram 17.3

in which opposing arrows represent inverse dualities.

Theorem 17.14. *Assume* (μ), *let G be a self-slender abelian group, and let* $X = K(G, 1)$. *There is a commutative diagram*

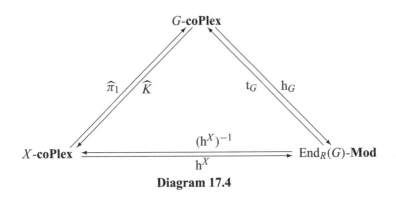

Diagram 17.4

in which opposing arrows represent inverse dualities.

Remark 17.15. It is worth noting that these diagrams begin with an abelian group. Moreover, the vertices of these triangles represent categories from ring theory, homology theory, and the topology of point set topological spaces.

Remark 17.16. A right R-module G is self-small if it is finitely generated as an R-module or if its additive structure $(G, +)$ is a self-small abelian group. At the time of this writing, the only known examples of self-slender right R-modules G are those G whose additive structure $(G, +)$ is a self-slender abelian group. The *torsion-free* abelian group G is self-slender iff G does not contain copies of $\mathbf{Q}$, $\prod_{\mathbf{N}} \mathbf{Z}$, or the p-adic integers $\widehat{\mathbf{Z}}_p$ for some rational prime p, [59, Theorem 95.3].

17.5 Three diamonds

17.5.1 A diagram for an abelian groups

The goal of this section is to construct Diagrams (17.5), (17.6), and (17.7).

Recall that **SAb** is the category of sequences of abelian groups. Let **SKAb** be the category of sequences $(\ldots, X_2, X_1)$ of topological spaces where $X_k = K(A_k, 1)$ and where A_k takes values in the category **Ab** of abelian groups. Maps in **SKAb** are $(\ldots, [f_2], [f_1])$ where $[f_k]$ is the homotopy equivalence class of the continuous map $f_k : X_k \longrightarrow X'_k$. We established in Theorems 17.7 and 17.10 that the category equivalences $\widehat{\pi}_1, \widehat{\pi}_1^*, \widehat{K}$, and $\widehat{K}^*$ exist.

For each abelian group A the functorially constructed topological space $K(A, 1)$ has the property that

$$\pi_p(K(A, 1)) \cong \begin{cases} A & \text{if } p = 1 \text{ and} \\ 0 & \text{for all integers } p > 1. \end{cases} \tag{17.4}$$

Define a functor

$$K(\cdot) : \mathbf{SAb} \longrightarrow \mathbf{SKAb}$$

by

$$K(\ldots, A_2, A_1) = (\ldots K(A_2, 1), K(A_1, 1))$$

for each $S = (\ldots, A_2, A_1) \in \mathbf{SAb}$.

Lemma 17.17. *The functors* $\Pi_1(\cdot)$ *and* $K(\cdot)$ *are inverse category equivalences.*

Proof: By (17.4) we have

$$\Pi_1 \circ K(S) \cong S$$

for each sequence $S \in \mathbf{SAb}$. Since $K(A, 1)$ is unique to A up to homotopy, we have

$$K \circ \Pi_1(\ldots, X_2, X_1) \sim (\ldots, X_2, X_1)$$

for $(\ldots, X_2, X_1) \in \mathbf{SKAb}$. This completes the proof.

The commutative triangle Diagram (17.1) is the triangle at the top of Diagram (17.5), and the commutative triangle Diagram (17.3) forms the bottom triangle in Diagram (17.5).

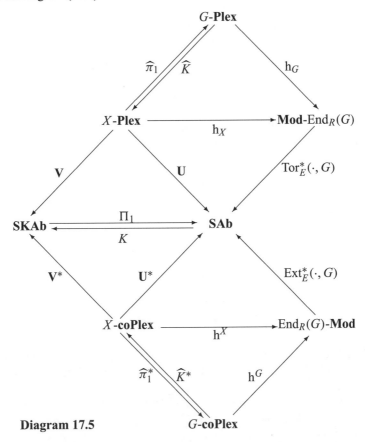

Diagram 17.5

As in Chapter 6 we let

$$\mathrm{Tor}_E^*(\cdot, G) = (\ldots, \mathrm{Tor}_E^3(\cdot, G), \mathrm{Tor}_E^2(\cdot, G), \mathrm{Tor}_E^1(\cdot, G))$$

$$\mathrm{Ext}_E^*(\cdot, G) = (\ldots, \mathrm{Ext}_E^3(\cdot, G), \mathrm{Ext}_E^2(\cdot, G), \mathrm{Ext}_E^1(\cdot, G)).$$

Each functor has image in the category **SAb**.

We define the functor

$$\mathbf{U}(\cdot) = \mathrm{Tor}_E^*(\mathrm{h}_X(\cdot), G) : X\text{-}\mathbf{Plex} \longrightarrow \mathbf{SAb}.$$

Define the functor **V** in Diagram (17.5),

$$\mathbf{V}(\cdot) = K \circ \mathbf{U}(\cdot) : X\text{-}\mathbf{Plex} \longrightarrow \mathbf{SKAb}.$$

Then $\mathbf{U}(\cdot)$ and $\mathbf{V}(\cdot)$ are the unique functors that make the top half of Diagram (17.5) commute.

Proposition 17.18. *For each X-plex C*

$$h_X(\mathcal{C}) = H_0 \circ \mathrm{Hom}(G, \widehat{\pi}_1(\mathcal{C})) = \mathrm{coker}\,\mathrm{Hom}(G, \lambda_1^*).$$

Similarly, do not assume (μ). Define the unique functors $\mathbf{U}^*(\cdot)$ and $\mathbf{V}^*(\cdot)$ that make the lower half of Diagram (17.5) commute. We need not assume (μ) since coker $\mathrm{Hom}(\lambda_1, G)$ exists for any map λ_1. See page 138.

Remark 17.19. We can fill in a blank place in Diagram (17.5) as follows. Let **Bim** be the set of $\mathrm{End}_R(G)$-$\mathrm{End}_R(G)$-bimodules M such that

$$\mathrm{Tor}_E^*(M, G) \cong \mathrm{Ext}_E^*(M, G).$$

Then **Bim** and the canonical inclusion functors

$$\iota : \mathbf{Bim} \longrightarrow \mathbf{Mod}\text{-}\mathrm{End}_R(G)$$

and

$$\jmath : \mathbf{Bim} \longrightarrow \mathrm{End}_R(G)\text{-}\mathbf{Mod}$$

embed in Diagram (17.5) in a way that makes Diagram (17.5) commute. It is interesting that such an unnatural isomorphism should arise in such a natural context.

Theorem 17.20. *Suppose that G is an abelian group, and let $X = K(G, 1)$. There is a commutative diagram, Diagram (17.5), of additive categories and functors, in which opposing arrows denote inverse functors.*

Remark 17.21. All of the arrows in Diagrams (17.1) and (17.5) are functors. This property does not apply to the diagram constructed in Diagram (16.1), where some of the arrows are not functors.

17.5.2 Self-small and self-slender

If we assume (μ), then self-slender groups G exist. Assume that G is self-small and self-slender, then we will produce the more detailed diagram, Diagram (17.6). We begin with Diagram (17.5).

Since G is self-small, there is a commutative triangle Diagram (17.2) of categories and category equivalences. This triangle forms the top commutative triangle in Diagram (17.6). Similarly, assuming (μ) and assuming that G is self-slender, Diagram (17.4) forms the bottom commutative triangle in Diagram (17.6).

The vertical maps in Diagram (17.6) are the homology functors

$$H_*^P(\cdot) : G\text{-}\mathbf{Plex} \longrightarrow \mathbf{SAb} \quad \text{and}$$
$$H_*^c(\cdot) : G\text{-}\mathbf{coPlex} \longrightarrow \mathbf{SAb}$$

of algebraic complexes. If G is self-small, then Lemma 11.19 states that for each integer $k > 0$

$$\text{Tor}^k_{\text{End}_R(G)}(\text{h}_G(\cdot), G) = H^P_k(\cdot),$$

so that the inclusion of the homology functor $H^P_*(\cdot)$ in Diagram (17.6) preserves the commutativity of the diagram.

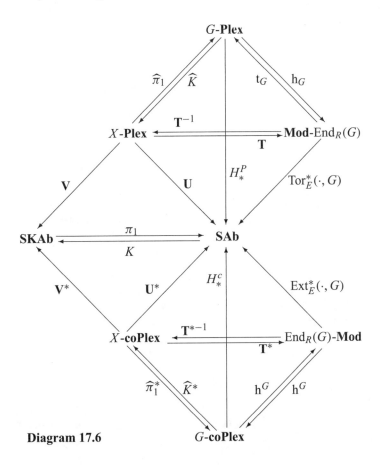

Diagram 17.6

Assuming (μ), and assuming that G is self-slender, page 132 contains the identity

$$\text{Ext}^k_E(\text{h}^G(\cdot), G) = H^c_k(\cdot)$$

for each integer $k > 0$. Thus we can insert the vertical homology functor $H^c_*(\cdot)$ into the bottom diamond of Diagram (17.6) and preserve the commutativity of the diagram.

Theorem 17.22. *Assume (μ), and let G be a self-small and a self-sender abelian group. There is a commutative diagram (17.6) of additive categories and functors, in which opposing arrows represent inverse category equivalences.*

Remark 17.23. Self-small and self-slender abelian groups occur naturally when studying $\mathrm{End}_R(G)$ and G. For instance [15] shows that self-small abelian groups occur naturally when studying **Mod**-$\mathrm{End}_R(G)$, while the work in [70] shows that self-slender abelian groups occur naturally when studying $\mathrm{End}_R(G)$-**Mod**. Reduced torsion-free finite-rank abelian groups are both self-small and self-slender. If M is a righht R-module and if $M, +$ is an rtffr group then M is self-small and self-slender.

Corollary 17.24. *Assume (μ) and assume that G is a rtffr group. There is a commutative diagram (17.6) of additive categories and functors, in which opposing arrows represent inverse category equivalences.*

17.5.3 Coherent complexes

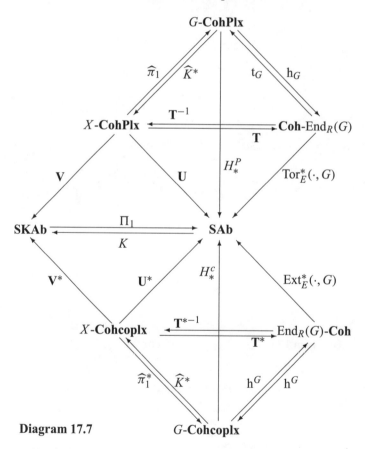

Diagram 17.7

Let us examine which parts of Diagrams (17.2) and (17.6) are retained when we remove the self-small and self-slender assumptions from the abelian group G.

Review coherent right $\mathrm{End}_R(G)$-modules and coherent G-plexes from the section beginning on page 138.

An X-plex $\mathcal{C}$ is *coherent* if each term X_k of $\mathcal{C}$ is a $K(Q_k, 1)$ space for some direct summand Q_k of $\oplus_n G$ for some integer $n = n(\mathcal{C}, k)$. The category X-**CohPlx** is the full subcategory of X-**Plex** whose objects are coherent X-plexes. We say that the X-complex $\mathcal{W}$ is a *coherent X-coplex* if each term W_k of $\mathcal{W}$ is a $K(Q_k, 1)$ space for some direct summand Q_k of $\oplus_n G$ for some integer $n = n(\mathcal{W}, k)$. The full subcategory of X-**coPlex** whose objects are coherent G-coplexes is denoted by X-**Cohcoplx**.

Theorem 17.25. *Let G be an abelian group and let $X = K(G, 1)$. There is a commutative triangle*

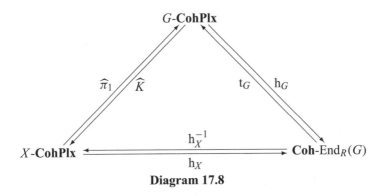

Diagram 17.8

of categories and category equivalences.

Proof: The proof is left as an exercise for the reader.

Dualizing the above process yields a different commutative triangle. Notice the absence of the hypotheses in the following theorem.

Theorem 17.26. *Let G be an abelian group and let $X = K(G, 1)$. There is a commutative triangle*

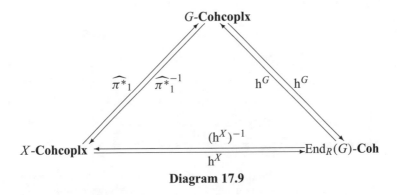

Diagram 17.9

of categories and inverse category equivalences.

Substituting Diagrams (17.8) and (17.9) into Diagram (17.7) yields a commutative diagram.

17.6 Prism diagrams

The diagrams above suggest that there is a connection between topology and algebra that can be expressed as category equivalences and dualities between categories determined by an abelian group. We will further examine this relationship by showing that for each abelian group G and $K(G, 1)$ space X there is a commutative triangle determined by X and a commutative triangle determined by G whose vertices are categorically equivalent.

Unless stated explicitly, we make no assumptions other than that G is an abelian group and $X = K(G, 1)$.

The functors $\widehat{\pi}_1(\cdot)$, $\widehat{K}(\cdot)$, $h_X(\cdot)$, $h_G(\cdot)\,\pi_1$, and $K(\cdot)$ have been defined.

Define the arrow χ on the right side of Diagram (17.10) so that the triangle commutes.

$$\chi(\cdot) = \mathrm{Tor}_E^*(\cdot, G) \circ h_G(\cdot).$$

By the commutativity of Diagram (17.1), the upper triangle of Diagram (17.10) is commutative. Define $\tau(\cdot)$ so that Diagram (17.10) commutes.

$$\tau(\cdot) = K \circ \mathrm{Tor}_E^*(\cdot, G).$$

Then define $\sigma(\cdot)$ so that Diagram (17.10) commutes.

$$\sigma(\cdot) = \tau \circ h_X(\cdot).$$

Theorem 17.27. *Let G be an abelian group.*

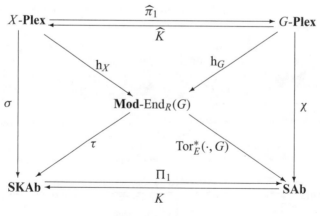

Diagram 17.10

There is a commutative diagram Diagram (17.10) of categories and functors, in which opposing arrows represent inverse category equivalences.

By dualizing the above process we construct Diagram (17.11).

Theorem 17.28. *Let G be an abelian group.*

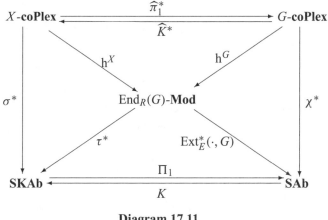

Diagram 17.11

There is a commutative diagram Diagram (17.11) of categories and functors, in which opposing arrows represent inverse category equivalences.

In the case where G is self-small or self-slender we can be specific about χ and χ^*.

Corollary 17.29. *Let G be an abelian group and let $X = K(G, 1)$.*

1. *Assume that G is self-small. Then $\mathrm{h}_G(\cdot)$ and $\mathrm{h}_X(\cdot)$ are category equivalences and $\chi(\cdot) \sim H_*^P(\cdot)$, the homology functor.*
2. *Assume (μ) and assume that G is self-slender. Then $\mathrm{h}^G(\cdot)$ and $\mathrm{h}^X(\cdot)$ are dualities, and $\chi^*(\cdot) \sim H_*^c(\cdot)$.*

Proof: Parts 1 and 2 follow from Diagrams (17.2) and (17.4). This completes the proof.

Remark 17.30. One of the first observations we can make about Diagrams (17.10) and (17.11) is that the triangle with vertices X-**Plex**, **Mod**-$\mathrm{End}_R(G) = $ **Mod**-$\mathrm{End}_C(X)$, and **SKAb** is one of topology, while the triangle with vertices G-**Plex**, **Mod**-$\mathrm{End}_R(G)$, and **SAb** is one of algebra. Because the horizontal arrows in Diagram (17.10) represent inverse functors, we see that the triangle of topology is the same as the triangle of algebra. The same can be said of the triangles in Diagram (17.11).

17.7 Direct sums

We will, as we did in Section 16.6, apply our techniques to the uniqueness of decomposition of abelian groups and topological spaces.

Let **S** be a category in which $\cong$ is an equivalence relation, and let **X** be a category in which $\sim$ is an equivalence relation. We say that $\mathbf{X}/\!\!\sim$ is *a complete set of topological invariants for* **S** *modulo* $\cong$ if there is a functorial bijection $\Sigma : \mathbf{S}/\!\!\cong \longrightarrow \mathbf{X}/\!\!\sim$.

Let **KAb** be *the category of $K(A, 1)$ spaces for some abelian group A*. Let **Ab** denote the *category of abelian groups*.

Theorem 17.31. *1. **KAb**/ $\sim$ is a complete set of topological invariants for **Ab** modulo isomorphism.*

*2. **Ab**/ $\cong$ is a complete set of algebraic invariants for **KAb** modulo homotopy equivalence.*

Proof: (a) The bijection $\Sigma : \textbf{Ab}/\cong \longrightarrow \textbf{KAb}/\sim$ is induced by the functor $K(\cdot)$ in Diagram (17.1) when we identify **Ab** with the set of sequences $(\dots, 0, 0, A_1)$, and we identify **KAb** with the sequences $(\dots, \{x\}, \{x\}, X_1)$.

(b) The bijection $\Sigma : \textbf{KAb}/\sim \longrightarrow \textbf{Ab}/\cong$ is induced by the functor $\pi_1(\cdot)$ in Diagram (17.1). This completes the proof.

Corollary 17.32. **KAb**/$\sim$ *is a complete set of topological invariants for* **Ab** *modulo isomorphism.*

Remark 17.33. Using the complete set of topological invariants **KAb**/$\sim$ of **Ab** we can characterize topologically any direct sum decomposition property in **Ab**. We demonstrate with a couple of our favorites.

Let $Y \in$ **KAb**. We say that Y is *K-indecomposable* if given $U, V \in$ **KAb** such that $Y \sim U \times V$ then U or V contracts to a point. We say that Y has a *unique K-decomposition* if

1. $Y = \prod_{i \in \mathcal{I}} X_i$ for some finite set $\{X_i \mid i \in \mathcal{I}\} \subset$ **KAb** of K-indecomposable spaces, (this product is called a **K-*decomposition of Y***), and
2. $Y = \prod_{j \in \mathcal{J}} Y_j$ for some finite set $\{Y_j \mid j \in \mathcal{J}\} \subset$ **KAb** of K-indecomposable spaces, then there is a bijection $\Sigma : \mathcal{I} \longrightarrow \mathcal{J}$ such that $X_i \sim Y_{\Sigma(i)}$ for each $i \in \mathcal{I}$.

We say that A has a *unique decomposition* if

1. $A = \bigoplus_{i \in \mathcal{I}} B_i$ for some finite set $\{B_i \mid i \in \mathcal{I}\}$ of indecomposable abelian groups, and
2. If $A = \bigoplus_{j \in \mathcal{J}} C_j$ for some finite set $\{C_j \mid j \in \mathcal{J}\}$ of indecomposable abelian groups, then there is a bijection $\Sigma : \mathcal{I} \longrightarrow \mathcal{J}$ such that $B_i \cong C_{\Sigma(i)}$ for each $i \in \mathcal{I}$.

Lemma 17.34. *Let A be an abelian group and let Y be a $K(A, 1)$ space.*

1. If $A = B \oplus C$ then $Y \sim K(B, 1) \times K(C, 1)$.
2. If $Y \sim U \times V$ for some $K(B, 1)$ and $K(C, 1)$ spaces U and V, respectively, then $A \cong B \oplus C$.
3. A is an indecomposable abelian group iff Y is K-indecomposable.

Proof: 1. Suppose that $A = B \oplus C$ as abelian groups. By [69, Theorem 1B.5], $K(B, 1) \times K(C, 1)$ is a $K(B \oplus C, 1) = K(A, 1)$ space. Hence $Y \sim K(B, 1) \times K(C, 1)$.

2. Suppose that $Y = U \times V$ for some $K(B, 1)$ and $K(C, 1)$ spaces U and V, respectively. By [69, Theorem 1B.5], Y is a $K(B \oplus C, 1)$ space, so that $A = \pi_1(Y) = B \oplus C$.

Part 3 follows from parts 1 and 2. This proves the lemma.

Theorem 17.35. *Let A be an abelian group and let Y be a K(A, 1) space. Then A has a unique decomposition iff Y has a unique K-decomposition.*

Proof: One proceeds as in Section 16.6. We sketch a proof. Suppose that A has a unique decomposition $\oplus_{i \in \mathcal{I}} B_i$ for some finite set $\{B_i \mid i \in \mathcal{I}\}$ of indecomposable abelian groups. Let $U_i = K(B_i, 1)$. Then by Lemma 17.34.3, U_i is K-indecomposable, and

$$Y = K(A, 1) = K\left(\bigoplus_{\mathcal{I}} B_i, 1\right) \sim \prod_{\mathcal{I}} K(B_i, 1) = \prod_{\mathcal{I}} U_i.$$

Another K-decomposition $Y \sim \prod_{j \in \mathcal{J}} V_j$ corresponds under the equivalence $K(\cdot)$ to a decomposition $\oplus_{\mathcal{J}} C_j$ of A. That is, $V_j \sim K(C_j) = K(C_j, 1)$. Since A has the unique finite direct sum decomposition $\oplus_{\mathcal{I}} B_i$ we may assume without loss of generality that $\mathcal{J} = \mathcal{I}$ and that $C_i = B_i$ for $i \in \mathcal{I}$. Then $V_i \sim K(C_i) = K(B_i) \sim U_i$, hence Y has a unique K-decomposition. The proof of the converse is found by reversing the above argument in the obvious way. This completes the proof.

Corollary 17.36. *Let A be an abelian group and let Y be a K(A, 1) space. Then Y has a unique K-decomposition if A is in one of the following sets.*

1. *A is a finitely generated abelian group.*
2. *A is a finite direct sum of subgroups of* **Q**.
3. *A is a divisible (=injective) abelian group of finite rank.*

Proof: By theorems found in [9], A possesses a unique decomposition if it falls into one of these sets. So by the previous result Y possesses a unique K-decomposition.

A space $Y \in \mathbf{KAb}$ has the *cancellation property in* **KAb** if given $Y \times U \sim Y \times V$ for some $U, V \in \mathbf{KAb}$, then $U \sim V$. We say that the abelian group A has the *cancellation property* if given $A \oplus B \cong A \oplus C$ for some abelian groups B and C then $B \cong C$.

Theorem 17.37. *Let A be an abelian group and let Y be a K(A, 1) space. Then A has the cancellation property iff Y has the cancellation property in* **KAb**.

Proof: The category equivalence $\pi_1(\cdot)$ takes finite Cartesian products in **KAb** to finite direct sums in **Ab**. Moreover $\pi_1(\cdot)$ has inverse $K(\cdot)$ on the objects of **Ab**. The rest of the proof is done in the by now familiar manner. This completes the proof.

Corollary 17.38. *Let A be an abelian group and let Y be a K(A, 1) space. Then Y has the cancellation property in* **KAb** *if A is in one of the following sets.*

1. *A is a finitely generated abelian group.*
2. *A is a divisible (=injective) abelian group.*
3. *A is a finite direct sum of pure subgroups of the* **Z**-*adic integers.*

Proof: Theorems found in [9, 59] state that groups A in these classes have the cancellation property. Then by the previous theorem Y has the cancellation property in **KAb**.

Recall that the group A has the *power cancellation property* if for each integer $n > 0$ and abelian group B, $A^n \cong B^n$ implies that $A \cong B$. Recall that A has a Σ-*unique decomposition* if for each integer $n > 0$, A^n has a unique finite direct sum decomposition.

The topological space Y has the *power cancellation property* in **KAb** if for each integer $n > 0$ and topological space $Z \in$ **KAb**, $Y^n \sim Z^n$ implies that $Y \sim Z$. We say that Y has Σ-*unique decomposition* in **KAb** if for each integer $n > 0$, Y^n has unique K-decomposition.

Then the proof of Theorem 17.35 can be used to prove the following three results.

Theorem 17.39. *Let A be an abelian group and let $Y = K(A, 1)$. Then A has the cancellation property iff Y has the cancellation property in* **KAb**.

Theorem 17.40. *Let A be an abelian group and let $Y = K(A, 1)$. Then A has the power cancellation property iff Y has the power cancellation property in* **KAb**.

Theorem 17.41. *Let A be an abelian group and let $Y = K(A, 1)$. Then A has a Σ-unique decomposition iff Y has a Σ-unique decomposition in* **KAb**.

Remark 17.42. The method given in the above section provides a systematic way of characterizing any direct sum decomposition property of an abelian group in topological terms.

17.8 Algebraic number fields

The above applications to direct sum decompositions of abelian groups and the work in Section 2.6 can be combined to give a nice classification of the unique factorization property in algebraic number fields.

Let G and H be abelian groups. Recall that G is *locally isomorphic to H* if given an integer $n > 0$ there is an integer $m > 0$ relatively prime to n, and maps $f : G \to H$ and $g : H \to G$ such that $fg = gf = m \cdot 1$. See Chapter 2 and [46] for properties of *locally isomorphic* abelian groups.

Let **k** be an algebraic number field, let $\overline{E}$ be the ring of algebraic integers in **k**, and let

$$\Omega(\overline{E}) = \{\text{abelian groups } G \,|\, \text{End}(G) \cong \overline{E}\}.$$

The *class number of* **k**, denoted by

$$h(\mathbf{k})$$

is the number of isomorphism classes of fractional right ideals of $\overline{E}$.

Given an abelian group G the *class number of G* is the number of isomorphism classes of groups H that are locally isomorphic to G. The class number of G is denoted by

$$h(G).$$

The spaces X and Y are *power homotopic* iff there is an integer $n > 0$ such that $X^n \sim Y^n$. Given a topological space X the *class number of X in* **KAb** is the number of homotopy classes of topological spaces Y that are power isomorphic to X in **KAb**. The *class number of X* is denoted by

$$h(X).$$

The abelian group G has $\Sigma(h)$-*unique decomposition* if $h > 0$ is the least integer that has the following property. There is a finite set $\{G_1, \ldots, G_h\}$ of indecomposable groups G_i such that for each $n > 0$, if $G^n = H_1 \oplus \cdots \oplus H_t$ for some integer t and indecomposable abelian groups $H_1, \ldots, H_t$ then for each $i = 1, \ldots, t$ there is an integer $1 \le j(i) \le h$ such that $H_i \cong G_{j(i)}$.

The topological space X has $\Sigma(h)$-*unique decomposition in* **KAb** if $h > 0$ is the least integer with the following property. There is a finite set $\{X_1, \ldots, X_h\}$ of K-indecomposable spaces X_i such that for each $n > 0$, if $X^n \sim Y_1 \oplus \cdots \oplus Y_t$ for some integer t and K-indecomposable spaces $Y_1, \ldots, Y_t$ in **KAb** then for each $i = 1, \ldots, t$ there is an integer $1 \le j(i) \le h$ such that $Y_i \sim X_{j(i)}$.

Two sequences s_n and t_n are said to be *asymptotically equal* if $\lim_{n \to \infty} s_n/t_n = 1$.

Theorem 17.43. *Let* **k** *be an algebraic number field of degree f, and let $h > 0$ be an integer. Let $L(p)$ and $h(G(p))$ be the integers defined in Property 3.14. The following are equivalent.*

1. $h(\mathbf{k}) = h$.
2. *The sequence $\{L(p)h(G(p)) \mid$ rational primes $p\}$ is asymptotically equal to the sequence $\{hp^{f-1} \mid$ rational primes $p\}$.*
3. $h(G) = h$ *for each $G \in \Omega(\overline{E})$.*
4. $h(G) = h$ *for some $G \in \Omega(\overline{E})$.*
5. *For each $G \in \Omega(\overline{E})$ there are exactly h isomorphism classes of groups H such that $G^n \cong H^n$ for some integer $n > 0$.*
6. *For some $G \in \Omega(\overline{E})$ there are exactly h isomorphism classes of groups H such that $G^n \cong H^n$ for some integer $n > 0$.*
7. *Each $G \in \Omega(\overline{E})$ has $\Sigma(h)$-unique decomposition.*
8. *Some $G \in \Omega(\overline{E})$ has $\Sigma(h)$-unique decomposition.*
9. *Given $G \in \Omega(\overline{E})$, $h(X) = h$ for each $K(G, 1)$-space X.*
10. *For some $G \in \Omega(\overline{E})$, $h(X) = h$ for each $K(G, 1)$-space X.*
11. *Given $G \in \Omega(\overline{E})$, each $K(G, 1)$-space has $\Sigma(h)$-unique decomposition in* **KAb**.
12. *For some $G \in \Omega(\overline{E})$, each $K(G, 1)$-space has $\Sigma(h)$-unique decomposition in* **KAb**.

Proof: $1 \Leftrightarrow 2$ follows from Theorem 3.15.

$1 \Leftrightarrow 3$. Let $G \in \Omega(\overline{E})$. Then $h(G) = h(\overline{E}) = h(\mathbf{k})$ by [39, Corollary 3.2]. This proves $1 \Leftrightarrow 3$.

$3 \Rightarrow 5$. Let $G \in \Omega(\overline{E})$. Let $\{(G_1), \ldots, (G_h)\}$ be the set of isomorphism classes of groups G_i that are locally isomorphic to G, and let $\{(H_i) \mid i \in I\}$ be a set of isomorphism classes of groups H such that $G^n \cong H^n$ implies $G \cong H$. By Warfield's theorem 2.1, for each $(G_i) \in \{(G_1), \ldots, (G_h)\}$ there is an integer n such that $G^n \cong G_i^n$. Then $\{(G_1), \ldots, (G_h)\} \subset \{(H_i) \mid i \in I\}$. On the other hand, let $(H_i) \in \{(H_i) \mid i \in I\}$. Then $G^n \cong H^n$ for some integer $n > 0$. By Warfield's theorem 2.1, G is locally isomorphic to H_i, so that $\{(G_1), \ldots, (G_h)\} \supset \{(H_i) \mid i \in I\}$. Hence $\{(G_1), \ldots, (G_h)\} = \{(H_i) \mid i \in I\}$, so that $h =$ the number of isomorphism classes of groups H such that $G^n \cong H^n$ for some integer $n > 0$. This proves part 5.

$5 \Rightarrow 7$. Let $G \in \Omega(\overline{E})$. By part 5 there is a set $\{(G_1), \ldots, (G_h)\}$ of isomorphism classes of groups H such that $G^n \cong H^n$ for some integer $n > 0$. Let $n > 0$ be an integer, and let $G^n \cong H_1 \oplus \cdots \oplus H_t$ for some integer $t > 0$ and some indecomposable groups $H_1, \ldots, H_t$. Since $\overline{E} = \text{End}(G)$ is an integral domain, Corollary 2.12 states that each H_i is locally isomorphic to G, so $H_i \cong G_{j(i)}$ for some $1 \leq j(i) \leq h$. Thus there is a minimal integer $0 < h' \leq h$ and a set $\{(K_1), \ldots, (K_{h'})\}$ of isomorphism classes (K_i) of indecomposable groups such that for each $i = 1, \ldots, t$, there is an integer $1 \leq j(i) \leq h'$ such that $H_i \cong K_{j(i)}$.

Specifically, given $(H) \in \{(G_1), \ldots, (G_h)\}$, then there is an integer $n > 0$ such that $G^n \cong H^n$. By our choice of $\{(K_1), \ldots, (K_{h'})\}$, $(H) \in \{(K_1), \ldots, (K_{h'})\}$, hence $\{(G_1), \ldots, (G_h)\} \subset \{(K_1), \ldots, (K_{h'})\}$, whence $h \leq h'$. That is, $h = h'$, which proves part 7.

$7 \Rightarrow 3$. Let $G \in \Omega(\overline{E})$. Suppose that G has $\Sigma(h)$-unique decomposition. There is a set $\{(K_1), \ldots, (K_h)\}$ of isomorphism classes (K_i) such that for each integer $n > 0$, if $G^n \cong H_1 \oplus \cdots \oplus H_t$ for some integer m and indecomposable groups $H_1, \ldots, H_t$ then for each $i = 1, \ldots, t$ there exists a $1 \leq j(i) \leq h$ such that $H_i \cong K_{j(i)}$. Let $\{(G_1), \ldots, (G_{h(G)})\}$ be a complete set of isomorphism classes of groups H that are locally isomorphic to G. Suppose that H is locally isomorphic to G. By Warfield's theorem 2.1, there is an integer $n > 0$ such that $G^n \cong H^n$. Then $H \cong K_j$ for some $1 \leq j \leq h$, so that $\{(G_1), \ldots, (G_{h(G)})\} \subset \{((K_1), \ldots, (K_h)\}$. Consequently, $h(G) \leq h$.

Let $(K) \in \{(K_1), \ldots, (K_h)\}$. By the minimality of h, there is an integer $n > 0$ and a direct sum decomposition $G^n \cong K \oplus H_2 \oplus \cdots \oplus H_t$. Since $\text{End}(G) = \overline{E}$ is a commutative integral domain, Corollary 2.12 states that K is locally isomorphic to G, so there exists a $1 \leq j \leq h(G)$ such that $K \cong G_j$. It follows that $h \leq h(G)$, and hence that $h = h(G)$. Thus we have proved that $1 \Leftrightarrow 3 \Rightarrow 5 \Rightarrow 7 \Rightarrow 3$.

$3 \Rightarrow 4 \Rightarrow 6 \Rightarrow 8 \Rightarrow 3$. By Butler's theorem 1.10 [46, Theorem I.2.6], there is a $G \in \Omega(\overline{E})$. Then part 3 implies part 4. The rest follows as in $3 \Rightarrow 5 \Rightarrow 7 \Rightarrow 3$.

$3 \Rightarrow 9$. Let $G \in \Omega(\overline{E})$ and let X be a $K(G, 1)$-space. Recall the functors $\widehat{K}(\cdot)$ and $\widehat{\pi}(\cdot)$ from Theorem 17.7. Let $\{(G_1), \ldots, (G_{h(G)})\}$ be a complete set of isomorphism classes of groups H that are locally isomorphic to G. Let $\{(X_1), \ldots, (X_{h(X)})\}$ be a complete set of the homotopy classes of topological spaces Y that are power homotopic to X. Let $H \in \{(G_1), \ldots, (G_{h(G)})\}$. By Warfield's theorem 2.1, there is an integer n such that $G^n \cong H^n$. Then

$$X^n \sim \widehat{K}(G)^n \sim \widehat{K}(G^n) \sim \widehat{K}(H^n) \sim \widehat{K}(H)^n,$$

which shows us that X is power isomorphic to $\widehat{K}(H)$. Thus $\widehat{K}(\cdot) : G\text{-}\mathbf{Plex} \to X\text{-}\mathbf{Plex}$ induces a map $f : \{(G_1), \ldots, (G_{h(G)})\} \to \{(X_1), \ldots, (X_{h(X)})\}$ such that

$f(G_i) = (\widehat{K}(G_i))$. Similarly $\widehat{\pi}_1(\cdot) : X\text{-}\mathbf{Plex} \to G\text{-}\mathbf{Plex}$ induces a map $g : \{(X_1), \ldots, (X_{h(X)})\} \to \{(G_1), \ldots, (G_{h(G)})\}$ such that $g(X_i) = (\widehat{\pi}_1(X_i))$. Since $\widehat{K}(\cdot)$ and $\widehat{\pi}_1(\cdot)$ are inverse category equivalences, one shows that f and g are inverse bijections. Thus $h(G) = h(X)$. This proves part 9.

$9 \Rightarrow 11$. Let $G \in \Omega(\overline{E})$ and let X be a $K(G, 1)$-space. Let $\{(X_1), \ldots, (X_{h(X)})\}$ be a complete set of homotopy classes of topological spaces X_i that are power homotopic to X. Let $n > 0$ be an integer, and suppose that $X^n \sim Y_1 \times \cdots \times Y_t$ for some indecomposable topological spaces $Y_1, \ldots, Y_t$. Then the category equivalence $\widehat{\pi}_1 : X\text{-}\mathbf{Plex} \to G\text{-}\mathbf{Plex}$ (Theorem 17.7) takes the Cartesian product of K-indecomposable spaces Y_i in $\mathbf{KAb}$ to a direct sum of indecomposable abelian groups

$$G^n \cong \widehat{\pi}_1(X)^n \cong \widehat{\pi}_1(X^n) \cong \widehat{\pi}_1(Y_1) \oplus \cdots \oplus \widehat{\pi}_1(Y_t).$$

It follows from Corollary 2.12 that since $\text{End}(G) \cong \overline{E}$ is a commutative integral domain, and since each $\widehat{\pi}_1(Y_i)$ is indecomposable, each $\widehat{\pi}_1(Y_i)$ is locally isomorphic to G. By Warfield's theorem 2.1, there is an integer $m > 0$ such that

$$\widehat{\pi}_1(X^m) \cong G^m \cong \widehat{\pi}_1(Y_i)^m \cong \widehat{\pi}_1(Y_i^m),$$

and since $\widehat{\pi}_1(\cdot)$ is a category equivalence,

$$X^m \sim Y_i^m$$

for each $i = 1, \ldots, t$. Then for each $i = 1, \ldots, t$ there is a $1 \leq j(i) \leq h(X)$ such that $Y_i \sim X_{j(i)}$. Thus X has $\Sigma(h)$-unique decomposition in $\mathbf{KAb}$ for some integer $0 \leq h \leq h(X)$.

Let $\{(U_1), \ldots, (U_h)\}$ be a set of homotopy classes of topological spaces associated with the definition of $\Sigma(h)$-unique decomposition for X. There are $h(X)$ homotopy classes $(X_1), \ldots, (X_{h(X)})$, and by arguing above with Warfield's theorem 2.1 there are integers $m, m_1, \cdots, m_{h(X)} > 0$ such that

$$X^m \sim \prod_{i=1}^{h(X)} X_i^{m_i}.$$

Thus $\{(X_1), \ldots, (X_{h(X)})\} \subset \{(U_1), \ldots, (U_h)\}$, and so $h(X) \leq h$. Hence $h(X) = h$, which proves part 11.

$11 \Rightarrow 5$. Proceed in the by now familiar pattern, using the category equivalences $\widehat{\pi}_1$ and $\widehat{K}$ to show that the $\Sigma(h)$-unique decomposition of G follows from the $\Sigma(h)$-unique decomposition of the $K(G, 1)$-space X.

This proves $3 \Rightarrow 9 \Rightarrow 11 \Rightarrow 5$. The equivalences $9 \Leftrightarrow 10$ and $11 \Leftrightarrow 12$ are clear. This completes the logical cycle.

By restricting the value of the integer $h > 0$ in the above theorem to $h = 1$ we have charactertized the algebraic number fields $\mathbf{k}$ with unique factorization.

Theorem 17.44. *Let* **k** *be an algebraic number field with degree* f. *The following are equivalent.*

1. **k** *has unique factorization.*
2. *The sequence* $\{L(p)h(G(p)) \mid$ *rational primes* $p\}$ *is asymptotically equal to* $\{p^{f-1} \mid$ *rational primes* $p\}$.
3. *Each* $G \in \Omega(\overline{E})$ *has the power cancellation property.*
4. *Some* $G \in \Omega(\overline{E})$ *has the power cancellation property.*
5. *Each* $G \in \Omega(\overline{E})$ *has* Σ-*unique decomposition.*
6. *Some* $G \in \Omega(\overline{E})$ *has* Σ-*unique decomposition.*
7. *Given* $G \in \Omega(\overline{E})$, *each* $K(G, 1)$-*space has the power cancellation property in* **KAb**.
8. *For some* $G \in \Omega(\overline{E})$, *each* $K(G, 1)$-*space has the power cancellation property in* **KAb**.
9. *Given* $G \in \Omega(\overline{E})$, *each* $K(G, 1)$-*space has* Σ-*unique decomposition in* **KAb**.
10. *For some* $G \in \Omega(\overline{E})$, *each* $K(G, 1)$-*space has* Σ-*unique decomposition in* **KAb**.

In particular, let **k** be a quadratic number field, let $\overline{E}$ be the ring of algebraic integers in **k**, and for each rational prime p let $E(p) = \mathbf{Z} + p\overline{E}$. Let $G(p)$ be a reduced torsion-free rank two abelian group such that $\text{End}(G(p)) \cong E(p)$. These groups exist by Butler's theorem 1.10 [46, Theorem I.2.6]. Let $L(p) = \text{card}(u(\overline{E})/u(E(p)))$, where $u(R)$ is the groups of units in the ring R. For an abelian group H let $h(H)$ be the number of isomorphism classes of groups L that are locally isomorphic to H.

Theorem 17.45. *Let* **k** *be an quadratric number field. The following are equivalent.*

1. **k** *has unique factorization.*
2. *The sequence* $\{L(p)h(G(p)) \mid$ *rational primes* $p\}$ *is asymptotically equal to the sequence of rational primes.*
3. *Each strongly indecomposable reduced torsion-free rank two abelian group* G *such that* $\text{End}(G) \cong \overline{E}$ *has the power cancellation property.*
4. *Some strongly indecomposable reduced torsion-free rank two abelian group* G *such that* $\text{End}(G) \cong \overline{E}$ *has the power cancellation property.*
5. *Each strongly indecomposable reduced torsion-free rank two abelian group* G *such that* $\text{End}(G) \cong \overline{E}$ *has* Σ-*unique decomposition.*
6. *Some strongly indecomposable reduced torsion-free rank two abelian group* G *such that* $\text{End}(G) \cong \overline{E}$ *has* Σ-*unique decomposition.*
7. *Given a strongly indecomposable reduced torsion-free rank two abelian group* G *such that* $\text{End}(G) \cong \overline{E}$, *each* $K(G, 1)$-*space has the power cancellation property in* **KAb**.
8. *For some strongly indecomposable reduced torsion-free rank two abelian group* G *such that* $\text{End}(G) \cong \overline{E}$, *each* $K(G, 1)$-*space has the power cancellation property in* **KAb**.
9. *Given a strongly indecomposable reduced torsion-free rank two abelian group* G *such that* $\text{End}(G) \cong \overline{E}$, *each* $K(G, 1)$-*space has* Σ-*unique decomposition in* **KAb**.

10. *For some strongly indecomposable reduced torsion-free rank two abelian group*
    G such that $\text{End}(G) \cong \overline{E}$, *each $K(G, 1)$-space has Σ-unique decomposition in*
    KAb.

Consider some of the more popular consequences of the properties used above.
In Theorem 17.43, for a given integer $h > 0$, we have characterized those algebraic
number fields **k** with class number $h(\mathbf{k}) = h$. Determining the class number of an
algebraic number field is a problem that goes back to Gauss. Specifically, one wishes
to write a list of the quadratic number fields **k** such that $h(\mathbf{k}) = 1$, or equivalently
such that **k** has unique factorization. See theorem 17.45.1.

Theorem 17.45.2 involves a sequence $\{s_p \mid$ primes $p\}$ of integers that is asymptotically equal to the sequence of rational primes. Thus $\{s_p \mid$ primes $p\}$ is related to the
Prime Number Theorem, although we do not explore the relationship here.

Theorem 17.45.3 and 17.45.5 are about direct sum decompositions of strongly
indecomposable torsion-free rank two abelian groups. These properties, called power
cancellation and Σ-unique decomposition, will be hard because they are related to
hard problems in number theory. We will also apply our methods to the study of
more general direct sum decompositions of abelian groups. Consequently, we give a
complete set of topological invariants for abelian groups up to isomorphism, and we
show how one would characterize any direct sum property of an abelian group.

Theorem 17.45.7 and 17.45.9 are about the classic $K(G, 1)$-spaces from point set
topology. We give decomposition properties of $K(G, 1)$-spaces for which $\text{End}(G) \cong$
$\overline{E}$ to provide a topological characterization of the unique factorization property in the
quadratic number field **k**.

Remark 17.46. The category equivalence $K(\cdot) : \mathbf{Ab} \longrightarrow \mathbf{KAb}$ shows us that we can
draw a picture of, or sculpt a figure of, an abelian group. This is different from what
students of abelian group theory were told in the 1970s. They were philosophically
advised that torsion-free abelian groups are like clouds without definite shape or
boundary.

Remark 17.47. Over the past 35 years certain aspects of abelian group theory have
blurred with the study of modules over general associative rings. Diagrams (17.1),
(17.5), (17.8), and (17.10) represent an idea in abelian group theory that is not readily
extended to modules over more general rings. The key observation to see this is that a
homotopy group is not in general a module over a ring $R \neq \mathbf{Z}$. Theorems exist that construct a topological space with specified homotopy group or fundamental group, but
there is no theorem that produces a space with the more general specified homotopy
"module." Therefore Diagrams (17.1), (17.5), (17.8), and (17.10) seem to represent a
concept in abelian group theory that will be unique to abelian groups for some time.

17.9 Exercises

1. Construct the commutative diagram (17.12).
2. Assume (μ), and suppose that G is self-small and self-slender. Construct a version
   of Diagram (17.12) that takes advantage of the category equivalences implied by
   the self-small hypothesis.

3. Construct a version of Diagram (17.12) that uses G-Cohplx and X-Cohplx.

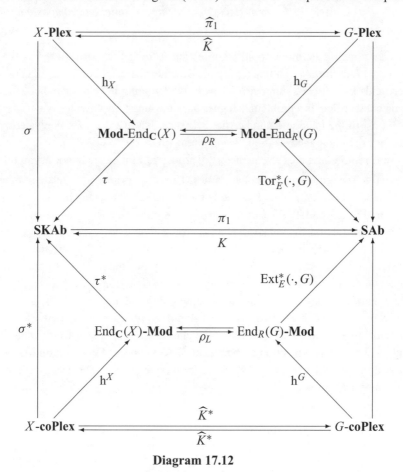

Diagram 17.12

4. We say that G has the *double cancellation property* if given abelian groups H and K such that $G \oplus G \oplus H \cong G \oplus G \oplus K$ then $H \cong K$. Characterize those abelian groups with the double cancellation property.

5. We say that G has *power cancellation of degree n* if given an abelian group H such that $G^n \cong H^n$ then $G \cong H$. Characterize those abelian groups that possess the power cancellation property of degree n.

6. We say that G has Σ-*unique decomposition* if G^n has unique direct sum decomposition for each integer $n > 0$. Characterize those abelian groups that possess the Σ-unique decomposition.

7. Generalize the above three exercises.

8. An abelian group A is divisible if $nA = A$ for each integer $n > 0$. Characterize the divisible groups A in topological terms.

9. Fill in the details of Theorem 17.43.

10. Let $h > 0$ be an integer. Give examples of groups G that possess $\Sigma(h)$-unique decomposition.

17.10 Problems for future research

1. Add to the lists in Theorems 17.44 and 17.45.
2. C. F. Gauss suggests that there are at most finitely many quadratic number fields **k** such that $h(\mathbf{k}) = 1$. Reformulate this conjecture in terms of abelian groups and then find a finite list of equivalence classes of abelian groups A with $h(A) = 1$.
3. Let $h > 0$ be an integer. Which groups G have $\Sigma(h)$-unique decomposition.

18

Marginal isomorphisms

18.1 Ore localization

Your author considers cancellation and uniqueness of decomposition results to be some of the most beautiful results in mathematics. Of these the most used is certainly the Azumaya–Krull–Schmidt theorem 1.4. This theorem states essentially that if a right R-module G is the direct sum

$$G \sim G_1 \oplus \cdots \oplus G_t, \tag{18.1}$$

where $t > 0$ is an integer and in which $G_1, \ldots, G_t$ are right R-modules with local endomorphism rings, then up to the isomorphism of the direct summands, (18.1) is the only direct sum decomposition for G into decomposables. Furthermore, G satisfies the internal cancellation property

$$G \sim H \oplus K \cong H \oplus L \Rightarrow K \cong L$$

for right R-modules $H, K, L \in \mathbf{P}_o(G)$.

Under the hypothesis that $\operatorname{End}_R(G)$ possesses a semi-primary classical right ring of quotients Q_G, the goal of this chapter is to prove an Azumaya-like uniqueness of decomposition of G. We introduce a means of comparison called *margimorphism* which leads us to define *marginal summands* of the right R-module $G^{(n)}$ for integers $n > 0$. Margimorphism is a generalization of isomorphism that was first sudied in [37]. While some of our results conclude that margimorphism is a coarser comparison of right R-modules than isomorphism, there are some results that show that isomorphism follows from the weaker form of comparison. For instance, we introduce functors $\mathbf{QH}_G(\cdot)$ and $\mathbf{B}_G(\cdot)$ such that marginal summands H and H' of $G^{(n)}$ are margimorphic iff $\mathbf{QH}_G(H) \cong \mathbf{QH}_G(H')$ as finitely generated projective right Q_G-modules iff $\mathbf{B}_G(H) \cong \mathbf{B}_G(H')$ as semi-simple right Q_G-modules. Ultimately, we define a class of modules $\mathbf{C}$ that is closed under the margimorphism classes of its objects for which each $H \in \mathbf{C}$ possesses a unique decomposition *à la* Azumaya–Krull–Schmidt. Each object in $\mathbf{C}$ satisfies the internal cancellation property defined above. Furthermore, we classify the modules G for which Q_G is semi-simple Artinian. That is, we classify those G such that $\operatorname{End}_R(G)$ possesses a semi-simple Artinian classical right ring of quotients.

The ring theory that we use is a generalization to the noncommutative setting of localization theory in commutative rings. For example, for a torsion-free finite-rank abelian group G, the classical ring of quotients of $\text{End}_{\mathbf{Z}}(G)$ is the finite-dimensional $\mathbf{Q}$-algebra $\text{End}_{\mathbf{Z}}(G) \otimes \mathbf{Q}$. We will assume that G is a right R-module for some ring R and that $\text{End}_R(G)$ possesses a *semi-primary* classical right ring of quotients Q_G. That is, Q_G is a ring that is semi-simple modulo the nilpotent Jacobson radical $\mathcal{J}(Q_G)$. Localizing to form Q_G is not commutative in any sense of the word. The generalization from commutative tools to the noncommutative setting will account for most of our effort.

18.1.1 Preliminary concepts and examples

Let E be a ring. An element $c \in E$ is *right regular* if

$$\text{for each } x \in E, cx = 0 \Rightarrow x = 0.$$

An element $c \in E$ is *regular* if it is left and right regular. The set of regular elements in E is denoted by $\mathcal{C}(E)$. The *right Ore property* for regular elements in a ring E states that

> Given a $c \in \mathcal{C}(E)$ and an $x \in E$ there exists a
>
> $d \in \mathcal{C}(E)$ and a $y \in E$ such that $xd = cy$.

The ring Q is a *classical right ring of quotients* of E provided that

1. Each regular element of E is a unit in Q, and
2. Each element of Q is of the form xc^{-1} for some $x \in E$ and some regular $c \in E$.

If item 2 is satisfied then E is called a *right order in Q*. Item 2 is equivalent to

2′ For each $q \in Q$ there is a regular $c \in E$ such that $qc \in E$.

The field of rational numbers $\mathbf{Q}$ is the classical ring of quotients of the integers $\mathbf{Z}$. The field of rational functions $\mathbf{Q}(x)$ over $\mathbf{Q}$ is the classical field of quotients of the ring $\mathbf{Z}[x]$ of polynomials over $\mathbf{Z}$. **Ore's theorem** states that E possesses a classical right ring of quotients Q iff $\mathcal{C}(E)$ satisfies the *right Ore property*.

The ideal J is *nil* if each element of J is nilpotent, and J is *nilpotent* if $J^n = 0$ for some integer $n > 0$. Nil ideals are important to our investigations because *idempotents lift modulo nil ideals*. (See [7, Proposition 27.1].) Idempotents lift modulo J if $(a + J)^2 = a + J$ in E/J implies that there is an $e^2 = e \in E$ such that $e + J = a + J$.

The ring Q is *semi-primary* if $Q/\mathcal{J}(Q)$ is semi-simple Artinian and $\mathcal{J}(Q)$ is nilpotent. We present some properties of semi-primary rings that we will use without reference. See [7, Section 28] for details. Let Q be semi-primary. Since idempotents lift modulo the nilpotent ideal $\mathcal{J}(Q)$ and since $Q/\mathcal{J}(Q)$ is semi-simple Artinian, Q is a *semi-perfect* ring.

Let E be a ring. The right E-module M has a *unique decomposition* if

1. $M = M_1 \oplus \cdots \oplus M_t$ for some integer $t > 0$ and indecomposable right E-modules $M_1, \ldots, M_t$, and
2. $M = M_1' \oplus \cdots \oplus M_s'$ for some integer $s > 0$ and some indecomposable right E-modules $M_1', \ldots, M_s'$ then $s = t$ and there is a permutation π of $\{1, \ldots, t\}$ such that $M_i \cong M_{\pi(i)}'$ for each $i = 1, \ldots, t$.

For example, let E be a semi-perfect ring. An indecomposable projective right E-module has a local endomorphism ring so that the finitely generated projective right E-modules satisfy the Azumaya–Krull–Schmidt theorem 1.4. Specifically, each $P \in \mathbf{P}_o(E)$ has a unique decomposition $P = P_1 \oplus \cdots \oplus P_t$ for some integer $t > 0$ and indecomposable projective right E-modules $P_1, \ldots, P_t$. Semi-primary rings are semi-perfect, so finitely generated projective right E-modules over a semi-primary ring E possesses a unique decomposition.

Example 18.1. Right Artinian rings are semi-primary.

Example 18.2. The ring $\begin{pmatrix} \mathbf{Q} & 0 \\ \mathbf{Q}^{(\aleph_o)} & \mathbf{Q} \end{pmatrix}$ is semi-primary but not right Artinian.

Example 18.3. This is an example of a local ring E in which $\mathcal{J}(E)$ is a nil ideal but not nilpotent. For each integer $n > 0$ let T_n denote the ring of lower triangular $n \times n$ matrices over $\mathbf{Q}$. Then $\mathcal{J}(T_n)^n = 0$ but $\mathcal{J}(T_n)^{n-1} \neq 0$. Then $J = \oplus_{n>0} \mathcal{J}(T_n)$ is a nil ideal in the ring $R = \prod_{n>0} T_n$ but not a nilpotent ideal. Let $E = J + \mathbf{Q}1_R$. Then E is a local ring, $\mathcal{J}(E) = J$ is a nil but not nilpotent ideal. Thus, although E is a local ring with nil Jacobson radical, E is not semi-primary.

Example 18.4. The endomorphism ring $\mathrm{End}(G)$ of an rtffr group G possesses an Artinian classical (left and right) ring of quotients

$$\boxed{\mathbf{Q}\mathrm{End}(G) = \mathbf{Q} \otimes_{\mathbf{Z}} \mathrm{End}(G)}$$

called the *quasi-endomorphism ring of G*. However, this example does not capture the flavor of the type of localization we will encounter in this section.

Example 18.5. Let x, y be elements such that $yx - xy = 1$ and let $E = \mathbf{Q}[x, y]$. Then E is a Noetherian domain called a *Weyl algebra*. See [72, Corollary 1.3.16] for details. Evidently E is not commutative and it is well known that E possesses a classical (right and left) ring of quotients Q. Specifically, Q is a division ring. One of the strange properties possessed by E is that 0 and E are the only ideals in E. Thus in generalizing commutative results we cannot assume a complicated lattice of ideals in E.

Example 18.6. The following example is due to A. V. Jategaonkar [72]. There is a local principal right ideal domain E whose *right* ideals form a chain. In particular,

E possesses a classical right ring of quotients Q. Furthermore, Q is a division ring. On the other hand, E contains an infinite direct sum

$$I_1 \oplus I_2 \oplus I_3 \oplus \cdots$$

of nonzero *left* ideals I_i of E. The reader can show that such a domain does not satisfy the left Ore property. Thus Q is not the classical left ring of quotients of E.

18.1.2 Noncommutative localization

In this subsection we will investigate the localization theory necessary to our investigation of margimorphisms and direct sum decompositions. This localization will lead us to investigate the localization of $\text{End}_E(P)$ for finitely generated projective right E-modules P.

We will need to know some noncommutative localization theory.

Lemma 18.7. *Let E be a right order in the ring Q. If $c \in E$ is right regular, then c is right regular in Q.*

Proof: Say that $c \in E$ is right regular and let $cq = 0$ for some $q \in Q$. There is a $d \in E$ and $x \in E$ such that $q = xd^{-1}$. Then $0 = cq = cqd = cx$ and the right regularity of c in E shows us that $x = 0$. Hence $q = 0$. Thus c is right regular in Q.

Lemma 18.8. *Assume that E has a semi-primary classical right ring of quotients Q.*

1. *$\mathcal{J}(Q) \cap E = \mathcal{N}(E)$.*
2. *Each right regular element in Q is a unit.*

Proof: 1. This is an exercise for the reader.

2. Let $c \in Q$ be a right regular element. Since Q is semi-primary Q possesses the descending chain condition on principal right ideals, [7, Theorem 28.4]. Then the chain

$$cQ \supset c^2Q \supset c^3Q \supset \cdots$$

of principal right ideals in Q is eventually constant, so that $c^kQ = c^{k+1}Q$ for some integer $k > 0$. Since c^k is right regular, $Q = cQ$, so that $cd = 1_Q$ for some right regular element $d \in Q$. In a similar manner, $dc' = 1_Q$ for some $c' \in Q$, so that $c = c(dc') = c'$. Hence c is a unit in Q. This completes the proof.

Lemma 18.9. *Assume that E has a semi-primary classical right ring of quotients Q.*

1. *Right regular elements in E are left regular.*
2. *$Q/\mathcal{J}(Q)$ is the semi-simple classical right ring of quotients of $E/\mathcal{N}(E)$.*
3. *$c \in E$ is regular iff $c + \mathcal{N}(E)$ is regular in $E/\mathcal{N}(E)$.*

Proof: 1. Let $c \in E$ be a right regular element. By Lemma 18.7, c is right regular in Q. Since Q is semi-primary Lemma 18.8.2 states that c is a unit in Q. Hence c is left regular in E.

2. By Lemma 18.8.1, $\mathcal{N}(E) = \mathcal{J}(Q) \cap E$, so there is a natural embedding $E/\mathcal{N}(E) \subset Q/\mathcal{J}(Q)$.

Let c be a regular element in E. By part 1, c is a unit of Q, so that c maps to a unit $\bar{c}$ in $Q/\mathcal{J}(Q)$. Hence $\bar{c}$ is regular in $E/\mathcal{N}(E)$.

Let $\bar{x} = x + \mathcal{J}(Q) \in Q/\mathcal{J}(Q)$. Since Q is the classical right ring of quotients of E there is a regular element $c \in E$ such that $xc \in E$, so that $\bar{x}\bar{c} \in E/\mathcal{N}(E)$ for some regular element $\bar{c}$ in $E/\mathcal{N}(E)$. We say that $E/\mathcal{N}(E)$ is a right order in $Q/\mathcal{J}(Q)$.

Let $c + \mathcal{N}(E) = \bar{c}$ be a regular element in $E/\mathcal{N}(E)$ for some $c \in E$. Because $E/\mathcal{N}(E)$ is a right order in $Q/\mathcal{J}(Q)$, $\bar{c}$ is a (right) regular element in the semi-simple ring $Q/\mathcal{J}(Q)$. Hence $\bar{c}$ is a unit in $Q/\mathcal{J}(Q)$. That is, each regular element $\bar{c}$ of $E/\mathcal{N}(E)$ is a unit in $Q/\mathcal{J}(Q)$. Hence $Q/\mathcal{N}(Q)$ is the semi-simple Artinian classical right ring of quotients of $E/\mathcal{N}(E)$.

3. We leave this proof as an exercise for the reader.

We investigate the classical ring of quotients of eEe for $e^2 = e \in E$. This requires a series of preliminary lemmas. The reader can find the following result in [7, Corollaries 27.7, 28.6 and Proposition 28.11], but try to prove it yourself.

Lemma 18.10. *Let Q be a semi-primary ring, let $e^2 = e \in Q$, and let $n > 0$ be an integer.*

1. $\mathrm{Mat}_n(Q)$ is semi-primary.
2. eQe is semi-primary.

Proof: 1. We know that

$$\mathcal{J}(\mathrm{Mat}_n(Q)) = \mathrm{Mat}_n(\mathcal{J}(Q))$$

and it follows readily that $\mathcal{J}(\mathrm{Mat}_n(Q))$ is a nilpotent ideal. If the reader has never seen it before, they should try proving that

$$\frac{\mathrm{Mat}_n(Q)}{\mathcal{J}(\mathrm{Mat}_n(Q))} = \mathrm{Mat}_n(Q/\mathcal{J}(Q)).$$

Thus $\mathrm{Mat}_n(Q)$ is semi-simple Artinian modulo its nilpotent Jacobson radical $\mathrm{Mat}_n(\mathcal{J}(Q))$, whence $\mathrm{Mat}_n(Q)$ is semi-primary.

2. We have observed that Q is a semi-perfect ring in which $\mathcal{J}(Q)$ is nilpotent. By [7, Corollary 27.7], eQe is semi-perfect so that $eQe/\mathcal{J}(eQe)$ is semi-simple Artinian. The reader can show that $\mathcal{J}(eQe) = e\mathcal{J}(Q)e$ and that this ideal in eQe is nilpotent. Thus eQe is semi-primary.

Theorem 18.11. *Assume that E possesses a semi-primary classical right ring of quotients Q. Then $\mathrm{Mat}_n(Q)$ is the semi-primary classical right ring of quotients of $\mathrm{Mat}_n(E)$ for each integer $n > 0$.*

Proof: We sketch a proof. By Lemma 18.10, $\mathrm{Mat}_n(Q)$ is a semi-primary ring. The ring $\mathrm{Mat}_n(E)$ is a right order in $\mathrm{Mat}_n(Q)$. (This is a nice exercise in the induction property associated with the right Ore property in E.) Then a right regular $c \in \mathrm{Mat}_n(E)$ is a (right regular =) unit in the semi-primary ring $\mathrm{Mat}_n(Q)$. This is what we had to prove.

The purpose behind the next two results is that we want to be able to deduce that the semi-simple classical right ring of quotients of eEe is eQe. This technical result is designed to allow us some commutativity in our use of regular elements in E.

Lemma 18.12. *Suppose that E is a semi-prime ring, let $e^2 = e \in E$, and let $c \in eEe$. Then c is right regular in eEe iff $c + (1 - e)$ is right regular in E.*

Proof: Suppose that $c \in eEe$ and that $c + (1-e)$ is right regular in E. If $x = exe \in eEe$ and if $cx = 0$ then

$$(c + (1 - e))x = cx + (1 - e)x = cx + (1 - e)(exe) = 0$$

so that $x = 0$. Thus c is right regular in eEe.

Conversely, suppose that c is right regular in eEe and let $(c + (1 - e))x = 0$ for some $x \in E$. Since $c = ece$ we have that

$$(1 - e)x = -cx = -(ece)x \in (1 - e)E \cap eE = 0.$$

Then $c(exEe) = (ece)(xEe) = c(xEe) = 0$. Since $c \in eEe$ is right regular, $exEe = 0$, and since E is semi-prime,

$$exEe = 0 \Rightarrow (exE)^2 = 0 \Rightarrow exE = 0 \Rightarrow ex = 0.$$

Thus $x = ex + (1 - e)x = 0$, which proves that $c + (1 - e)$ is right regular in E. This completes the proof.

The following is referred to in the literature as the Faith–Utumi theorem. See [31, Theorem 10.15] or [33] for a proof. We say that

$$Q = \mathrm{Mat}_n(D)$$

if there are elements e_{ij}, $1 \leq i,j \leq n$ in Q called *matrix units* and a subring $D \subset Q$ such that

1. $1 = e_{11} + \cdots + e_{nn}$,
2. $e_{ij}e_{k\ell} = \delta_{jk}e_{i\ell}$ where δ is the Kronecker delta,
3. $e_{ii}d = de_{ii}$ and $e_{ij}d = de_{ij} = 0$ for each integer $1 \leq i \neq j \leq n$ and each $d \in D$, and
4. Given $q \in Q$ there are $d_{ij} \in D$ such that $q = \sum_{ij} e_{ij}d_{ij}$.

Faith–Utumi Theorem *Suppose E is a ring perhaps without unit. Suppose that Q is the simple Artinian classical right ring of quotients of E. Write $Q = \mathrm{Mat}_t(D)$ for* *some division ring D. Then $E \cap D$ contains a domain F such that*

1. *D is the classical right ring of quotients of F,*
2. *$\mathrm{Mat}_t(F) \subset E$,*
3. *Q is the classical right ring of quotients of $\mathrm{Mat}_t(F)$, and*
4. *$Q = \{xc^{-1} \,\big|\, x \in E \text{ and } 0 \neq c \in F\}$.*

Lemma 18.13. *Suppose that E possesses a semi-simple Artinian classical right ring of quotients Q, let $Q = Q_1 \times \cdots \times Q_t$ for some simple Artinian rings $Q_1, \ldots, Q_t$, and let $e^2 = e \in Q$. Then*

1. *Q_i is the simple Artinian classical right ring of quotients of $E \cap Q_i$.*
2. *For each $x \in eQe$ there is a regular $c \in E$ such that*

    (a) *ece is regular in $E \cap eQe$,*
    (b) *$c = ece \oplus (1 - e)c(1 - e)$, and*
    (c) *$xc \in E \cap eQe$.*

3. *eQe is the semi-simple Artinian classical right ring of quotients of $E \cap eQe$.*

Proof: 1. Let e_i be the unique central idempotent in Q such that $e_i Q = Q_i$. Since Q is the classical right ring of quotients of E there is a regular $c \in E$ such that $e_i c \in E$ for each $i = 1, \ldots, t$. Then

$$cE \subset e_1 cE \times \cdots \times e_t cE \subset E.$$

We will show that Q_i is the simple classical right ring of quotients of $e_i cE$. (This is a ring without unit.)

Let $y \in Q_i$. By the right Ore property in E, there is a regular $d \in cE$ such that $(yc)d \in cE$. Since $y \in Q_i$,

$$y(e_i c)(e_i d) = (e_i y)(cd) = (yc)d \in e_i cE.$$

Since cd is a unit in Q, $(e_i c)(e_i d)$ is a unit in Q_i, so that $(e_i c)(e_i d)$ is regular in $e_i cE$. Thus $e_i cE$ is a right order in Q_i.

Let d_1 be a right regular element in $e_1 cE$. Since $e_1 cE$ is a right order in $e_1 Q$, d_1 is a right regular element in $e_1 Q$ (Lemma 18.7). Hence d_1 is a unit in the simple Artinian ring $e_1 Q$, whence $e_1 Q$ is the simple Artinian classical right ring of quotients of $e_1 cE$.

Inasmuch as $e_i cE \subset E \cap Q_i \subset Q_i$, Q_i is the simple Artinian classical right ring of quotients of $E \cap Q_i$.

2. Let $e^2 = e \in Q$. By part 1 we can assume without loss of generality that Q is the simple Artinian classical right ring of quotients of E. Since Q is simple Artinian, there is a set of matrix units

$$e_{ij} \text{ such that } 1 \leq i, j \leq t$$

for Q and an integer $s \leq t$ such that

$$e = e_{11} \oplus \cdots \oplus e_{ss}. \tag{18.2}$$

Let $x = exe \in eQe$. By the Faith–Utumi Theorem (page 317) (and using the notation therein) there is a regular $c \in F \subset E$ such that $xc \in E$. Because $c \in F$ is a diagonal matrix,

$$e_{ii}c = e_{ii}ce_{ii} \in \text{Mat}_n(F) \subset E \text{ for all } i = 1, \ldots, t,$$

so that $ec = ece \in E \cap eQe$ and

$$c = ece \oplus (1 - e)c(1 - e) = ec \oplus (1 - e)c.$$

Furthermore, because c is a unit in Q,

$$(ece)(ec^{-1}e) = (ec)(c^{-1}e) = e,$$

so that ece is a unit in eQe. Then ece is a regular element in $E \cap eQe$ such that $x(ece) = x(ec) = xc \in E \cap eQe$. This proves part 2.

3. By Lemma 18.10.2, eQe is a semi-primary ring. To see that that eQe is semi-simple, let $I \subset eQe$ be a right ideal such that $I^2 = 0$. Then

$$(IQ)^2 = I(QI)Q = eIe(QeIe)Q = (eIe)^2Q = I^2Q = 0.$$

Because Q is semi-simple, $IQ = 0 = I$. Hence eQe is semi-simple. By part 2, $E \cap eQe$ is a right order in the semi-simple Artinian ring eQe. Thus eQe is the classical right ring of quotients of $E \cap eQe$. This completes the proof.

Theorem 18.14. *Assume that E possesses a semi-primary classical right ring of quotients Q. If $e^2 = e \in E$ then $eQe/eJ(Q)e$ is the semi-simple Artinian classical right ring of quotients of $eEe/eN(E)e$.*

Proof: Since Q is the semi-primary classical right ring of quotients of E, Lemma 18.9.2 shows us that $Q/J(Q)$ is the semi-simple classical right ring of quotients of $E/N(E)$. By Lemma 18.13.3,

$$eQe/eJ(Q)e \cong \bar{e}(Q/J(Q))\bar{e}$$

is then the semi-simple classical right ring of quotients of

$$\frac{E}{N(E)} \cap \bar{e}\frac{Q}{J(Q)}\bar{e} = \frac{eEe + J(Q)}{J(Q)} = \frac{eEe}{eN(E)e}.$$

This completes the proof.

Corollary 18.15. *Suppose that E possesses a semi-simple Artinian classical right ring of quotients Q, and let $e^2 = e \in Q$. Let $M \subset V$ be right E-modules such that $V = VQ = MQ$. Then for each $x \in Ve$ there is a regular $c \in E$ such that*

1. $c = ece \oplus (1 - e)c(1 - e)$, *and*
2. ece *is a unit in eQe,*
3. $ece \in E \cap eQe$,
4. $xc \in M \cap Ve$.

Proof: Without loss of generality we assume that Q is a simple Artinian classical right ring of quotients for E. Choose matrix units e_{ij} and a division ring D such that $\mathrm{Mat}_n(D) = Q$. Assume without loss of generality that $e = e_{11} + \cdots + e_{ss}$ for some integer $s \leq t$. By the Faith–Utumi Theorem (page 317) there is a domain $F \subset D$ such that $\mathrm{Mat}_n(F) \subset E$ and each $q \in Q$ has the form xc^{-1} for some $x \in E$ and $c \in F$.

Let $x \in Ve$. There is a regular $c \in F$ such that $ec \in E$ and $xc \in M$. Since $c \in F$, $c = e_{11}ce_{11} + \cdots + e_{tt}ce_{tt}$, so that $c = ece \oplus (1 - e)c(1 - e)$. This proves part 1.

Since c is a unit in Q, one can use part 1 to show that $(ece)^{-1} = ec^{-1}e$ in eQe. Notice that by our choice of $c \in F$, $ece = ec \in E$, so that $ece \in E \cap eQe$. This proves parts 2 and 3.

Furthermore, $xc = (xe)c = x(ece) \in M \cap Ve$. This proves part 4 and completes the proof.

Lemma 18.16. *Let E be a ring that possesses a semi-primary classical right ring of quotients Q. Let $e^2 = e \in E$ and let $x = xe \in Q$. There is a regular element $c = ece \oplus (1 - e)c(1 - e) \in E$ such that $xc \in E$.*

Proof: Let $\bar{x}$ denote the image of an element x modulo $\mathcal{J}(Q)$.

Let $y_{-1} = 0$, let $x = xe = x_0 = y_0 \in Ee$, and let $c_0 = 1$.

Proceed inductively. Assume that for some integer $k \geq 0$, there are elements $y_k \in e\mathcal{J}(Q)^k e$, $x_k \in Ee$, and a regular element $c_k \in E$ such that

$[k_1] \; c_k = ec_k e \oplus (1 - e)c_k(1 - e) \in E,$
$[k_2] \; ec_k = ec_k e$ is regular in eEe and a unit in $eQe,$
$[k_3] \; y_k = y_{k-1}c_k - x_k \in \mathcal{J}(Q)^k e \,.$

The induction step begins. Evidently $\bar{e}^2 = \bar{e} \in \dfrac{E + \mathcal{J}(Q)}{\mathcal{J}(Q)}$, and by Lemma 18.9, the ring $Q/\mathcal{J}(Q)$ is the semi-simple classical right ring of quotients of $\dfrac{E + \mathcal{J}(Q)}{\mathcal{J}(Q)}$.

Because $c \in E$ is regular iff $c \in \dfrac{E + \mathcal{J}(Q)}{\mathcal{J}(Q)}$ is regular, and by Corollary 18.15, there is a regular element $c_{k+1} \in E$ such that

$[k + 1_1] \; \bar{c}_{k+1} = \overline{ec}_{k+1}\bar{e} \oplus (1 - \bar{e})\bar{c}_{k+1}(1 - \bar{e}) \in \dfrac{E + \mathcal{J}(Q)}{\mathcal{J}(Q)},$

$[k + 1_2] \; \overline{ec}_{k+1}\bar{e}$ is a unit in $\dfrac{eQe + \mathcal{J}(Q)}{\mathcal{J}(Q)},$

$[k + 1_3] \; \bar{y}_k \bar{c}_{k+1} = (\bar{y}_k \bar{e})\bar{c}_{k+1} = \bar{y}_k(\overline{ec}_{k+1}\bar{e})$ which is in

$$\dfrac{(E \cap \mathcal{J}(Q)^k e) + \mathcal{J}(Q)^{k+1}e}{\mathcal{J}(Q)^{k+1}e}.$$

Since $c_{k+1}, e \in E$, we assume without loss of generality that

$[k + 1_1'] \; c_{k+1} = ec_{k+1}e \oplus (1 - e)c_{k+1}(1 - e) \in E.$

Since units lift modulo the Jacobson radical, item $[k + 1_2]$ implies that $ec_{k+1}e$ is a unit in eQe. Then

$[k + 1'_2]$ $ec_{k+1}e$ is regular in eEe and a unit in eQe.

Since $y_k \in \mathcal{J}(Q)^k e$, and by $[k + 1_1]$ and $[k + 1_3]$

$$\bar{y}_k \bar{c}_{k+1} = (\bar{y}_k \bar{c}_{k+1})\bar{e} \in \frac{(E \cap \mathcal{J}(Q)^k e) + \mathcal{J}(Q)^{k+1} e}{\mathcal{J}(Q)^{k+1} e}.$$

Hence there is an

$$x_{k+1} \in E \cap \mathcal{J}(Q)^k e \subset Ee$$

such that

$[k + 1'_3]$ $y_{k+1} = y_k c_{k+1} - x_{k+1} \in \mathcal{J}(Q)^{k+1} e.$

This completes the induction process.

Because Q is semi-primary there is an integer $n > 0$ such that $\mathcal{J}(Q)^n = 0$. By item $[k'_3]$ (with $k = n$),

$$y_n = y_{n-1} c_n - x_n \in \mathcal{J}(Q)^n e = 0.$$

Hence there are regular elements $c_1, \ldots, c_n \in E$ such that

$$0 = y_{n-1} c_n - x_n$$
$$= (\cdots (x_0 c_1 - x_1) c_2 - \cdots) c_n - x_n$$
$$= x_0(c_0 \cdots c_n) - X$$

for some $X \in E$. Let $c = c_0 \cdots c_n$. By $[k_1]$,

$$c = ece \oplus (1 - e)c(1 - e)$$

is a regular element in E, and by our choice of $x_k \in Ee$ for $k \geq 0$, we have

$$x_k(c_k \cdots c_n) = (x_k e)(c_k \cdots c_n) = x_k e(c_k \cdots c_n)e \in Ee.$$

Since X is a sum of the $x_k(c_k \cdots c_n)$,

$$xc = x_0(c_0 \cdots c_n)e = X \in Ee.$$

Item $[k_2]$ implies that $c = c_1 \cdots c_n$ is regular in E, and $[k_1]$ implies that $c = ece \oplus (1 - e)c(1 - e)$. This completes the proof.

Theorem 18.17. *Assume that E has a semi-primary classical right ring of quotients Q. If $e^2 = e \in E$ then eQe is the semi-primary classical right ring of quotients of eEe.*

Proof: Regard eEe as a subring of eQe with unit e. By Lemma 18.7 it suffices to show that eEe is a right order in eQe. Let $x \in eQe$. By Lemma 18.16 there is a regular $c = ece \oplus (1 - e)c(1 - e) \in E$ such that ece is regular in eEe and $xc \in E$. Then

$$x(ece) = (exe)(ece) = e(xc)e \in eEe,$$

so that eEe is a right order in eQe. This completes the proof.

Some examples will illustrate the hypotheses of the above theorem.

Example 18.18. Let x be an indeterminant and let $E = \begin{pmatrix} \mathbf{Z}[x] & 0 \\ \mathbf{Z} & \mathbf{Z} \end{pmatrix}$ where $\mathbf{Z}$ is a $\mathbf{Z}[x]$-module by identifying $\mathbf{Z} \cong \mathbf{Z}[x]/(x)$. We note that E is a Noetherian ring such that $\mathcal{J}(E) = \begin{pmatrix} 0 & 0 \\ \mathbf{Z} & 0 \end{pmatrix}$ is nilpotent. But $x + \mathcal{J}(E)$ is a regular element of $E/\mathcal{J}(E)$ that does not lift to a regular element of E.

Moreover, E possesses a semi-perfect classical ring of quotients $Q = \begin{pmatrix} S & 0 \\ Q & Q \end{pmatrix}$, where $S = \mathbf{Q}[x]_{(x)}$ is formed by localizing $\mathbf{Q}[x]$ at the (maximal) ideal (x).

Example 18.19. Let $Q = \begin{pmatrix} S & 0 \\ Q & Q \end{pmatrix}$ as in the previous example. Then $\mathcal{J}(Q) = \begin{pmatrix} xS & 0 \\ Q & 0 \end{pmatrix}$ so that $\mathcal{J}(Q)$ is not a nil ideal.

Example 18.20. Let E and Q be as in the previous two examples. Let $e^2 = e \in E$ be such that $eE = \begin{pmatrix} \mathbf{Z}[x] & 0 \\ 0 & 0 \end{pmatrix}$. Then $eEe = \mathbf{Z}[x]$ is a commutative Noetherian integral domain, and $S = eQe$ is a commutative localization of the integral domain eEe. However, eQe is not the classical ring of quotients of eEe.

18.2 Marginal isomorphisms

We fix an associative ring R and a right R-module G. Throughout the sequel let us agree to an additional hypothesis and notation associated with G. When needed we will explicitly state that

> $\operatorname{End}_R(G)$ possesses a semi-primary classical right ring of quotients Q_G.

For instance, G satisfies these conditions if $\operatorname{End}_R(G)$ is a right Artinian ring, if $\operatorname{End}_R(G)$ is a semi-prime right Goldie ring (e.g., a semi-prime right Noetherian ring), or if G is a reduced torsion-free group of finite rank. Specifically, each right regular element of $\operatorname{End}_R(G)$ is a unit in Q_G.

18.2.1 Margimorphism and localizations

Let

$$a : G \longrightarrow G' \quad \text{and} \quad b : G' \longrightarrow G$$

be R-module maps. The pair (a, b) is called a *marginal isomorphism pair* in R or we say that (a, b) is a *margimorphism pair* if $ab \in \text{End}_R(G')$ and $ba \in \text{End}_R(G)$ are regular endomorphisms. In this case a and b are called *margimorphisms* and we say that *G and G' are margimorphic* or that *G is margimorphic to G'*. The reader will prove that the relation *is margimorphic to* is an equivalence relation on any set of right R-modules. Examples of margimorphisms include isomorphisms of modules and quasi-isomorphisms of reduced torsion-free abelian groups. For instance, if $\text{End}_R(G)$ and $\text{End}_R(G')$ are domains then any pair (a, b) of maps as above such that $ab \neq 0 \neq ba$ is a margimorphism pair.

Let G, H, K be right R-modules. If G is margimorphic to $H \oplus K$ then we say that *$H \oplus K$ is a marginal decomposition of G*, and that *H is a marginal summand of G*. Direct summands of modules are marginal summands. Let

$$\mathbf{QP}_o(G) = \{H \in \mathbf{Mod}\text{-}R \,\big|\, H \oplus H' \text{ is margimorphic to } G^{(n)}$$
$$\text{for some right } R\text{-module } H' \text{ and for some integer } n \geq 0\}$$

For example, if $R = G = E$ is a right Noetherian domain and if I, J are right ideals in G then $I \oplus J$ is a marginal decomposition of $G \oplus G$.

Define an additive functor

$$\mathbf{QH}_G(\cdot) : \mathbf{Mod}\text{-}R \longrightarrow \mathbf{Mod}\text{-}Q_G$$

by

$$\mathbf{QH}_G(\cdot) = \text{Hom}_R(G, \cdot) \otimes_{\text{End}_R(G)} Q_G.$$

Since Q_G is a flat left $\text{End}_R(G)$-module $\mathbf{QH}_G(\cdot)$ is a left exact functor.

Our immediate goal is to prove that margimorphic R-modules have isomorphic classical right rings of quotients. The first few lemmas are technical.

Lemma 18.21. *Let G and G' be right R-modules, let (a, b) be a margimorphism pair for G and G', and let $c \in \text{End}_R(G)$ be a regular element. Then acb is a regular element in $\text{End}_R(G')$.*

Proof: Let $c \in \text{End}_R(G)$ be a regular element, and suppose that $(acb)x = 0$ for some $x \in \text{End}_R(G')$. Then $(ba)c(bxa) = 0$. Since ba and c are regular in $\text{End}_R(G)$, $bxa = 0 = (ab)x(ab)$, and since ab is regular in $\text{End}_R(G')$, $x = 0$. Hence acb is right regular in $\text{End}_R(G')$. Similarly, acb is left regular in $\text{End}_R(G')$, which proves the lemma.

Lemma 18.22. *Assume that* $\text{End}_R(G)$ *possesses a classical right ring of quotients* Q_G. *Suppose that* $G^{(n)}$ *and* G' *are margimorphic right R-modules for some integer* $n > 0$. *There is a natural embedding of rings*

$$\text{End}_R(G') \longrightarrow \text{End}_{Q_G}(\mathbf{QH}_G(G')) : \phi \longmapsto \mathbf{QH}_G(\phi).$$

Proof: First we work on $G^{(n)}$. Because $\mathbf{H}_G(\cdot)$ is a faithful functor on $\mathbf{P}_o(G)$, (Theorem 1.14), the ring mapping

$$\text{End}_G(G^{(n)}) \longrightarrow \text{End}_{\text{End}_R(G)}(\mathbf{H}_G(G^{(n)})) : \psi \longmapsto \mathbf{H}_G(\psi)$$

is a ring embedding. Moreover, let

$$K = \{\phi \in \mathbf{H}_G(G^{(n)}) \, | \, \phi \otimes 1 = 0\}.$$

Since Q_G is the classical localization of $\text{End}_R(G)$, K is the $\text{End}_R(G)$-submodule of $\mathbf{H}_G(G^{(n)})$ of elements x such that $xc = 0$ for some regular $c \in \text{End}_R(G)$. Thus $K = 0$, whence the canonical map

$$\text{End}_{\text{End}_R(G)}(\mathbf{H}_G(G^{(n)})) \longrightarrow \text{End}_{Q_G}(\mathbf{QH}_G(G^{(n)})) : \psi \longmapsto \psi \otimes 1$$

is a ring embedding. Then the composition

$$\text{End}_R(G^{(n)}) \longrightarrow \text{End}_{Q_G}(\mathbf{QH}_G(G^{(n)})) : \psi \longmapsto \mathbf{QH}_G(\psi) \qquad (18.3)$$

of these maps is an embedding of rings.

Now, because $\mathbf{QH}_G(\cdot)$ is an additive functor there is a ring mapping

$$\text{End}_R(G') \longrightarrow \text{End}_{Q_G}(\mathbf{QH}_G(G')) : \phi \longmapsto \mathbf{QH}_G(\phi).$$

Let $\phi \in \text{End}_R(G')$ and suppose that $\mathbf{QH}_G(\phi) = 0$. Then $b\phi a \in \text{End}_R(G^{(n)})$ and

$$\mathbf{QH}_G(b\phi a) = \mathbf{QH}_G(b)\mathbf{QH}_G(\phi)\mathbf{QH}_G(a) = 0.$$

By the injectivity of (18.3), $b\phi a = 0 = (ab)\phi(ab)$, and since $ab \in \text{End}_R(G')$ is a regular element, $\phi = 0$. This proves the lemma.

As a consequence of the next result the existence of a classical right ring of quotients is preserved by margimorphism.

Lemma 18.23. *Let* G *and* G' *be margimorphic right R-modules. If* $\text{End}_R(G)$ *possesses a classical right ring of quotients* Q_G *then* $\text{End}_R(G')$ *possesses a classical right ring of quotients* $Q_{G'}$.

Proof: We will prove that the regular elements of $\text{End}_R(G')$ satisfy the right Ore condition. We begin with a margimorphism pair $(a : G \longrightarrow G', b : G' \longrightarrow G)$. Let $x, c \in \text{End}_R(G')$ where c is regular. By Lemma 18.21, acb is a regular element in $\text{End}_R(G)$, and by hypothesis the regular elements of $\text{End}_R(G)$ satisfy the right Ore condition, so there is a $y \in \text{End}_R(G)$ and a regular $d \in \text{End}_R(G)$ such that

$(bxa)d = (bca)y$. Then $(ab)x(adb) = (ab)(cayb) \in \text{End}_R(G')$ and ab is regular, so $x(adb) = c(ayb)$. By Lemma 18.21, adb is regular in $\text{End}_R(G')$, so that the regular elements of $\text{End}_R(G')$ satisfy the right Ore condition. Hence $\text{End}_R(G')$ has a classical right ring of quotients $Q_{G'}$. This completes the proof.

If (a, b) is a margimorphism pair for G and G' then ab is regular in $\text{End}_R(G')$ and ba is regular in $\text{End}_R(G)$. Therefore, ab is a unit in $Q_{G'}$ and ba is a unit in Q_G, whenever these classical right rings of quotients exist.

The following result states that margimorphisms are isomorphisms when Q_G exists.

Lemma 18.24. *Let G, G' be right R-modules such that $\text{End}_R(G)$ possesses a classical right ring of quotients Q_G. If $(a : G \longrightarrow G', b : G' \longrightarrow G)$ is a margimorphism pair, then the induced maps*

$$\mathbf{QH}_G(a) : \mathbf{QH}_G(G) \longrightarrow \mathbf{QH}_G(G') \text{ and}$$

$$\mathbf{QH}_G(b) : \mathbf{QH}_G(G') \longrightarrow \mathbf{QH}_G(G)$$

are isomorphisms of right Q_G-modules.

Proof. Assume that (a, b) is a margimorphism pair for G and G'. An application of the left exact functor $\mathbf{QH}_G(\cdot)$ yields the inclusions

$$\mathbf{QH}_G(ba)Q_G = \mathbf{QH}_G(ba)\mathbf{QH}_G(G) \subset \mathbf{QH}_G(G) = Q_G.$$

Since $\mathbf{QH}_G(ba)$ is left multiplication on $\mathbf{QH}_G(G) \cong Q_G$ by the unit $ba \in Q_G$, it follows that $\mathbf{QH}_G(ba)$ is an isomorphism on Q_G. Since $\mathbf{QH}_G(ba) = \mathbf{QH}_G(b)\mathbf{QH}_G(a)$, $\mathbf{QH}_G(b)$ is a surjection and $\mathbf{QH}_G(a)$ is an injection.

Observe that $\mathbf{QH}_G(ab) : \mathbf{QH}_G(G') \longrightarrow \mathbf{QH}_G(G')$ is left multiplication by ab. Suppose that $\mathbf{QH}_G(b)x = 0$ for some $x \in \mathbf{QH}_G(G')$. Then $\mathbf{QH}_G(ab)x = 0 = (ab)x$. There is a regular $d \in \text{End}_R(G)$ such that $xd \in H_G(G')$ so that $(ab)(xd) = 0 = (ab)(xd)b$. Since $(xd)b \in \text{End}_R(G')$, and since ab is regular in $\text{End}_R(G')$, $xdb = 0 = xd(ba)$. Since ba and d are units in Q_G, $xd(ba) = 0 = x$. Thus $\mathbf{QH}_G(b)$ is an injection, whence $\mathbf{QH}_G(b)$ is an isomorphism. Subsequently, $\mathbf{QH}_G(a)$ is an isomorphism, and this completes the proof.

Theorem 18.25. *Let G be a right R-module such that $\text{End}_R(G)$ possesses a classical right ring of quotients Q_G. If G and G' are margimorphic, then Q_G is the classical right ring of quotients of $\text{End}_R(G')$.*

Proof. Let (a, b) be a margimorphism pair for G and G'. By Lemmas 18.22 and 18.23, $\text{End}_R(G')$ possesses a classical right ring of quotients $Q_{G'}$, and there is an embedding of rings

$$q : \text{End}_R(G') \longrightarrow \text{End}_{Q_G}(\mathbf{QH}_G(G')) : \phi \longmapsto \mathbf{QH}_G(\phi).$$

By Lemma 18.24,

$$\mathbf{QH}_G(a) : \mathbf{QH}_G(G) \longrightarrow \mathbf{QH}_G(G')$$

is an isomorphism of right $\mathrm{End}_R(G)$-modules, so that

$$\mathrm{End}_{Q_G}(\mathbf{QH}_G(G')) \cong \mathrm{End}_{Q_G}(\mathbf{QH}_G(G)) \qquad (18.4)$$

via the map that sends

$$x \longmapsto \mathbf{QH}_G(a)^{-1} x \mathbf{QH}_G(a)$$

for each $x \in \mathrm{End}_{Q_G}(\mathbf{QH}_G(G'))$. We can then identify rings

$$\mathrm{End}_{Q_G}(\mathbf{QH}_G(G')) \cong \mathrm{End}_{Q_G}(\mathbf{QH}_G(G)) = \mathrm{End}_{Q_G}(Q_G) = Q_G. \qquad (18.5)$$

Thus, to prove that Q_G is the classical right ring of quotients of $\mathrm{End}_R(G')$, it suffices to prove that $\mathrm{End}_{Q_G}(\mathbf{QH}_G(G'))$ is the classical right ring of quotients of $\mathrm{End}_R(G')$.

Let $c \in \mathrm{End}_R(G')$ be regular. We claim that c is a unit in $\mathrm{End}_{Q_G}(\mathbf{QH}_G(G'))$. By Lemma 18.21, bca is regular in $\mathrm{End}_R(G)$, so that

$$\mathbf{QH}_G(bca) = \mathbf{QH}_G(b)\mathbf{QH}_G(c)\mathbf{QH}_G(a)$$

is a unit in the classical ring of quotients Q_G. By Lemma 18.24, $\mathbf{QH}_G(a)$ and $\mathbf{QH}_G(b)$ are isomorphisms, so that $\mathbf{QH}_G(c)$ is also an isomorphism. Inasmuch as left multiplication by c is the map $\mathbf{QH}_G(c)$, c is a unit in $\mathrm{End}_{Q_G}(\mathbf{QH}_G(G'))$, as claimed.

Let $x \in \mathrm{End}_{Q_G}(\mathbf{QH}_G(G'))$. We have

$$\mathbf{QH}_G(a)^{-1} x \mathbf{QH}_G(a) \in Q_G.$$

Since Q_G is the classical right ring of quotients of $\mathrm{End}_R(G)$ and since $\mathbf{QH}_G(\phi)$ is left multiplication by ϕ for each $\phi \in \mathrm{End}_R(G)$, there is a regular $d \in \mathrm{End}_R(G)$ such that

$$\mathbf{QH}_G(a)^{-1} x \mathbf{QH}_G(ad) = [\mathbf{QH}_G(a)^{-1} x \mathbf{QH}_G(a)] \mathbf{QH}_G(d)$$

$$\in \mathrm{End}_R(G).$$

From the ring embedding q it follows that

$$x(adb) = x\mathbf{QH}_G(adb) = \mathbf{QH}_G(a)[\mathbf{QH}_G(a)^{-1} x \mathbf{QH}_G(ad)]\mathbf{QH}_G(b)$$

is in $q(\mathrm{End}_R(G')) = \mathrm{End}_R(G')$. By Lemma 18.21, adb is a regular element in $\mathrm{End}_R(G')$, so that $\mathrm{End}_R(G')$ is a right order in $\mathrm{End}_{Q_G}(\mathbf{QH}_G(G'))$. Therefore, $\mathrm{End}_{Q_G}(\mathbf{QH}_G(G'))$ is the classical right ring of quotients of $\mathrm{End}_R(G')$. Given (18.4) and (18.5), the proof is complete.

An example shows that in some cases margimorphisms are isomorphisms.

Example 18.26. Let

$$R = G = E = \left\{ \begin{pmatrix} x & 0 \\ y & x \end{pmatrix} \,\Big|\, x \in \mathbf{Z}, y \in \mathbf{Q}/\mathbf{Z} \right\}.$$

The reader will show that each element $x \neq \pm 1$ of E is a zero divisor. Then $E = Q_G$ is its own classical ring of quotients. By Theorem 18.25, if G is margimorphic to G' then E is the classical ring of quotients of $\mathrm{End}_R(G')$, whence $\mathrm{End}_R(G) = \mathrm{End}_R(G')$ and so $G \cong G'$. It follows that margimorphisms of G are isomorphisms.

Example 18.27. Let $E = \mathbf{Z}[x, y]$ be the polynomial ring in noncommuting indeterminants x and y. This is a domain E that does not have a classical right ring of quotients or a classical left ring of quotients. By Corner's theorem [59] there is an abelian group G such that $\mathrm{End}_{\mathbf{Z}}(G) \cong E$. The results of this section do not apply to G or E.

Example 18.28. Let E be a right Noetherian domain that is not left Goldie. Then E possesses a classical right ring of quotients Q which is a division ring and which is not a left quotient ring of E. Let I and J be nonzero right ideals in E. We will show that $I \oplus J$ is margimorphic to $E \oplus E$. Let $G = E \oplus E$. Then $\mathrm{End}_E(G)$ possesses a simple Artinian classical right ring of quotients $\mathrm{Mat}_2(Q)$. Hence, with $G' = I \oplus J$, Theorem 18.25 states that $\mathrm{End}_E(G')$ possesses a simple Artinian classical right ring of quotients $\mathrm{Mat}_2(Q)$. Let $(x, y) \in G'$ be such that $x \neq 0 \neq y$ and let $a : G \longrightarrow G'$ be defined by $a(s, t) = (xs, yt)$ for each $(s, t) \in G$. Let $b : G' \longrightarrow G$ be the inclusion map. Then a and b are injections so that ab and ba are (right) regular elements in $\mathrm{Mat}_2(Q)$, hence they are regular endomorphisms. Therefore G and G' are margimorphic.

18.2.2 Marginal summands

The right R-module H is a *marginal summand* of G if G is margimorphic to $H \oplus H'$ for some right R-module H'. The following results are the next steps toward our goal of treating marginal summands of G like simple modules.

Lemma 18.29. *Let G be a right R-module such that $\mathrm{End}_R(G)$ possesses a semi-primary classical right ring of quotients Q_G. If for a given integer $n > 0$, $G^{(n)}$ is margimorphic to G' then $\mathrm{Mat}_n Q_G$ is the semi-primary classical right ring of quotients of $\mathrm{End}_R(G^{(n)})$ and $\mathrm{End}_R(G')$.*

Proof: Since Q_G is the semi-primary classical right ring of quotients of $\mathrm{End}_R(G)$, Theorem 18.11 shows that $\mathrm{Mat}_n(Q_G)$ is the semi-primary classical right ring of quotients of $\mathrm{End}_R(G^{(n)})$. Because $G^{(n)}$ is margimorphic G', Theorem 18.25 states that $\mathrm{Mat}_n(Q_G)$ is the semi-primary classical right ring of quotients of $\mathrm{End}_R(G')$.

Theorem 18.30. *Suppose that $\mathrm{End}_R(G)$ has a semi-primary classical right ring of quotients Q_G, suppose that $G^{(n)}$ is margimorphic to $G' = H \oplus H'$ for some integer $n > 0$, and let $e^2 = e \in \mathrm{End}_R(G')$ be such that $e(G') = H$. Then*

$$\mathbf{QH}_G(G^{(n)}) \cong \mathbf{QH}_G(H) \oplus \mathbf{QH}_G(H') \text{ and}$$
$$\mathbf{QH}_G(H) \cong eQ_G^{(n)}$$

as right Q_G-modules.

Proof: Suppose that $\text{End}_R(G)$ possesses a semi-primary classical right ring of quotients Q_G. Let $n > 0$ be an integer, let $(a : G^{(n)} \longrightarrow G', b : G' \longrightarrow G^{(n)})$ be a margimorphism pair, let $G' = H \oplus H'$, and let $e^2 = e \in \text{End}_R(G')$ be such that $eG' = H$ and $(1 - e)G' = H'$. By Lemmas 18.24 and 18.29, $\textbf{QH}_G(a) : \textbf{QH}_G(G^{(n)}) \longrightarrow \textbf{QH}_G(G')$ is an isomorphism, so that

$$\textbf{QH}_G(G^{(n)}) \cong \textbf{QH}_G(G') = \textbf{QH}_G(H \oplus H') \cong \textbf{QH}_G(H) \oplus \textbf{QH}_G(H').$$

An application of $\cdot \otimes_{\text{End}_R(G)} Q_G$ to the equation $e\textbf{H}_G(G') = \textbf{H}_G(H)$ yields the natural identity

$$e\textbf{QH}_G(G') = \textbf{QH}_G(H). \qquad (18.6)$$

Furthermore, the fact that $\textbf{QH}_G(a)$ is an isomorphism implies that the map

$$\text{End}_{Q_G}(\textbf{QH}_G(G')) \longrightarrow \text{End}_{Q_G}(\textbf{QH}_G(G^{(n)}))$$

defined by

$$\phi \longmapsto \textbf{QH}_G(a)^{-1}\textbf{QH}_G(\phi)\textbf{QH}_G(a)$$

is an isomorphism of rings. Then

$$eQ_G^{(n)} \cong \textbf{QH}_G(a)^{-1}\textbf{QH}_G(e)\textbf{QH}_G(a)\textbf{QH}_G(G^{(n)})$$
$$= \textbf{QH}_G(a)^{-1}\textbf{QH}_G(e)\textbf{QH}_G(G')$$
$$\cong \textbf{QH}_G(e)\textbf{QH}_G(G')$$
$$= e\textbf{QH}_G(G').$$

Then by (18.6), $eQ_G^{(n)} \cong \textbf{QH}_G(H)$, which completes the proof of the theorem.

Theorem 18.31. *Suppose that* $\text{End}_R(G)$ *has a semi-primary classical right ring of quotients* Q_G, *and suppose that* $H \in \textbf{QP}_o(G)$. *Then*

$$\text{End}_{Q_G}(\textbf{QH}_G(H))$$

is naturally the semi-primary classical right ring of quotients of $\text{End}_R(H)$.

Proof: Suppose that $H \oplus H'$ is margimorphic to $G^{(n)}$, and let $e^2 = e \in \text{End}_R(H \oplus H')$ be such that $e(H \oplus H') = H$. By Theorem 18.11, $\text{Mat}_n(Q_G)$ is the semi-primary classical right ring of quotients of $\text{Mat}_n(\text{End}_R(G)) = \text{End}_R(G^{(n)})$, and by Theorem 18.25, $\text{Mat}_n(Q_G)$ is the semi-primary classical right ring of quotients of $\text{End}_R(H \oplus H')$. Theorem 18.17 then states that $e\text{End}_R(H \oplus H')e$ has a semi-primary classical right ring of quotients $e\text{Mat}_n(Q_G)e$. It is clear that

$$e\text{End}_R(H \oplus H')e = \text{End}_R(H)$$

so $\text{End}_R(H)$ has semi-primary classical right ring of quotients $e\text{Mat}_n(Q_G)e$.

Moreover, Theorem 18.30 states that $\mathbf{QH}_G(H) \cong eQ_G^{(n)}$ so there is an isomorphism of rings

$$e\mathrm{Mat}_n(Q_G)e \cong \mathrm{End}_{Q_G}(eQ_G^{(n)}) \cong \mathrm{End}_{Q_G}(\mathbf{QH}_G(H)).$$

Hence $\mathrm{End}_{Q_G}(\mathbf{QH}_G(H))$ is the semi-primary classical right ring of quotients $\mathrm{End}_R(H)$. The reader can show that the embedding is given by

$$q : \mathrm{End}_R(H) \longrightarrow \mathrm{End}_{Q_G}(\mathbf{QH}_G(H)) : \phi \longrightarrow \mathbf{QH}_G(\phi).$$

This completes the proof.

Corollary 18.32. *Suppose that* $\mathrm{End}_R(G)$ *has a semi-primary classical right ring of quotients* Q_G, *and suppose that for some integer* $n > 0$, $G^{(n)}$ *is margimorphic to* $H \oplus H'$. *If* $e^2 = e \in \mathrm{End}_R(H \oplus H')$ *is such that* $e(H \oplus H') = H$ *then*

$$e\mathrm{Mat}_n(Q_G)e$$

is naturally the semi-primary classical right ring of quotients of $\mathrm{End}_R(H)$.

18.2.3 Marginal summands as projectives

Recall that $\mathbf{QP}_o(G) = \{H \in \mathbf{Mod}\text{-}R \mid H \text{ is a marginal summand of } G^{(n)} \text{ for some integer } n > 0\}$. Using the localization theory developed in the previous subsection we will prove the following result.

Theorem 18.33. *Suppose that* G *is a right* R-*module such that* $\mathrm{End}_R(G)$ *possesses a semi-primary classical right ring of quotients* Q_G, *and let* $H, K \in \mathbf{QP}_o(G)$. *Then*

1. $\mathbf{QH}_G(H)$ *is a finitely generated projective right* Q_G-*module and*
2. H *is margimorphic to* K *iff* $\mathbf{QH}_G(H) \cong \mathbf{QH}_G(K)$ *as right* Q_G-*modules.*

Proof: 1. Theorem 18.30 shows us that if H is a marginal summand of $G^{(n)}$ then $\mathbf{QH}_G(H)$ is a direct summand of $\mathbf{QH}_G(G^{(n)}) \cong Q_G^{(n)}$. This proves part 1.

2. The proof of part 2 is a series of lemmas.

Lemma 18.34. *Let* G *be a right* R-*module such that* $\mathrm{End}_R(G)$ *possesses a semi-primary classical right ring of quotients* Q_G. *Let* $a : G \longrightarrow G'$ *and* $b : G' \longrightarrow G$ *be* R-*module maps. If* $\mathbf{QH}_G(a)$ *and* $\mathbf{QH}_G(b)$ *are isomorphisms then* (a, b) *is a margimorphism pair for* G *and* G'.

Proof: Suppose that

$$\mathbf{QH}_G(a) : \mathbf{QH}_G(G) \longrightarrow \mathbf{QH}_G(G') \text{ and}$$
$$\mathbf{QH}_G(b) : \mathbf{QH}_G(G') \longrightarrow \mathbf{QH}_G(G)$$

are isomorphisms of right Q_G-modules. By Lemma 18.22, $ab \in \mathrm{End}_R(G')$ lifts to a unit $\mathbf{QH}_G(ab) \in \mathrm{End}_{Q_G}(\mathbf{QH}_G(G'))$, so that ab is regular in $\mathrm{End}_R(G')$. Similarly, ba

is regular in $\text{End}_R(G)$, so (a, b) is a margimorphism pair for G and G'. This proves the lemma.

Lemma 18.35. *Let G be a right R-module and suppose that $\text{End}_R(G)$ has a semi-primary classical right ring of quotients Q_G. If $G^{(h)}$ is margimorphic to G' and if $G^{(k)}$ is margimorphic to G'', then*

1. $\mathbf{QH}_G(G^{(h)}) \cong \mathbf{QH}_G(G')$ *as right Q_G-modules.*
2. $G^{(h+k)}$ *is margimorphic to $G' \oplus G''$.*

Proof: Since $\text{End}_R(G)$ possesses a semi-primary classical right ring of quotients Q_G, Lemma 18.29 states that the rings $\text{End}_R(G^{(h)})$, $\text{End}_R(G^{(h+k)})$, $\text{End}_R(G')$, and $\text{End}_R(G'')$ possess semi-primary classical right rings of quotients.

  1. Let (a, b) be a margimorphism pair for $G^{(h)}$ and G'. By Lemma 18.22 and Theorem 18.25,

$$\text{End}_{Q_G}(\mathbf{QH}_G(G^{(h)}))$$

is the semi-primary classical right ring of quotients of $\text{End}_R(G^{(h)})$ via the embedding $\phi \longmapsto \mathbf{QH}_G(\phi)$. Then the regular element ba in $\text{End}_R(G^{(h)})$ is sent to the unit

$$\mathbf{QH}_G(ba) \in \text{End}_{Q_G}(\mathbf{QH}_G(G^{(h)})).$$

Similarly, $\mathbf{QH}_G(ab)$ is a unit in $\text{End}_{Q_G}(\mathbf{QH}_G(G'))$. Then $\mathbf{QH}_G(a) : \mathbf{QH}_G(G^{(h)}) \longrightarrow \mathbf{QH}_G(G')$ is an isomorphism.

  2. Let $(a : G^{(h)} \longrightarrow G', b : G' \longrightarrow G^{(h)})$ and $(a' : G^{(k)} \longrightarrow G'', b' : G'' \longrightarrow G^{(k)})$ be margimorphism pairs. By part 1 and its proof, $\mathbf{QH}_G(a)$, $\mathbf{QH}_G(b)$, $\mathbf{QH}_G(a')$, $\mathbf{QH}_G(b')$ are isomorphisms, whence

$$(a \oplus a' : G^{(h+k)} \longrightarrow G' \oplus G'', \quad b \oplus b' : G' \oplus G'' \longrightarrow G^{(h+k)})$$

is a pair of maps such that

$$\mathbf{QH}_G(a \oplus a')\mathbf{QH}_G(b \oplus b') = \mathbf{QH}_G(ab) \oplus \mathbf{QH}_G(a'b') \quad \text{and}$$

$$\mathbf{QH}_G(b \oplus b')\mathbf{QH}_G(a \oplus a') = \mathbf{QH}_G(ba) \oplus \mathbf{QH}_G(b'a')$$

are isomorphisms. Then by Lemma 18.34, $(a \oplus a', b \oplus b')$ is a margimorphism pair, and hence $G^{(h+k)}$ is margimorphic to $G' \oplus G''$. This completes the proof of the lemma.

Lemma 18.36. *Let G be a right R-module and suppose that $\text{End}_R(G)$ has a semi-primary classical right ring of quotients Q_G. Let $H, K \in \mathbf{QP}_o(G)$. There is an integer $m > 0$ and a right R-module L such that $H \oplus K \oplus L$ is margimorphic to $G^{(m)}$.*

Proof: There are integers $h, k > 0$ and right R-modules H', K' such that $H \oplus H'$ is margimorphic to $G^{(h)}$ and $K \oplus K'$ is margimorphic to $G^{(k)}$. By part 2 of the previous lemma,

$$(H \oplus H') \oplus (K \oplus K') \cong H \oplus K \oplus (H' \oplus K')$$

is margimorphic to $G^{(h+k)}$.

We are now ready to continue the **Proof of Theorem 18.33.2.**

2. Suppose that we are given $H, K \in \mathbf{QP}_o(G)$. By Theorem 18.31, $\mathrm{End}_R(H)$ and $\mathrm{End}_R(K)$ possess semi-primary classical right rings of quotients $\mathrm{End}_{Q_G}(\mathbf{QH}_G(H))$ and $\mathrm{End}_{Q_G}(\mathbf{QH}_G(K))$. Suppose that there is a margimorphism pair $(a : H \longrightarrow K, b : K \longrightarrow H)$ of right R-modules. Then the regular $ba \in \mathrm{End}_R(H)$ maps to a unit $\mathbf{QH}_G(b)\mathbf{QH}_G(a) = \mathbf{QH}_G(ba) \in \mathrm{End}_{Q_G}(\mathbf{QH}_G(H))$. Similarly, $\mathbf{QH}_G(a)\mathbf{QH}_G(b) = \mathbf{QH}_G(ab)$ is a unit in $\mathrm{End}_{Q_G}(\mathbf{QH}_G(K))$ so that

$$\mathbf{QH}_G(a) : \mathbf{QH}_G(H) \longrightarrow \mathbf{QH}_G(K)$$

is an isomorphism.

Conversely, assume that $u : \mathbf{QH}_G(H) \longrightarrow \mathbf{QH}_G(K)$ is an isomorphism of right Q_G-modules. By Lemma 18.36, there is an integer m and a right R-module L such that $G^{(m)}$ is margimorphic to

$$G' = H \oplus K \oplus L.$$

If $e^2 = e, f^2 = f \in \mathrm{End}_R(G')$ are such that $eG' = H$ and $fG' = K$ then $\mathbf{QH}_G(H) \cong eQ_G$ and $\mathbf{QH}_G(K) \cong fQ_G$ by Theorem 18.30. Thus we can identify u with $u = fue \in \mathrm{End}_R(G')$. By Lemma 18.16 there is a regular $c \in \mathrm{End}_R(G')$ such that $c = ece \oplus (1 - e)c(1 - e)$ and $uc = f(uc)e \in \mathrm{Hom}_R(H, K)$. Similarly, there is a regular $d = fdf \oplus (1 - f)d(1 - e)$ such that $u^{-1}d = eu^{-1}df \in \mathrm{Hom}_R(K, H)$. Then $(\mathbf{QH}_G(uc), \mathbf{QH}_G(u^{-1}d))$ is easily seen to be a pair of isomorphisms, so by Lemma 18.34, $(uc, u^{-1}d)$ is a margimorphism pair for H and K. This completes the proof of Theorem 18.33.

Example 18.37. This example will demonstrate the necessity of the semi-primary hypothesis in Theorem 18.33. Let R be the local ring $R = \mathbf{Q}[x]_{(x)}$, and let $G = R \oplus \mathbf{Q}$ be a right R-module. Then $\mathrm{End}_R(G) = Q_G = \begin{pmatrix} R & 0 \\ \mathbf{Q} & \mathbf{Q} \end{pmatrix}$. The reader will show that each regular element in $\mathrm{End}_R(G)$ is a unit. Let $H = R$ as a direct summand of G. Then $(1_H : H \longrightarrow H, x : H \longrightarrow H)$ is a margimorphism pair. But $\mathbf{QH}_G(x)$, left multiplication by x, is not an isomorphism since x is not a unit in the local integral domain (not a field) $R = \mathrm{End}_R(R) = \mathrm{End}_{Q_G}(\mathbf{QH}_G(R))$.

18.2.4 Projective Q_G-modules

Let

> $\mathbf{P}_o(Q_G) = $ the category of finitely generated projective right Q_G-modules.

Theorem 18.38. *Let G be a right R-module such that $\mathrm{End}_R(G)$ has a semi-primary classical right ring of quotients Q_G. There is a faithful left exact functor*

$$\mathbf{QH}_G(\cdot) : \mathbf{QP}_o(G) \longrightarrow \mathbf{P}_o(Q_G)$$

such that, for each $H, K \in \mathbf{QP}_o(G)$,

1. *H is margimorphic to K iff $\mathbf{QH}_G(H) \cong \mathbf{QH}_G(K)$.*
2. *$\mathrm{End}_{Q_G}(\mathbf{QH}_G(H))$ is the semi-primary classical right ring of quotients of $\mathrm{End}_R(H)$.*
3. *For each $P \in \mathbf{P}_o(Q_G)$ there is an $H \in \mathbf{QP}_o(G)$ such that $\mathbf{QH}_G(H) \cong P$.*
4. *$\mathbf{QH}_G(H) \cong P \oplus P'$ iff H is margimorphic to $K \oplus K'$ for some R-modules K and K' such that $\mathbf{QH}_G(K) \cong P$ and $\mathbf{QH}_G(K') \cong P'$ as right Q_G-modules.*

Proof: Let $H, K \in \mathbf{P}_o(G)$.

Since $\mathbf{H}_G(\cdot)$ and $\cdot \otimes_{\mathrm{End}_R(G)} Q_G$ are left exact functors $\mathbf{QH}_G(\cdot)$ is left exact.

Theorem 18.33 shows us that $\mathbf{QH}_G(\cdot) : \mathbf{QP}_o(G) \longrightarrow \mathbf{P}_o(Q_G)$ is a well-defined additive functor. Let $\phi : H \longrightarrow K$ be a map in $\mathbf{QP}_o(G)$ such that $\mathbf{QH}_G(\phi) = 0$. By Lemma 18.36 we may assume without loss of generality that $G^{(n)}$ is margimorphic to $G' = H \oplus K \oplus L$ for some integer $n > 0$ and a right R-module L. We may then consider ϕ to be a mapping $G' \longrightarrow G'$. By Lemma 18.22, $\mathbf{QH}_G(\phi) = 0$ implies that $\phi = 0$. Hence $\mathbf{QH}_G(\cdot) : \mathbf{QP}_o(G) \longrightarrow \mathbf{P}_o(Q_G)$ is a faithful functor.

1. By Theorem 18.33, H is margimorphic to K iff $\mathbf{QH}_G(H) \cong \mathbf{QH}_G(K)$.

2. Theorem 18.31 shows us that $\mathrm{End}_R(\mathbf{QH}_G(H))$ is the semi-primary classical right ring of quotients of $\mathrm{End}_R(H)$.

4. Part 4 follows immediately from parts 1 and 3.

Part 3 is important enough to have its own number. Since $\mathbf{QH}_G(H)$ is a finitely generated projective right Q_G-module for each $H \in \mathbf{QP}_o(G)$, it is natural to ask if each projective right Q_G-module can be written as $\mathbf{QH}_G(H)$ for some $H \in \mathbf{QP}_o(G)$.

Theorem 18.39. *Let G be a right R-module such that $\mathrm{End}_R(G)$ has a semi-primary classical right ring of quotients Q_G. If $P \oplus P' \cong Q_G^{(n)}$ for some integer $n > 0$, then $G^{(n)}$ is margimorphic to $H \oplus H'$ for some G-generated $H, H' \in \mathbf{QP}_o(G)$ such that $\mathbf{QH}_G(H) \cong P$ and $\mathbf{QH}_G(H') \cong P'$.*

Proof: Let $n > 0$ be an integer and assume that

$$P \oplus P' = \mathbf{QH}_G(G^{(n)}) = F$$

as right Q_G-modules. There are idempotents

$$e, e' \in \mathrm{End}_{Q_G}(F)$$

such that $e \oplus e' = 1$, $eF = P$ and $e'F = P'$. Since $\mathrm{End}_{Q_G}(F)$ is naturally the classical right ring of quotients of $\mathrm{End}_R(G^{(n)})$ (Theorem 18.31), there is a regular $c \in \mathrm{End}_R(G^{(n)})$ such that

$$ec, e'c \in \mathrm{End}_R(G^{(n)}).$$

Since $G^{(n)}$ is a right R-module and since $ee' = 0$,

$$G' = ecG^{(n)} \oplus e'cG^{(n)} \subset G^{(n)}.$$

Let $a : G^{(n)} \longrightarrow G'$ be the map defined by $a(x) = c(x)$, and let $b : G' \longrightarrow G^{(n)}$ be the inclusion map. For each $\phi \otimes 1 \in \mathbf{QH}_G(G^{(n)})$ we have

$$\mathbf{QH}_G(ba)(\phi \otimes 1) = (ba)\phi \otimes 1 = (c\phi) \otimes 1 \in \mathbf{QH}_G(G^{(n)}),$$

which shows that $\mathbf{QH}_G(ba)$ is the isomorphism of left multiplication by c. Subsequently, $\mathbf{QH}_G(b)$ is a surjection. But b is inclusion and $\mathbf{QH}_G(\cdot)$ is left exact so $\mathbf{QH}_G(b)$ is an injection, whence $\mathbf{QH}_G(b)$ and $\mathbf{QH}_G(ba) = \mathbf{QH}_G(b)\mathbf{QH}_G(a)$ are isomorphisms. It follows that $\mathbf{QH}_G(a)$ is an isomorphism. Then by Lemma 18.34, $G^{(n)}$ is margimorphic to G'. We will identify $\mathbf{QH}_G(G') = \mathbf{QH}_G(G^{(n)})$.

Therefore,

$$H = ecG^{(n)} \text{ and } H' = e'cG^{(n)}$$

are objects in $\mathbf{QP}_o(G)$ such that

$$eG' = eH + eH' = eecG^{(n)} + ee'cG^{(n)} = H,$$

so that $\mathbf{H}_G(H) = e\mathbf{H}_G(G')$. Because we identify $\mathbf{QH}_G(G') = \mathbf{QH}_G(G^{(n)})$ we have

$$\mathbf{QH}_G(H) = e\mathbf{QH}_G(G') = e\mathbf{QH}_G(G^{(n)}) = eF = P.$$

Similarly, $\mathbf{QH}_G(H') = P'$. This completes the proof of Theorem 18.39 and thus we have proved Theorem 18.38.3.

18.3 Uniqueness of direct summands

18.3.1 Totally indecomposable modules

We say that the R-module G is *totally indecomposable* if each R-module that is margimorphic to G is indecomposable. Examples will give us some intuition about this strong notion of indecomposability.

Example 18.40. 1. Each field $\mathbf{k}$ is totally indecomposable as a right $\mathbf{k}$-module.
2. Let R be a domain such that R satisfies the right Ore condition. These domains exist in abundance. See [72]. Then R is totally indecomposable as a right R-module.

Example 18.41. Let G be a right R-module such that $\mathrm{End}_R(G)$ is a (not necessarily commutative) right Noetherian domain. (For example, choose $R = G$ to be a right Noetherian domain.) By Goldie's theorem there is a division ring Q_G such that Q_G is the classical right ring of quotients of $\mathrm{End}_R(G)$. If G is margimorphic to $H \oplus H'$, then by Theorem 18.25, Q_G contains $e^2 = e$ such that $e(H \oplus H') = H$. Since the nonzero elements in Q_G are units, $e = 1$ or $e = 0$, and hence $H = 0$ or $H' = 0$. Thus G is totally indecomposable.

Example 18.42. Let (x) denote the ideal generated by the indeterminant x in $\mathbf{Z}[x]$, let

$$R = G = \mathrm{End}_R(G) = \left\{ (p,q) \,\middle|\, p,q \in \mathbf{Z}[x] \text{ and } p - q \in (x) \right\},$$

$$H = \left\{ (xp, 0) \,\middle|\, p \in \mathbf{Z}[x] \right\}, \quad \text{and}$$

$$H' = \left\{ (0, xq) \,\middle|\, q \in \mathbf{Z}[x] \right\}.$$

The right R-module G is indecomposable since, as the reader will show, $\mathrm{End}_R(G)$ has no idempotents e other than $0, 1$. But G is not totally indecomposable since if we let $\jmath$ be the inclusion map then it is readily seen that the pair

$$(x : G \longrightarrow H \oplus H', \quad \jmath : H \oplus H' \longrightarrow G)$$

is a margimorphism pair of right R-modules. Observe that

$$Q_G = \mathbf{Q}(x) \times \mathbf{Q}(x)$$

is the semi-prime Artinian classical ring of quotients of $\mathrm{End}_R(G)$, where $\mathbf{Q}(x)$ is the ring of rational functions with coefficients in $\mathbf{Q}$.

Example 18.43. Compare this example with Theorem 18.46. Let $p \in \mathbf{Z}$ be a prime and let

$$R = G = \mathrm{End}_R(G) = \left\{ \begin{pmatrix} x & 0 \\ y & x \end{pmatrix} \,\middle|\, x \in \mathbf{Z}_p, y \in \mathbf{Z}_p/p\mathbf{Z}_p \right\}.$$

Then R is a local commutative ring. Since $R/\mathcal{J}(R) \cong \mathbf{Z}/p\mathbf{Z}$ and since $\mathrm{ann}_R(\mathcal{J}(R)) \neq 0$, $R = Q_R$ is its own local classical ring of quotients. We note that if G is margimorphic to $H \oplus H'$ then $G \cong H \oplus H'$ since regular elements in R are units. Let $e^2 = e \neq 0$ be such that $e(H \oplus H') = H$. Since R is local, $e = 1$ or $e = 0$, hence $G = H$, whence G is totally indecomposable. Moreover since $p \in \mathcal{J}(R)$ is not nilpotent, R is not semi-primary.

18.3.2 Morphisms of totally indecomposables

The ring E is *local* if E has a unique maximal right ideal iff the nonunits of E form an ideal. We leave it to the reader to show that if E is a local ring and if $\mathcal{J}(E)$ consists of zero divisors then E is totally indecomposable. A result due to J. D. Reid shows that the reduced torsion-free finite rank abelian group G is totally indecomposable (= *strongly indecomposable*) iff

$$\mathbf{Q}\mathrm{End}(G) = \mathrm{End}_{\mathbf{Z}}(G) \otimes \mathbf{Q}$$

is a local finite-dimensional $\mathbf{Q}$-algebra. See [9, 59]. The above examples suggest that Q_G is local when G is totally indecomposable. Notice that in the next result Q_G is not necessarily a semi-primary ring.

Theorem 18.44. *Let G be a right R-module such that $\text{End}_R(G)$ possesses a classical right ring of quotients Q_G. If Q_G is a local ring then G is totally indecomposable.*

Proof: Suppose that Q_G is a local ring, and assume that G is margimorphic to $H \oplus H'$. Lemma 18.24 implies that

$$
\begin{aligned}
Q_G &= \mathbf{QH}_G(G) \\
&\cong \mathbf{QH}_G(H \oplus H') \\
&= \mathbf{QH}_G(H) \oplus \mathbf{QH}_G(H').
\end{aligned}
$$

Since Q_G is a local ring, one summand, say $\mathbf{QH}_G(H')$, is equal to 0. By Theorem 18.38, $\mathbf{QH}_G(\cdot) : \mathbf{QP}_o(G) \longrightarrow \mathbf{P}_o(Q_G)$ is a faithful functor, so that $Q_G(H') = 0$ implies that $H' = 0$. We conclude that G is totally indecomposable. This completes the proof.

Example 18.45. We construct a right R-module G such that Q_G exists, Q_G is not a local ring, but such that G is totally indecomposable. Let

$$
R = G = E = \left\{ \begin{pmatrix} x & 0 \\ y & x \end{pmatrix} \,\Big|\, x \in \mathbf{Z}_6, y \in \mathbf{Z}_6/6\mathbf{Z}_6 \right\}.
$$

Then

$$
\mathcal{J}(E) = \left\{ \begin{pmatrix} x & 0 \\ y & x \end{pmatrix} \,\Big|\, x \in 6\mathbf{Z}_6, y \in \mathbf{Z}_6/6\mathbf{Z}_6 \right\},
$$

so that $E/\mathcal{J}(E) \cong \mathbf{Z}/6\mathbf{Z}$, which is not a field. That is, E is not a local ring. Furthermore, E is indecomposable since $E/\mathcal{N}(E) \cong \mathbf{Z}_6$ is indecomposable. The reader will show that $c = \begin{pmatrix} x & 0 \\ y & x \end{pmatrix}$ is a regular element in E iff $x \notin 2\mathbf{Z}_6$ and $x \notin 3\mathbf{Z}_6$. But then a regular c is regular modulo $\mathcal{J}(E)$, which makes c a unit modulo $\mathcal{J}(E)$. Since units lift modulo $\mathcal{J}(E)$, regular elements of E are units of E. That is, E is a quotient ring. It follows that any margimorphism for E is an isomorphism for E. Then because it is indecomposable, E is totally indecomposable.

In the presence of a semi-primary classical right ring of quotients there is an interesting connection between totally indecomposable R-modules and local rings.

Theorem 18.46. *Let G be a right R-module such that $\text{End}_R(G)$ has a semi-primary classical right ring of quotients Q_G. The following are equivalent for G.*

1. *Q_G is a local ring.*
2. *The nonregular elements of $\text{End}_R(G)$ form a nil ideal.*
3. *G is totally indecomposable.*

Proof: $1 \Rightarrow 2$. Suppose that Q_G is local, and let $x \in \text{End}_R(G)$ be a nonregular element. Then x is not a unit of Q_G. Since Q_G is local $x \in \mathcal{J}(Q_G)$, and since Q_G is semi-primary x is nilpotent. This proves part 2.

$2 \Rightarrow 3$. Assume to the contrary that G is margimorphic to $G' = H \oplus H'$ where $H \neq 0 \neq H'$. Then by Theorem 18.31, Q_G contains the nonzero idempotents $e, e' \in \text{End}_R(G')$ such that $eG = H$ and $e'G' = H'$. By Lemma 18.13.2 there is a regular $c \in \text{End}_R(G)$ such that $ec, e'c \in \text{End}_R(G)$. Then ec and $e'c$ are nonregular elements in $\text{End}_R(G)$ whose sum is regular and not nilpotent. Hence part 2 is false.

$3 \Rightarrow 1$. Let G be totally indecomposable and let $e, e' \in Q_G$ be nonzero idempotents such that $e \oplus e' = 1$. Then $Q_G = eQ_G \oplus e'Q_G$ as right Q_G-modules. By Theorem 18.38.4 there are right R-modules $H \neq 0 \neq H' \in \mathbf{QP}_o(G)$ such that $\mathbf{QH}_G(H) \cong eQ_G$ and $\mathbf{QH}_G(H') \cong e'Q_G$. It follows that

$$\mathbf{QH}_G(G) \cong Q_G$$
$$\cong eQ_G \oplus e'Q_G$$
$$\cong \mathbf{QH}_G(H) \oplus \mathbf{QH}_G(H')$$
$$\cong \mathbf{QH}_G(H \oplus H'),$$

so that G is margimorphic to $H \oplus H'$ by Theorem 18.38.1. Since G is totally indecomposable, $H = 0$ or $H' = 0$. Thus $e = 1$ or $e' = 1$. That is, Q_G is a semi-primary ring whose only idempotents are $0, 1$. Such a ring is local. This proves part 1 and completes the logical cycle.

The totally indecomposable right R-modules in $\mathbf{QP}_o(G)$ are characterized in the following manner.

Theorem 18.47. *Let G be a right R-module such that $\text{End}_R(G)$ possesses a semi-primary classical right ring of quotients Q_G, and let $H \in \mathbf{QP}_o(G)$. The following are equivalent.*

1. *H is totally indecomposable.*
2. *$\mathbf{QH}_G(H)$ is an indecomposable projective right Q_G-module.*
3. *$\text{End}_{Q_G}(\mathbf{QH}_G(H))$ is a local ring.*

Proof: $1 \Leftrightarrow 2$. Assume part 1, and let $\mathbf{QH}_G(H) = P \oplus P'$ for some projective right Q_G-modules P and P'. By Theorem 18.38.4 there are $K, K' \in \mathbf{QP}_o(G)$ such that $P \cong \mathbf{QH}_G(K)$ and $P' \cong \mathbf{QH}_G(K')$, so that

$$\mathbf{QH}_G(H) \cong P \oplus P'$$
$$\cong \mathbf{QH}_G(K) \oplus \mathbf{QH}_G(K')$$
$$\cong \mathbf{QH}_G(K \oplus K').$$

Theorem 18.38.1 then states that H is margimorphic to $K \oplus K'$, so that by part 1, one summand, say K', is zero. Then $P' = \mathbf{QH}_G(K') = 0$ and so $\mathbf{QH}_G(H)$ is indecomposable. This proves part 2.

Conversely, assume that part 1 is false. Then H is margimorphic to $K \oplus K'$ for some right R-modules $K \neq 0 \neq K'$. By Theorem 18.38.1,

$$\mathbf{QH}_G(H) \cong \mathbf{QH}_G(K) \oplus \mathbf{QH}_G(K'),$$

where $\mathbf{QH}_G(K) \neq 0 \neq \mathbf{QH}_G(K')$. Thus $\mathbf{QH}_G(H)$ is decomposable. That is, part 2 is false.

$1 \Leftrightarrow 3$ follows from Theorems 18.31 and 18.46.

Example 18.48. Let

$$R = G = E = \left\{ \begin{pmatrix} p(x) & 0 \\ q(x) & p(x) \end{pmatrix} \middle| p(x) \in \mathbf{Q}[x], q(x) \in \mathbf{Q}(x)/\mathbf{Q}[x] \right\},$$

where $\mathbf{Q}(x)$ is the ring of rational functions over $\mathbf{Q}$. Then each nonconstant element $r(x)$ in E is a zero divisor in E, so that E is its own classical ring of quotients. But E is not semi-primary. The reader can prove that G is totally indecomposable but E is not local.

We close this subsection by showing that the margimorphism of two totally indecomposable objects can be established with relatively little work.

Theorem 18.49. *Let G and G' be totally indecomposable right R-modules such that $\mathrm{End}_R(G)$ possesses a semi-primary classical right ring of quotients Q_G. Then G is margimorphic to G' iff there are maps $a : G \longrightarrow G'$ and $b : G' \longrightarrow G$ such that either ba or ab is not nilpotent.*

Proof: Suppose that G and G' are not margimorphic and let $a : G \longrightarrow G'$ and $b : G' \longrightarrow G$ be two R-module maps. One of ab, ba is not regular, say ba. By Theorem 18.46, ba is nilpotent, say $(ba)^\ell = 0$. Then

$$(ab)^{\ell+1} = a(ba)^\ell b = 0,$$

so that ab is nilpotent.

The converse is obvious. This completes the proof.

18.3.3 Semi-simple marginal summands

Our next goal is to prove that $\mathbf{QH}_G(\cdot)$ can be used to change marginal summands of G into semi-simple right Q_G-modules.

Theorem 18.50. *Let G be a right R-module and suppose that $\mathrm{End}_R(G)$ possesses a semi-primary classical right ring of quotients Q_G. The functor $\mathbf{QH}_G(\cdot)$ induces a bijective correspondence*

$$\Delta : \begin{cases} \{[H] \,\big|\, H \in \mathbf{QP}_o(G)\} \longrightarrow \{(P) \,\big|\, P \in \mathbf{P}_o(Q_G)\} \\ [H] \longmapsto (\mathbf{QH}_G(H)) \end{cases}$$

from the set of margimorphism classes $[H]$ *of* $H \in \mathbf{QP}_o(G)$ *onto the set of isomorphism classes* (P) *of finitely generated projective right* Q_G-*modules.*

Proof: The proof is a verification, using Theorem 18.38, that the assignment $H \mapsto \mathbf{QH}_G(H)$ defines the stated bijection. We leave the details as an exercise for the reader.

Recall that in a semi-prime right Goldie ring E,

$$P \cong P \oplus P' \Rightarrow P' = 0$$

for each $P \in \mathbf{P}_o(E)$.

Lemma 18.51. *Let* E *be a ring such that* $E/\mathcal{N}(E)$ *is semi-prime right Goldie, and let*

$$\boxed{\mathbf{A}_E(M) = M \otimes_E E/\mathcal{N}(E) \cong M/M\mathcal{N}(E)}$$

for right E-*modules* M. *Then* $\mathbf{A}_E(\cdot) \; : \; \mathbf{P}_o(E) \longrightarrow \mathbf{P}_o(E/\mathcal{N}(E))$ *is a full additive functor such that*

1. *For each* $\overline{P} \in \mathbf{P}_o(E/\mathcal{N}(E))$ *there is a* $P \in \mathbf{P}_o(E)$ *such that* $\mathbf{A}_E(P) = \overline{P}$.
2. $\mathbf{A}_E(P) \cong \mathbf{A}_E(P')$ *iff* $P \cong P'$.

Proof: The fullness of $\mathbf{A}_E(\cdot)$ follows from the projective property of $P \in \mathbf{P}_o(E)$. The rest follows from the fact that idempotents lift modulo a nil ideal (see [9, Corollary 9.6], [7, Proposition 17.18]), and Nakayama's Theorem 1.1. We leave the details as an exercise for the reader.

The next result is somewhat of a surprise. We will show that (marginal) direct summands of G correspond to semi-simple Q_G-modules in a functorial manner.

Theorem 18.52. *Let* G *be a right* R-*module such that* $\mathrm{End}_R(G)$ *possesses a semi-primary classical right ring of quotients* Q_G.

1. *The functor*

$$\boxed{\mathbf{B}_G(\cdot) = \mathbf{A}_{Q_G} \circ \mathbf{QH}_G(\cdot) = \mathbf{H}_G(\cdot) \otimes_{\mathrm{End}_R(G)} Q_G/\mathcal{J}(Q_G)}$$

induces a functorial bijective correspondence

$$\{[H] \, \big| \, H \in \mathbf{QP}_o(G)\} \longrightarrow \{(M) \, \big| \, M \text{ is a finitely generated}$$

$$\text{semi-simple right } Q_G\text{-module }\}$$

from the set of margimorphism classes $[H]$ *of* $H \in \mathbf{QP}_o(G)$ *onto the set of isomorphism classes* (M) *of finitely generated semi-simple right* Q_G-*modules,* M.
2. *If* H *is margimorphic to* $K_1 \oplus K_2$, *then* $\mathbf{B}_G(H) \cong \mathbf{B}_G(K_1) \oplus \mathbf{B}_G(K_2)$ *as semi-simple right* Q_G-*modules.*

3. *If $\mathbf{B}_G(H) \cong M_1 \oplus M_2$ for some semi-simple right Q_G-modules M_1 and M_2, then*
   H is margimorphic to $K_1 \oplus K_2$ for some right R-modules K_1 and K_2 such that
   $\mathbf{B}_G(K_1) \cong M_1$, and $\mathbf{B}_G(K_2) \cong M_2$.
4. *H is a totally indecomposable right R-module iff $\mathbf{B}_G(H)$ is a simple right*
   Q_G-module.

Proof: 1. By Theorem 18.50 and Lemma 18.51 the functors $\mathbf{QH}_G(\cdot)$ and $\mathbf{A}_G(\cdot)$ induce
bijections, so their composition $\mathbf{B}_G(\cdot)$ induces a functorial bijection as indicated in
part 1.

2. Follows from part 1 and Theorem 18.38.

3. Suppose that $\mathbf{B}_G(H) \cong M_1 \oplus M_2$. By part 1 there are K_1 and $K_2 \in \mathbf{QP}_o(G)$ such
that $\mathbf{B}_G(K_1) \cong M_1$ and $\mathbf{B}_G(K_2) \cong M_2$. Then

$$\mathbf{B}_G(H) \cong M_1 \oplus M_2 \cong \mathbf{B}_G(K_1 \oplus K_2)$$

which by the bijection in part 1 implies that H is margimorphic to $K_1 \oplus K_2$.

4. Follows from parts 1, 2, and 3. This completes the proof of the theorem.

18.3.4 Jónsson's theorem and margimorphisms

Torsion-free finite-rank abelian groups G and G' are *quasi-isomorphic* if they are
margimorphic abelian groups. The interested reader can prove what the abelian
groupist already knows. The torsion-free finite-rank abelian group G is quasi-
isomorphic to the torsion-free abelian group G' iff there are maps $a : G \longrightarrow G'$ and
$b : G' \longrightarrow G$ such that ab and ba are multiplication by (possibly different) nonzero
integers iff G is isomorphic to a subgroup of finite index in G'. Quasi-isomorphism
is the motivation for our study of margimorphism pairs.

Like margimorphism, quasi-isomorphism can be viewed as isomorphism in the
appropriate additive category. Let $\mathbf{QAb}$ be the category whose objects are the torsion-
free finite rank abelian groups and whose homsets are the groups

$$\mathbf{QHom_Z}(\cdot, \cdot) = \mathrm{Hom_Z}(\cdot, \cdot) \otimes_\mathbf{Z} \mathbf{Q}.$$

The category $\mathbf{QAb}$ is the inspiration for the functor $\mathbf{QH}_G(\cdot)$ and the category $\mathbf{QP}_o(G)$.
For instance, J. D. Reid (see [9, 59]) shows that two *rtffr* groups are quasi-isomorphic
iff they are isomorphic in the category $\mathbf{QAb}$. Compare this with Theorem 18.38.1.
Given $G \in \mathbf{QAb}$ then $\mathrm{End_Z}(G)$ has an Artinian classical ring of quotients

$$\mathbf{QEnd}(G) = \mathrm{End_{QAb}}(G) = \mathrm{End_Z}(G) \otimes \mathbf{Q}$$

called the *quasi-endomorphism ring* of G. The quasi-endomorphism ring is a finite-
dimensional ($=$ Artinian $=$ semi-primary) $\mathbf{Q}$-algebra. This is the motivation for our
standing hypothesis about $\mathrm{End}_R(G)$ and Q_G. Furthermore, $G \in \mathbf{QAb}$ is totally
indecomposable (referred to in the abelian group literature as *strongly indecomposable*) iff G is indecomposable in $\mathbf{QAb}$ iff $\mathbf{QEnd_Z}(G)$ is a local ring. Compare this
with Theorem 18.46. Thus $\mathbf{QAb}$ effectively motivates further discussion of topics
surrounding margimorphisms and totally indecomposable right R-modules.

One of the properties of **QAb** that is especially appealing is that the direct sum decompositions of objects in **QAb** satisfy a strong uniqueness property. Thus $G \in$ **QAb** can be written as $G \cong G_1 \oplus \cdots \oplus G_t$ in **QAb** for some integer $t > 0$ and some indecomposable objects G_i, and if $G = G'_1 \oplus \cdots \oplus G'_s$ in **QAb** for some integer $s > 0$ and some indecomposable objects G'_j then $s = t$ and there is a permutation π of $\{1, \ldots, t\}$ such that $G_i \cong G'_{\pi(i)}$ in **QAb**. From the Azumaya–Krull–Schmidt theorem 1.4 in **QAb** and from the fact that groups G and G' are isomorphic in **QAb** iff they are quasi-isomorphic groups, we prove Jónsson's theorem. This theorem started the investigation of maps between groups that induce a comparison of groups that is coarser than isomorphism.

Theorem 18.53. [59, Jónsson's Theorem 92.5] *Let $G \in$ **QAb**.*

1. *There are a positive integer t and strongly indecomposable torsion-free finite rank abelian groups $G_1, \ldots, G_t$ such that G is quasi-isomorphic to $G_1 \oplus \cdots \oplus G_t$.*
2. *If G is quasi-isomorphic to $G'_1 \oplus \cdots \oplus G'_s$ for some integer s and strongly indecomposable abelian groups $G'_1, \ldots, G'_s$ then $s = t$ and there is a permutation π of $\{1, \ldots, t\}$ such that G_i is quasi-isomorphic to $G'_{\pi(i)}$ for each $i \in \{1, \ldots, t\}$.*

We will search for a similar theorem among the right R-modules for which $\mathrm{End}_R(G)$ possesses a semi-primary classical right ring of quotients.

Theorem 18.54. *Let G be a right R-module such that $\mathrm{End}_R(G)$ possesses a semi-primary classical right ring of quotients Q_G. There is an integer $t > 0$ and totally indecomposable right R-modules $G_1, \ldots, G_t$ such that*

$$G \text{ is margimorphic to } G_1 \oplus \cdots \oplus G_t \tag{18.7}$$

with the following properties. Let $H \in \mathbf{QP}_o(G)$.

1. *H is margimorphic to $H_1 \oplus \cdots \oplus H_s$ for some $H_1, \ldots, H_s \in \{G_1, \ldots, G_t\}$.*
2. *If H is margimorphic to $H'_1 \oplus \cdots \oplus H'_r$ for some integer r and some totally indecomposable right R-modules $H'_1, \ldots, H'_r$, then $r = s$ and there is a permutation π of $\{1, \ldots, s\}$ such that H_i is margimorphic to $H'_{\pi(i)}$ for each $i \in \{1, \ldots, s\}$.*
3. *If $K, K', L \in \mathbf{QP}_o(G)$ and if $L \oplus K$ is margimorphic to $L \oplus K'$ then K is margimorphic to K'.*

Proof: Since Q_G is a semi-primary ring there is a positive integer t and indecomposable cyclic projective right Q_G-modules $P_1, \ldots, P_t$ such that

$$Q_G = P_1 \oplus \cdots \oplus P_t.$$

By Theorems 18.38.3 and 18.47 there are totally indecomposable right R-modules $G_1, \ldots, G_t \in \mathbf{QP}_o(G)$ such that $\mathbf{QH}_G(G_i) \cong P_i$ for $i = 1, \ldots, t$. Because $\mathbf{QH}_G(\cdot)$ is an additive functor,

$$\mathbf{QH}_G(G) = Q_G$$
$$= P_1 \oplus \cdots \oplus P_t$$
$$\cong \mathbf{QH}_G(G_1) \oplus \cdots \oplus \mathbf{QH}_G(G_t)$$
$$\cong \mathbf{QH}_G(G_1 \oplus \cdots \oplus G_t),$$

and by Lemma 18.36, $G_1 \oplus \cdots \oplus G_t \in \mathbf{QP}_o(G)$. Then G is margimorphic to $G_1 \oplus \cdots \oplus G_t$ by Theorem 18.38.1.

1. Suppose that $H \oplus H'$ is margimorphic to $G^{(n)}$ for some positive integer n. An application of $\mathbf{QH}_G(\cdot)$ to (18.7) and another pair of appeals to Theorem 18.38.1 show us that

$$\mathbf{QH}_G(H) \oplus \mathbf{QH}_G(H') \cong \mathbf{QH}_G(H \oplus H')$$
$$\cong \mathbf{QH}_G(G^{(n)})$$
$$\cong \mathbf{QH}_G(G_1 \oplus \cdots \oplus G_t)^{(n)}$$
$$\cong \mathbf{QH}_G(G_1)^{(n)} \oplus \cdots \oplus \mathbf{QH}_G(G_t)^{(n)}.$$

By our choice of G_i, $\mathbf{QH}_G(G_i) = P_i$ is an indecomposable projective right R-module over the semi-primary ring Q_G, so by Lemma 1.3, $\mathbf{QH}_G(G_i)$ has a local endomorphism ring. Then by the Azumaya–Krull–Schmidt theorem 1.4,

$$\mathbf{QH}_G(H) \cong \mathbf{QH}_G(G_1)^{(h_1)} \oplus \cdots \oplus \mathbf{QH}_G(G_t)^{(h_t)}$$
$$\cong \mathbf{QH}_G(G_1^{(h_1)} \oplus \cdots \oplus G_t^{(h_t)})$$

for some integers $h_1, \ldots, h_t \geq 0$. Theorem 18.38.1 then implies that

$$H \text{ is margimorphic to } G_1^{(h_1)} \oplus \cdots \oplus G_t^{(h_t)}.$$

Renumber the totally indecomposable direct summands in the direct sum $G_1^{(n_1)} \oplus \cdots \oplus G_t^{(n_t)}$ to show that there is an integer s and totally indecomposable right R-modules $H_1, \ldots, H_s$ that are isomorphic to elements of $\{G_1, \ldots, G_t\}$ and such that

$$H \text{ is margimorphic to } H_1 \oplus \cdots \oplus H_s. \tag{18.8}$$

This proves part 1.

2. Write H as in (18.8). Suppose that H is margimorphic to $H_1' \oplus \cdots \oplus H_r'$ for some positive integer r and totally indecomposable right R-modules $H_1', \ldots, H_r'$. By Theorem 18.47, $\mathbf{QH}_G(H_i)$ and $\mathbf{QH}_G(H_j')$ are indecomposable projective right Q_G-modules, so by the Azumaya–Krull–Schmidt theorem 1.4, $s = r$ and $\mathbf{QH}_G(H_i) \cong \mathbf{QH}_G(H_{\pi(i)}')$ for some permutation π of the indices $\{1, \ldots, s\}$. Theorem 18.38.1 then shows us that H_i is margimorphic to $H_{\pi(i)}'$ for each $i \in \{1, \ldots, s\}$.

3. Let $K, K', L \in \mathbf{QP}_o(G)$ and suppose that $L \oplus K$ is margimorphic to $L \oplus K'$. By Theorem 18.38, $\mathbf{QH}_G(K), \mathbf{QH}_G(K'), \mathbf{QH}_G(L)$ are projective right Q_G-modules

such that

$$\mathbf{QH}_G(L) \oplus \mathbf{QH}_G(K) \cong \mathbf{QH}_G(L \oplus K)$$
$$\cong \mathbf{QH}_G(L \oplus K')$$
$$\cong \mathbf{QH}_G(L) \oplus \mathbf{QH}_G(K').$$

Since Q_G is semi-primary, the Azumaya–Krull–Schmidt theorem 1.4 implies that $\mathbf{QH}_G(K) \cong \mathbf{QH}_G(K')$. Therefore, K is margimorphic to K' by Theorem 18.38.1. This completes the proof of the theorem.

Notice that Jónsson's theorem for torsion-free groups of finite rank is an immediate consequence of Theorem 18.54.

Corollary 18.55. *Let G be a right R-module such that* $\mathrm{End}_R(G)$ *possesses a semi-primary classical right ring of quotients* Q_G. *There is a positive integer t and totally indecomposable right R-modules $G_1, \ldots, G_t$ such that*

1. *G is margimorphic to $G_1 \oplus \cdots \oplus G_t$.*
2. *If G is margimorphic to $G'_1 \oplus \cdots \oplus G'_s$ for some integer s and some totally indecomposable right R-modules $G'_1, \ldots, G'_s$ then $s = t$ and there is a permutation π of $(1, \ldots, t)$ such that G_i is margimorphic to $G'_{\pi(i)}$ for each $i \in \{1, \ldots, t\}$.*

18.4 Nilpotent sets and margimorphism

In this section we will use the machinery surrounding margimorphism to discuss theorems on the uniqueness of direct sum decompositions of right R-modules up to isomorphism.

A set $\{H_1, \ldots, H_s\}$ is *nilpotent* if for each $i \neq j \in \{1, \ldots, s\}$ each composition $H_i \longrightarrow H_j \longrightarrow H_i$ of R-module maps is a nilpotent element of $\mathrm{End}_R(H_i)$. We should view nilpotency of sets as the diametric opposite of margimorphic modules. If $\{H, K\}$ is a nilpotent set, then there cannot be a margimorphism pair (a, b) for H and K since no nilpotent element is regular.

Recall the functor $\mathbf{B}_G(\cdot)$ and Theorem 18.52. The first result shows us that margimorphism and nilpotent sets are compatible ideas.

Lemma 18.56. *Let G be a right R-module such that* $\mathrm{End}_R(G)$ *has a semi-primary classical right ring of quotients* Q_G. *Let $H, K \in \mathbf{QP}_o(G)$, and assume that $\{H, K\}$ is a nilpotent set. If H' is margimorphic to H then $\{H', K\}$ is a nilpotent set.*

Proof: Let $(a : H \longrightarrow H', b : H' \longrightarrow H)$ be a margimorphism pair. We note that because Q_G exists and is semi-primary, Q_H exists and is semi-primary (Theorem 18.31). Then by Theorem 18.25, $\mathrm{End}_R(H')$ has a semi-primary classical right ring of quotients $Q_{H'}$, and $Q_{H'} \cong Q_H$.

Let $\psi_1 : H' \longrightarrow K$ and $\psi_2 : K \longrightarrow H'$ be two R-module maps. Then $\psi_1 a : H \longrightarrow K$ and $b\psi_2 : K \longrightarrow H$. Since $\{H, K\}$ is a nilpotent set, $b\phi\psi_2\psi_1 a$ is a nilpotent endomorphism of H for each $\phi : H' \longrightarrow H'$. Then there is an integer $n > 0$ such that $(b\phi\psi_2\psi_1 a)^n = 0$ so that

$$0 = a(b\phi\psi_2\psi_1 a)^n b = ab(\phi\psi_2\psi_1 ab)^n.$$

Since ab is regular $(\phi\psi_2\psi_1 ab)^n = 0$, and since ϕ was arbitrarily chosen,

$$\psi_1\psi_2 ab \in \mathcal{N}(\mathrm{End}_R(H')).$$

Because ab is a unit in $Q_{H'}$, $ab + \mathcal{J}(Q_{H'})$ is a unit in $Q_{H'}/\mathcal{J}(Q_{H'})$, so that ab maps to a regular element in $\mathrm{End}_R(H')/\mathcal{N}(\mathrm{End}_R(H'))$. Thus $\psi_2\psi_1 \in \mathcal{N}(\mathrm{End}_R(H'))$. Similarly, $\psi_1\psi_2 \in \mathcal{N}(\mathrm{End}_R(K))$ and therefore $\{H', K\}$ is a nilpotent set. This completes the proof.

Theorem 18.57. *Let G be a right R-module such that $\mathrm{End}_R(G)$ possesses a semi-primary classical right ring of quotients Q_G, and let $H, K \in \mathbf{QP}_o(G)$. The following are equivalent.*

1. *$\{H, K\}$ is not a nilpotent set.*
2. *Some totally indecomposable marginal summand $L \neq 0$ of H is margimorphic to a marginal summand of K.*
3. *Some nonzero simple Q_G-submodule of $\mathbf{B}_G(H)$ is isomorphic to a Q_G-submodule of $\mathbf{B}_G(K)$.*
4. *$\mathrm{Hom}_{Q_G}(\mathbf{B}_G(H), \mathbf{B}_G(K)) \neq 0$.*

Proof. $4 \Leftrightarrow 3$ follows immediately from Shur's lemma and the fact that $\mathbf{B}_G(H)$ and $\mathbf{B}_G(K)$ are semi-simple right Q_G-modules.

$3 \Rightarrow 2$. Suppose that some simple Q_G-submodule M of $\mathbf{B}_G(H)$ is isomorphic to a simple Q_G-submodule of $\mathbf{B}_G(K)$. Since $\mathbf{B}_G(H)$ and $\mathbf{B}_G(K)$ are semi-simple right Q_G-modules,

$$M \oplus N \cong \mathbf{B}_G(H) \quad \text{and} \quad M \oplus P \cong \mathbf{B}_G(K)$$

for some semi-simple Q_G-modules N and P. By Theorem 18.52.1 there is an $L \in \mathbf{QP}_o(G)$ such that $\mathbf{B}_G(L) = M$. Since M is simple, Theorem 18.52.3 and 18.52.4 imply that L is totally indecomposable marginal direct summand of H and K.

$2 \Rightarrow 1$. By Lemma 18.56 we can assume without loss of generality that L is a proper nonzero direct summand of H and K. Let

$$j_X : L \longrightarrow X \quad \text{and} \quad \pi_X : X \longrightarrow L$$

be the canonical injection and projection for $X \in \{H, K\}$ with direct summand L. Then $j_K \pi_H : H \longrightarrow K$ and $j_H \pi_K : K \longrightarrow H$ are maps such that

$$(j_H \pi_K)(j_K \pi_H) = j_H (\pi_K j_K)\pi_H = j_H 1_K \pi_H = j_H \pi_H,$$

which is an idempotent, not nilpotent, endomorphism of H. Since L is a proper nonzero direct summand of H, $j_H \pi_H \neq 0, 1$. Hence $\{H, K\}$ is not a nilpotent set.

$1 \Rightarrow 4$. Assume that $\{H, K\}$ is not a nilpotent set. There are R-module maps

$$\psi_H : H \longrightarrow K \text{ and } \psi_K : K \longrightarrow H$$

such that $\psi_K \psi_H$, say, is not a nilpotent endomorphism. Since $\mathbf{QH}_G(\cdot)$ is a faithful functor, $\mathbf{QH}_G(\psi_K \psi_H)$ is not a nilpotent endomorphism of $\mathbf{QH}_G(H)$.

By hypothesis, $\mathrm{End}_R(G)$ possesses a semi-primary classical right ring of quotients Q_G. Since $\mathcal{J}(Q_G)$ is nilpotent, any Q_G-module map $\phi : \mathbf{QH}_G(H) \longrightarrow \mathbf{QH}_G(H)\mathcal{J}(Q_G)$ is nilpotent. Then

$$\text{image } \mathbf{QH}_G(\psi_K \psi_H) \not\subset \mathbf{QH}_G(H)\mathcal{J}(Q_G).$$

Inasmuch as

$$\mathbf{B}_G(H) \cong \mathbf{QH}_G(H)/\mathbf{QH}_G(H)\mathcal{J}(Q_G),$$

we see that

$$\mathbf{B}_G(\psi_K \psi_H) : \mathbf{B}_G(H) \longrightarrow \mathbf{B}_G(H)$$

is a nonzero endomorphism of the semi-simple module $\mathbf{B}_G(H)$. Then

$$0 \neq \mathbf{B}_G(\psi_H) \in \mathrm{Hom}_{Q_G}(\mathbf{B}_G(H), \mathbf{B}_G(K)).$$

This proves part 4, and completes the logical cycle.

Corollary 18.58. *Let G be a right R-module such that $\mathrm{End}_R(G)$ possesses a semi-primary classical right ring of quotients Q_G, and let $H, K \in \mathbf{QP}_o(G)$. The following are equivalent.*

1. $\{H, K\}$ is a nilpotent set.
2. If L is a totally indecomposable marginal summand of H and of K then $L = 0$.

Proof: Apply Theorem 18.57.1 and 18.57.2.

Corollary 18.59. *Let G be a right R-module such that $\mathrm{End}_R(G)$ possesses a semi-primary classical right ring of quotients Q_G, and let $\{H_1, \ldots, H_t\} \subset \mathbf{QP}_o(G)$. The following are equivalent.*

1. $\{H_1, \ldots, H_t\}$ is a nilpotent set.
2. Let $1 \leq i \neq j \leq t$. If L is a totally indecomposable marginal summand of H_i and a marginal summand of H_j then $L = 0$.

Proof: This follows immediately from Corollary 18.58 and the fact that $\{H_1, \ldots, H_t\}$ is a nilpotent set iff $\{H_i, H_j\}$ is a nilpotent set for all integers $1 \leq i \neq j \leq t$.

An application of the above corollary will prove the next theorem.

Theorem 18.60. *Let G be a right R-module such that $\mathrm{End}_R(G)$ possesses a semi-primary classical right ring of quotients Q_G. Let*

$$\{K_1, \ldots, K_t\} \subset \mathbf{QP}_o(G)$$

be a set of totally indecomposable right R-modules. The following are equivalent.

1. *$\{K_1, \ldots, K_t\}$ is a nilpotent set.*
2. *If $1 \leq i \neq j \leq t$ then K_i is not margimorphic to K_j.*

Proof. $1 \Rightarrow 2$. Suppose that $K_i, K_j \in \mathbf{QP}_o(G)$ are totally indecomposable right R-modules that form a nilpotent set for $i \neq j$. Since K_i and K_j are totally indecomposable, Theorem 18.52 implies that $\mathbf{B}_G(K_i)$ and $\mathbf{B}_G(K_j)$ are simple modules. By Corollary 18.59, $\mathbf{B}_G(K_i)$ is not isomorphic to $\mathbf{B}_G(K_j)$. Then by Corollary 18.57, K_i and K_j are margimorphic. This proves part 2.

$2 \Rightarrow 1$. Suppose that K_i and K_j are not margimorphic for some $i \neq j$. Theorem 18.52 implies that $\mathbf{B}_G(K_i)$ and $\mathbf{B}_G(K_j)$ are nonisomorphic simple right Q_G-modules. Then by Corollary 18.59, $\{K_i, K_j\}$ is not a nilpotent set. This proves part 1 and completes the logical cycle.

We will have occasion to use the following relationship between nilpotent sets and commutativity. Note the lack of the hypothesis that $\mathrm{End}_R(G)$ has a classical right ring of quotients.

Lemma 18.61. *Let G be a right R-module, suppose that $G = H_1 \oplus H_2$, and let $e_1^2 = e_1$ be the idempotent corresponding to H_1. Then $\{H_1, H_2\}$ is a nilpotent set iff e_1 is central modulo $\mathcal{N}(\mathrm{End}_R(G))$.*

Proof. Suppose that $\{H_1, H_2\}$ is a nilpotent set, and let $\phi \in \mathrm{End}_R(G)$. Then $e_1\phi = e_1\phi e_1 \oplus e_1\phi e_2$ and $\phi e_1 = e_1\phi e_1 \oplus e_2\phi e_1$. We claim that $e_1\phi e_2 \in \mathcal{N}(\mathrm{End}_R(G))$.

Let $\psi \in \mathrm{End}_R(G)$. Since $\{H_1, H_2\}$ is a nilpotent set, $e_2\psi e_1\phi e_2$ is a nilpotent element in $\mathrm{End}_R(H_2)$. Then for large enough integer $m > 0$,

$$(\psi(e_1\phi e_2))^{m+1} = (\psi e_1\phi e_2)(e_2\psi e_1\phi e_2)^m = 0.$$

As claimed, $e_1\phi e_2 \in \mathcal{N}(\mathrm{End}_R(G))$. Similarly, $e_2\phi e_1 \in \mathcal{N}(\mathrm{End}_R(G))$.

It follows that

$$e_1\phi - \phi e_1 = e_1\phi e_2 - e_2\phi e_1 \in \mathcal{N}(\mathrm{End}_R(G)),$$

and hence e_1 is central modulo $\mathcal{N}(\mathrm{End}_R(G))$.

Conversely, suppose that e_1 and $e_2 = 1 - e_1$ are central modulo $\mathcal{N}(\mathrm{End}_R(G))$, and let $\psi_1 : H_1 \longrightarrow H_2$ and $\psi_2 : H_2 \longrightarrow H_1$ be R-module maps. By supposition,

$$0 = e_1e_2\psi_1 \equiv e_1\psi_1 e_2 \bmod \mathcal{N}(\mathrm{End}_R(G)),$$

so that $\psi_2\psi_1 = \psi_2(e_2\psi_1 e_1)$ is nilpotent. Thus $\{H_1, H_2\}$ is a nilpotent set. This completes the proof.

18.5 Isomorphism from margimorphism

Recall the functor $\mathbf{A}_G(\cdot)$ introduced in Lemma 18.51. In this section we will use $\mathbf{A}_G(\cdot)$ to functorially translate the margimorphism of marginal summands of G into the uniqueness of direct summands of G. Specifically, we deduce isomorphism of right R-modules from the margimorphism of right R-modules. We should consider this deduction in light of the fact that most modern mathematicians would not expect any useful relationship between the isomorphism class of G and the margimorphism class of G. Observe the lack of the quotient ring hypothesis in the next theorem.

Theorem 18.62. *Let G be a right R-module, and suppose that*

$$G = H \oplus H' = G_1 \oplus \cdots \oplus G_t,$$

where $\{G_1, \ldots, G_t\}$ is a nilpotent set of (not necessarily indecomposable) right R-modules. Then there is a nilpotent set $\{H_1, \ldots, H_t\}$ such that $G_i \cong H_i \oplus H_i'$ for some right R-modules $H_1', \ldots, H_t'$ and such that $H \cong H_1 \oplus \cdots \oplus H_t$.

Proof: Define $\mathbf{A}_G(\cdot)$ as in Lemma 18.51. The composite functor

$$\mathbf{A}_G \circ \mathbf{H}_G(\cdot) : \mathbf{Mod}\text{-}R \longrightarrow \mathbf{Mod}\text{-}\mathrm{End}_R(G)/\mathcal{N}(\mathrm{End}_R(G))$$

is defined by

$$\mathbf{A}_G \circ \mathbf{H}_G(\cdot) = \mathrm{Hom}_R(G, \cdot) \otimes_{\mathrm{End}_R(G)} \mathrm{End}_R(G)/\mathcal{N}(\mathrm{End}_R(G)).$$

Notice that

$$\mathbf{A}_G \circ \mathbf{H}_G(G) = \mathrm{End}_R(G)/\mathcal{N}(\mathrm{End}_R(G)).$$

For each $x \in \mathrm{End}_R(G)$ we let

$$\bar{x} = \mathbf{A}_G(x) = x + \mathcal{N}(\mathrm{End}_R(G)) \in \mathbf{A}_G \circ \mathbf{H}_G(G).$$

Let f be the idempotent in $\mathrm{End}_R(G)$ corresponding to the direct summand H of G. Then $\mathbf{H}_G(H) = f\mathbf{H}_G(G)$ so that

$$\mathbf{A}_G \circ \mathbf{H}_G(H) = \mathbf{H}_G(H)/\mathbf{H}_G(H)\mathcal{N}(\mathrm{End}_R(G))$$

$$= \bar{f}\mathbf{A}_G \circ \mathbf{H}_G(G).$$

Because $\{G_1, \ldots, G_t\}$ is a nilpotent set, the idempotents $e_1, \ldots, e_t \in \mathrm{End}_R(G)$ associated with the direct sum $G_1 \oplus \cdots \oplus G_t$ are central modulo $\mathcal{N}(\mathrm{End}_R(G))$ (Lemma 18.61). Then $e_i f$ maps to an idempotent $\overline{e_i f}$ in $\mathbf{A}_G \mathbf{H}_G(G)$. Moreover,

$$\bar{e}_i = \bar{e}_i(\bar{f} \oplus \overline{(1-f)})$$

$$= \overline{e_i f} \oplus \bar{e}_i \overline{(1-f)}.$$

Hence the isomorphisms

$$\mathbf{A}_G \circ \mathbf{H}_G(G_i) \cong \bar{e}_i \mathbf{A}_G \circ \mathbf{H}_G(G)$$
$$= \bar{e}_i \bar{f} \mathbf{A}_G \circ \mathbf{H}_G(G) \oplus \bar{e}_i (\overline{1 - f}) \mathbf{A}_G \circ \mathbf{H}_G(G).$$

are natural. By the Arnold–Lady theorem 1.13 and Lemma 18.51, there are $H_i, H_i' \in \mathbf{P}_o(G)$ such that

$$\bar{e}_i \bar{f} \mathbf{A}_G \circ \mathbf{H}_G(G_i) \cong \mathbf{A}_G \circ \mathbf{H}_G(H_i) \text{ and}$$
$$\bar{e}_i(1 - \bar{f}) \mathbf{A}_G \circ \mathbf{H}_G(G_i) \cong \mathbf{A}_G \circ \mathbf{H}_G(H_i'),$$

so that

$$\mathbf{A}_G \circ \mathbf{H}_G(G_i) \cong \bar{e}_i \bar{f} \mathbf{A}_G \circ \mathbf{H}_G(G) \oplus \bar{e}_i(\overline{1 - f}) \mathbf{A}_G \circ \mathbf{H}_G(G)$$
$$\cong \mathbf{A}_G \circ \mathbf{H}_G(H_i) \oplus \mathbf{A}_G \circ \mathbf{H}_G(H_i')$$
$$\cong \mathbf{A}_G \circ \mathbf{H}_G(H_i \oplus H_i').$$

The Arnold–Lady theorem 1.13 and Lemma 18.51 imply that $G_i \cong H_i \oplus H_i'$.

Furthermore, since the $\bar{e}_i$ are central idempotents such that $\bar{1} = \bar{e}_1 \oplus \cdots \oplus \bar{e}_t$, we have a direct sum of orthogonal idempotents

$$\bar{f} = \bar{e}_1 \bar{f} \oplus \cdots \oplus \bar{e}_t \bar{f},$$

which corresponds to a direct sum of modules

$$\mathbf{A}_G \circ \mathbf{H}_G(H) = \bar{f} \mathbf{A}_G \circ \mathbf{H}_G(G)$$
$$= \bar{e}_1 \bar{f} \mathbf{A}_G \circ \mathbf{H}_G(G) \oplus \cdots \oplus \bar{e}_t \bar{f} \mathbf{A}_G \circ \mathbf{H}_G(G)$$
$$\cong \mathbf{A}_G \circ \mathbf{H}_G(H_1) \oplus \cdots \oplus \mathbf{A}_G \circ \mathbf{H}_G(H_t)$$
$$\cong \mathbf{A}_G \circ \mathbf{H}_G(H_1 \oplus \cdots \oplus H_t).$$

As above, $H \cong H_1 \oplus \cdots \oplus H_t$. The reader will prove as an exercise that since H_i is a direct summand of G_i for each $i = 1, \ldots, t$, $\{H_1, \ldots, H_t\}$ is a nilpotent set. This completes the proof.

Corollary 18.63. *Let G be a right R-module such that $\text{End}_R(G)$ possesses a semi-primary classical right ring of quotients Q_G. Suppose that*

$$G = H \oplus H' \cong G_1 \oplus \cdots \oplus G_t$$

for some right R-modules $G_1, \cdots, G_t$ such that

$$\text{Hom}_R(G_i, G_j) = 0 \text{ for each } 1 \leq i < j \leq t.$$

Then there is a set $\{H_1, \ldots, H_t, H'_1, \ldots, H'_t\}$ *such that*

1. $G_i \cong H_i \oplus H'_i$ *for each* $1 \le i \le t$ *and*
2. $H \cong H_1 \oplus \cdots \oplus H_t$.

Proof: The reader can show that $\{G_1, \ldots, G_t\}$ is a nilpotent set, and then apply Theorem 18.62.

In the next result we show that some of the above properties for G are passed onto $H \in \mathbf{P}_o(G)$.

Corollary 18.64. *Suppose there is a nilpotent set* $\{G_1, \cdots, G_t\}$ *of indecomposables* G_i, *and a direct sum decomposition* $G = G_1 \oplus \cdots \oplus G_t$. *Then given* $H \in \mathbf{P}_o(G)$, *there is a nilpotent set* $\{H_1, \ldots, H_r\}$ *of (not necessarily indecomposable) right R-modules such that* $H = H_1 \oplus \cdots \oplus H_r$.

Proof: Since $H \in \mathbf{P}_o(G)$, there is a positive integer n and a right R-module H' such that $G^{(n)} \cong H \oplus H'$. The reader will prove as an exercise that $\{G_1^{(n)}, \ldots, G_t^{(n)}\}$ is a nilpotent set. Then an appeal to Theorem 18.62 completes the proof.

The following is an interesting exercise for the reader. An idempotent e is indecomposable if $e = f \oplus f'$ for some idempotents f and f' implies that $f = 0$ or $f' = 0$.

Lemma 18.65. *Let R be a ring that contains indecomposable nonzero central idempotents* $e_1, \cdots, e_t$ *such that* $e_1 \oplus \cdots \oplus e_t = 1$. *If f is a nonzero indecomposable idempotent in R, then* $f = e_i$ *for some integer* $i \in \{1, \ldots, t\}$.

The uniqueness of direct sum decompositions is thought to be a very sensitive property possessed by very few modules. Margimorphism must be viewed as a very coarse measure of a module. For example, each right ideal in a domain is margimorphic to the domain. The following theorem is interesting because it starts with an hypothesis about a marginal direct sum decomposition about G and concludes with a unique direct sum decomposition for G. That is, from knowing that G is margimorphic to $K \oplus K'$ we conclude that G is uniquely written as $G = H \oplus H'$. This is an interesting marriage of the sensitive with the coarse.

Theorem 18.66. *Let G be a right R-module such that* $\mathrm{End}_R(G)$ *has a semi-primary classical right ring of quotients* Q_G. *Furthermore, suppose that*

$$G \text{ is margimorphic to } K_1 \oplus \cdots \oplus K_r,$$

where $\{K_1, \ldots, K_r\}$ *is a nilpotent set of nonzero totally indecomposable right R-modules.*

1. $G = G_1 \oplus \cdots \oplus G_t$ *for some nilpotent set* $\{G_1, \ldots, G_t\}$ *of indecomposable right R-modules.*

2. *If also $G \cong G'_1 \oplus \cdots \oplus G'_s$ for some indecomposable right R-modules $G'_1, \ldots, G'_s$ then $s = t$ and there is a permutation π of $\{1, \ldots, t\}$ such that $G_i \cong G'_{\pi(i)}$ for each $i \in \{1, \ldots, t\}$.*

3. *If H, L, and L' are right R-modules such that $G \cong H \oplus L \cong H \oplus L'$ then $L \cong L'$.*

Proof: 1. Since Q_G is semi-primary there is an integer t and a finite set of orthogonal indecomposable idempotents $\{e_1, \ldots, e_t\} \subset \operatorname{End}_R(G)$ such that $1 = e_1 \oplus \cdots \oplus e_t$. Then

$$G = e_1 G \oplus \cdots \oplus e_t G,$$

and since e_i is indecomposable, $e_i G$ is an indecomposable right R-module. Let $e_i G = G_i$.

For the sake of contradiction, assume that $K \neq 0$ is a totally indecomposable marginal summand of G_i and of G_j for some $i \neq j$. Since G is margimorphic to $G_i \oplus G_j \oplus X$ for some right R-module X, G is margimorphic to $K^{(2)} \oplus Y$ for some right R-module Y. By Theorem 18.54, Y is margimorphic to $Y_1 \oplus \cdots \oplus Y_r$ for some $Y_i \in \{K_1, \ldots, K_t\}$. Thus $K_1 \oplus \cdots \oplus K_t$ is margimorphic to $K^{(2)} \oplus Y_1 \oplus \cdots \oplus Y_r$. By the uniqueness of the marginal direct sum decomposition $K_1 \oplus \cdots \oplus K_t$ for G (Theorem 18.54), K is margimorphic to at least two totally indecomposable modules in $\{K_1, \cdots, K_r\}$. Say, for example, that K is margimorphic to K_1 and to K_2. However, the set $\{K_1, \ldots, K_r\}$ is nilpotent, so Theorem 18.60 implies that K_1 is not margimorphic to K_2. This is the contradiction that we sought, and this proves part 1.

2. Since $G = G_1 \oplus \cdots \oplus G_t$ for some nilpotent set $\{G_1, \cdots, G_t\}$, $1 = e_1 \oplus \cdots \oplus e_t$, where the e_i are indecomposable nonzero idempotents in $\operatorname{End}_R(G)$. Let $\bar{e}_i$ be the idempotent image of e_i in $\mathbf{A}_G \circ \mathbf{H}_G(G) = \operatorname{End}_R(G)/\mathcal{N}(\operatorname{End}_R(G))$. By Lemma 18.61 the $\bar{e}_i$ are central modulo $\mathcal{N}(\operatorname{End}_R(G))$.

Let

$$G = G'_1 \oplus \cdots \oplus G'_s$$

for some indecomposable R-modules $G'_1, \cdots, G'_s$. The associated nonzero idempotents $\bar{f}_1, \cdots, \bar{f}_s$ are indecomposable. Then by Lemma 18.65, after a permutation of the subscripts and an induction on t, $s = t$ and $\bar{e}_1 = \bar{f}_1, \ldots, \bar{e}_t = \bar{f}_t$.

It follows that

$$\mathbf{A}_G \circ \mathbf{H}_G(G_i) = \bar{e}_i \mathbf{A}_G \circ \mathbf{H}_G(G)$$
$$= \bar{f}_i \mathbf{A}_G \circ \mathbf{H}_G(G)$$
$$= \mathbf{A}_G \circ \mathbf{H}_G(G'_i),$$

so that by Theorem 1.13 and Lemma 18.51,

$$G_i \cong G'_i \text{ for all } i = 1, \ldots, t.$$

3. Suppose that $G = H \oplus L \cong H \oplus L'$. Write $H = H_1 \oplus \cdots \oplus H_p$ and $L = L_1 \oplus \cdots \oplus L_s$ as direct sums of nonzero indecomposables. Let (X) denote the isomorphism class

of X. By part 2 the lists

$$(H_1), \ldots, (H_p), (L_1), \ldots, (L_s) = (G_1), \ldots, (G_t)$$

are the same. Then $p + s = t$, and since $\{(G_1), \ldots, (G_t)\}$ is a nilpotent set

$$\{(H_1), \ldots, (H_p)\} \cap \{(L_1), \ldots, (L_s)\} = \emptyset.$$

Similarly, write $L' = L'_1 \oplus \cdots \oplus L'_q$ and note that

$$\{(H_1), \ldots, (H_p)\} \cup \{(L'_1), \ldots, (L'_s)\} = \{(G_1), \ldots, (G_t)\}$$

and

$$\{(H_1), \ldots, (H_p)\} \cap \{(L'_1), \ldots, (L'_q)\} = \emptyset.$$

It follows that $p + s = t = r + q$ whence $s = q$. Since the complement of $\{(H_1), \ldots, (H_p)\}$ is unique in $\{(G_1), \ldots, (G_t)\}$,

$$\{(L_1), \ldots, (L_s)\} = \{(L'_1), \ldots, (L'_s)\}.$$

Hence, $L_i \cong L'_{\pi(i)}$ for some permutation π of $\{1, \ldots, s\}$, so that

$$L \cong L_1 \oplus \cdots \oplus L_s \cong L'.$$

This completes the proof.

With the above theorem it would be useful to have an external condition equivalent to the existence of a nilpotent set.

Theorem 18.67. *Let G be a right R-module such that $\text{End}_R(G)$ possesses a semi-primary classical right ring of quotients Q_G. The following are equivalent.*

1. *There is a nilpotent set $\{K_1, \ldots, K_r\}$ of nonzero totally indecomposable right R-modules such that G is margimorphic to $K_1 \oplus \cdots \oplus K_r$.*
2. *$Q_G / \mathcal{J}(Q_G)$ is a product of division rings.*

Proof: $2 \Rightarrow 1$. Assume that $Q_G / \mathcal{J}(Q_G)$ is a product of division rings

$$Q_G / \mathcal{J}(Q_G) = D_1 \times \cdots \times D_r.$$

Then $(D_1), \ldots, (D_r)$ is a complete list of the distinct isomorphism classes of simple Q_G-modules. By Theorem 18.52, for each D_i there is a totally indecomposable marginal direct summand K_i of G such that

$$D_i \cong \mathbf{B}_G(K_i).$$

Since $D_i \not\cong D_j$ for $i \neq j \in \{1, \ldots, r\}$, Theorem 18.52.1 shows us that K_i is not margimorphic to K_j for $i \neq j$. Then by Theorem 18.60, $\{K_1, \ldots, K_r\}$ is a nilpotent set. Furthermore, because

$$\mathbf{B}_G(G) = Q_G/\mathcal{J}(Q_G)$$
$$= D_1 \oplus \cdots \oplus D_r$$
$$\cong \mathbf{B}_G(K_1) \oplus \cdots \oplus \mathbf{B}_G(K_r)$$
$$\cong \mathbf{B}_G(K_1 \oplus \cdots \oplus K_r),$$

Theorem 18.52.1 states that

$$G \text{ is margimorphic to } K_1 \oplus \ldots \oplus K_r.$$

$1 \Rightarrow 2$. Assume that there is a nilpotent set $\{K_1, \ldots, K_r\}$ of nonzero totally indecomposable right R-modules such that G is margimorphic to $K_1 \oplus \cdots \oplus K_r$. We claim that $(\mathbf{B}_G(K_1)), \ldots, (\mathbf{B}_G(K_r))$ is a complete list of distinct isomorphism classes of simple right Q_G-modules.

By Theorem 18.52, the assignment

$$\Omega : \{[K_1], \ldots, [K_r]\} \to \{(M) \mid M \text{ is a semi-simple right } Q_G\text{-module}\}$$

that sends the margimorphism class $[K]$ of $K \in \mathbf{QP}_o(G)$ to the isomorphism class (M) of the semi-simple right Q_G-module $\mathbf{B}_G(K)$, is well defined. Theorem 18.52.1 shows us that Ω is an injection. Since K_i is totally indecomposable, Theorem 18.52.4 states that $\mathbf{B}_G(K_i)$ is a simple right Q_G-module. Thus the image of Ω is in $\{(M) \mid M$ is a simple right Q_G-module$\}$. Let M be a simple right Q_G-module. By Theorem 18.52.1 and 18.52.4, there is a totally indecomposable marginal summand K of G such that $M \cong \mathbf{B}_G(K)$. The uniqueness of decomposition Theorem 18.54.2 then implies that K is margimorphic to one of the modules $K_1, \ldots, K_r$. Thus Ω is a bijection. This proves our claim.

Consequently,

$$Q_G/\mathcal{J}(Q_G) = \mathbf{B}_G(G)$$
$$= \mathbf{B}_G(K_1 \oplus \cdots \oplus K_r)$$
$$= \mathbf{B}_G(K_1) \oplus \cdots \oplus \mathbf{B}_G(K_r)$$

is a direct sum of distinct simple modules. Shur's Lemma then shows us that

$$\text{Hom}_{Q_G}(\mathbf{B}_G(K_i), \mathbf{B}_G(K_j)) = 0 \text{ for each } i \neq j \in \{1, \ldots, r\}$$

so that

$$Q_G/\mathcal{J}(Q_G) = \text{End}_{Q_G}(\mathbf{B}_G(K_1)) \times \cdots \times \text{End}_{Q_G}(\mathbf{B}_G(K_r))$$

is a product of division rings. This completes the proof.

Corollary 18.68. *Let G be a right R-module such that $\mathrm{End}_R(G)$ possesses a semi-primary classical right ring of quotients Q_G. If $Q_G / \mathcal{J}(Q_G)$ is a product of division rings, then*

1. *$G = G_1 \oplus \cdots \oplus G_t$ for some nilpotent set $\{G_1, \ldots, G_t\}$ of indecomposable right R-modules.*
2. *If also $G = G'_1 \oplus \cdots \oplus G'_s$ for some indecomposable right R-modules $G'_1, \ldots, G'_s$ then $s = t$ and after a permutation of the subscripts $G_i \cong G'_i$ for each $i = 1, \ldots, t$.*
3. *If H, L, and L' are right R-modules such that $G = H \oplus L \cong H \oplus L'$, then $L \cong L'$.*

Proof: Apply Theorems 18.66 and 18.67.

A set $\{K, L\}$ of R-modules is a *rigid set* if $\mathrm{Hom}_R(K, L) = \mathrm{Hom}_R(L, K) = 0$. Clearly, a rigid set is a nilpotent set.

Corollary 18.69. *Let G be a right R-module such that $\mathrm{End}_R(G)$ possesses a classical right ring of quotients Q_G that is a product of division rings. Then*

1. *$G = G_1 \oplus \cdots \oplus G_t$ for some rigid set $\{G_1, \ldots, G_t\}$ of indecomposable right R-modules. (That is, $\mathrm{Hom}_R(G_i, G_j) = 0$ for each $i \neq j \in \{1, \ldots, t\}$.)*
2. *If also $G = G'_1 \oplus \cdots \oplus G'_s$ for some indecomposable right R-modules $G'_1, \ldots, G'_s$, then $s = t$ and after a permutation of the subscripts $G_i \cong G'_i$ for each $i = 1, \ldots, t$.*
3. *If H, L, and L' are right R-modules such that $G = H \oplus L \cong H \oplus L'$, then $L \cong L'$.*

The next result shows us that under our standing hypotheses G is margimorphic to a direct sum of modules each of which has the uniqueness of decomposition listed in Theorem 18.66.

Corollary 18.70. *Let G be a right R-module such that $\mathrm{End}_R(G)$ possesses a semi-primary classical right ring of quotients Q_G. Then*

$$G \text{ is margimorphic to } B_1 \oplus \cdots \oplus B_p$$

for some right R-modules $B_1, \ldots, B_p$ such that each B_i has the uniqueness of decomposition property described in Theorem 18.66.2.

Proof: By Theorem 18.54.1 there are totally indecomposable right R-modules $G_1, \ldots, G_t$ such that G is margimorphic to $G_1 \oplus \cdots \oplus G_t$. Choose any disjoint subsets

$$I_1, \ldots, I_p \subset \{G_1, \ldots, G_t\}$$

such that

$$I_1 \cup \cdots \cup I_p = \{G_1, \ldots, G_t\}$$

and such that for each j no two R-modules in I_j are margimorphic. Then by Theorem 18.60, I_j is a nilpotent set. By setting $B_j = \oplus\{K|K \in I_j\}$, G is margimorphic to

$$G_1 \oplus \cdots \oplus G_t = B_1 \oplus \cdots \oplus B_p$$

because the I_j are pairwise disjoint sets. By Theorem 18.66, each B_j enjoys the uniqueness of decomposition property given in Theorem 18.66.2. This completes the proof.

Example 18.71. Let G be any abelian group such that $\mathbf{QEnd}(G) = D_1 \times \cdots \times D_t$ for some finite-dimensional division $\mathbf{Q}$-algebras $D_1, \ldots, D_t$. There are many such groups. Then by Theorem 18.66.2, G has unique decomposition.

Example 18.72. Let G be any abelian group such that $\mathbf{QEnd}(G)$ is the full ring of $t \times t$ lower triangular matrices over the algebraic number field $\mathbf{k}$. Then by Theorem 18.66.2, G has unique decomposition.

Example 18.73. This example is due to A. L. S. Corner [59, Theorem 90.2]. Let $t \geq 2$ be an integer. There is a group G whose endomorphism ring E has finite index in

$$\mathrm{Mat}_2(\mathbf{Z}) \times \underbrace{\mathbf{Z} \times \cdots \times \mathbf{Z}}_{t}.$$

The group has the property that there are several ways to write G as a direct sum of indecomposable groups. Specifically for each partition $(a_1, \ldots, a_r)$ of $2 + t$ there is a direct sum

$$G = A_1 \oplus \cdots \oplus A_r,$$

where each A_i is indecomposable and $\mathrm{rank}(A_i) = a_i$ for each $i = 1, \ldots, r$.

Philosophically, we conclude from the above results that uniqueness of decomposition fails when we introduce pairs of margimorphic totally indecomposable marginal summands into G. Thus, while we might know that G satisfies the conditions of Theorem 18.66 and thus has a unique direct sum decomposition, we cannot draw the same conclusions for $G \oplus G$ unless there are other strong hypotheses on G.

18.6 Semi-simple endomorphism rings

In this section we characterize in terms of totally indecomposable marginal summands the right R-modules G whose endomorphism ring has a semi-simple Artinian classical right ring of quotients Q_G. The ring E is *semi-prime* if $\mathcal{N}(E) = 0$ or equivalently if $I^2 \neq 0$ for any nonzero (right) ideal $I \subset E$. We say that E is a *prime* ring if $IJ \neq 0$ for any nonzero (right) ideals $I, J \subset E$.

The ring E is *semi-simple Artinian* if

$$E = M_1^{(n_1)} \oplus \cdots \oplus M_t^{(n_t)}$$

for some integers $t, n_1, \ldots, n_t > 0$ and some simple right E-modules $M_1, \ldots, M_t$ such that $M_i \not\cong M_j$ for $i \neq j$. Let us investigate this a little. Since the M_i are distinct Shur's lemma shows us that

$$\text{Hom}_E(M_i, M_j) = 0 \text{ for each } i \neq j \in \{1, \ldots, t\}$$

and that $D_i = \text{End}_E(M_i)$ is a division ring for each $i = 1, \ldots, t$. Thus

$$
\begin{aligned}
E &\cong \text{End}_E(M_1^{(n_1)} \oplus \cdots \oplus M_t^{(n_t)}) \\
&\cong \text{End}_E(M_1^{(n_1)}) \times \cdots \times \text{End}_E(M_t^{(n_t)}) \\
&\cong \text{Mat}_{n_1}(D_1) \times \cdots \times \text{Mat}_{n_t}(D_t).
\end{aligned}
$$

The reader can prove the converse. Subsequently, E is semi-simple Artinian iff E is a finite product of matrix rings over division rings.

The ring E is *simple Artinian* if

$$E = M^{(n)}$$

for some simple right E-module M and integer $n > 0$. Therefore, E is simple Artinian iff $E = \text{Mat}_n(D)$ for some division ring D and some integer $n > 0$.

We have arrived at the goal of this subsection. A ring E is a *right Ore domain* if for each nonzero $x, y \in E$, (a) $xy \neq 0$ and (b) there are nonzero $c, d \in E$ such that $xc = yd$. The ring E is a right Ore domain iff E possesses a classical right ring of quotients Q that is a division ring.

Theorem 18.74. *Let G be a right R-module. The following are equivalent.*

1. *$\text{End}_R(G)$ has a semi-simple Artinian classical right ring of quotients Q_G.*
2. *There are integers $t, n_1, \ldots, n_t$ and submodules $K_1, \ldots, K_t$ of G such that*
   (a) *G is margimorphic to $K_1^{(n_1)} \oplus \cdots \oplus K_t^{(n_t)}$.*
   (b) *$\{K_1, \ldots, K_t\}$ is a rigid set. That is, $\text{Hom}_R(K_i, K_j) = 0$ for each $i \neq j \in \{1, \ldots, t\}$.*
   (c) *$\text{End}_R(K_i)$ is a right Ore domain for each $i \in \{1, \ldots, t\}$.*

Proof: $1 \Rightarrow 2$. Suppose that Q_G is semi-simple Artinian. We write

$$Q_G = M_1^{(n_1)} \oplus \cdots \oplus M_t^{(n_t)}$$

for some integers $t, n_1, \ldots, n_t > 0$ and simple Q_G-modules $M_1, \ldots, M_t$ such that $M_i \not\cong M_j$ for $i \neq j$. By Theorem 18.38.3, there are $K_1, \ldots, K_t \in \mathbf{QP}_o(G)$ such that

$$\mathbf{QH}_G(K_i) \cong M_i$$

for each $i \in \{1, \ldots, t\}$, and by Lemma 18.36, $K_1^{(n_1)} \oplus \cdots \oplus K_t^{(n_t)}$ is in $\mathbf{QP}_o(G)$. Moreover, because

$$\mathbf{QH}_G(G) \cong Q_G$$
$$= M_1^{(n_1)} \oplus \cdots \oplus M_t^{(n_t)}$$
$$\cong \mathbf{QH}_G(K_1^{(n_1)} \oplus \cdots \oplus K_t^{(n_t)}).$$

Theorem 18.38.1 implies that G is margimorphic to the direct sum $G' = K_1^{(n_1)} \oplus \cdots \oplus K_t^{(n_t)}$. This is part 2(a).

By Shur's lemma, $\mathrm{Hom}_{Q_G}(M_i, M_j) = 0$ for $i \neq j$, and $\mathbf{QH}_G(\cdot)$ is a faithful functor by Theorem 18.38, so

$$\mathrm{Hom}_R(K_i, K_j) \subset \mathrm{Hom}_{Q_G}(\mathbf{QH}_G(K_i), \mathbf{QH}_G(K_j))$$
$$= \mathrm{Hom}_{Q_G}(M_i, M_j)$$
$$= 0.$$

This is part 2(b).

By Theorem 18.38.2,

$$D_i = \mathrm{End}_{Q_G}(M_i) = \mathrm{End}_{Q_G}(\mathbf{QH}_G(K_i))$$

is the classical right ring of quotients of $\mathrm{End}_R(K_i)$. Since M_i is a simple right R-module D_i is a division ring, and thus $\mathrm{End}_R(K_i)$ is a right Ore domain. This proves part 2(c), and so proves part 2.

$2 \Rightarrow 1$. Assume part 2. By part 2(a), G is margimorphic to

$$G' = K_1^{(n_1)} \oplus \cdots \oplus K_t^{(n_t)}.$$

Suppose that we have shown that $\mathrm{End}_R(G')$ has a semi-simple Artinian classical right ring of quotients $Q_{G'}$. Then by Theorem 18.25, $Q_{G'}$ is the semi-simple Artinian classical ring of quotients of $\mathrm{End}_R(G)$. Thus to prove that $\mathrm{End}_R(G)$ has a semi-simple Artinian classical right ring of quotients Q_G it suffices to show that $\mathrm{End}_R(G')$ has a semi-simple Artinian classical right ring of quotients.

By part 2(b), $\mathrm{Hom}_R(K_i, K_j) = 0$ for each $i \neq j$, so that

$$\mathrm{End}_R(G') = \mathrm{End}_R(K_1^{(n_1)}) \times \cdots \times \mathrm{End}_R(K_t^{(n_t)})$$
$$= \mathrm{Mat}_{n_1}(\mathrm{End}_R(K_1)) \times \cdots \times \mathrm{Mat}_{n_t}(\mathrm{End}_R(K_t)).$$

Fix $i \in \{1, \ldots, t\}$. Since $\mathrm{End}_R(K_i)$ is a right Ore domain (part 2(c)), $\mathrm{End}_R(K_i)$ has a classical right ring of quotients D_i that is a division ring. The simple Artinian ring $\mathrm{Mat}_{n_i}(D_i)$ is thus the classical right ring of quotients of $\mathrm{Mat}_{n_i}(\mathrm{End}_R(K_i))$

(Theorem 18.11). We conclude that

$$\mathrm{Mat}_{n_1}(D_1) \times \cdots \times \mathrm{Mat}_{n_t}(D_t)$$

is the semi-simple Artinian classical right ring of quotients of $\mathrm{End}_R(G')$. Given our reductions, we have proved part 1. This completes the proof.

By restricting the above theorem to the case where $t = 1$, we see that condition Theorem 18.74.2b is vacuous. Thus with $t = 1$ in the above theorem we have characterized those R-modules G for which Q_G is a simple Artinian ring.

Theorem 18.75. *Let G be a right R-module. The following are equivalent.*

1. $\mathrm{End}_R(G)$ *has a simple Artinian classical right ring of quotients Q_G.*
2. G *is margimorphic to $K^{(n)}$ for some integer n and right R-module K such that $\mathrm{End}_R(K)$ is a right Ore domain.*
3. *(a)* $\mathrm{End}_R(G)$ *has a semi-primary classical right ring of quotients.*
   (b) If H is a fully invariant G-generated E-submodule of G, then H is margimorphic to G.

Proof: $1 \Leftrightarrow 2$ follows from Theorem 18.74 and the above comments.

$1 \Rightarrow 3$. Assume part 1. Part 3(a) follows from the fact that a simple Artinian ring is a semi-primary ring.

Let $H \neq 0 \subset G$ be a fully invariant G-generated R-submodule. Since Q_G is simple Artinian, there is a projective Q_G-module P such that

$$\mathbf{QH}_G(G) = \mathbf{QH}_G(H) \oplus P.$$

By Theorem 18.38.3, there is an $H' \in \mathbf{QP}_o(G)$ such that $\mathbf{QH}_G(H') \cong P$. Also, by Lemma 18.36, $H \oplus H' \in \mathbf{QP}_o(G)$. Then

$$\begin{aligned}
\mathbf{QH}_G(G) &= \mathbf{QH}_G(H) \oplus P \\
&= \mathbf{QH}_G(H) \oplus \mathbf{QH}_G(H') \\
&= \mathbf{QH}_G(H \oplus H'),
\end{aligned}$$

so that by Theorem 18.38.1, G is margimorphic to

$$G' = H \oplus H'.$$

By Theorem 18.25, Q_G is the simple Artinian classical right ring of quotinets of $\mathrm{End}_R(G')$. The simplicity of Q_G implies that $\mathrm{End}_R(G')$ is a prime ring. Choose $e^2 = e \in \mathrm{End}_R(G')$ such that $e(G') = H$. Then $\mathbf{H}_G(H) = e\mathbf{H}_G(G')$. By Lemma 18.13 there is a regular $c \in \mathrm{End}_R(G')$ such that $c = ece + (1-e)c(1-e)$ and $ce, c(1-e) \in \mathrm{End}_R(G)$. Since H is fully invariant in G, $\mathbf{H}_G(H) = e\mathbf{H}_G(G')$ is an ideal in $\mathrm{End}_R(G')$.

Then

$$[c(1-e)\mathrm{End}_R(G')]\mathbf{H}_G(H) = c(1-e)e\mathbf{H}_G(H)$$
$$= 0.$$

Since $\mathrm{End}_R(G')$ is a prime ring $c(1-e)\mathrm{End}_R(G') = 0$. Because c is a unit in Q_G, $c(1-e) = 0 = 1-e$ so that $H' = 0$. That is, G is margimorphic to H. This proves part 3(b), and this proves part 3.

$3 \Rightarrow 1$. Assume part 3, and let $I, J \subset \mathrm{End}_R(G)$ be nonzero ideals such that $JI = 0$. Since I is an ideal of $\mathrm{End}_R(G)$, $IG \subset G$ is a fully invariant G-generated submodule of G. By part 3(b) there is a margimorphism pair (a, b) for G and IG such that $aG \subset IG \subset G$. Then a is a regular endomorphism of G. Applying J, we see that $JaG \subset (JI)G = 0$, so that $Ja = 0$. Since a is regular, $J = 0$, and so $\mathrm{End}_R(G)$ is prime. Since $\mathcal{J}(Q_G) \cap \mathrm{End}_R(G)$ is a nilpotent ideal in the prime ring $\mathrm{End}_R(G)$, $\mathcal{J}(Q_G) \cap \mathrm{End}_R(G) = 0 = \mathcal{J}(Q_G)$. Thus $Q_G = Q_G/\mathcal{J}(Q_G)$ is a prime, semi-simple Artinian ring. Hence Q_G is simple Artinian. This proves part 1 and completes the logical cycle.

Next is a result whose proof will require the reader to cross-pollinate Theorem 18.74 with Theorem 18.75.

Corollary 18.76. *The following are equivalent for a right R-module G.*

1. *$\mathrm{End}_R(G)$ possesses a classical right ring of quotients that is a finite product of division rings.*
2. *G is margimorphic to $K_1 \oplus \cdots \oplus K_t$ for some rigid set $\{K_1, \ldots, K_t\}$ of right R-modules such that $\mathrm{End}_R(K_i)$ is a right Ore domain for each $i \in \{1, \ldots, t\}$.*

The theorems involving the existence of semi-simple Artinian classical rings of quotients rest on an understanding of when $\mathrm{End}_R(K)$ is a right Ore domain. We give a characterization of modules G for which $\mathrm{End}_R(G)$ is a right Ore domain.

Theorem 18.77. *The following are equivalent for the right R-module G.*

1. *$\mathrm{End}_R(G)$ is a right Ore domain.*
2. *(a) $\mathrm{End}_R(G)$ possesses a semi-primary classical right ring of quotients Q_G, and*
   (b) G is margimorphic to each nonzero G-generated submodule of G.

Proof: $1 \Rightarrow 2$. Suppose that $\mathrm{End}_R(G)$ is a right Ore domain. Then Q_G exists and it is a division ring, a prime semi-primary ring. This proves part 2(a). By Theorem 18.75(3b), part 2(b) holds.

$2 \Rightarrow 1$. Assume part 2. By part 2(a), Q_G exists and is a semi-primary ring. By part 2(b) and Theorem 18.75, Q_G is a simple Artinian ring, say $Q_G = \mathrm{Mat}_n(D)$ for some division ring D and integer $n \geq 1$.

Suppose that $Q_G = P \oplus P'$ for some nonzero indecomposable right Q_G-module P. By Theorem 18.39 there are G-generated $H, H' \in \mathbf{QP}_o(G)$ such that $P \cong \mathbf{QH}_G(H)$

and $P' = \mathbf{QH}_G(H')$. Then by part 2(b), G is margimorphic to H and to $H \oplus H'$. Theorem 18.38.1 then shows us that

$$P \cong \mathbf{QH}_G(H)$$
$$\cong \mathbf{QH}_G(G)$$
$$\cong \mathbf{QH}_G(H \oplus H')$$
$$\cong \mathbf{QH}_G(H) \oplus \mathbf{QH}_G(H').$$

Since $P \neq 0$ is indecomposable, $\mathbf{QH}_G(H') = 0$ and since $\mathbf{QH}_G(\cdot)$ is a faithful functor, $H' = 0$. Hence $Q_G = P$ is indecomposable. As required by part 1, Q_G is a division ring. This completes the proof.

Corollary 18.78. *Let G be a right R-module such that $\mathrm{End}_R(G)$ possesses a semi-primary classical right ring of quotients Q_G. Then $\mathrm{End}_R(G)$ is a right Ore domain iff G is margimorphic to each nonzero G-generated submodule of G.*

Example 18.79. Let $\mathbf{Z} \subset X \subset \mathbf{Q}$ be a subgroup such that $\mathrm{Hom}(X, \mathbf{Z}) = 0$ while $\mathrm{End}(X) = \mathbf{Z}$. For example, the group

$$X = \left\{ \frac{n}{m} \,\middle|\, n \in \mathbf{Z} \text{ and } m \in \mathbf{Z} \text{ is cube-free} \right\}$$

will do. Further choose a subgroup $\mathbf{Z} \subset Y \subset X$ such that $\mathrm{Hom}(X, Y) = \mathrm{Hom}(Y, \mathbf{Z}) = 0$. For example, let

$$Y = \left\{ \frac{n}{m} \,\middle|\, m, n \in \mathbf{Z} \text{ where } m \text{ is square-free} \right\}.$$

Let

$$G = \mathbf{Z} \oplus X.$$

Then Y is G-generated but Y is not a (quasi-) summand of G by the Baer–Kulikov–Kaplansky theorem [59]. Observe that

$$\mathrm{End}(G) = \begin{pmatrix} \mathbf{Z} & 0 \\ \mathbf{Z} & \mathbf{Z} \end{pmatrix}$$

is not semi-prime but that $\mathbf{Q}\mathrm{End}(G) = Q_G$ is an Artinian (= semi-primary) ring with nonzero nilradical. Thus $\mathrm{End}(G)$ does not possess a semi-simple Artinian classical right ring of quotients even though G is a direct sum of groups whose endomorphism rings are commutative Noetherian integral domains.

18.7 Exercises

Let R and E be rings, let G, H, K be right R-modules, and let ϕ be an R-module homomorphism. We assume that E has a classical right ring of quotients Q.

1. Let E be a Noetherian semi-prime ring and assume that E is tffr. Show that $E = E_d \times E_o$ where E_d is the divisible subgroup of E and where E_o is a rtffr ring.

2. Let Q be a semi-perfect ring. ($Q/\mathcal{J}(Q)$ is semi-simple Artinian and idempotents lift modulo $\mathcal{J}(Q)$.) If P is a finitely generated projective right Q-module then $\text{End}_Q(P)$ is semi-perfect.

3. Let Q be a semi-primary ring.

    (a) Q is a *semi-perfect* ring. (See the previous exercise.)
    (b) If $n > 0$ is an integer, then $\text{Mat}_n(Q)$ is semi-primary.
    (c) If $e^2 = e \in Q$ then eQe is a semi-primary ring.

4. If each element of Q has the form xc^{-1} for some $x \in E$ and regular $c \in E$, then $\mathcal{J}(Q) \cap E = \mathcal{N}(E)$.

5. Suppose that E has a semi-primary classical right ring of quotients Q. Let $\mathcal{N}(E) = \mathcal{J}(Q) \cap E$. Then $c \in E$ is a regular element iff $c + \mathcal{N}(E)$ is a regular element in $E/\mathcal{N}(E)$.

6. Suppose that E is a subring of a semi-simple Artinian ring Q. If for each $x \in Q$ there is a regular $c \in E$ such that $xc \in E$, then each right regular element of E is a unit of Q. Hence right regular elements of E are left regular.

7. Let Q be a semi-simple Artinian ring and let $e^2 = e \in Q$. Then eQe is a semi-simple Artinian ring.

8. Suppose that Q is a semi-local ring. Show that $\mathcal{J}(\text{Mat}_n(Q)) = \text{Mat}_n(\mathcal{J}(Q))$ and that

$$\text{Mat}_n(Q)/\mathcal{J}(\text{Mat}_n(Q)) \cong \text{Mat}_n(Q/\mathcal{J}(Q)).$$

That is, $\text{Mat}_n(Q)$ is a semi-local ring.

9. Give an example of a right Artinian ring Q and an $e^2 = e \in Q$ such that eQe is not a right Artinian ring.

10. Let Q be the classical right ring of quotients of E. Show that Q is a flat left E-module.

11. If $e^2 = e, f^2 = f \in \text{End}_R(G)$ and if $eG = H$ and $fG = K$ then $e\text{End}_R(G)f = \text{Hom}_R(K, H)$.

12. Let $e^2 = e \in E$. Show that $\mathcal{N}(eEe) = e\mathcal{N}(E)e$ and that $\mathcal{J}(eEe) = e\mathcal{J}(E)e$.

13. A subring E of Q is a *right order in Q* if to each $x \in Q$ there are $y \in E$ and a regular $c \in E$ such that $x = yc^{-1}$.

    Suppose that $\text{End}_R(G)$ has a semi-primary classical right ring of quotients Q_G, and suppose that G is margimorphic to $H \oplus H'$. If $e^2 = e \in \text{End}_R(H \oplus H')$ is such that $e(H \oplus H') = H$, then $\text{End}_R(H)$ is a right order in $e\text{Mat}_n(Q_G)e = \text{End}_{Q_G}(\mathbf{QH}_G(H))$.

14. Let $\mathbf{k}$ be a field and let $Q = \begin{pmatrix} \mathbf{k} & 0 \\ \mathbf{k}^{(\aleph_o)} & \mathbf{k} \end{pmatrix}$. Prove that Q is a semi-primary ring such that $\mathcal{J}(Q)^2 = 0$ but Q is not right Artinian.

15. Suppose that E is a right order in S and let $x_1, \ldots, x_t \in S$. Show that there is a regular element $c \in E$ such that $x_1 c \ldots, x_t c \in E$.

16. Let E be a right order in S and let $c \in E$ be a regular element. Show that given $x \in S$ there is a regular element $d \in cE$ such that $xd \in E$.

17. If $e^2 = e \in E$ then show that $\text{Hom}_E(eE, e\mathcal{N}(E)) = e\mathcal{N}(E)e$.
18. If $\{G_1, G_2\}$ is a nilpotent set and if $G_i = H_i \oplus H_i'$ then $\{H_1, H_2\}$ is a nilpotent set.
19. If $\{G_1, G_2\}$ is a nilpotent set and if G_i is margimorphic to $H_i \oplus H_i'$ for each $i = 1, 2$, then $\{H_1, H_2\}$ is a nilpotent set.
20. If $\{G_1, G_2\}$ is a nilpotent set and if n_1 and n_2 are positive integers then $\{G_1^{(n_1)}, G_2^{(n_2)}\}$.
21. If $\{G_1, G_2, \ldots, G_t\}$ is a nilpotent set then $\{G_1, G_2 \oplus \cdots \oplus G_t\}$ is a nilpotent set.
22. Prove Theorem 18.52.
23. Let G be a right R-module such that $\text{End}_R(G)$ has a semi-primary classical right ring of quotients Q_G. Let $H \in \mathbf{QP}_o(H)$. Show that H is totally indecomposable iff $\mathbf{QH}_G(H)$ is an indecomposable right Q_G-module.

18.8 Problems for future research

Let R be an associative ring with identity, let G be a right R-module, let $\text{End}_R(G)$ be the endomorphism ring of G, let $\mathcal{Q}$ be a G-plex, let $\mathcal{P}$ be a projective resolution of right $\text{End}_R(G)$-modules.

1. Prove Theorem 18.17 under the hypothesis that Q is right perfect, and not just semi-primary.
2. The *maximal right ring of quotients* $Q_r(E)$ of a ring E is defined as follows. Let $\bar{E}$ be the injective hull of the right E-module E. Then $Q_r(E)$ is the ring of left $\text{End}_E(\bar{E})$-module endomorphisms of $\bar{E}$. Generalize Chapter 7 to right R-modules G such that $Q_r(\text{End}_R(G))$ is semi-primary.
3. Find $Q_r(\text{End}(G))$ for rtffr groups G.
4. Characterize $\mathcal{N}(\text{End}_R(G))$; specifically, given conditions on which $\mathcal{N}(\text{End}_R(G))$ is T-nilpotent.
5. Study those G such that $\text{End}_R(G)$ is a simple right Noetherian domain or prime ring.
6. Give internal conditions that imply two modules are margimorphic.
7. Give conditions under which $\text{End}_R(G)$ has a (semi-primary) classical right ring of quotients.
8. Give more general conditions under which G has a unique decomposition, or internal cancellation.
9. Study the properties of direct sum decompositions of modules of the form $G \oplus G$.
10. Study nilpotent sets of R-modules. Find an internal method for determining that a pair of R-modules form a nilpotent set.
11. Determine those totally indecomposable right R-modules G that have the following properties.
    (a) If $G^{(n)} = H \oplus K$ then $H = H_1 \oplus \cdots \oplus H_t$ for some R-modules H_i such that G and H_i are closely related. (Perhaps they are margimorphic.)
    (b) If H is closely related to G then $G \cong H$.
    (c) If $G^{(n)} = H \oplus K \cong H \oplus L$ then $K \cong L$.
12. In future research, take advantage of the fact that $H \in \mathbf{QP}_o(G)$ can be treated as simple Q_G-modules up to margimorphism.

13. Generalize Jónsson's theorem and Theorem 18.54 to include the maximal right ring of quotients of $\text{End}_R(G)$.

14. Characterize those right R-modules G such that $Q_G = \begin{pmatrix} \mathbf{Q} & 0 \\ \mathbf{Q} & \mathbf{Q} \end{pmatrix}$ is the classical right ring of quotients of $\text{End}_R(G)$.

15. Characterize those right R-modules G such that for some division ring D, $Q_G = \begin{pmatrix} D & 0 \\ D & D \end{pmatrix}$ is the classical right ring of quotients of $\text{End}_R(G)$.

16. Generalize the previous problem to rings Q_G of $n \times n$ lower triangular matrices over a division ring D.

17. Generalize the previous problem to a ring of 2×2 matrices $Q_G = \begin{pmatrix} D_1 & 0 \\ V & D_2 \end{pmatrix}$, where V is a D_1-D_2-bimodule.

18. Let $D_1, \ldots, D_t$ be a list of division algebras and let V_{ij} be a list of $D_i - D_j$-vector spaces. Characterize those right R-modules G whose endomorphism ring is a right order in the ring of $t \times t$ matrices (V_{ij}). This assumes that this matrix defines a ring.

Bibliography

1. U. Albrecht, *Locally A-projective abelian groups and generalizations*, Pac. Jr. Math. **141**, (2), (1990), 209–228.
2. U. Albrecht, *Endomorphism rings of faithfully flat abelian groups*, Resultate der Mathematik, **17**, (1990), 179–201.
3. U. Albrecht, *Baer's Lemma and Fuch's Problem 84a*, Trans. Am. Math. Soc. **293**, (1986), 565–582.
4. U. Albrecht, *Faithful abelian groups of infinite rank*, Proc. Am. Math. Soc. **103**, (1), (1988), 21–26.
5. U. Albrecht; H. P. Goeters, *A dual to Baer's Lemma*, Proc. Am. Math. Soc. **105**, (1989), 217–227.
6. H. M. K Angad-Gaur, *The homological dimension of a torsion-free group of finite rank as a module over its ring of endomorphisms*, Rend. Sem. Mat. Univ. Padova **57**, (1977), 299–309.
7. F. W. Anderson; K. R. Fuller, *Rings and Categories of Modules*, Graduate Texts in Mathematics **13**, Springer-Verlag, New York-Berlin, (1974).
8. D. M. Arnold, *Abelian Groups and Representations of Finite Partially Ordered Sets*, Canadian Mathematical Society: Books in Mathematics, Springer, New York, (2000).
9. D. M. Arnold, *Finite Rank Abelian Groups and Rings*, Lecture Notes in Mathematics **931**, Springer-Verlag, New York, (1982).
10. D. M. Arnold, *Endomorphism rings and subgroups of finite rank torsion-free abelian groups*, Rocky Mt. Jr. Math. **12**, (2), (1982), 241–256.
11. D. M. Arnold, *A finite global Azumaya Theorem in additive categories*, Proc. Am. Math. Soc. **91**, (1), May, (1984), 25–29.
12. D. Arnold J. Hausen, *Modules with the summand intersection property*, Comm. Algebra **18**, (1990), 519–528.
13. D. M. Arnold; R. Hunter; F. Richman, *Global Azuamya Theorems in additive categories*, Jr. Pure Appl. Algebra **16**, (1980), 223–242.
14. D. M. Arnold; L. Lady, *Endomorphism rings and direct sums of torsion-free abelian groups*, Trans. Am. Math. Soc. **211**, (1975), 225–237.
15. D. M. Arnold; C. E. Murley, *Abelian groups A such that* $\mathrm{Hom}(A, \cdot)$ *preserves direct sums of copies of A*, Pac. Jr. Math. **56**, (1), (1975), 7–20.

16. D. M. Arnold; R. S. Pierce; J. D. Reid; C. Vinsonhaler; W. Wickless, *Torsion-free abelian groups of finite rank projective as modules over their endomorphism rings*, Jr. Algebra **71**, (1), July (1981), 1–10.

17. M. F. Atiyah; I. G. Macdonald, *Introduction to Commutative Algebra*, Addison-Wesley Publishing Co., Reading, MA–Sydney, (1969).

18. G. Azumaya, *Some aspects of Fuller's Theorem*, Lecture Notes in Mathematics **700**, Springer-Verlag, New York–Berlin, (1979), 34–45.

19. H. Bass, *Algbraic K-Theory*, W. A. Benjamin, Inc., New York–Amsterdam, (1968).

20. R. A. Beaumont; R. S. Pierce, *Subrings of algebraic number fields*, Acta Sci. Math. (Szeged), **22**, (1961), 202–216.

21. R. A. Beaumont; R. S. Pierce, *Torsion-free rings*, Illinois Jr. Math., **5**, (1961), 61–98.

22. A. L. S. Corner, *Every countable torsion-free ring is an endomorphism ring*, Proc. London Math. Soc. **13**, (1963), 23–33.

23. A. L. S. Corner; R. Gobel, *Prescribing endomorphism algebras*, Proc. London Math. Soc. **50**, (3), (1985), 447–494.

24. R. Colpi, *Some remarks on equivalences between categories of modules*, Comm. Algebra **18**, (1990), 1935–1951.

25. R. Colpi, *Tilting modules and *-modules*, Comm. Algebra **21**, (1993), 1095–1102.

26. R. Colpi; C. Menini, *On the structure of *-modules*, Jr. Algebra **158**, (1993), 400–419.

27. G. D'Este, *Some remarks on representable equivalences*, Topics in Algebra – Banach Center Publications **26**, (1990), 223–232.

28. A. Dress, *On the decomposition of modules*, Bull. Am. Math. Soc. **75**, (1969), 984–986.

29. M. Dugas; T. G. Faticoni, *Cotorsion-free groups cotorsion as modules over their endomorphism rings*, Abelian Groups, Lecture Notes in Pure and Applied Mathematics, Marcel Dekker, New York, (1993), 111–127.

30. M. Dugas; A. Mader; C. Vinsonhaler, *Large E-rings exist*, Jr. Algebra, **108**, (1987), 88–101.

31. C. Faith, *Algebra I: Algebra and Categories*, Springer-Verlag, New York–Berlin, (1974).

32. C. Faith, *Algebra II: Ring Theory*, Springer-Verlag, New York–Berlin, (1976).

33. C. Faith, *Orders in simple Artinian rings*, Trans. Am. Math. Soc. **114**, (1965), 61–64.

34. C. Faith; S. Page, *FPF Ring Theory*, London Mathematical Society Lecture Note Series **88**, Cambridge University Press, Cambridge, (1984).

35. C. Faith, E. A. Walker, *Direct sum representations of injective modules*, Jr. Algebra **5**, (1967), 203–221.

36. T. G. Faticoni, *Dimensions and Endomorphisms*, submitted (2008).

37. T. G. Faticoni, *Marginal isomorphisms*, Jr. Algebra, **319**, (2008), 4575–4620.

38. T. G. Faticoni, *Class number of an abelian group*, Jr. Algebra, **314**, (2007), 978–1008.

39. T. G. Faticoni, *Addendum: Class number of an abelian group*, Jr. Algebra, **314**, (2007), 1009–1010.

40. T. G. Faticoni, *L-Groups, Modules, and Abelian Groups*, De Gruyter, (2008), 339–349.

41. T. G. Faticoni, *Conductor of an abelian group*, Jr. Algebra, **321**, (2009), 1977–1996.

42. T. G. Faticoni, *A duality for self-slender modules*, Comm. Algebra, Dec., **35**, (12), (2007), 4175–4182.

43. T. G. Faticoni, *Addendum: Diagrams of abelian groups*, Bull. Aust. Math. Soc., **80**, (2009), 65–68.

44. T. G. Faticoni, *Diagrams of abelian groups*, to appear in Bull. Aust. Math. Soc., **80**, (2009), 38–64.

45. T. G. Faticoni, *Torsion-free Free Finite Rank Groups as Point Set Topological Spaces*, Abelian Groups, Rings, Modules, and Homological Algebra, Chapman and Hall/CRC, Boca-Raton, (2006), Chapter 10.

46. T. G. Faticoni, *Direct Sum Decompositions of Torsion-Free Finite Rank Groups*, Chapman and Hall/CRC, Boca-Raton, (2007).

47. T. G. Faticoni, *Direct Sums and Refinement*, Comm. Algebra **27**, (1), (1999), 451–464.

48. T. G. Faticoni, *Modules over endomorphism rings as homotopy classes*, In *Abelian Groups and Modules*, A. Facchini, C. Menini, eds., Mathematics and its Applications, **343**, Kluwer Academic Publishers, (1995), 163–183.

49. T. G. Faticoni, *Torsion-free groups torsion as modules over their endomorphism rings*, Bull. Aust. Math. Soc., **50**, (2), (1994), 177–195.

50. T. G. Faticoni, *Categories of Modules over Endomorphism Rings*, Memoirs Am. Math. Soc. **492**, (May, 1993).

51. T. G. Faticoni, *The endlich Baer splitting property*, Pac. Jr. Math, **157**, (2), (1991), 225–240.

52. T. G. Faticoni, *Gabriel filters on the endomorphism ring of a torsion-free abelian group*, Comm. Algebra **18**, (9), (1990), 2841–2883.

53. T. G. Faticoni, *On the Lattice of right ideals of the endomorphism ring of an abelian group*, Bull. Aust. Math. Soc. **38**, (2), (1988), 273–291.

54. T. G. Faticoni, *Each countable reduced torsion-free commutative ring is a pure subring of an E-ring*, Comm. Algebra **15**, (12), (1987), 2545–2564.

55. T. G. Faticoni, *On quasi-projective covers*, Trans. Am. Math. Soc., **278**, (1), (1983), 101–111.

56. T. G. Faticoni; H. Pat Goeters, *Examples of torsion-free groups flat as modules over their endomorphism rings*, Comm. Algebra **19**, (1), (1991), 1–28.

57. T. G. Faticoni; H. Pat Goeters, *On torsion-free Ext*, Comm. Algebra **16**, (9), (1988), 1853–1876.

58. T. G. Faticoni; P. Schultz, *Direct sum decompositions of acd groups with primary regulating quotient*, Abelian Groups and Modules, Lecture Notes in Pure and Applied Mathematics, **182**, (May 1993), Marcell Dekker, (1996).

59. L. Fuchs, *Infinite Abelian Groups I, II*, Academic Press, New York–London, (1969, 1970).

60. L. Fuchs; L. Salce, *Uniserial modules over valuation rings*, Jr. Algebra **85**, (1983), 14–31.

61. K. Fuller, *Density and equivalence*, Jr. Algebra **29**, (1974), 528–550.

62. J. L. Garcia Hernandez; M. Saorin, *Endomorphism rings and category equivalences*, Jr. Algebra **127**, (1989), 182–205.

63. J. L. Garcia Hernandez; Jr. L. Gomez Pardo, *Hereditary and semi-hereditary endomorphism rings*, Lecture Notes in Mathematics **1197**, Springer-Verlag, New York–Berlin, (1986), 83–89.

64. J. L. Garcia Hernandez; Jr. L. Gomez Pardo, *On the endomorphism rings of quasi-projective modules*, Math. Z. **196**, (1987), 87–108.

65. J. L. Garcia Hernandez; Jr. L. Gomez Pardo, *Closed submodules of free modules over the endomorphism ring of a quasi-injective module*, Comm. Algebra **16**, (1), (1988), 115–137.

66. J. L. Gomez Pardo, *Endomorphism rings and dual modules*, Jr. Algebra **130**, (2), (1990), 477–493.

67. J. L. Gomez Pardo, *Counterinjective modules and duality*, Jr. Pure Appl. Algebra **61**, (1989), 165–179.

68. J. L. Gomez Pardo; Jr. M. Hernandez, *Coherence of endomorphism rings*, Arch. Math. **48**, (1987), 40–52.

69. A. Hatcher, *Algebraic Topology*, Cambridge University Press, Cambridge, (2002).

70. M. Huber; R. B. Warfield, Jr., *Homomorphisms between cartesian powers of an abelian group*, Lecture Notes in Mathematics **874**, Springer-Verlag, New York–Berlin, (1981), 202–227.

71. S. Jain, *Flat and FP-injectivity*, Proc. Am. Math. Soc. **41**, (2), (1973), 437–442.

72. A. V. Jategaonkar, *Localization in Noetherian Rings*, London Math. Soc. Lecture Note Series **98**, (1986).

73. I. Kaplansky, *Commutative Rings*, Allyn and Bacon Inc., Boston–New York, (1970).

74. I. Kaplansky, *Infinite Abelian Groups*, The University of Michigan Press, Ann Arbor, Michigan, (1954).

75. I. Kaplansky, *Fields and Rings*, The University of Chicago Press, Chicago–London, (1969).

76. S. M. Khuri, *Modules with regular, perfect, Noetherian or Artinian endomorphism rings*, Lecture Notes in Mathematics **1448**, Springer-Verlag, New York–Berlin, (1989), 7–18.

77. P. A. Krylov, *Torsion-free abelian groups with hereditary rings of endomorphisms*, Algebra and Logic **27**, (3), (1989), 184–190.

78. P. A. Krylov, A. V. Mikhalev, A. A. Tuganbaev, *Endomorphism Rings of Abelian Groups*, Kluwer Academic Publishers, (2003).

79. R. Kuebler; J. Reid, *On a paper of Richman and Walker*, Rocky Mt. Jr. Math. **5**, (4), (1975), 585–592.

80. E. L. Lady, *Nearly isomorphic torsion-free abelian groups*, Jr. Algebra **35**, (1975), 235–238.

81. E. L. Lady, *Torsion-free Modules over Dedekind Domains*, manuscript.

82. S. Lang, *Algebraic Number Theory*, Addison-Wesley, New York, (1970).

83. L. S. Levy, *Direct-sum cancellation and genus class groups*, In *Methods in Module Theory*, G. Abrams, Jr. Haefner, K. M. Rangaswamy eds., Lecture Notes in Pure and Applied Mathematics, **140**, (1993), 203–218.

84. A. Mader, *Almost Completely Decomposable Abelian Groups*, Proceedings of the Meeting on Abelian Groups and Modules held at the University of Padua, June, 1994.

85. A. Mader, *Almost Completely Decomposable Groups*, volume 13, Gordon and Breach Science Publishers, France–Germany, (2000).

86. A. Mader; C. Vinsonhaler, *Torsion-free E-modules*, Jr. Algebra, **108**, (1987), 88–101.

87. C. Murley, *The classification of certain classes of torsion-free abelian groups*, Pac. Jr. Math. **40**, (1972), 647–665.

88. C. Menini; D. Orsatti, *Representable equivalences between categories of modules and applications*, Rend. Sem. Mat. Univ. Padova **82**, (1982), 203–231.

89. G. P. Niedzwecki; J. Reid, *Abelian groups projective over their endomorphism rings*, Jr. Algebra **159**, (1993), 139–149.

90. K. C. O'Meara; C. Vinsonhaler, *Separative cancellation and multiple isomorphism in torsion-free abelian groups*, Jr. Algebra **221**, (1999), 536–550.

91. A. T. Paras, *Abelian groups as Noetherian modules over their endomorphism rings*, In Abelian Group Theory and Related Topics, Contemporary Mathematics, **171**, R. Gobel, P. Hill, W. Lieber, eds., AMS, Providence RI, (1994), 325–332.

92. J. Reid, *Abelian groups finitely generated over their endomorphism rings*, Lecture Notes in Mathematics **874**, Springer-Verlag, New York–Berlin, (1981), 41–52.

93. J. D. Reid, *On quasi-decompositions of torsion-free abelian groups*, Proc. Am. Math. Soc. **13**, (1961), 550–554.

94. I. Reiner, *Maximal Orders*, Academic Press Inc., New York, (1975).

95. F. Richman, *An extension of the theory of completely decomposable torsion-free abelian groups*, Trans. Am. Math. Soc. **279**, (1), (1983), 175–185.

96. F. Richman; E. A. Walker, *Primary abelian groups as modules over their endomorphism rings*, Math. Z. **89**, (1965), 77–81.

97. J. Rotman, *An Introduction to Homology Theory*, Pure and Applied Mathematics **85**, Academic Press, New–York–San Francisco–London, (1979).

98. M. Sato, *Fuller's theorem on equivalence*, Jr. Algebra **52**, (1978), 274–284.

99. J. Seltzer, *A cancellation criterion for finite rank torsion-free abelian groups*, Proc. Am. Math. Soc. **94**, (1985), 363–368.

100. L. Small, *Semi-hereditary rings*, Bull. Am. Math. Soc. **73**, (1966), 656–658.

101. J. T. Stafford, *Noetherian full quotient rings*, Proc. London Math. Soc. (3), **44**, (1982), 385–404.

102. B. Stenstrom, *Rings of Quotients: An Introduction to Ring Theory*, Die Grundlehren Band **217**, Springer-Verlag, New York–Berlin, (1975).

103. I. N. Stewart, D. O. Tall, *Algebraic Number Theory*, 2nd edn., Chapman and Hall/CRC, Boca-Raton–New York, (1987).

104. R. G Swan, *Projective modules over group rings and maximal orders*, Ann. Math. (2) **76**, (1962), 55–62.

105. J. Trlifaj, *Every *-module is finitely generated*, Jr. Algebra **169**, (1994), 392–398.

106. C. Vinsonhaler, *Torsion-free abelian groups quasi-projective over their endomorphism rings II*, Pac. Jr. Math. **74**, (1978), 261–265.

107. C. Vinsonhaler, *The divisible and E-injective hulls of a torsion-free group*, In *Abelian Groups and Modules, Proceedings of the Udine Conference*, Udine, April 1984, R. Göbel, C. Metelli, D. Orsatti, L. Salce, eds., CISM **287**, Springer-Verlag, Wein–New York, (1984), 163–181.

108. C. Vinsonhaler, W. Wickless, *Homological dimension of completely decomposable groups*, In *Abelian Groups*, Lecture Notes in Pure and Applied Mathematics **146**, ed. L. Fuchs, R. Göbel, (1993), 247–258.

109. C. Vinsonhaler, W. Wickless, *Balanced projective and cobalanced injective torsion-free groups of finite rank*, Acta Math. Hung. **46** (3–4), (1985), 217–225.

110. C. Vinsonhaler, W. Wickless, *Injective hulls of torsion-free abelian groups as modules over their endomorphism rings*, Jr. Algebra, **58** (1), (1979), 64–69.

111. C. Vinsonhaler, W. Wickless, *Torsion-free abelian groups quasi-projective over their endomorphism rings*, Pac. Jr. Math. **70**, (2), (1977), 1–9.

112. R. B. Warfield, Jr., *Countably generated modules over commutative Artinian rings*, Pac. Jr. Math. **60**, (2), (1975), 289–302.

113. R. Wisbauer, *Foundations of Module and Ring Theory*, Gordon and Breach Science Publishers, New York, (1991).

114. H. Zassenhaus, *Orders as endomorphism rings of modules of the same rank*, Jr. London Math. Soc. **42**, (1967), 180–182.

115. B. Zimmerman-Huisgen, *Endomorphism rings of self-generators*, Pac. Jr. Math. **61**, (1), (1975), 587–602.

Index